THOMPSON'S

Core Textbook
of Anatomy

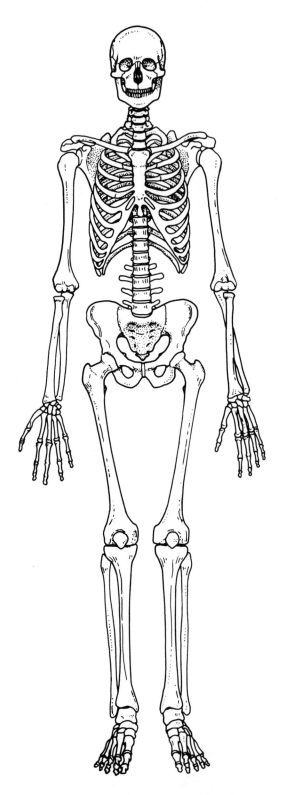

The human skeleton with pronated left forearm.

THOMPSON'S

Core Textbook of Anatomy

Second Edition

Elizabeth J. Akesson, M.Sc.
Assistant Professor, Department of Anatomy
University of Toronto, Faculty of Medicine
Toronto, Ontario, Canada

Jacques A. Loeb, M.B.B.S. (Hons), F.R.C.S., F.R.C.S.(C), F.A.C.S.
Formerly Lecturer in Anatomy
University of London
Formerly Chairman, Department of Surgery
Oakville Trafalgar Memorial Hospital
Presently Consultant in Surgery
Department of Anatomy
University of Toronto, Faculty of Medicine
Toronto, Ontario, Canada

Linda Wilson-Pauwels, B.Sc.A.A.M., A.O.C.A.
Associate Chairman
Department of Art as Applied to Medicine
University of Toronto, Faculty of Medicine
Toronto, Ontario, Canada

J. B. LIPPINCOTT
COMPANY
Philadelphia

Grand Rapids London
New York   Sydney
St. Louis   Tokyo
San Francisco

Acquisitions Editor: Lisa McAllister
Developmental Editor: Richard Winters
Manuscript Editors: Leslie E. Hoeltzel and Lynda Kenny
Indexer: Barbara Littlewood
Designer: Anita Curry
Design Coordinator: Ellen C. Dawson
Production Supervisor: Charlene C. Squibb
Production Manager: Carol A. Florence
Cover Design: Mark A. James
Compositor: TAPSCO, Inc.
Printer/Binder: The Murray Printing Company

Second Edition

1   3   5   6   4   2

Library of Congress Cataloging-in-Publication Data

Thompson, James S. (James Scott), 1919–
    [Core textbook of anatomy]
    Thompson's core textbook of anatomy.—2nd ed./Elizabeth J.
Akesson, Jacques A. Loeb, Linda Wilson-Pauwels.
        p.   cm.
    Rev. ed. of: Core textbook of anatomy/J.S. Thompson. c1977.
    Bibliography: p.
    Includes index.
    ISBN 0-397-50849-2
    1. Anatomy, Human.   I. Akesson, E. J.   II. Loeb, Jacques A.
III. Wilson-Pauwels, Linda.   IV. Title.   V. Title: Core textbook of
anatomy.
    [DNLM: 1. Anatomy.   QS 4 T473c]
QM23.2.T48 1990
611—dc19
DNLM/DLC
for Library of Congress   88-23088
                                                                      CIP

The authors and publisher have exerted every effort to ensure that drug
selection and dosage set forth in this text are in accord with current
recommendations and practice at the time of publication. However, in view
of ongoing research, changes in government regulations, and the constant
flow of information relating to drug therapy and drug reactions, the reader is
urged to check the package insert for each drug for any change in indications
and dosage and for added warnings and precautions. This is particularly
important when the recommended agent is a new or infrequently
employed drug.

To Tom, Joan, and Hugh

Preface to the Second Edition

Why have we revised *Thompson's Core Textbook of Anatomy*? The original edition, written by Professor J. S. Thompson, was produced as a concise textbook of anatomy. Students and healthcare practitioners who have this first edition emphasize the importance of an abbreviated textbook as a valuable study guide and reference book. Since Professor Thompson's death, further contractions have occurred in the first-year curricula of many medical schools, including that of the University of Toronto. Curricula have been revised to expose the student of anatomy to a course of study that eliminates attention to minute details but, by means of a series of lectures and human dissection, stresses functional anatomy and its clinical relevance. In this edition we have tried to meet these objectives. By eliminating unnecessary details, by focusing on clinical basics, and by emphasizing the regional approach, the *Core Textbook of Anatomy* is designed to meet the needs of today's students and practitioners in all healthcare professions.

One obvious departure from the first edition is a reordering of the sequence of the body regions to be studied. This corresponds to the order in which our students dissect the body. There are some changes in terminology that correspond to changes in the anglicized version of the Nomina Anatomica; many new illustrations have been added, previous illustrations have been revised, and some have been enhanced by the use of color. Surface anatomy photographs, radiographs, and modern imaging techniques are used in some parts of the book, not to be comprehensive but to show readers the application of anatomy relevant to the various healthcare professions.

We are indebted to the many individuals who have helped with this edition and apologize to any whose names have been omitted. We are particularly grateful to Drs. Ian Taylor, Ming Lee, and Ken McCuaig for the exchange of ideas during informal discussions; to Mr. Steve Toussaint for providing many dissected specimens that were used as models for our illustrations; to Mr. Douglas Stoddard, our medical student photographic model; and to Tom, Joan, and Hugh, our respective spouses, who got to hate the sight of the backs of our heads as we sat at the computer keyboard or drawing board but who endured!

Finally we should like to thank Dr. Margaret W. Thompson for allowing us to revise the book, Mr. David Barnes of the J. B. Lippincott Company, who invited us to write the second edition of the *Core Textbook of Anatomy*, and the editorial staff of the J. B. Lippincott Company for their assistance with the preparation of the book.

E. J. A., J. A. L., L. W.-P.

From the Preface to the First Edition

This book is written for the medical student who, in today's shortened courses, must understand the basic principles of gross anatomy and retain sufficient knowledge to facilitate later study and clinical practice.

It is based upon many years of teaching medical students and seeing their problems in the course of close association in the laboratory. As a set of mimeographed notes, it was used by students over a period of five or six years, and it covers the material that might reasonably be learned in a course of 150 to 200 hours. The notes have been modified and corrected; clinical notes have been added to emphasize the relevance of the areas studied. The terminology, with few exceptions, is an anglicized version of that used in the internationally recognized *Nomina Anatomica*. Simplification has been stressed in order to promote easy understanding in a short time.

The detailed drawings found in standard atlases of gross anatomy are often difficult to understand. Therefore the illustrations in this book consist almost entirely of simplified line drawings which are intended more for comprehension than for the exhaustive depiction of structures and relationships. The book is designed to be used in conjunction with an atlas in which the form and the detailed relationships of structures can be visualized.

In general the drawings of bilaterally symmetrical structures are for those on the right side. However, occasionally a drawing of the left side is included to stimulate the realization that it is all too easy to become a "one-sided" doctor. Students must train themselves to understand the mirror-image relationship between the two sides of the body.

One shock that is in store for beginning students is the discovery that there are many variations of "normal" form and structure. These anomalies vary in frequency for different organs or parts of the body, and the student or practitioner must be aware of their existence. However, limitations of space required that most variations be ignored by the author. Consequently, although the student will know what to expect in the great majority of cases, he or she should realize that variations from normal can occur.

Towards the back of the book will be found a short list of suggested reading and references. Students may find these references helpful if they wish to have more information on particular topics. On occasion the individual books are referred to in the body of the text.

J. S. T.

Contents

THOMPSON'S

Core Textbook
of Anatomy

An Introduction to Gross Anatomy

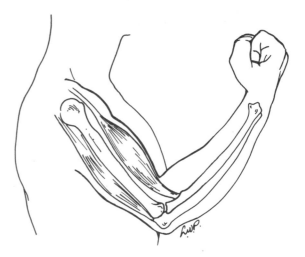

TERMS AND CONCEPTS THE STUDENT SHOULD UNDERSTAND

The human body is a complicated structure and needs to be described in standard terms that can be understood by all practitioners and students.

ANATOMICAL POSITIONS

The ***anatomical position*** is the "standardized" position of the body from which any part may be related to any other part through the use of defined descriptive terms. It is used throughout medical and paramedical practice and can be learned best by the student assuming this position. The anatomical position is as follows:

> The individual is standing erect.
> The face, eyes, and whole body are directed forward.
> The hands are by the side with the palms facing forward.
> The heels are together; the feet are pointed forward so that the great toes touch and are parallel to each other.

In this position the eyes are ***lateral*** to the bridge of the nose, the lips are ***anterior*** to the incisor teeth, the tongue is ***posterior*** to the incisor teeth, the incisor teeth are ***medial*** to the angles of the mouth, and the nose is ***superior*** to the mouth. Also, the hands are distal to the forearms and the forearms are proximal to the hands.

The terms ***proximal*** and ***distal*** are also used in describing portions of nerves and blood vessels and should be taken to mean "toward their origin," that is, proximal, or "away from their origin," that is, distal. For example, the distal part of a spinal nerve is that portion farthest away from the spinal cord. Confusion may arise because the student begins his studies by dissecting the cadaver lying on its back and practitioners examine patients in a similar position. Remember, therefore, that the recumbent body is lying on its posterior surface with its anterior surface presenting; the head end is superior and the foot end is inferior. The body presenting in this manner is said to be in the ***supine*** position and is in the ***prone*** position when turned over to lie on its face.

PLANES

Certain planes of the body are defined and should be clearly understood. In each case the description applies to the body in the anatomical position.

Figure 1-1.
Planes of the body.

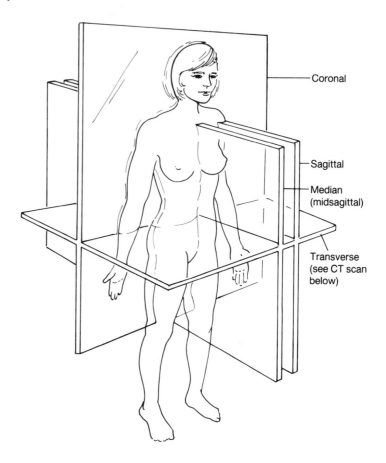

Coronal

Sagittal

Median
(midsagittal)

Transverse
(see CT scan
below)

The sagittal and coronal planes refer to the positions of the sagittal and coronal sutures of the skull (Fig. 1-1).

> The ***median plane*** is that vertical plane that bisects the body into right and left halves.
> A ***sagittal plane*** is any plane parallel to the median plane.
> A ***coronal plane*** is any vertical plane at right angles to the median plane.
> A ***transverse plane*** or ***horizontal plane*** is any plane at right angles to both the median and coronal planes.

Since the advent of computed tomography, anatomists are making an effort to think of and to teach a cross-sectional perspective of the body; unless otherwise defined, cross-section is synonymous with transverse plane (Fig. 1-2)

MOVEMENTS

Movements of a joint may be around any one (or several) of three axes. The following descriptions are applicable to most joints, but elaborations and modifications will be introduced in the sections discussing individual joints.

Flexion-Extension.   In general, ***flexion*** of a joint, usually in a sagittal plane, reduces the angle between two bones and ***extension*** increases it; for example, bending (flexion) and straightening (extension) of the elbow (Fig. 1-3). In some joints this description leads to confusion and spe-

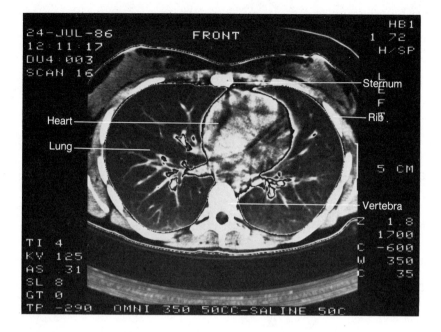

Figure 1-2.
Computed tomogram (cross-section)
of thorax (CT scan).

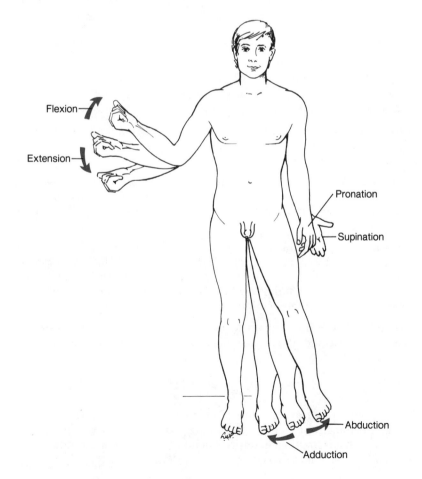

Figure 1-3.
Body movements related to the
anatomical position.

Figure 1-4.
Section of compact bone.

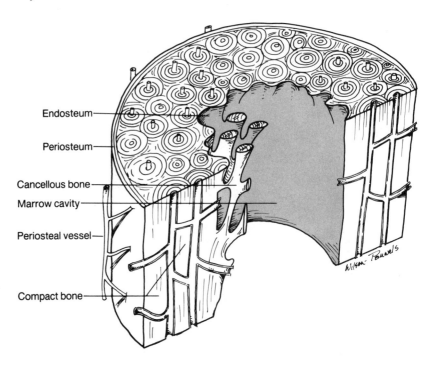

Endosteum

Periosteum

Cancellous bone

Marrow cavity

Periosteal vessel

Compact bone

cial terms are used, for example, ***dorsiflexion*** and ***plantarflexion*** at the ankle joint.

Abduction-Adduction. In general, ***abduction*** (ab=from, ducere=lead) takes the structure away from the long axis of the body (or a limb) and ***adduction*** (ad=to) brings the part back toward the long axis (Fig. 1-3).

Circumduction. A combination of flexion-extension and abduction-adduction leads to a conical movement called ***circumduction***. Rotation is not a component of circumduction.

Rotation. A bone may rotate around its long axis. ***Medial rotation*** brings the anterior surface of the bone or limb closer to the median plane, whereas ***lateral rotation*** takes the anterior surface away from the median plane.

Pronation-Supination (Fig. 1-3). Used in reference to movement of the forearm around its long axis. The palm faces forward in supination and backward in pronation. Based on morphology and comparative anatomical concepts, supination and pronation also occur in the foot and will be detailed in the appropriate section.

Inversion-Eversion. Relates to movements of the foot. ***Inversion*** directs the sole toward the median plane, and ***eversion*** directs it away from the median plane.

BONE

Bone is the supporting structure of the body, consisting primarily of a network of connective tissue fibers interspersed with cells lying in a ground substance that is impregnated with calcium salts to produce rigidity. Bone is usually divided into two types, based on density (Fig. 1-4):

> ***Compact Bone***: dense bone usually forming a firm outer shell around a central mass of cancellous bone.
>
> ***Cancellous Bone***: bone that consists of a mass of spicules between which is bone marrow. This marrow may be active in blood cell formation (red marrow) or may be inert and fatty (yellow marrow). In most long bones, the central portion of the shaft is hollow and is filled with yellow marrow.

Bones may also be classified by shape:

> ***Long bones*** are tubular with a shaft and two expanded ends (*e.g.,* humerus).
>
> ***Short bones*** are cuboidal in shape (*e.g.,* bones of the wrist).
>
> ***Flat bones*** consist of two plates of compact bone separated by cancellous bone (*e.g.,* ilium).
>
> ***Irregular bones*** have various shapes (*e.g.,* facial bones).
>
> ***Sesamoid bones*** are round or oval and are found within tendons.

As a living tissue capable of growth and repair, bone requires an adequate blood and nerve supply. Nutrient arteries enter the bone through foramina that can be easily identified on some bones.

Bone Growth

Bone grows circumferentially by the deposition of new bone under the periosteum with simultaneous resorption of endosteal bone. Bone lengthens at each end. Growth occurs at the epiphyseal plates on both sides of the plate. The shaft of the bone is the ***diaphysis***, where bone forms first, in the ***primary center*** of ossification. The part of the diaphysis adjacent to the epiphysis is called the ***metaphysis***. The ***epiphyseal plate*** separates the diaphysis from the ***epiphysis*** in which the ***secondary center*** of ossification occurs. When the epiphyseal plate ossifies, growth stops at that end of the bone (Fig. 1-5).

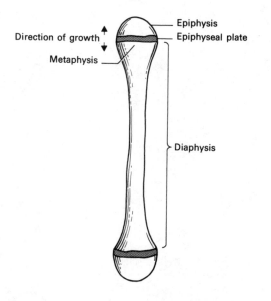

Figure 1-5.
Growth, in length, in a typical long bone.

Bone Age

Calcification (ossification) occurs in the epiphyses at different times that are characteristic for each bone and its epiphyses. The process usually starts before birth or shortly after it. Completion of ossification of the epiphyses occurs somewhat earlier in females, and thus female bones tend to be shorter than those of the male.

CLINICAL NOTE

Radiologic examination will show the appearance and degree of ossification of the secondary centers and the completion of the process of ossification of the bone; the approximate age of the patient may be judged from these appearances (Fig. 1-6). Details of the age of these occurrences are difficult to remember and are best looked up in an appropriate reference book (e.g., Gray's Anatomy). If an epiphyseal plate is damaged by trauma, unequal growth, with resultant deformity of the bone, may occur.

The last epiphysis of a bone to ossify is called the growing end of the bone. In the upper limb the upper end of the humerus and lower ends of the radius and ulna are the growing ends, whereas in the lower limb the lower femoral epiphysis and the upper tibial epiphysis are the first to appear and the last to ossify so that the femur grows from its lower end and the tibia from its upper end.

Because the upper end of the humerus is the growing end, if an amputation through the humerus becomes necessary in childhood, the shaft of the humerus must be left short relative to the soft tissue covering its end at the stump to prevent subsequent growth in length of the bone from

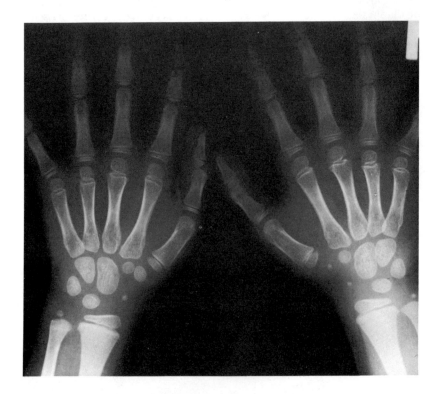

Figure 1-6.
Radiograph of child's hand showing ossification of carpal bones and unfused epiphyses in the long bones. Clear spaces between shaft and epiphyses are cartilage.

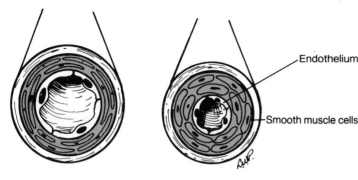

Figure 1-7.
Contraction of smooth muscle in wall
of blood vessel.

Relaxed: larger lumen
and thinner wall

Contracted: narrowed lumen
and thicker wall

*outstripping the available soft tissue. Similar considerations would apply
to amputations through the tibia in childhood.*

MUSCLES

There are three types of muscle:

> ***Striated muscle***, so-called because of its microscopic appearance,
> is under the conscious control of the somatic nervous system
> and is referred to as ***voluntary muscle***.
>
> ***Smooth muscle*** is not striated and is under the control of the
> autonomic nervous system without conscious control or input
> and may be called ***involuntary***. It forms the walls of most
> hollow viscera.
>
> ***Cardiac muscle*** is a special type of striated muscle found in the
> heart and can originate and maintain its own rhythmic
> contractions.

A muscle contracts by shortening the length of the cells that make
up its substance. Thus when smooth muscle contracts it will thicken the
wall and narrow the lumen of a hollow viscus (Fig. 1-7).

When striated (skeletal) muscle contracts it will move its insertion
(distal end) toward its origin (proximal end). The distinction between
origin and insertion is largely a matter of convention and semantics. Iso-
metric contraction can occur when the muscle fibers are made to contract
by a conscious effort but the attachments are prevented from moving by an
external resistance or by deliberate contraction of an opposing muscle
(*e.g.,* as displayed by ***body builders***) (Fig. 1-8).

A muscle can contract about 35% of the length of its fibers. A ***motor
unit*** is one motor nerve cell and the muscle cells it supplies (see Fig. 1-
18). The fewer the number of muscle cells in a motor unit, the more dis-
creet and precise will be the movement.

Prime mover is the name given to the muscle that produces the desired movement. An ***antagonist*** will oppose the prime mover, and a ***synergist*** prevents unwanted movement at a joint over which the prime mover
passes. In reality any muscle can perform the above-mentioned functions,
and a smoothly executed, skilled movement requires the action of the
prime mover with simultaneous relaxation of the antagonist and the stabilizing influence of the synergist (Fig. 1-8).

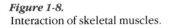

Figure 1-8.
Interaction of skeletal muscles.

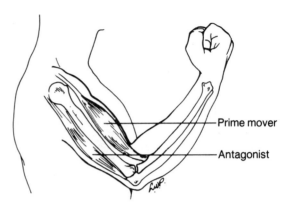

Hold a pencil in the act of writing and note that the fine movements are carried out by the muscles that flex and extend the digits while the wrist extensors and flexors steady the wrist.

Fixators steady joints that are closer to the skeletal axis than the origin of the prime mover. Muscles that fix the shoulder and elbow relative to the axial skeleton enable the fingers and wrists to function in discrete movements.

When a motor unit is stimulated by an impulse arising in the motor nerve cell, the response of the muscle fiber will be total, that is, there will be maximal contraction of all the muscle fibers supplied by that motor nerve—the "all or none principle" of response.

It should be obvious from this that a powerful muscular effort will employ all, or most, of the motor units of a muscle whereas a weak effort is the result of partial use of available units.

Specific muscle actions will be described herein but are by no means as well understood as their descriptions imply.

We can study a muscle's action by postulating the movement that would result from the approximation of its insertion to its origin, or we can force a movement against resistance and determine by palpation which muscle contracts. Try flexing your elbow against resistance and see how the biceps brachii muscle of the arm stands out. We can stimulate a muscle by electricity and watch it twitch or we can apply electrodes to muscles and record their electrical activities during certain movements or in various postures. This last advance of *electromyography* has yielded much useful information and shattered many old anatomists' deductive illusions.

JOINTS

There are several different types of joints or *articulations* in the body; their only common feature is that two or more bones are held together in some manner. Rarely, a cartilaginous structure may take the place of bone, for example, where the ends of ribs join the sternum.

Joints are generally classified in terms of the materials that join the bones (Fig. 1-9).

Bony Joints

In some places two bones have joined together to make a single bone (*e.g.*, the conversion of the two frontal bones in the fetus and child to the single adult form) so that no true joint remains; this is called a *synostosis*.

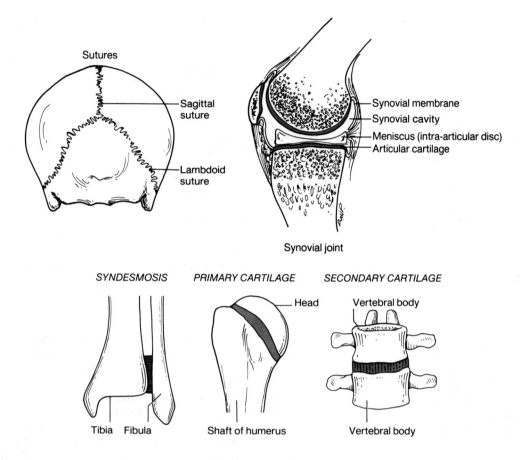

Fibrous Joints

Suture. The bones articulate with each other and are held together by a thin layer of fibrous tissue. The edges of the bones are usually reciprocally irregular. Sutures are found only in the skull.

Syndesmosis. The bony surfaces are held together by fibrous tissue, but the irregularities of the bones are less marked than in sutures, for example, lower tibiofibular joint.

Cartilaginous Joints

Primary (synchondrosis). The component bones are held together by hyaline cartilage as in the junction between the diaphysis and epiphysis of a long bone.

Secondary (symphysis). Here, fibrocartilage (and ligaments) join the bones. For example, the joints between the ***bodies of vertebrae*** or the ***pubic symphysis*** are secondary cartilaginous joints.

Synovial Joints

These are the most common joints in the body and, because of their special structure, normally allow free movements between the bones they join.

Figure 1-9.
Diagrammatic examples of different joint types.

A
ARTICULAR DISC

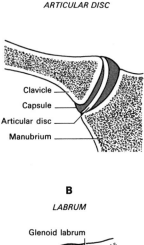

Clavicle
Capsule
Articular disc
Manubrium

B
LABRUM

Glenoid labrum

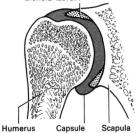

Humerus   Capsule   Scapula

C
TENDON

Capsule

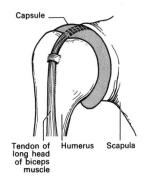

Tendon of   Humerus   Scapula
long head
of biceps
muscle

Figure 1-10.
Special features of certain joints:
(**A**) sternoclavicular joint; (**B**) and
(**C**) shoulder joint.

Characteristics of Synovial Joints (Fig. 1-10). A synovial joint has certain typical features.

The bone ends are covered by hyaline ***articular cartilage***.
There is a joint ***cavity***, but this is more potential than real.
A capillary layer of ***synovial fluid*** fills the cavity. When a joint is inflamed, the amount of fluid may increase.

The joint is surrounded by a ***capsule*** of fibrous tissue that may be thickened in places to form ***intrinsic ligaments***. Both capsule and ligaments are of major importance in maintaining the stability of the joint and the bones in their normal relationships.
Extrinsic ligaments are connective tissue bands that hold bones together but are separated from the capsule.
Synovial membrane lines the whole cavity of the joint except over the hyaline, articular cartilages. This synovial membrane is made of connective tissue lined by flattened cells that secrete synovial fluid, which lubricates the joint and has a surface-tension-like effect in keeping the articular cartilages in apposition.

Some synovial joints contain additional structures:

Articular disc. This fibrocartilaginous disc adapts incongruent surfaces of bone and permits movements around more than one axis (*e.g.,* sternoclavicular joint and knee joint). The disc is attached to one or both bones of the joint (Fig. 1-10).
Labrum. This ***fibrocartilaginous ring*** deepens an articular surface. The glenoid labrum of the shoulder joint and the acetabular labrum of the hip joint are two examples (Fig. 1-10).
Tendons. In some joints a tendon passes through the joint cavity, surrounded by synovial membrane, for example, the tendon of the long head of the biceps muscle in the shoulder joint.

Types of Synovial Joints. Several classifications of synovial joints are based on the shape of the bony surfaces or the various movements possible at the joint, or both. One classification is simply ***uniaxial, biaxial***, or ***multiaxial*** depending on the number of axes around which the joint can be moved. However, the most usual classification is based on the shape of the joint (Fig. 1-11).

Ginglymus Joint. This is a hinge joint, for example, the humeroulnar joint (uniaxial).

Pivot Joint. One bone rotates around its long axis; for example, at the superior radioulnar joint the radius rotates (uniaxial).

Condyloid Joint (knuckle-like). This joint has two axes at right angles to each other (biaxial). The joint surfaces are usually oval, for example, between the head of one of the four medial metacarpals and the base of its proximal phalanx.

Saddle-shaped Joint. This is seen in the carpometacarpal joint of the thumb (biaxial).

Ball and Socket Joint. A part of a sphere fits into a hollow, for example, the hip and shoulder joints (multiaxial).

Plane Joint. This allows gliding motions. The joint between two carpal bones is an example (usually uniaxial).

Innervation of Joints. The nerves that pass over a joint and supply the muscles that move the joint send sensory fibers to the joint and the skin overlying it (Hilton's law). The sensations transmitted are those of pain and position.

Position sense or ***proprioception*** is an interesting phenomenon that enables the body to determine the position and degree of angulation of a joint. It is also a feature of skin sensation when it interprets deformity of the skin surface by pressure applied to it. For example, put your hand in your pocket and take out a coin and identify it without looking. The texture of the coin, the feel of its surface and rim, and its weight are measures of skin proprioception; its shape is recognized by the positions your finger and thumb joints adopt in grasping it. This is joint proprioception. Position sense of joints also tells you whether you are walking on an even or uneven surface, stepping upward, and so forth.

CLINICAL NOTE

When a joint becomes inflamed it may fill with an excess of synovial fluid, and this swelling can be visible where the capsule is subcutaneous, for example, around the knee. If the inflammation causes pain, the muscles around the joint will go into spasm, and later they will waste in bulk (atrophy) quite rapidly so that a lot of physiotherapy will be required when the acute phase of inflammation has subsided to remobilize the joint and to build up the wasted muscles.

BLOOD VESSELS

There are three principal types of blood vessels in the body. ***Arteries*** carry blood away from the heart. ***Capillaries*** form a microscopic network of vessels joining arteries and veins. Through the thin walls of capillaries various substances and cells pass from the blood to the tissue fluids. Tissue fluids are returned to the circulation by way of capillaries and lymph vessels (Fig. 1-12). Veins return the blood to the heart. All blood vessels are lined by delicate flattened cells that form the inner lining, the ***endothelium*** (see Fig. 1-7).

Arteries have relatively thick walls and appear round in cross-section (Fig. 1-13). The larger arteries contain much elastic tissue in their walls to absorb the pulse wave that is produced by each heart beat. In smaller arteries the elastic tissue is largely replaced by smooth muscle. Major arteries supply particular parts of the body, but there are some variations in individual bodies.

Veins may be identified by the fact that they are usually flaccid and thin walled and, in the cadaver, may contain clotted blood. In general,

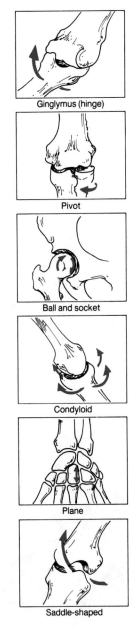

Ginglymus (hinge)

Pivot

Ball and socket

Condyloid

Plane

Saddle-shaped

Figure 1-11.
Types of synovial joints.

Figure 1-12.
Lymphatic drainage of interstitial fluid
from capillary bed.

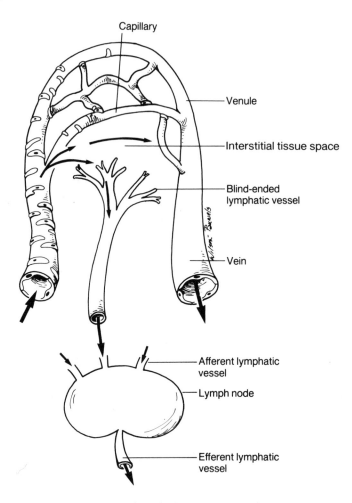

Capillary

Venule

Interstitial tissue space

Blind-ended
lymphatic vessel

Vein

Afferent lymphatic
vessel

Lymph node

Efferent lymphatic
vessel

veins accompany arteries and, with the exception of some major veins and some subcutaneous veins, rarely have names of their own. They are referred to as **venae comitantes** of a particular artery. The venous system displays considerable variations and often does not follow textbook descriptions.

Venous pressure is much lower than arterial pressure, and venous return of blood to the heart depends, to some extent, on the contraction of surrounding muscles. For example, in the lower limb the veins in the deep layers are surrounded by the skeletal muscles which, when they contract, squeeze the venous blood both upward and downward; however, downward flow is prevented by the presence of valves that allow the blood to flow only in an upward direction, that is, toward the heart (Fig. 1-13). Subcutaneous veins contain similar valves, and the student can observe these on his own forearm.

LYMPHATIC SYSTEM

The lymphatic system provides for drainage of tissue fluids back to the venous system and also provides the main **immune mechanism** of the body. Tiny particles of foreign protein from an infection or an implanted piece of tissue drain through the lymph vessels to regional lymph nodes. This alerts the immune mechanism to the presence of foreign substances, and special cells of the lymphoid system produce **antibodies** or form **killer lymphocytes** directed against the foreign substance.

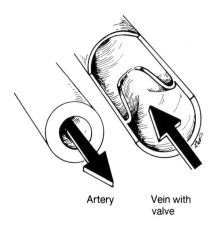

Artery   Vein with
               valve

Figure 1-13.
Differences between artery and vein.

The antibody can be carried to the tissues by the bloodstream; this is known as the humeral mechanism of the immune reaction. For transplanted tissue, lymphoid cells are important in the actual rejection of the graft.

The lymphatic system is the only normal means by which protein in the tissue fluids is conducted back into the circulatory system. It also carries fat from the intestines to the bloodstream. For a more detailed description of the immune system and the specific lymph cells involved, the student should consult the appropriate texts on immunology and histology.

Lymph Vessels

The lymph vessels start as blind capillaries in the tissue spaces; these capillaries, lined by endothelium, join to become larger lymph-collecting vessels that eventually drain into the regional *lymph nodes* of the area (see Fig. 1-12). The walls of the larger lymph vessels, ducts, or trunks contain smooth muscle.

The vessels that drain to a lymph node are called *afferent* vessels. Vessels that leave lymph nodes and carry the lymph to the next group of nodes are called *efferent* vessels (Fig. 1-14). This progression from node to node continues in several steps until the lymph vessels become large

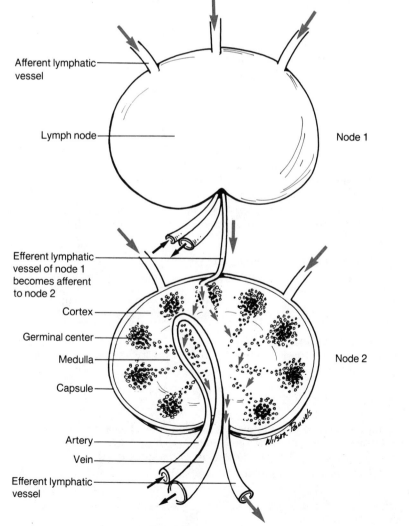

Figure 1-14.
Lymph flow.

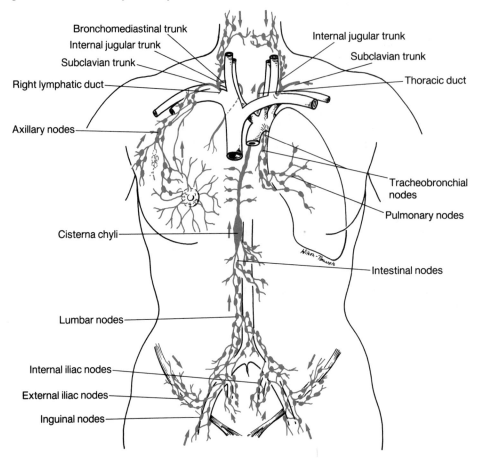

Bronchomediastinal trunk
Internal jugular trunk
Subclavian trunk
Right lymphatic duct
Axillary nodes
Internal jugular trunk
Subclavian trunk
Thoracic duct
Tracheobronchial nodes
Pulmonary nodes
Cisterna chyli
Intestinal nodes
Lumbar nodes
Internal iliac nodes
External iliac nodes
Inguinal nodes

Figure 1-15.
An overview of the lymphatic drainage of the body.

enough to be called ***lymphatic trunks***. These trunks join to form either the ***thoracic duct***, which drains into the venous system at the junction of the left internal jugular and subclavian veins, or the ***right lymphatic duct***, which joins the corresponding veins on the right side. In general, the thoracic duct drains the whole body except for the right upper limb, the right half of the thoracic cavity, and the right side of the head and neck (Fig. 1-15).

Superficial Lymph Vessels. The vessels of the skin form a fine network on the deep surface of the epithelium; they then join to form slightly larger vessels that eventually pass inward to join the deeper vessels in the deep fascia. The superficial vessels tend to follow the course of the superficial veins of the skin. Larger lymph vessels contain valves.

Deep Lymph Vessels. The deep vessels run in the deep fascia, again paralleling major blood vessels. Visceral lymphatics drain to regional nodes grouped around the origins of the major arteries.

Lymph Node

Lymph nodes consist of an aggregation of encapsulated lymphatic tissue and vary in size from a few millimeters to about 2 cm in diameter. When they respond to infection or deposits of malignant cells, they can become very large and easily palpable.

Structure of Lymph Nodes (see Fig. 1-14). A lymph node has a hilus at which its blood supply enters and leaves. The various afferent lymphatic vessels enter at the periphery and the efferent lymphatic vessels leave at the hilus to become the afferent vessels of the next group of nodes in the ascending chain of lymphatic drainage.

The node has a cortex and a medulla, through which connective tissue trabeculae run, passing from the hilus to the periphery. The cortex contains collections of lymphatic cells at various stages of maturation arranged in lymphatic follicles with germinal centers. In the medulla the cells are arranged in cords. ***Macrophages*** are found on the trabeculae and can remove foreign matter from the lymph. See the carbon deposits in pulmonary and mediastinal lymph nodes, derived from the inhalation of hydrocarbon combustion products in the atmosphere.

CLINICAL NOTE

Lymph nodes can become enlarged when they contain secondary (metastatic) deposits of tumor cells or when they are the primary focus of malignant disease arising in the lymphatic system, for example, ***leukemia*** *and* ***Hodgkin's disease***. *They may also enlarge in infections of their drainage territory. Some specific parasitic infections, occurring mostly in tropical countries, may block lymph nodes to the extent that they cannot drain their regional territories, and massive enlargement of limbs, owing to accumulation of protein laden-tissue fluids, produces the grotesque features of* ***elephantiasis***.

The precise locations of lymph nodes and their areas of drainage will be described, in detail, in the regional chapters of this book.

Lymph

Lymph is a translucent, watery fluid with a specific gravity of about 1.015. It contains many lymphocytes, protein, and, in the vessels draining the intestines, some fat.

Functions of Lymphatic System

1. The drainage of tissue fluid and protein into the venous system.
2. The absorption and transportation of fat from the small bowel to the portal venous system.
3. The protection of the body by providing an immune system for the rejection of foreign protein or cells.

General Plan of Lymphatic Drainage of the Body (Fig. 1-15)

Superficial Lymph Vessels of the skin and subcutaneous tissue drain three distinct areas of the body.

1. The ***scalp***, ***face***, and ***neck*** drain through ***cervical*** nodes to the right lymphatic duct or the thoracic duct.
2. The ***upper limb*** and ***trunk*** (anterior and posterior surfaces) above the level of the umbilicus drain to the ***axillary nodes*** and thence to the right lymphatic duct or thoracic duct.
3. The skin of the ***lower limb***, ***perineum***, ***external genitalia***, the ***trunk*** below the umbilicus (anterior and posterior surfaces),

the ***buttocks***, and the lower half of the ***anal canal*** drain to the ***inguinal*** nodes.

Deep Vessels.   The deep vessels drain parts of the body deep to the deep fascia and usually follow the course of the blood vessels.

NERVES

The nervous system may be divided, by function, into two major portions: somatic and autonomic. The ***somatic*** nervous system consists of ***sensory*** and ***motor*** components.

The ***somatic sensory system*** transmits ordinary sensations of touch, pain, temperature, or position from sensory receptors (Fig. 1-16). The special senses, such as taste, sight, and smell, are not usually considered to belong exclusively to either somatic or visceral systems. The ***somatic motor*** system allows voluntary movement by causing contraction of striated muscle.

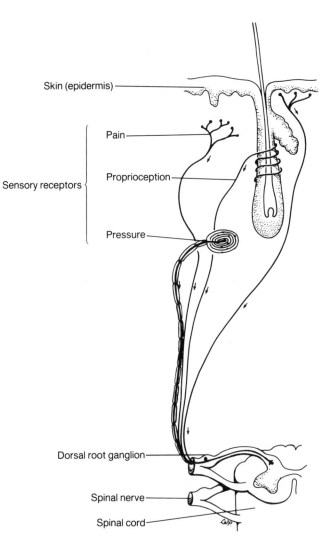

Figure 1-16.
Sensory receptors of the skin.

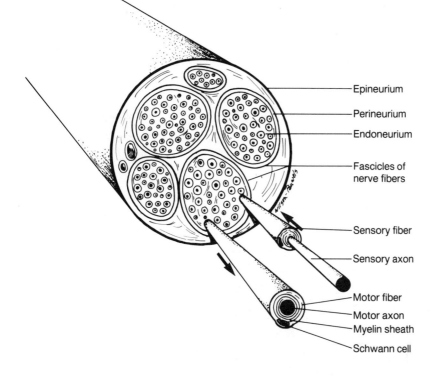

Figure 1-17.
Cross-section of a peripheral nerve.

Epineurium
Perineurium
Endoneurium
Fascicles of nerve fibers
Sensory fiber
Sensory axon
Motor fiber
Motor axon
Myelin sheath
Schwann cell

The ***autonomic nervous system*** consists of **sympathetic** and ***para-sympathetic*** components, both of which are motor to smooth (involuntary) muscle, glands, blood vessels and so forth. There are, in addition, sensory fibers that serve the internal organs (*e.g.,* those that transmit the pain of smooth muscle colic). These visceral sensory pathways have not been determined with absolute certainty, but it is generally believed that the sensory neurons reside in the ***dorsal root ganglia*** of spinal nerves and in the sensory ganglia of cranial nerves and that the peripheral processes of these neurons are included in the substance of the autonomic "motor" nerves. Further details of these pathways and the concept of visceral pain are found in Chapters 12 and 16.

Structure of Nerves

A peripheral nerve, as a definable entity in the body, consists of many parallel ***axons*** that carry impulses to and from the central nervous system. Somatic axons are covered by an "insulating" layer of ***myelin*** and a sheath of ***Schwann*** (neurilemma) cells. Several axons are bound together by a connective tissue ***perineurium*** to form a bundle. Many such bundles are held together by a surrounding ***epineurium*** to form a typical nerve. The term ***endoneurium*** refers to a connective tissue sheath that surrounds one axon, its myelin sheath and neurilemma (Fig. 1-17).

An ***axon*** is an elongated process of a nerve cell, and a ***dendrite*** is a shorter process. Dendrites carry impulses to the nerve cell body, and axons carry impulses away from the cell body. This makes the peripheral part of

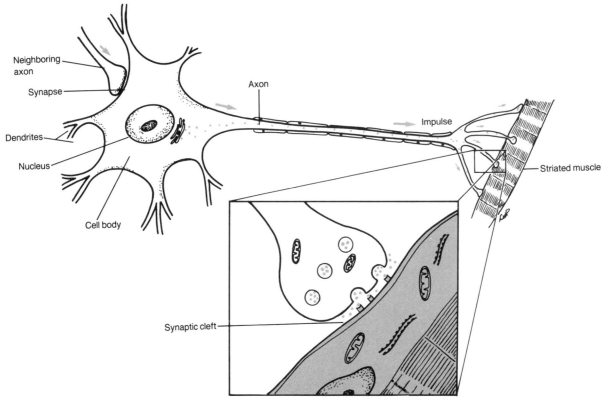

Motor end plate

Figure 1-18.
Typical motor neuron.

a sensory nerve fiber a dendrite, but, because of its extreme length, it is called an "axon." Figure 1-18 shows a typical motor neuron and its processes. A ***synapse*** (Fig. 1-18) is a special junction at which the axons of one neuron transmit an impulse to the dendrites, cell body, or axon of another neuron.

A typical sensory neuron has a T-shaped process that permits an impulse to pass from the periphery to the cell body, located in a sensory ganglion, and into the central nervous system without synapsing.

A typical ***spinal nerve*** comes from the spinal cord in two groups of rootlets that unite to form two roots. The sensory rootlets emerge from the posterolateral aspect of the cord, and the motor rootlets arise from its anterolateral aspect. The sensory (dorsal) ganglion is seen on the posterior root, and both roots join to form the spinal nerve in the ***intervertebral foramen*** (Fig. 1-19).

Immediately after passing through the intervertebral foramen the spinal nerve divides into a ***posterior primary ramus*** and an ***anterior primary ramus***. At, or very near, the point of division all spinal nerves are connected to the sympathetic trunk by a ***gray ramus communicans***. In the region of the first thoracic to the second lumbar nerves, a ***white ramus communicans*** connects the spinal nerve to a sympathetic ganglion (Fig. 1-19). This results in both anterior and posterior primary rami containing somatic motor and sensory fibers and sympathetic (autonomic) components (see Fig. 16-1). Details of the ***sympathetic nervous system*** are given in Chapter 16.

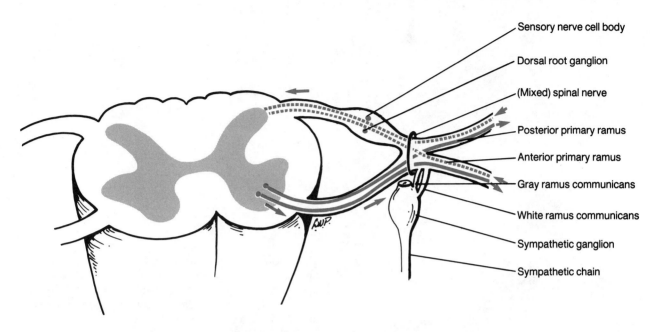

The ***posterior primary rami*** supply somatic motor, sensory, and sympathetic fibers to the muscles and skin of the back, from the midline to approximately the lateral borders of the ***erector spinae*** group of muscles. The ***anterior primary rami*** supply motor, sensory, and sympathetic fibers to the anterior and lateral muscles and skin of the neck and trunk and to all of the muscles and skin of the limbs, including limb muscles that have migrated to the back (superficial muscles of the back).

Figure 1-19.
A typical thoracic (mixed) spinal nerve.

The Skin and Subcutaneous Structures of the Pectoral Region

This short chapter will describe skin and subcutaneous structures in general terms and then will proceed to describe the breast, the pectoral muscles and fascia.

The pectoral muscles are integral parts of the upper limb but must be mentioned here because they cover the wall of the thorax, which is the subject of the next chapter. They will be mentioned again, in greater detail, in Chapter 35 where the upper limb will be introduced.

The entire surface of the body is covered by skin that consists of a layer of epithelium, deep to which are several layers of connective tissue. The epithelium forms the *epidermis*, and the first layer of connective tissue forms the corium or *dermis*. These two layers together comprise the *skin*. The other layers of connective tissue form the *tela subcutanea* or *superficial fascia*, and the *deep fascia* that surrounds muscles (Fig. 2-1).

Muscles lie deep to the deep fascia, and, where bone is not covered by muscle, the deep fascia will usually be attached to bone.

THE SKIN

As mentioned above, skin consists of two layers: the epidermis and the dermis.

EPIDERMIS

The epidermis consists of a layer of *epithelium* arranged in four strata. The cells formed in the deepest layer (germinativum) become flattened and hardened (keratinized) as they are pushed toward the surface, where they eventually flake off. The epidermis has a relatively smooth surface, except where sweat glands and hair follicles are present and over the hands, feet, fingers, and toes, where it is ridged to form the patterns of fingerprints, palm prints, and so forth. In addition the skin exhibits flexure lines (skin creases) opposite joints and tension lines, which divide the surface into

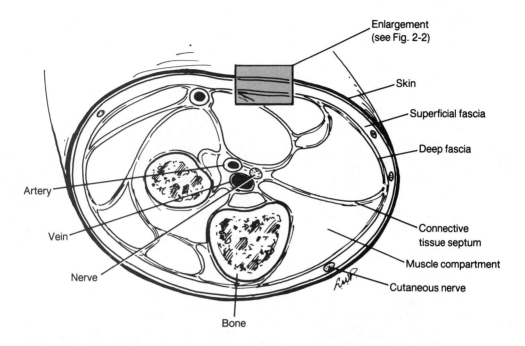

Enlargement
(see Fig. 2-2)

Skin

Superficial fascia

Deep fascia

Connective
tissue septum

Muscle compartment

Cutaneous nerve

Artery

Vein

Nerve

Bone

Figure 2-1.
Diagrammatic cross-section of limb.

large numbers of polygonal patterns. The deep surface of the epidermis is very uneven, and the dermis projects into it in many places (Fig. 2-2).

Epithelium consists of closely packed cells separated by very little intercellular material. The cells are generally cuboidal in shape, but the cube may be elongated to form ***columnar cells*** or flattened to form ***squamous cells***. Because the squamous cells predominate, in multiple layers, particularly near the surface, the epithelium is called ***stratified squamous*** (Fig. 2-3).

In some areas, particularly those exposed to pressure, the skin will thicken and produce hard calluses composed of dense layers of keratinized squamous cells.

Note that fingernails and toenails are composed of specially compacted layers of degenerated, keratinized squamous cells and that the nail beds from which they grow are integral parts of the skin.

Skin coloring is influenced by ***melanin***, a dark brown pigment that resides in the germinative layer of the skin and is produced by cells called ***melanoblasts***. ***Carotene***, an orange pigment similar to vitamin A, is found in the stratum corneum, the dermis, and the subcutaneous fat.

DERMIS

The dermis is a layer of ***connective tissue*** that is densely packed and forms a base for the epidermis. Its superficial surface is irregular, and projections of connective tissue push into the epidermis. The skin is supplied by fine blood vessels and nerves that reside in the dermis.

Connective tissue consists of several kinds of cells embedded in a ***matrix*** of gelatinous material through which pass many ***connective tissue fibers***. These fibers may be ***elastic*** or ***collagenous***; the latter, when chemically tanned, are the substance of leather. Elastic fibers, as the name implies, have the ability to resume their original shape after being stretched, and account for the elasticity of the skin, particularly in the young. Pick up

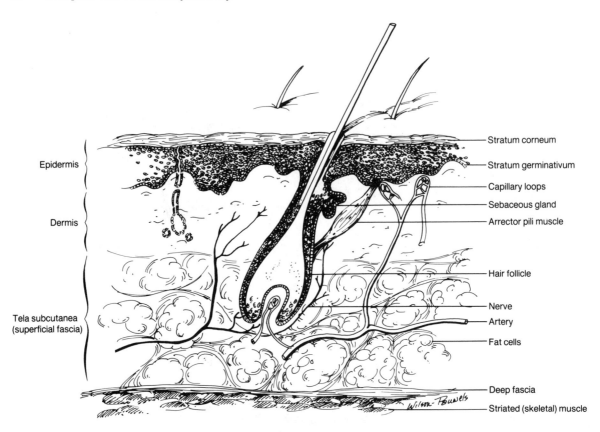

Figure 2-2.
Skin and subcutaneous tissue.

the skin on the back of your hand and watch it recoil to its original shape upon release. Compare this with the skin on your grandmother's hand.

The cells of the dermis are *fibroblasts*, *fat cells*, *melanoblasts*, and many others. *Nerves*, *blood vessels*, and *lymph vessels* run through the connective tissue.

The functions of the skin are varied and important. The skin is the "interface" between the body and the outside world. It provides the following:

1. Protection: (a) *mechanical*, from abrasions and other harmful forces; (b) from *fluid loss*, which can be enormous in cases of severe burns.
2. Heat Regulation: (a) *sweat glands* allow the body to be cooled by the evaporation of sweat from the skin surface; (b) blood vessels, by dilating or constricting, control the amount of blood that is in contact with the surface, and therefore control the amount of heat lost by the body.
3. Sensation: Superficial nerves and their end-organs, or corpuscles,

Figure 2-3.
Epithelia.

Stratified squamous   Simple squamous   Cuboidal   Columnar

permit the recognition of touch, pain, temperature, pressure, and vibration (see Fig. 1-16 for some typical cutaneous sensory nerve endings).

APPENDAGES OF THE SKIN

The skin has several appendages:

1. Nails (mentioned above).
2. Sweat glands (sudorific glands) that produce sweat and consist of epithelium-lined, convoluted tubules that reside below the dermis and empty through ducts that open on the epidermis.
3. Hair follicles located below the dermis, producing body hair, the distribution of which is self-evident (see Fig. 2-2).
4. Sebaceous glands that empty their oily secretions into hair follicles (see Fig. 2-2).

SUPERFICIAL FASCIA

Deep to the dermis lies more connective tissue, but its fibers blend with that of the dermis so that no distinct plane of cleavage can be found. The subcutaneous tissue contains fat cells and varies in thickness with the obesity of the individual. This layer is called the ***tela subcutanea*** or, more commonly, the ***superficial fascia***. Note that fat is present in the superficial fascia throughout the body except in the eyelids and the penis.

Functions of Superficial Fascia
1. It serves as a place of storage for water and fat.
2. It forms a layer of insulation protecting the body from heat loss.
3. It provides mechanical protection from pressure and blunt injuries.
4. It provides a pathway for vessels and nerves.
5. It provides a site for the location of a special organ, the breast.
6. In certain areas of the body it contains a thin layer of muscle, especially in the neck and face where it contains the striated ***muscles of facial expression***.
7. It provides a smooth contour for the outline of the body.
8. In some places, *e.g.*, the point of the elbow, it contains ***bursae*** that are potential spaces, lined by synovial membrane, allowing the skin to glide smoothly over bone and affording the skin some protection against pressure.

DEEP FASCIA

This fascia is composed of a fibrous tissue layer interposed between the superficial fascia and muscles. Its fibers are continuous with those of the superficial fascia and the connective tissue between the muscle fibers. It is a tough structure that allows muscles to move freely over adjacent structures. It contains muscles within their own compartments, as will be seen when the limbs are described (see Fig. 2-2).

Functions of Deep Fascia
1. It permits free play of muscles.
2. It carries nerves and blood vessels.
3. It fills spaces between muscles.
4. It provides attachments for some muscles.

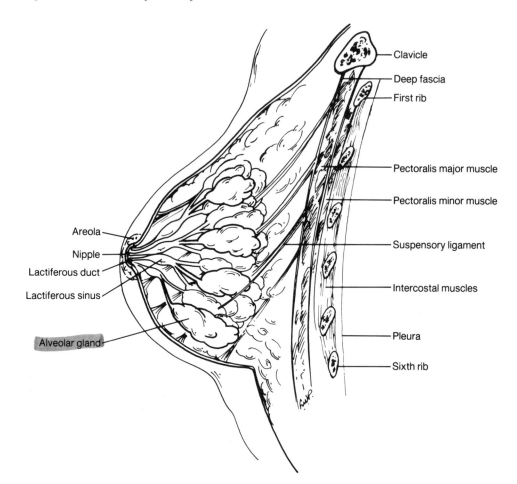

Clavicle

Deep fascia

First rib

Pectoralis major muscle

Pectoralis minor muscle

Suspensory ligament

Intercostal muscles

Pleura

Sixth rib

Areola

Nipple

Lactiferous duct

Lactiferous sinus

Alveolar gland

Figure 2-4.
Active breast (sagittal section).

THE BREAST

The ***breast*** or ***mammary gland*** consists of modified sweat gland tissue located in the superficial fascia (Fig. 2-4). In the male and in preadolescent females it is rudimentary, consisting only of the nipple and a few ducts.

At puberty, in the female, the glandular volume and complexity increase considerably, as does the amount of fat within its stroma. The contour and size of the female breast show considerable individual variations, but the location and size of its base are fairly constant, covering the area of the second to sixth ribs and extending upward and laterally into the subcutaneous tissue of the axilla (armpit), to form the ***axillary tail***.

The resting female breast consists of ducts, a few alveolar glands grouped into lobes, and much fibrous and fatty stroma. Some of the fibrous tissue runs from the deep fascia to the skin and areola in strands that appear to support the breast. These were called the ***suspensory ligaments*** by Astley Cooper. The lobes empty onto the nipple through about 15 to 20 ***lactiferous ducts***. The nipple, which contains erectile tissue, is surrounded by pigmented skin, the ***areola***. During pregnancy these alveolar glands become greatly enlarged.

The arterial supply of the breast is from the ***lateral thoracic artery*** (a branch of the axillary artery), from branches from the second to fifth intercostal arteries and from perforating branches from the ***internal thoracic artery***.

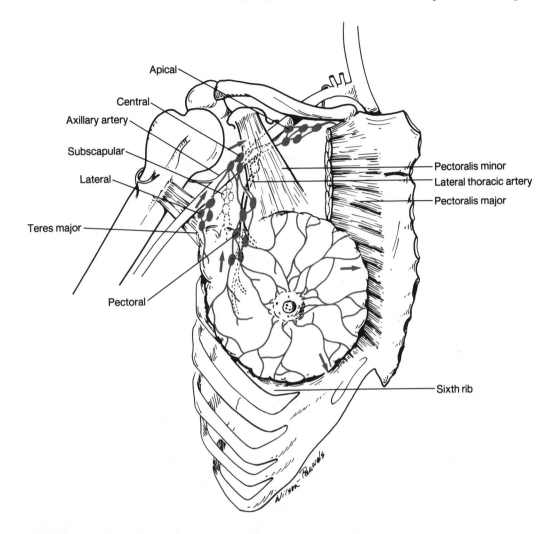

Figure 2-5.
Lymphatic drainage of right breast.

The ***lymphatic drainage*** of the breast is of great clinical importance in the spread of breast cancer (Fig. 2-5). There is a network of subcutaneous lymph vessels around the areola, and this, and the deeper lymph vessels of the gland, drain as follows:

The lateral half of the breast drains mainly into the pectoral, apical, and subscapular subgroups of the ***axillary nodes***. The medial half drains to the axilla and, possibly, to the opposite breast, and by vessels piercing the medial ends of the intercostal spaces, to the ***internal thoracic*** (parasternal) nodes.

The lower half drains into the axilla and to anterior abdominal wall vessels, and thence to mediastinal nodes in the thorax.

The upper half drains into the axillary nodes, specifically to the apical subgroup.

CLINICAL NOTE

1. *Cancer of the breast may invade the suspensory ligaments and shorten them so that the skin overlying the breast or the nipple, or both, may be retracted. If tumor tissue blocks the deeper lymph vessels within the breast tissue, lymph from*

*the skin cannot be drained away, and the skin will become puffy and pitted to resemble the texture of the skin of an orange (**peau d'orange**). The upper lateral quadrant is the most common site for breast cancer.*

2. *The nipple is a poor reference point owing to its variable position, even in the male.*

3. *During pregnancy the breast enlarges and secretes a milk-like fluid (**colostrum**); within a few days after birth the colostrum changes to milk.*

4. *Morphologically mammals have the potential to develop many breasts along the **milk line**, which runs from the pectoral region to the groin. In man, accessory breasts or nipples may be seen along this line. Such accessory breast tissue does not always drain into its own nipple so that the breast secretions have no escape during lactation and may become very swollen and painful. Such isolated islands of mammary tissue are not uncommon in the axillae.*

THE PECTORAL REGION

As mentioned at the beginning of this chapter, the pectoral region will be referred to, briefly, in order to serve as an introduction to the thorax, which will form the subject of the next five chapters.

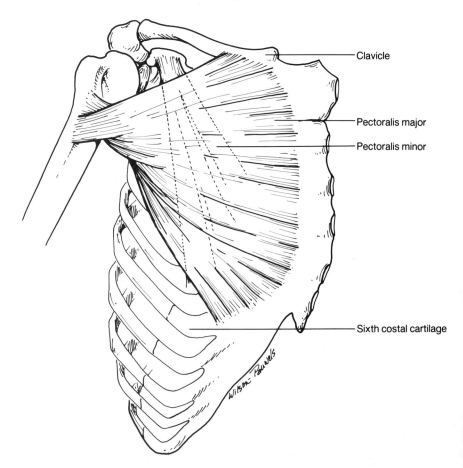

Figure 2-6.
Pectoralis major and minor muscles.

The only bone connecting the upper limb to the axial skeleton is the clavicle. It articulates with the sternum and the scapula, which have many muscular attachments to the axial and thoracic skeletons.

The anterior muscles in this group comprise the ***pectoral muscles***. They also form the anterior wall of the axilla (Fig. 2-6).

The pectoralis major, a large triangular muscle, arises from the sternum, clavicle, and the upper six ribs. It inserts into the upper end of the shaft of the humerus. It is covered by the ***pectoral fascia*** upon which the breast rests.

Behind the pectoralis major is the much smaller ***pectoralis minor*** muscle that arises from the third to fifth ribs and inserts into the coracoid process of the scapula.

The ***clavipectoral fascia*** is attached to the clavicle above, splits to enclose the pectoralis minor muscle, and is attached below to the fascia forming the floor of the axilla.

AXILLARY LYMPH NODES

The main group of lymph nodes of the upper limb are found in the axilla. The ***axillary nodes*** may be divided into five groups; the grouping is not necessarily distinct, and the location of individual nodes and groups of nodes varies.

Apical Group. The apical lymph nodes form a small group around the axillary artery at the very apex of the axilla. This group receives afferents from all the other axillary nodes. The efferents from the apical nodes join to form the ***subclavian lymph trunk***, which joins the ***jugular*** and ***bronchomediastinal trunk*** to form the ***right lymphatic trunk*** on the right side; on the left they join the ***thoracic duct***.

Central Group. The central group of nodes lies medial to the axillary artery deep to the pectoralis minor muscle. It receives afferents from the ***lateral***, ***pectoral***, and ***subscapular*** groups. Its efferents go to the apical group.

Lateral Group. This group of nodes is located lateral to the axillary artery near the lower border of teres major. They receive most of the lymph from the upper limb and drain into the ***central group***.

Subscapular Group. This group is situated along the subscapular vessels and receives afferents from the posterior surface of the chest wall and scapular region. Its efferents pass to the central group.

Pectoral Group. These nodes are found along the lateral thoracic artery at the lower border of pectoralis minor and are overlapped by the lower border of pectoralis major. They drain the anterior chest wall and the breast. Their efferents drain into the central and ***apical*** groups.

The Thoracic Wall

The thorax is the body cavity that contains the heart and its great vessels, the lungs and their respiratory passages, and the esophagus and other important structures. For the protection of these structures the thorax requires a strong, bony wall that will allow the essential movements of respiration. The bones of the thorax, particularly the ribs and their methods of articulation, are well suited to perform these dual tasks.

BONES OF THE THORACIC WALL

The bony skeleton of the thorax comprises the following (Fig. 3-1): 12 thoracic vertebrae; 12 pairs of ribs; and the sternum.

THE THORACIC VERTEBRAE

A typical vertebra consists of an anterior part, the body, and a posterior arch that is extended into lever-like processes and surrounds the vertebral foramen, which contains the spinal cord and its coverings (Fig. 3-2).

The body is connected to that of the vertebrae above and below by a fibrocartilaginous disc, the ***intervertebral disc*** (Fig. 3-3). In the aggregate all the vertebral bodies and their discs form a flexible pillar that bears a good deal of the total weight of the body. The foramina, combined, form the vertebral canal, which contains and protects the spinal cord. Each side of the arch contains half a notch, which, with that of the arch below, forms an intervertebral foramen for the passage of the spinal nerves.

The arch is projected from the posterior aspect of the body by two pedicles that continue into two laminae that fuse in the midline to form the spine of the vertebra.

Projecting from the junction of the laminae and pedicles are two laterally directed transverse processes for the attachment of muscles and two superior and two inferior articular facets that make synovial joints with the vertebrae above and below to permit the movements of flexion, extension, and rotation. (For further details of spinal movements, see Chapter 23.) The above remarks apply to vertebrae in general. Each area of the spine presents some typical variations; the most obvious of these, in the thoracic region, is the presence of ***costal facets*** for articulation with the ribs. These are located on the bodies, near the roots of the pedicles, and typically

3

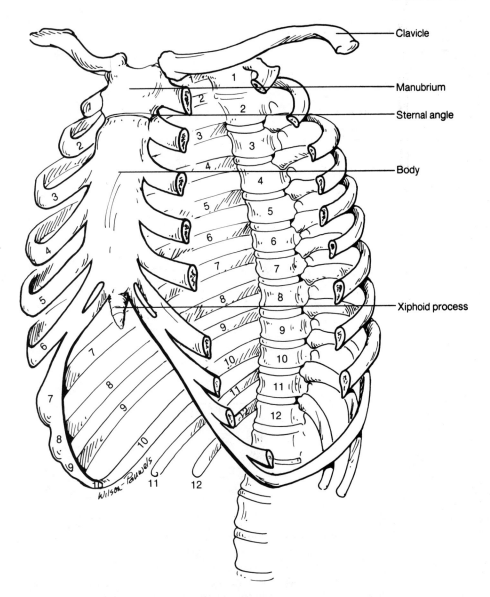

Clavicle

Manubrium

Sternal angle

Body

Xiphoid process

Wilson-Pauwels

there is a larger one on the upper part of the body for the numerically corresponding rib and a smaller one on the inferior aspect of the body of the vertebra above for the head of the rib below. The articular facets of the thoracic vertebrae face almost straight backward and forward so that little flexion/extension occurs in the thoracic spine; some rotation is possible.

The thoracic spine forms a curvature with its concavity forward.

Figure 3-1.
Thoracic cage.

CLINICAL NOTE

An abnormal lateral curvature can develop, not uncommonly, in children and is probably due to an imbalance of pull of the spinal muscles. This condition is called **scoliosis** *and can become quite severe and deforming if not treated during the child's growing phase (Fig. 3-4).*

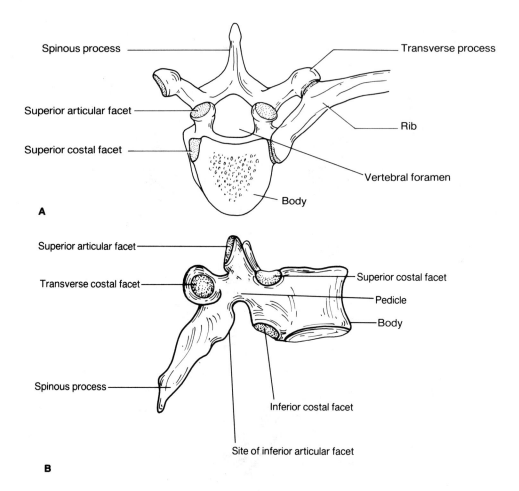

A

B

Figure 3-2.
Thoracic vertebra: (**A**) superior view;
(**B**) lateral view.

THE RIBS

The first seven ***true*** ribs attach by means of their own costal cartilages to the sternum. The next three ***false*** ribs attach to the sternum indirectly through the costal cartilage of the rib above. The last two ribs have no anterior attachment and are called ***floating*** ribs (see Fig. 3-1).

The 12th rib may be very short or absent; occasionally, a cervical rib may be present articulating with the seventh cervical vertebra behind and the first thoracic rib in front. A cervical rib may be incomplete, having a free end, or it may be connected to the first costal cartilage by a fibrous cord.

The typical rib (Fig. 3-5) has a ***head***, ***neck***, ***tubercle***, and ***shaft***. The head articulates with the body of its numerically equivalent vertebra and the one above, except for the first rib, which articulates only with the first thoracic vertebra, and the last two (three) ribs, which articulate with one vertebral body only.

The tubercle of the rib has a facet for articulation with the transverse process of its corresponding vertebra, and a rough, nonarticular area for ligamentous attachment to the transverse process. The last two ribs do not articulate with a transverse process. These ***costovertebral*** joints are syno-

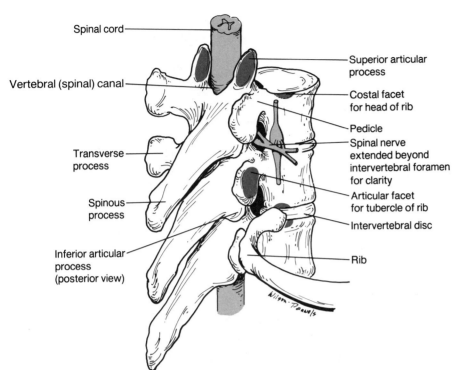

Spinal cord

Vertebral (spinal) canal

Transverse process

Spinous process

Inferior articular process (posterior view)

Superior articular process

Costal facet for head of rib

Pedicle

Spinal nerve extended beyond intervertebral foramen for clarity

Articular facet for tubercle of rib

Intervertebral disc

Rib

Figure 3-3.
Thoracic vertebrae illustrating articulation with a typical rib, vertebral canal, and spinal nerve exiting intervertebral foramen.

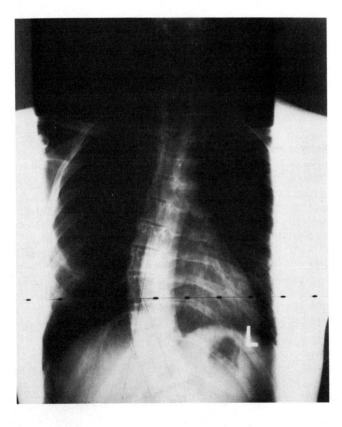

Figure 3-4.
Radiograph showing thoracic scoliosis. Lines are drawn on the radiograph to indicate the degree of curvature and to measure changes in the degree of curvature.

Figure 3-5.
Typical rib (posterior view).

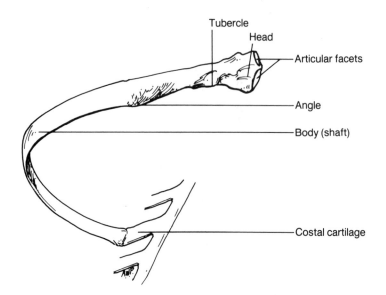

vial joints. Anteriorly, the shaft of the rib is joined to its costal cartilage, which, in turn, articulates with the sternum from the first to seventh ribs (Fig. 3-6). Ribs 8, 9, and 10 have costal cartilages that articulate with the costal cartilage above, and the last two ribs have no anterior junction but the free ends are capped by cartilage.

The first rib is flat with a sharp inner border; it slopes downward and forward, and the important nerve trunks and blood vessels passing from the neck to the upper limb cross its superior surface.

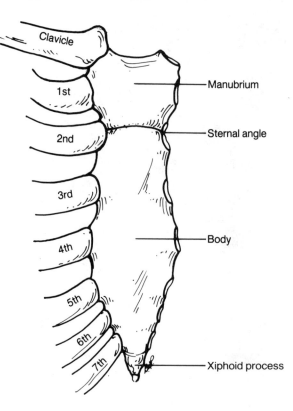

Figure 3-6.
Sternum illustrating facets for costal cartilages of ribs.

The shaft of a rib is twisted along its long axis, and the shaft bends sharply at the angle so that the lower border of a rib (except the second) cannot be placed on a flat surface. The importance of the angle and the torsion of the long axis will be appreciated when the movements of respiration are considered.

Each rib has a blunt upper border, but the lower border is sharp and projects below the costal groove.

CLINICAL NOTE

"Separation of a rib" is the injury in which a costal cartilage is dislocated from its rib. Fractures of ribs occur, most commonly, at their angles. There is a tendency for fractures, in general, to occur where the long axis of a bone changes direction. Another example is the clavicle, which tends to fracture at the junction of its lateral and middle thirds.

In severe crush injuries several ribs may be fractured at more than one level, thus producing an unstable segment of thoracic wall that gets sucked in and blown out in a paradoxical direction with movements of respiration, causing considerable hindrance (embarrassment is a common clinical term) to respiration and to the return of venous blood to the heart. This situation has to be remedied on an emergency basis by stabilization of the fragment, artificial ventilation of the lungs by a tube in the trachea (windpipe), or insertion of a tube into the thoracic cavity with drainage to an "underwater seal."

In infants and young children the ribs are more elastic and are less likely to fracture.

STERNUM

The sternum (Fig. 3-6) comprises three parts: The **manubrium** joins the body at the sternal angle, and the **xiphoid process** joins the lower end of the body. The first costal cartilage and the clavicle articulate with the manubrium, and the second costal cartilage joins the manubrium and the body at their junction, at the sternal angle (angle of Louis). The other five true ribs join the body by means of their costal cartilages.

CLINICAL NOTE

The first rib is difficult to palpate, but the sternal angle is prominent and easily palpable. Counting of ribs should start at the second costal cartilage, lateral to the sternal angle, and proceed downward and laterally.

MOVEMENTS OF THE THORACIC WALL

The chest increases in volume at each inspiration, causing air to be sucked into the lungs and facilitating the return of venous blood to the heart. The volume can increase in a **vertical diameter** by downward movement of the diaphragm (see below); in an **anteroposterior direction** by the raising of a rib, which, by virtue of its peculiar articulation with the vertebra at its head and tubercle, will elevate its anterior end and thus push the sternum forward; and in its **transverse diameter** by elevation of the bucket-handle-shaped ribs rotating their lower borders outward as the ribs are elevated (Fig. 3-7).

In expiration the elastic recoil of the chest wall allows the anteroposterior and transverse diameters to return to the resting position, and

Figure 3-7.
"Bucket-handle" action of rib during
inspiration.

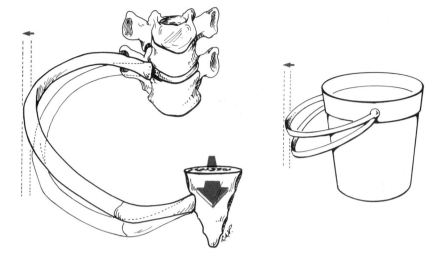

the diaphragm relaxes and is pushed back into its inactive position by the
pressure of the abdominal content and contraction of the abdominal wall
muscles.

THORACIC INLET AND OUTLET

The ***inlet*** of the thorax is bounded by the first thoracic vertebra, the first
ribs, and the upper border of the manubrium sterni.

The ***outlet*** is bounded by the 12th thoracic vertebra, the costal margins as far as the sternum, and the lower end of the xiphoid process.

The inlet is covered by a piece of cervical fascia, the ***suprapleural***
membrane, and the outlet is floored by the diaphragm.

MUSCLES OF THE THORACIC WALL

The muscles to be considered run from rib to rib, sternum to rib, or vertebra to rib (Fig. 3-8).

Serratus Posterior. The serratus posterior muscles, inferior and superior, are thin, flat muscles running from vertebrae to ribs.

Superior: The serratus posterior superior arises from the lower part of
the ligamentum nuchae, the spine of the seventh cervical vertebra, and the
spines of the upper two or three thoracic vertebrae. Inclining downward
and laterally it inserts into the outer surfaces of the second to fifth ribs
near their angles and elevates the ribs. The serratus posterior superior is
supplied by the anterior primary rami of the second to fifth intercostal
nerves.

Inferior: The serratus posterior inferior arises from the spines of the
lower two thoracic and upper two lumbar vertebrae, and, passing upward
and laterally, it is inserted into the lateral surfaces of the lower four ribs,
just lateral to their angles and depresses the lower four ribs. Its nerve supply is from the anterior primary rami of the 9th to 12th thoracic nerves.

External Intercostal. The external intercostal muscle passes from the
lower border of one rib to the upper border of the rib below, running in a

downward and forward direction (similar to the external oblique muscle of the abdomen). It extends from the tubercle of the rib to the costochondral junction, where it is replaced by fibrous tissue that extends to the sternum as the ***external intercostal membrane*** (see Figs. 3-8 and 3-11).

Internal Intercostal. The internal intercostal muscle passes from the floor of the costal groove of the rib above downward and backward to the upper border of the rib below (*i.e.*, at right angles to the external intercostal muscle). It extends from the sternum to the angles of the ribs and continues backward from this point as the ***internal intercostal membrane***.

Intercostalis Intimi. Once considered to be a separate, third layer similar to the transversus abdominis, the intercostalis intimi is now described as a deep layer of the internal intercostal muscle. It is separated from the internal intercostal muscle by the intercostal vessels and nerves that lie in the protection of the costal groove. The intercostalis intimi runs in the

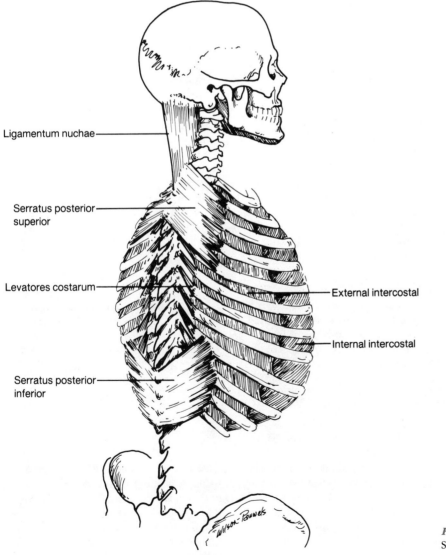

Ligamentum nuchae

Serratus posterior superior

Levatores costarum

External intercostal

Internal intercostal

Serratus posterior inferior

Figure 3-8.
Some muscles of respiration.

same direction as the internal intercostal muscle and occupies the middle two fourths of the intercostal space (see Fig. 3-11).

Subcostal. The subcostal muscles pass from the inside of the angle of the rib to the inner surface of the two ribs below (*i.e.*, they skip one rib.). They run in the same direction as the internal intercostals.

Transversus Thoracis. The transversus thoracis passes from the lower end of the posterior surface of the sternum upward and laterally to the second to sixth ribs. Its action would appear to be depression of those ribs.

Levatores Costarum. Each muscle arises from the posterior surface of a thoracic transverse process and inserts into the posterior surface of the tubercle of the rib below. It would appear to elevate the rib below (Fig. 3-8).

Much controversy surrounds the actions of the muscles of the thoracic wall in respiration. Theoretically, it would seem reasonable to suppose that if the first rib is fixed by the scalene muscles attached to it, contraction of the intercostal muscles below it would raise the second rib, which then becomes fixed and allows the third rib to be raised, and so on. This action would be assisted by the serratus posterior superior and by the levatores costarum.

The lower six ribs, which give attachment to the diaphragm and the abdominal muscles, are thought to be held steady by abdominal muscle contraction (fixator), thus allowing contraction of the diaphragmatic muscle to lower the diaphragm.

Now we have a paradox: we have assumed fixation of the first rib and postulated fixation of the lower six ribs so that contractions of the intercostal muscles would elevate ribs in sequence from above downward but depress them if looked at from the lower end. Electromyography has yielded conflicting results, showing variable degrees of activity in different intercostal spaces according to the depth of inspiration or the force of the expiratory effort. In deep, forced inspiration there seems to be activity in all of the intercostal muscles, possibly resulting in increased tension in the intercostal spaces. This keeps the ribs in normal alignment to each other and prevents sucking in or blowing out of intercostal space soft tissues.

CLINICAL NOTE

*Patients suffering from severe heart failure or respiratory difficulty can be seen to use **accessory muscles of respiration**. For example, they may be seen to hold onto a table to fix their upper limb girdle so that pectoralis major, serratus anterior, and some neck muscles can exert effort on their costal attachments.*

Forced expiration is assisted by contraction of the abdominal muscles.

INTERCOSTAL VESSELS

There are two sets of intercostal arteries (Fig. 3-9): anterior and posterior. Most ***posterior intercostal arteries*** arise from the thoracic aorta and run in the intercostal spaces toward the angles of the ribs and then hide in the costal groove between the pleura and the posterior intercostal membrane and, further forward, between the internal intercostal and the intercostalis intimi muscles. Each posterior intercostal artery gives off a collateral

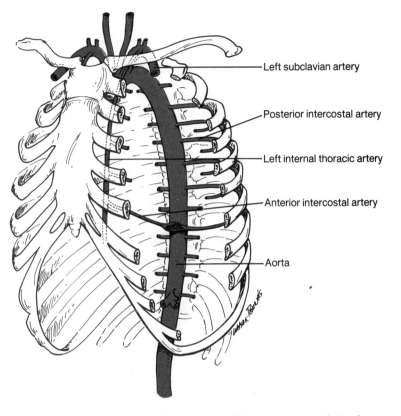

Figure 3-9.
Intercostal arteries.

Left subclavian artery

Posterior intercostal artery

Left internal thoracic artery

Anterior intercostal artery

Aorta

branch that runs along the upper border of the rib below. (The first two posterior intercostal arteries are branches of the superior costal branch of the costocervical branch of the subclavian artery.) At the anterior end of the intercostal spaces the posterior arteries anastomose (join with) the *anterior intercostal* branches of the *internal thoracic artery*.

The *internal thoracic artery* arises from the subclavian artery and descends about 1 cm lateral to the lateral border of the sternum, deep to the ribs, to the sixth intercostal space, where it divides into the *superior epigastric artery* (which enters the sheath of the rectus abdominis muscle) and the *musculophrenic artery*. The latter passes along the lower costal margin, supplies the diaphragm, and gives rise to the remaining anterior intercostal arteries.

Each intercostal artery has a corresponding *intercostal vein* that runs above the artery and drains into the *azygos vein* on the right side of the body and into the *hemiazygos veins* on the left (Fig. 3-10).

INTERCOSTAL NERVES

Each intercostal nerve comes from a thoracic spinal nerve and represents the *anterior primary ramus* of that thoracic spinal nerve. A typical thoracic spinal nerve gives off a *white ramus communicans* immediately after it exits from the intervertebral foramen. This ramus enters the corresponding sympathetic ganglion, and the ganglion returns a *gray ramus communicans* to the nerve (see Fig. 1-19).

The upper two thoracic nerves (T1 and T2) supply the upper limb and thorax, whereas the lower six intercostal nerves (T7 to T12) supply the thoracic and abdominal walls.

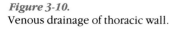

Figure 3-10.
Venous drainage of thoracic wall.

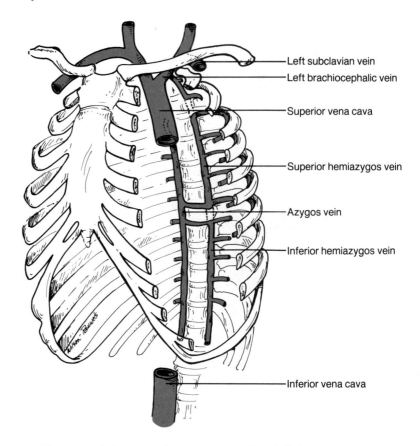

Left subclavian vein
Left brachiocephalic vein
Superior vena cava
Superior hemiazygos vein
Azygos vein
Inferior hemiazygos vein
Inferior vena cava

The seventh intercostal nerve, once it has left the intercostal space, runs upward and medially parallel to the lower costal margin to supply the area of the ***epigastrium***. The tenth intercostal nerve supplies the area of the umbilicus. These nerves, which supply abdominal muscles and skin, and the lower intercostal nerves must be considered when planning abdominal incisions.

Note that the intercostal nerves continue their downward and forward direction when they reach the abdomen, so that the 7th nerve supplies the epigastrium, the 10th the umbilical region, and the 12th the area just above the pubis. In their abdominal course they run between the muscle layers that are equivalent to the intercostal layers, that is, between the internal oblique and the transversus abdominis.

The typical intercostal nerve runs in the costal groove between the internal intercostal muscle and the intercostalis intimi. It lies below the intercostal artery, which lies below the intercostal vein. (This relationship of **v**ein, **a**rtery, and **n**erve [**VAN**] will be found to apply to most areas of the body.) Each intercostal nerve gives off a lateral branch to the skin, and a collateral branch that runs along the upper border of the rib below (Fig. 3-11).

Each intercostal nerve terminates in an ***anterior cutaneous branch*** that supplies the skin of the anterior surface of the thorax and the abdomen.

In addition to supplying the muscles of its own intercostal space and corresponding abdominal segment, the intercostal nerves supply the skin of their own area and the underlying ***parietal pleura*** and ***parietal peritoneum***.

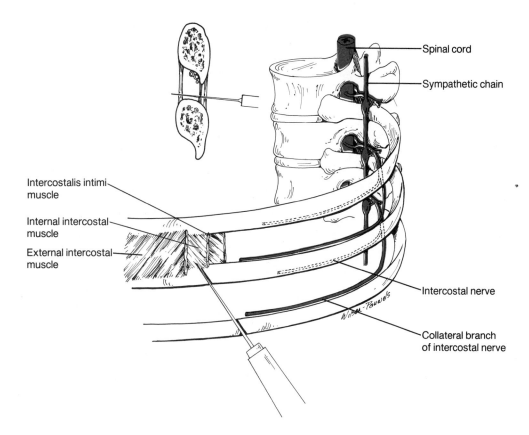

Spinal cord

Sympathetic chain

Intercostalis intimi muscle

Internal intercostal muscle

External intercostal muscle

Intercostal nerve

Collateral branch of intercostal nerve

CLINICAL NOTE

There is considerable overlap in the areas of cutaneous supply so that an attempt to anesthetize an intercostal space requires the injection of local anesthetic agent into the regions of the nerve above and the nerve below that space.

The intercostal VAN are protected in the costal groove near the lower end of the rib. If a needle has to be inserted into the chest, for example, for the withdrawal of fluid, it should be introduced close to the upper border of the rib, that is, well away from the costal groove (Fig. 3-11).

*If there is an obstruction to the flow of blood in the thoracic aorta, for example, **coarctation**, the circulation to the body wall and lower limbs has to take a circuitous route via vessels from the subclavian arteries anastomosing with intercostal arteries and vessels around the scapula. This increase in size of intercostal arteries causes deepening of the costal groove, which is recognizable on an x-ray as "notching" of the lower borders of some ribs.*

Figure 3-11.
Disposition of structures in an intercostal space.

THE DIAPHRAGM

The ***diaphragm*** is a dome-shaped structure consisting of a central tendon with peripheral striated muscle attaching the tendon to the sternum, ribs, and vertebral column (Figs. 3-12 and 3-13). The dome, which is often referred to as right dome and left dome, extends upward from the costal margin to about the level of the fifth intercostal space. Its precise level

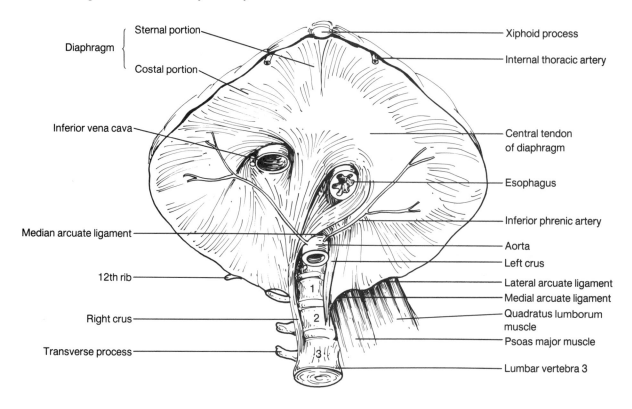

Diaphragm {
Sternal portion
Costal portion

Xiphoid process

Internal thoracic artery

Inferior vena cava

Central tendon of diaphragm

Esophagus

Inferior phrenic artery

Median arcuate ligament

Aorta

Left crus

12th rib

Lateral arcuate ligament

Medial arcuate ligament

Right crus

Quadratus lumborum muscle

Psoas major muscle

Transverse process

Lumbar vertebra 3

Figure 3-12.
Diaphragm from below.

cannot be marked because the diaphragm moves up and down during the respiratory cycle. It will rise as high as the fourth space in expiration and down to the sixth intercostal space in inspiration. The right dome sits at a higher level than the left because it is pushed upwards by the bulk of the liver, which resides below it. Although different parts and attachments of the diaphragm will be described, it must be realized that the diaphragm moves as a single unit.

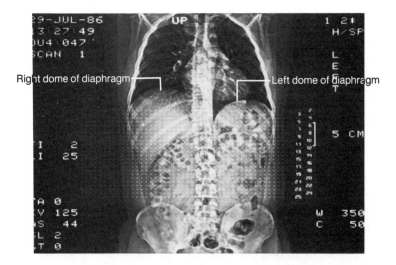

Right dome of diaphragm

Left dome of diaphragm

Figure 3-13.
CT scan (coronal plane) shows diaphragm separating thoracic and abdominal cavities.

Central Tendon

The ***central tendon*** is roughly trefoil in shape. It is a strong aponeurosis composed of interlacing fibrous tissue.

Muscular Portion

The muscular portion of the diaphragm, which inserts into the central tendon, has three different origins.

Sternal. Two muscular slips from the xiphoid process.

Costal. The costal origins consist of large slips of muscle arising from the deep surfaces of the lower six ribs and their costal cartilages at the costal margin, interdigitating with the slips of attachment of the transversus abdominis muscles.

Vertebral. The ***right crus*** arises from the upper three lumbar vertebrae and the ***left crus*** from the upper two lumbar vertebrae. In addition, the muscles of the diaphragm are attached to the arcuate ligaments.

There are five ***arcuate*** ligaments: one median and paired medial and lateral ligaments. The ***median*** arcuate ligament joins the two crura in the midline and arches over the entrance of the aorta into the abdomen. The ***medial*** arcuate ligament runs from the body of the first lumbar vertebra to its transverse process and arches over the psoas major muscle of the abdomen. The ***lateral*** arcuate ligament runs from the transverse process of the first lumbar vertebra to the 12th rib and arches over the quadratus lumborum muscle of the abdomen.

INNERVATION OF THE DIAPHRAGM

The diaphragm receives its motor innervation from the ***phrenic nerves***, which arise from the third, fourth, and fifth cervical anterior primary rami. In the embryo, part of the diaphragm develops in the septum transversum, which migrates from the neck region, taking its nerve supply with it. The sensory innervation is from the lower six intercostal nerves and from the phrenic nerves. The latter supply the central region of the diaphragm on its thoracic and abdominal aspects. Painful stimuli arising from the upper or lower surfaces of the central portion of the diaphragm will be felt in the shoulder tip regions (referred pain), since these areas are supplied by the same cervical nerves. Pain from the more peripheral parts of the diaphragm will be referred to the areas supplied by the six lower thoracic nerves and will be referred to the lower chest wall and the abdominal wall.

Diaphragmatic pain from, for example, inflammation of the diaphragmatic pleura (pleurisy) may present as abdominal pain and cause reflex spasm of abdominal wall muscles, resembling acute abdominal pain. It is a clinical aphorism that "examination of the painful abdomen begins in the chest."

BLOOD SUPPLY OF THE DIAPHRAGM

The arterial blood supply of the diaphragm comes from the phrenic branches of the aorta, the lower intercostal arteries, and the musculophrenic branches of the internal thoracic arteries (branches of the subclavian arteries).

The venous drainage is into the veins that correspond to the arteries and into the azygos and lower hemiazygos veins.

STRUCTURES PASSING THROUGH THE DIAPHRAGM

Aorta. The aorta passes from the thorax into the abdomen behind the median arcuate ligament in the midline at the level of the 12th thoracic vertebra. The thoracic duct and, usually, the azygos vein pass through the same opening.

Esophageal Opening. The esophageal opening is located about 2 cm to the left of the midline at the level of the tenth thoracic vertebra. It is located in the muscular tissue of the right crus, which loops around the esophagus and exerts a sling-like or "pinchcock" action on the lower end of the esophagus, helping to prevent reflux of gastric content into the esophagus. The esophageal tributaries of the left gastric vein, the esophageal branches of the left gastric artery, lymph channels from the upper stomach and lower esophagus, and the anterior and posterior gastric nerves (branches of the vagus nerves) accompany the esophagus through its opening (esophageal hiatus).

Vena Caval Opening. Unlike the aortic and esophageal openings, the passage for the inferior vena cava is through the ***tendinous*** part of the diaphragm, located at the level of the eighth thoracic vertebra just to the right of the midline. When the muscles of the diaphragm contract in deep respiration, the tension in the tendinous part of the diaphragm (which blends with the wall of the inferior cava) will keep the walls of the cava apart and facilitate return of its content to the heart. The vena caval opening also transmits the right phrenic nerve.

The ***sympathetic nerve trunks*** pass behind the medial arcuate ligaments, and the ***splanchnic nerves***, which are preganglionic sympathetic fibers from the thoracic region destined for the central abdominal plexuses, normally pass through the crura.

The superior epigastric artery, a terminal branch of the internal thoracic artery, passes through an opening between the sternal and costal fibers of the diaphragm.

FUNCTION OF THE DIAPHRAGM

The diaphragm contracts in inspiration, pushing the abdominal viscera inferiorly to allow increase in the vertical dimension of the thorax. The domes move more than the central tendon, which remains relatively fixed by its attachment to the ***mediastinum***. This can be confirmed by chest x-rays taken in inspiration and expiration. The aorta, passing behind the median arcuate ligament, is not subject to compression by diaphragmatic contractions, and the location and configuration of the inferior vena caval opening have already been mentioned above (see Fig. 3-12).

The sling-like arrangement of the right crus around the esophagus, as mentioned above, prevents gastric emptying (reflux) into the esophagus when the intra-abdominal pressure is raised during respiration.

DEVELOPMENT OF THE DIAPHRAGM

A brief digression into the territory of the embryologist seems appropriate at this point because the following clinical notes on diaphragmatic hernias require an understanding of how the diaphragm is formed.

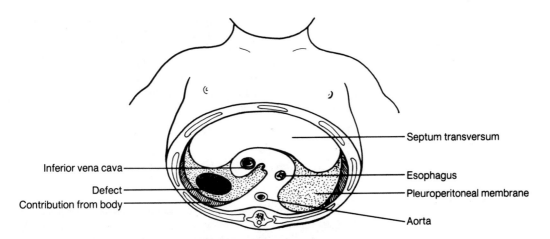

The diaphragm is formed by the fusion of the following embryonic elements (Fig. 3-14). The septum transversum forms the central tendon. The dorsal mesentery of the foregut, body wall mesoderm, and the pleuroperitoneal membranes form the muscular portions.

Figure 3-14.
Possible diaphragmatic defects in newborn.

CLINICAL NOTE: DIAPHRAGMATIC HERNIAS

A hernia (rupture) is the protrusion of the content of one body compartment, or cavity, into another compartment, through a normally occurring opening or through a traumatic opening between these compartments.

The diaphragm may rupture as a result of penetrating trauma or as the result of a severe compression injury to the thoracic or abdominal wall. In this case abdominal viscera are free to move into the chest cavity.

Nontraumatic hernias can occur through the following anatomical openings or embryonic defects:

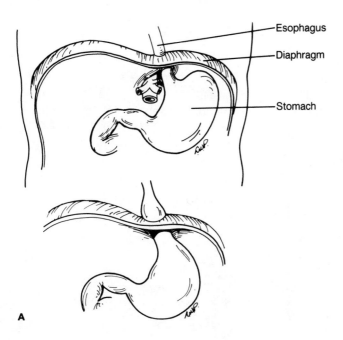

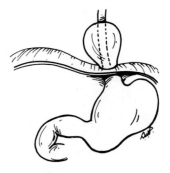

Figure 3-15.
The two types of hiatus hernias:
(**A**) sliding hiatus hernia;
(**B**) paraesophageal hernia.

A

B

1. *Rarely abdominal content will slip between the sternal and costal attachments of the diaphragm (foramen of Morgagni).*
2. *If the posteriorly placed pleuroperitoneal canal fails to be obliterated by the formation of the similarly named membrane, the potential for migration of large portions of abdominal viscera into the chest exists. This is a hernia through the foramen of Bochdalek, and it can cause severe respiratory difficulties in the newborn or manifest itself later in life (Fig. 3-14).*
3. *The central tendon may be deficient or absent.*
4. *Through the esophageal opening, the **esophageal hiatus**, resulting in a **hiatus hernia**.*

*Hiatus hernias are the most common of diaphragmatic hernias and include two types (Fig. 3-15): [1] a **sliding hernia** in which the esophageal hiatus is lax because of muscular weakness and because it has been stretched by the protrusion of the upper part of the stomach, aided and abetted by increased intra-abdominal pressure from obesity or pregnancy. In this hernia the cardia and upper part of the body of the stomach slide through the hiatus into the thorax. The gastroesophageal junction loses its normal angulation, and reflux of gastric secretions into the esophagus causes "heartburn," inflammation, ulceration, and eventual scarring with narrowing of the esophagus; and [2] a "rolling" or **paraesophageal** (Barrett's) hernia in which the gastroesophageal junction remains in place but the greater curvature part of the stomach herniates through the hiatus alongside the esophagus.*

Divisions of the Thorax

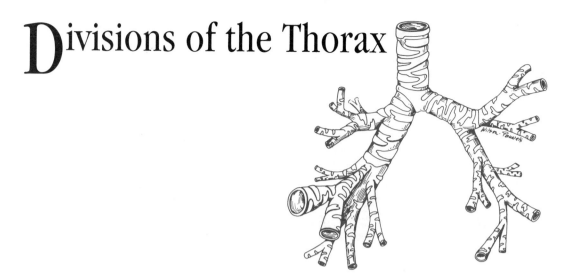

There are three major divisions of the thorax: the two ***pleural cavities*** (including the lungs) and the ***mediastinum*** (Fig. 4-1).

The ***pleural cavities*** are lined by a thin membrane, the ***parietal pleura***. It lines the inside of the thoracic wall and is reflected over the diaphragm and upward over the lateral aspect of the pericardium.

Because this is the first time that the reader has been confronted with the concept of a serous membrane, it is as well to elaborate on its nature and function, for similar principles apply to the peritoneum and the pericardium.

A serous membrane represents the remaining lining of a coelomic cavity that has become infolded, or invaginated, by the inward projection of developing viscera. Imagine a balloon into which a fist has been poked (Fig. 4-2); every contour of the fist will be covered by the inner layer, which becomes continuous with the outer layer around the wrist (in the diagram). The outer layer of this balloon then molds itself to the space in which it is itself contained. In the case of the thorax this outer layer will mold itself to the shape of the thoracic wall, the mediastinum, and the diaphragm. At the ***hilus*** (root) of the lung (which grew into this coelomic space) the outer layer will be continuous with the inner layer, just as it was around the diagrammatic wrist. The outer layer is called the ***parietal pleura*** (or peritoneum or pericardium), and the inner layer is the ***visceral pleura*** (or peritoneum or pericardium). The apposing surfaces of the parietal and visceral layers are lined by flattened cells (mesothelium) that produce a capillary layer of fluid which holds the two surfaces together by surface tension, just as two glass slides stick to each other when separated by a very thin layer of water. The space or ***pleural cavity*** (or peritoneal or pericardial cavity) is thus a potential space that will become a real space only if something is introduced (*e.g.*, more fluid, blood, or air) to break the "seal" of surface tension.

This concept of a coelomic space is so fundamental that the reader must try to grasp it at this stage. Subsequent descriptions of more complex spaces and configurations will be facilitated by this understanding.

Since the two layers are held together yet can slide freely upon each other (try it with glass slides), and since the membranes are connected to the parietal wall and the viscus by fibrous tissue, it must be obvious, at least with regard to the lungs, that when the thoracic wall expands the lung will also be "pulled out" into expansion. This creates a low pressure in the lungs, and air will be sucked in. When something breaks the "seal," the

4

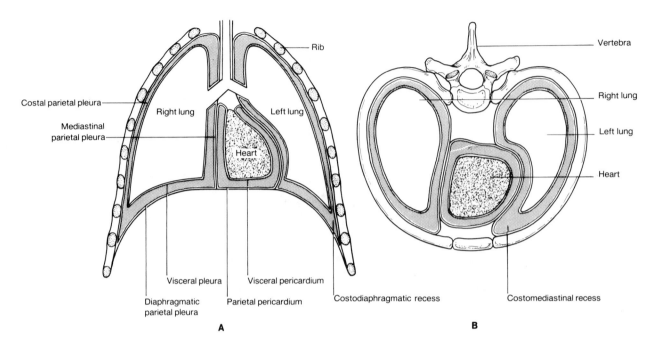

Figure 4-1.
Pleural and pericardial cavities:
(**A**) coronal section; (**B**) transverse
section.

elasticity of the lung will cause it to collapse away from the thoracic wall
and assume a volume of perhaps only 20% of its expanded state.

The *mediastinum* is the region between the right and left parietal
pleurae. Although the pleurae cover the mediastinum, they do not, by
definition, take part in it.

PLEURAL CAVITIES

The concept has been described above; the specifics follow.

THE PARIETAL PLEURA

The parietal pleura covers different parts of the walls and content of the
thorax and is named for the portions it covers. The *costal pleura* lines
the chest wall. The *mediastinal pleura* covers the mediastinum, and the
diaphragmatic pleura lies on the diaphragm. The *cervical pleura* (cu-
pula) projects into the root of the neck (because of the downward obliquity of the first rib) and can be described as rising to a point about 2.5 cm
above the center of the clavicle.

CLINICAL NOTE

*In this location the cervical pleura may be damaged by an injury or oper-
ation in the lower neck. Because of the shortness of the neck in infants
and young children the cervical pleura reaches a relatively higher level
and is even more vulnerable to injury in this age group.*

Lines of Pleural Reflection.   These markings are taught under the head-
ing Surface Anatomy, being the projections onto the body wall of deeper
lying structures. The parietal pleura reaches as high as the inner border of
the first rib into the neck, as described above. From this spot the reflection
of the parietal pleura can be followed around the chest wall (Fig. 4-3).

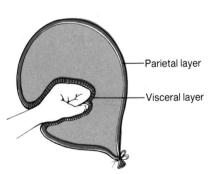

Figure 4-2.
Concept of a serous membrane lining a
body cavity.

On the left side the markings are as follows:

> From 2.5 cm above the middle of the clavicle to the second rib: anteriorly in the midline;
> to the 4th rib: anteriorly, in the midline;
> to the 6th rib: anteriorly, at lateral sternal margin;
> to the 8th rib: in midclavicular line;
> to the 10th rib: in midaxillary line;
> to the 12th rib: posteriorly to the edge of erector spinae muscles; in this position the pleura may lie below the level of the 12th rib.

On the right side the markings are as follows:

> From 2.5 cm above the middle of the clavicle to the second rib: anteriorly in the midline
> 2nd rib: midline
> 6th rib: midline
> 8th rib: anteriorly, midclavicular line
> 10th rib: midaxillary line
> 12th rib: posteriorly at the lateral border of erector spinae muscles (may reach below level of 12th rib).

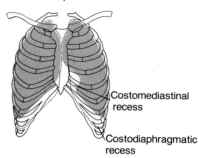

Figure 4-3.
Surface markings of lungs and pleura as drawn on the anterior aspect of thoracic cage.

VISCERAL PLEURA

The ***visceral pleura*** covers the lungs and is intimately adherent and fused to their surfaces. It provides them with a smooth shiny surface that permits free movement on the parietal pleura. The visceral pleura follows into the ***fissures*** of the lungs so that individual ***lobes*** of the lung are covered by pleura.

At the ***root (hilus)*** of the lung the visceral pleura becomes continuous with the parietal pleura.

PLEURAL RECESSES

In quiet respiration the lungs do not fill the available space in the pleural cavity. Therefore, there are areas where the pleural space is empty and parietal pleura is in contact with parietal pleura. These ***pleural recesses*** become almost filled with lung in very deep respiration and are as follows (see Fig. 4-3):

Costodiaphragmatic Recess. This recess is the depth of two ribs and their intercostal spaces and occurs between the lower margin of the lung and the attachment of the diaphragm to the ribs. In this area the diaphragmatic pleura is in contact with the ***costal pleura***.

Costomediastinal Recess. This recess is the area where the mediastinal pleura comes into contact with the costal pleura. This is made possible by the presence of a semicircular deficiency in the left lung anterior to the pericardium (cardiac notch).

THE LUNGS

The lungs (Fig. 4-4) are composed of ***alveoli*** (air sacks) and a supporting stroma in which run the blood vessels, lymphatics, nerves, and divisions of the air passages ***(bronchi)***. It is at the interface between the walls of the alveoli and the adjacent pulmonary capillaries that gaseous interchange takes place. The supporting stroma contains a good deal of elastic tissue which, when the pleural cavity is opened, will cause the lungs to

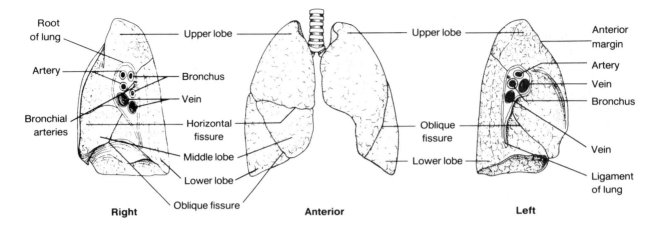

Right   **Anterior**   **Left**

Figure 4-4.
Lungs: Anterior and mediastinal surfaces.

collapse. Each lung has an abundant pulmonary circulation with a large **pulmonary artery** entering and two **pulmonary veins** draining each lung. The lung tissue itself is supplied by the bronchial arteries (see below). Each lung has a distinctive shape, but they have many common features:

Apex: The rounded superior pole of the lung lies immediately against the cervical pleura and rises as high as the neck of the first rib.

CLINICAL NOTE

Complete examination of the lung by auscultation (listening) and percussion must include the base of the neck above the middle third of the clavicle. There is no significant pleural recess at the apex.

Borders. The ***anterior border*** and the ***lower border*** of the lung are sharp and the ***posterior border*** is rounded and fits against the vertebral column.

Costal Surface. The costal surface is convex and fits against the thoracic wall; in the embalmed cadaver, because of the hardening of the tissues, it will show the markings of the ribs impressed upon it. These markings are not evident in living persons.

Mediastinal Surface. The mediastinal surface is applied against the mediastinum and is concave to fit against it. It contains the root of the lung (hilus) where the visceral and parietal pleura fit like a sleeve around the vessels and bronchi. The ***ligament*** of the lung is a downward prolongation of this sleeve toward the diaphragm. Passing through the root are the ***primary bronchus*** and the ***pulmonary artery and veins***, as well as the ***bronchial arteries*** and the ***autonomic nerves*** and ***lymphatics*** of the lung.

Diaphragmatic Surface. The diaphragmatic surface of each lung is concave to fit the diaphragm. This surface is spoken of as the ***base*** of the lung. The inferior margin of the lung descends into the costodiaphragmatic recess of the pleura only in very deep inspiration (see Fig. 4-3).

Differences Between the Lungs. The left lung differs from the right in having two lobes instead of three, the ***transverse fissure*** being absent. In addition the left lung has a ***cardiac notch*** on its anterior border to accommodate the bulge of the heart.

Caution. It is tempting to differentiate between the two lungs by the number of lobes or fissures, but these are not constant features. The correct orientation is achieved by identifying the apex, base, anterior and posterior borders, and the costal surface. The correct side to which the lung belongs becomes obvious.

Divisions of Lungs. The right lung is divided into three lobes by an *oblique fissure* and a *horizontal (transverse) fissure*. The lobes are *superior*, *middle*, and *inferior*. The left lung is divided by an *oblique fissure* into *superior* and *inferior* lobes. The lobulation has considerable clinical import. The superior lobe of the left lung has a projection below the cardiac notch, the *lingula* (tongue).

SURFACE MARKINGS OF LUNGS

The two lungs can be outlined on the surface of the thorax by the following markings. These markings are important in clinical medicine.

Right Lung. The apex of the right lung reaches the neck approximately 2.5 cm above the middle third of the clavicle. From there the outlines are as follows (see Fig. 4-3).

> Level of 2nd rib: at midline anteriorly
> Level of 4th rib: lateral margin of sternum anteriorly
> Level of 6th rib: midclavicular line
> Level of 8th rib: midaxillary line
> Level of 10th rib: lateral to 10th thoracic vertebra posteriorly.

Left Lung. The apex is at the neck of the first rib 2.5 cm above the middle third of the clavicle. From there the outlines are as follows (see Fig. 4-3).

> Level of 2nd rib: at midline anteriorly
> Level of 4th rib: edge of sternum
> Level of 6th rib: midclavicular line. There is a cardiac notch on the left side so that the margin runs almost at a right angle along the 4th rib and down to the 6th rib in the midclavicular line.
> Level of 8th rib: midaxillary line
> Level of 10th rib: lateral to 10th vertebra posteriorly

Note that these markings apply to quiet respiration; in deep inspiration they will move to approximate the markings of the parietal pleura.

SURFACE MARKINGS OF FISSURES

Normally the oblique fissure of each lung runs along a line that extends from the spine of the second thoracic vertebra around the chest to reach the sixth rib at the lateral border of the sternum. This is roughly the line along the medial border of the scapula when the hand is placed behind the head. The horizontal fissure of the right lung passes along the fourth rib from the lateral border of the sternum to meet the line of the oblique fissure.

PNEUMOTHORAX

If there is an injury to the thoracic wall or to the lung that permits air to pass into the pleural cavity, the elasticity of the lung will cause it to collapse, making the lung useless for the exchange of gases. This is a *pneumothorax* (Fig. 4-5A).

On occasion, a blowout can occur from a weak spot on the lung surface. This is called a *spontaneous pneumothorax*. If the hole in the lung

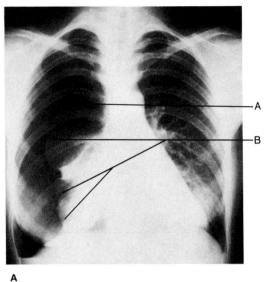

A

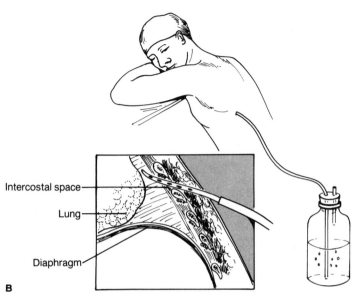

Intercostal space

Lung

Diaphragm

B

Figure 4-5.
A. Radiograph of a right pneumothorax. Note air in pleural cavity (*A*) and note margins of collapsed lung (*B*). **B.** Closed (tube) thoracotomy for drainage of air or fluid from pleural cavity.

or in the chest wall has a configuration that allows it to act like a valve, so that air can enter the pleural cavity during inspiration but is prevented, by the valve effect, from leaving during expiration, a condition called a ***tension pneumothorax*** results. This causes a buildup of pressure in the thoracic cavity, which can push the mediastinum to the opposite side and, eventually, embarrass the action of the uninvolved lung.

This extremely dangerous situation must be treated immediately by the insertion of a tube through an intercostal space. The other end of the tube is connected to an underwater seal in a bottle, which allows the trapped air to be expelled with each expiratory excursion without allowing outside air to reenter during inspiration (Fig. 4-5B).

If an accumulation of fluid in the pleural cavity occurs it will cause the lung to collapse to the extent of the volume occupied by the fluid. The clinical name is determined by the fluid involved. Blood produces a ***hemothorax***, pus a ***pyothorax***, and lymph (*e.g.,* from trauma to the thoracic duct) a ***chylothorax***. An accumulation of clear or straw-colored fluid may occur in pleurisy (inflammation of the pleura) or in malignancy of the lung. It might interest the student to know that in the days before chemotherapeutic agents for the treatment of ***pulmonary tuberculosis***, it was common practice to partially fill and, from time to time, refill the pleural cavity with air to permit partial collapse of the lung and allow it to "rest." This was an artificial ***pneumothorax***, which was kept up for months or years in some patients.

LUNG SEGMENTS

The trachea begins in the neck at the lower border of the cricoid cartilage and is about 11.5 cm long. In the thorax it divides at the ***carina*** into right and left main bronchi. The right bronchus is shorter, wider, and more directly in line with the trachea, whereas the left bronchus leaves the carina at a greater angle and is narrower. As a result of this, inhaled foreign bodies or secretions are more likely to enter the right bronchus and, in continuity of flow, the middle and lower lobes of the right lung (Fig. 4-6).

The trachea and main bronchi will be described in more detail later. The ***primary bronchi*** in turn divide into ***secondary bronchi*** that lead to the lobes of the lungs. These subdivide further into ***tertiary bronchi***, each of which supplies a segment of a lobe of a lung. Each tertiary bronchus with its pulmonary artery and vein is called a ***bronchopulmonary segment***, or segment for short.

These segments are of importance in surgery because lesions must be identified according to their segmental location, and the segments can be removed by surgery without encroaching on the remainder of the lung. The divisions of the segments are marked by intersegmental veins, which form a landmark in surgical resection of a segment.

In addition, the location and the direction of the tertiary bronchus in relation to each segment are important in establishing passive drainage of an infected segment. Physiotherapists rely on their knowledge of the locations of these segments, and the direction of their bronchi, to carry out postural drainage.

Right Lung. Ten segments.

Superior lobe: apical, anterior, posterior
Middle lobe: lateral, medial
Lower lobe: apical, anterior, lateral, medial, posterior

Left Lung. Nine segments.

Superior lobe: apicoposterior, anterior
Lingula (part of the superior lobe): superior, inferior
Inferior lobe: apical, anterior, lateral, posterior, medial

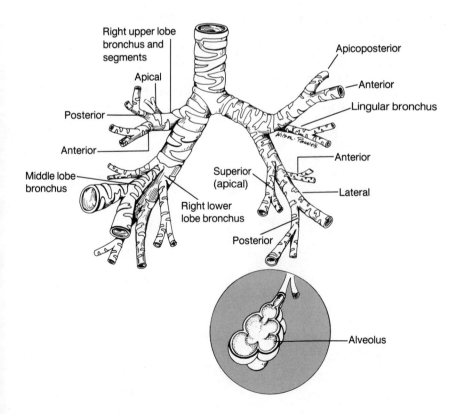

Figure 4-6.
Tracheal bifurcation and bronchial tree. **Inset:** alveolar sac and alveoli.

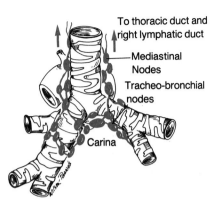

Figure 4-7.
Lymphatic drainage of lungs, bronchi, and trachea.

There is some controversy regarding the naming and the counting of the segments on the two sides. The list given above is standard for the right lung, but some clinicians and anatomists describe the superior (left) lobe as having the same apical, anterior, and posterior arrangements as on the right, and there is disagreement about the presence of a medial (cardiac) segment in the (left) inferior lobe (Fig. 4-6).

ROOT OF THE LUNG

The root (hilus) of the lung (see Fig. 4-4) contains the vessels, bronchi, and lymph nodes; it should be examined carefully.

Left Lung. The left lung hilus exhibits **veins, artery**, and **bronchus**, in that order from anterior to posterior. The **bronchial artery**, a branch of the aorta, lies close to the bronchus. Two veins will be found, one as the most inferior structure, one as the most anterior structure. The pulmonary artery lies above and anterior to the bronchus.

Right Lung. The two bronchi are posterior; one lies above the artery **(eparterial bronchus)**.

Both Lungs. The bronchial artery usually has divided into two branches by the time it reaches the root of the lung, so that two small arteries will be seen close to the bronchi. These arteries supply the bronchi and the **parenchyma** of the lung.

The **pulmonary plexus** of autonomic nerves surrounds the vessels of the root of each lung.

Lymph vessels drain the lungs and pass into the roots of the lungs, where they connect with lymph nodes at the lung root. Carbon deposits from inhaled pollutants will be seen on the surface of the lungs and will be aggregated in the nodes as a black deposit. The lymph nodes at the roots of the lungs are called the bronchopulmonary nodes; they drain into the tracheobronchial group, which is located at the division of the trachea, particularly in the inferior angle of the carina. Thence drainage proceeds to the mediastinal nodes and from these to the thoracic duct and the right lymphatic duct (Fig. 4-7).

The Heart and Mediastinum

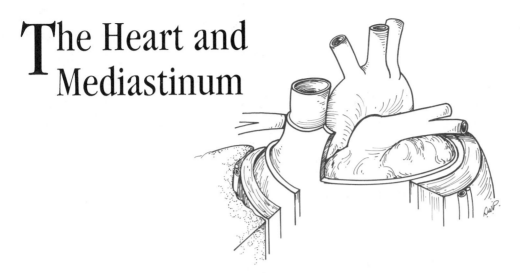

The **mediastinum** (Fig. 5-1) is that portion of the thoracic contents found between the two pleural cavities. Its main contents are the heart and great vessels, the trachea and bronchi, esophagus, vagus nerves, phrenic nerves, autonomic nerve plexuses, thoracic duct, thymus gland, and an abundance of lymph nodes.

It must be mentioned, at this stage, that the mediastinum that appears as a rigid structure in the cadaver is a flexible and somewhat mobile aggregate of structures in living persons.

DIVISIONS OF THE MEDIASTINUM

The mediastinum is divided into four parts mainly for ease of description and reference. The divisions are illustrated in Figure 5-1.

Superior Mediastinum. The superior mediastinum is that portion below the thoracic inlet and superior to an imaginary plane passing from the sternal angle to the lower border of the fourth thoracic vertebra.

The portion below the superior mediastinum is divided into three parts.

Anterior Mediastinum. This is the area between the sternum and the fibrous pericardium; it is very shallow and contains some fatty tissue and the thymus gland, which is prominent in children but vestigial in the mature adult.

Middle Mediastinum. This space contains the fibrous pericardium and its contents. The important structures are the heart and the great vessels that enter and leave it.

Posterior Mediastinum. This lies between the posterior aspect of the fibrous pericardium and the posterior wall of the thorax.

PERICARDIUM

The pericardium consists of a fibrous sack lined by two layers of serosal membrane, the **serous pericardium**, which is similar to the pleura (Fig. 5-2).

Fibrous Pericardium. The fibrous pericardium is a fibrous sack that encloses the serous pericardium and heart. It extends from the tendinous

Figure 5-1.
Divisions of the mediastinum.

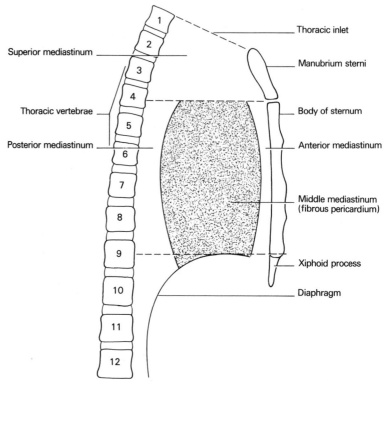

Thoracic inlet

Superior mediastinum

Manubrium sterni

Thoracic vertebrae

Body of sternum

Posterior mediastinum

Anterior mediastinum

Middle mediastinum
(fibrous pericardium)

Xiphoid process

Diaphragm

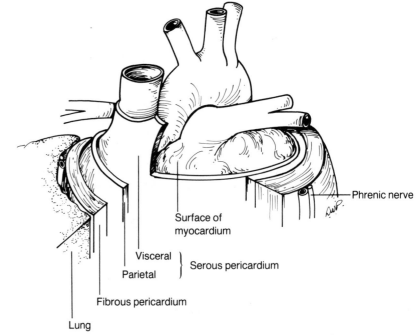

Phrenic nerve

Surface of
myocardium

Visceral
} Serous pericardium
Parietal

Fibrous pericardium

Lung

Figure 5-2.
Layers of the pericardium.

part of the diaphragm to which it is attached, to the level of the sternal angle. Its lower border is level with the xiphisternal junction. It extends about 1.5 cm to the right of the sternum and as far as 7.5 cm to the left of the midline at the fifth intercostal space.

The lungs and pleura wrap around the fibrous pericardium and intervene between it and the chest wall except for an area of the left anterior thoracic wall where the pericardium is in direct contact with the wall. This is the area of **cardiac dullness** in which the lack of resonance (due to the absence of air-containing lung tissue) can be elicited by **percussion**. In other words, this is the area of the **costomediastinal recess** of the pleura.

Above, the fibrous pericardium blends with the adventitia (outer coat of connective tissue) of the great vessels and the pretracheal fascia. Two **sternopericardial ligaments** attach the fibrous pericardium to the posterior surface of the sternum. The superior ligament passes to the upper end of the body of the sternum and the inferior ligament to its lower end.

The posterior relations of the fibrous pericardium are the trachea and its bifurcation, the esophagus and its plexus of autonomic nerves, and the thoracic aorta.

Although the surface markings of the fibrous pericardium may be described as extending from 1.5 cm lateral to the right border of the sternum, at the sternal angle, to the level of the xiphisternal joint, then horizontally to the left to the **apex beat** in the left fifth intercostal space 7 cm from the midline and hence to the left border of the sternal angle, it must be understood that the size and position of the heart and its covering pericardium exhibit much variation in living persons and are dependent on posture and individual physique.

CLINICAL NOTE

The tough, unyielding structure of the fibrous pericardium makes accumulations of fluid (e.g., blood from trauma or inflammatory exudate) extremely dangerous. They will compress the heart **(cardiac tamponade)** *and impair its function. Emergency aspiration of such fluids may be necessary. A needle can be introduced in an upward, lateral, and posterior direction from the left side of the xiphisternal junction to drain the sack.*

Serous Pericardium. The serous pericardium consists of visceral and parietal layers (*cf.* pleura). The **parietal pericardium** lines the inside of the fibrous pericardium. The **visceral pericardium (epicardium)** is closely applied to the surface of the heart and part of the great vessels. It presents a shiny surface and glides easily against the parietal pericardium. As was seen with the pleura, visceral and parietal pericardia are the continuous walls of a "balloon" into which the heart has been poked. The parietal and visceral layers are continuous with each other at the blending of the fibrous pericardium with the adventitia of the great vessels. Thus the origins of the great vessels are covered by a cuff of visceral pericardium. As was mentioned in reference to the pleura, the pericardial space is a potential space, containing a capillary layer of fluid.

Further understanding of the pericardium requires a description of the chambers of the heart and of the great vessels.

HEART

The human heart, unlike the traditional valentine, has a rather complicated shape, but the arrangements of the chambers and of the valves are of major importance and must be understood.

CHAMBERS OF THE HEART: GENERAL DESCRIPTION

The heart has four chambers, and their basic functions are as follows (Fig. 5-3):

Right Atrium. The right atrium receives venous blood from the whole of the body, except for the lungs, by means of the *superior* and *inferior venae cavae*. It then pumps the blood through the *right atrioventricular (tricuspid) orifice* into the right ventricle.

Right Ventricle. The right ventricle pumps the blood through the *pulmonary orifice* into the *pulmonary artery*, which carries it to the lungs.

Left Atrium. The left atrium receives oxygenated blood from the lungs by way of four *pulmonary veins*. It pumps the blood through the *left atrioventricular (mitral) orifice* into the left ventricle.

Left Ventricle. The left ventricle, the most muscular and powerful chamber, pumps the blood through the *aortic orifice* into the ascending aorta and hence to the rest of the body, except the lungs. The circulation of blood through the lungs for purposes of gaseous interchange is called the *pulmonary circulation*; that through the rest of the body is called the *systemic circulation*.

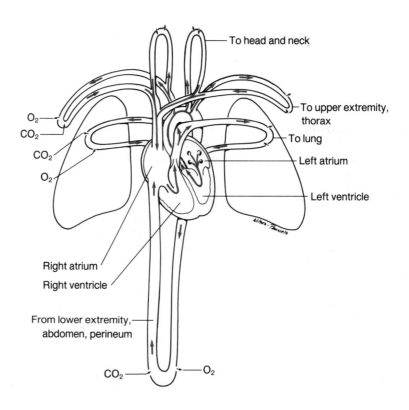

Figure 5-3.
Schematic overview of the chambers of the heart and vascular circulation.

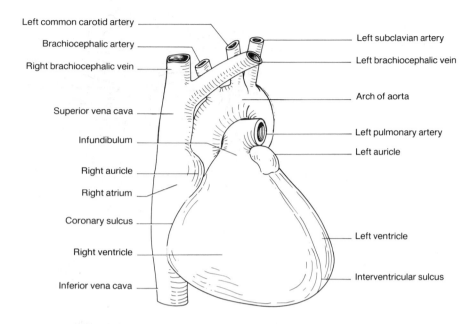

Note that the bronchial arteries, described in the previous chapter, supply the ***parenchyma*** of the lungs and are part of the systemic circulation.

HEART: GENERAL OUTLINE (Fig. 5-4)

Right Atrium. The right atrium forms the right border of the heart; the superior vena cava enters its upper aspect at the level of the third right costal cartilage, and the inferior vena cava enters it at its lower aspect opposite the level of the fifth intercostal space. The right atrium occupies about a quarter of the anterior aspect of the heart.

Right Ventricle. The right ventricle is shaped like a pyramid with a base that occupies the right two thirds of the inferior border of the heart and an apex that leads into the pulmonary artery at the level of the right third intercostal space.

Left Atrium. The left atrium forms the bulk of the posterior surface of the heart. Each "corner" of its posterior quadrangular surface is marked by the entrance of one of the four pulmonary veins (Fig. 5-5).

Auricles (Atrial Appendages). Each atrium has an anterior projection with relatively thick muscular walls. These appendages project forward and tend to hug the bases of the aorta and the pulmonary artery.

Left Ventricle. The left ventricle occupies the left third of the inferior border of the heart and forms the bulk of its left border from the apex, in the fifth intercostal space, to the beginning of the aorta, in the third intercostal space.

SURFACE ANATOMY OF THE HEART

The heart may be outlined on the anterior chest wall as follows (Fig. 5-6):
    The upper border is on a line from the second costal cartilage 3 cm

Figure 5-4.
Anterior view of heart and great vessels.

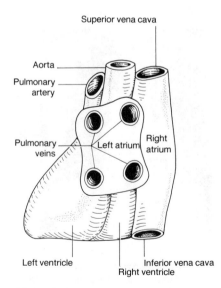

Figure 5-5.
Posterior view of heart and great vessels.

Figure 5-6.
Areas of auscultation of heart valve sounds. *A* = aortic; *T* = tricuspid; *P* = pulmonary; *M* = mitral.

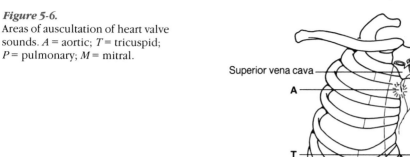

to the left of the midline to the right third costal cartilage 2 cm from the midline.

The right border runs from the right upper margin of the superior border to the fifth intercostal space just lateral to the border of the sternum. It has a slight convexity to the right.

The inferior border starts at the lower end of the right border and projects to the left fifth intercostal space in the midclavicular line (at the apex beat). The left border extends from the apex to the left margin of the superior border. The **apex** is formed by the left ventricle and is palpable in most subjects.

CLINICAL NOTE

*In some clinical conditions, such as high blood pressure, one or more chambers of the heart may be hypertrophied or dilated. This situation can be determined by percussion of the **cardiac dullness** and by noting a change in location of the apex beat, but a much better appreciation of the size of the heart and enlargements of specific chambers can be obtained by radiography. In a chest x-ray (posteroanterior view) the ratio of the width of the lower border of the heart to the width of the thoracic cage should be about 1:2. If the left atrium is enlarged it may produce a dent in the outline of the esophagus that can be visualized by the patient swallowing a radiopaque material (barium). Enlargement of the right atrium will cause visible bulging of the right border of the heart, as seen in an x-ray (Fig. 5-7).*

PERICARDIAL REFLECTIONS

Now that the student has an elementary concept of the shape and the chambers of the heart (see above text and diagrams), the reflections of the visceral serous pericardium (epicardium) may be considered. It must be remembered that the parietal pericardium assumes the shape of the interior of the fibrous pericardium and is essentially that of a hollow cone.

The visceral pericardium follows the shape of the heart and the great vessels at their "roots." To understand the reflections one must realize that

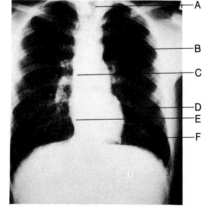

Figure 5-7.
Standard posteroanterior (PA) radiograph of the chest. Note air in trachea (*A*); aortic "knuckle" (*B*); margin of superior vena cava (*C*); margin of left ventricle (*D*); margin of right atrium (*E*); and cardiac apex (*F*).

in the "embryonic heart tube" one artery exited cranially and one vein entered it caudally. Although the veins and arteries and heart chambers become much more complicated with subsequent development, the basic principle of pericardial reflections remains; that is, the reflections occur where the visceral layer becomes continuous with the parietal layer at the entrances of the veins and the exits of the arteries.

The six veins must be surrounded by one sleeve of pericardium, and the two arteries must be surrounded by another sleeve of pericardium. These reflections appear complicated, but a reference to Figure 5-8 will show that the six veins plus the pericardium form the shape of an h and the arteries plus the pericardium form an oval.

SINUSES OF THE PERICARDIUM

Figure 5-8 shows that there is a ***transverse sinus*** that passes horizontally between the arterial and venous reflections of the pericardium.

The ***oblique sinus*** is a blind sack formed by the pericardial reflection around the veins. It is the cul-de-sac formed by the two uprights and the crossbar of the h.

CLINICAL NOTE

This description of the pericardial sinuses may seem somewhat esoteric, but the transverse sinus is of great importance to the cardiac surgeon. After the fibrous pericardium (and its inner lining, the parietal pericardium) have been opened and the heart exposed, the surgeon can pass a finger and a ligature through the transverse sinus between the arteries and the veins. Thus he can stop the circulation of blood into the arteries by tightening the ligature while he carries out surgery on the aorta or pulmonary artery.

The inferior vena cava (IVC) enters the fibrous pericardium immediately after piercing the tendinous part of the diaphragm. If the IVC has to be exposed within the thorax, the pericardium must be opened to permit access to its thoracic (terminal) part, which measures only about 2 cm.

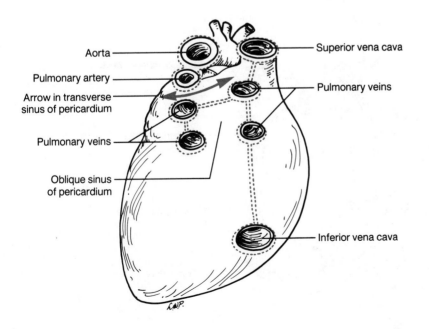

Figure 5-8.
Posterior aspect of heart showing the sinuses of the pericardium.

WALL OF THE HEART

The wall of the heart is basically composed of muscle. This ***cardiac muscle*** is a special type of striated muscle with two special properties: [1] It has the ability to generate spontaneous, rhythmic contractions; and [2] it can conduct a contractile impulse arising anywhere in its substance equally throughout the cardiac muscle. The rate of cardiac muscle contractions is generated in specialized heart muscle called the ***pacemaker***, and the rate of firing of the pacemaker and the force of contraction are modified by the autonomic nerve supply to the heart.

The muscles originate from a series of fibrous rings around the atrioventricular junctions and around the origins of the pulmonary artery and the aorta; these rings form the "fibrous" skeleton of the heart. The muscle loops out over the walls of the heart and returns to insert into the fibrous rings.

The wall of the heart consists of the following layers:

Visceral pericardium ***(epicardium)***
Myocardium (the cardiac muscle)
Endocardium (endothelial lining of the interior of the heart)

RIGHT ATRIUM

The right atrium receives venous blood from the superior and inferior venae cavae and from the ***coronary sinus***, which receives the blood drained by the cardiac veins from the heart itself (Fig. 5-9).

Superior Vena Cava. The superior vena cava is formed by the junction of the right and left ***brachiocephalic*** veins, which drain blood from the

Figure 5-9.
Interior of heart (arrows show direction of flow).

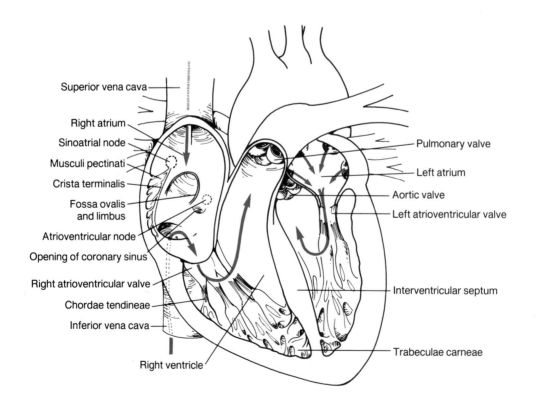

head and neck, the upper limbs and related areas, and some parts of the thoracic wall. This junction occurs behind the manubrium sterni to the right of the midline. The superior vena cava enters the right atrium at its upper end (at approximately the right third costal cartilage (Fig. 5-9).

Inferior Vena Cava. The inferior vena cava pierces the tendinous part of the diaphragm at the level of the eighth thoracic vertebra, somewhat to the right of the midline; it enters the pericardium immediately, and has a very short course (1-2 cm) within the pericardium before entering the lower part of the right atrium.

Right Auricle (Atrial Appendage). This is a thick walled, muscular, forward projection of the right atrium around the base of the aorta. The auricle has an irregular interior wall made of muscular ridges (***musculi pectinati***). The auricle forms a "backwater" in which the flow of blood is slow. If contractions of the auricle are irregular or weak, because of cardiac disease, clots can form in the auricle, and pieces of this clot may break off into the circulation and pass through the right ventricle to the lungs as emboli.

Sulcus Terminale. This groove on the external surface of the right atrium represents the junction between the sinus venosus and the primitive heart. It is represented on the inside of the atrium as a ridge, the ***crista terminalis*** (Fig. 5-9).

Wall of Right Atrium. The wall of the right atrium is thin, and anterolaterally the internal surface is raised in rough ridges, the ***musculi pectinati*** (Fig. 5-9).

Interior of Right Atrium

The interior of the right atrium shows numerous features (Fig. 5-9):

The superior vena cava enters its superior aspect and the inferior vena cava its inferior aspect. At the anterior border of the inferior vena caval opening there is a narrow crescent of tissue called the ***valve of the inferior vena cava***; it plays an important role in the fetal circulation.

Septum. The septum is the partition between the right and the left atria. It exhibits a depressed area, the ***fossa ovalis***, that represents the position of the embryonic ***foramen ovale*** between the two atria. A window is often discernible in the adult as a very small opening tucked superiorly under the edge of the ***limbus*** (raised rim around the fossa) of the fossa ovalis. It may admit the tip of a probe.

Note that the septum lies almost in the coronal rather than the sagittal plane so that the right atrium lies in front of rather than to the right of the left atrium.

CLINICAL NOTE

The foramen ovale may fail to close at birth, leading to the condition of a patent foramen ovale. This will be discussed later when the fetal circulation is described.

Crista Terminalis. The crista terminalis is a ridge on the right wall of the right atrium; it represents the junction between the sinus venosus and the primitive heart. The wall of the atrium is thin and smooth behind the crista terminalis, but anteriorly it is thickened and roughened by the musculi pectinati.

Figure 5-10.
Conducting system of the heart.

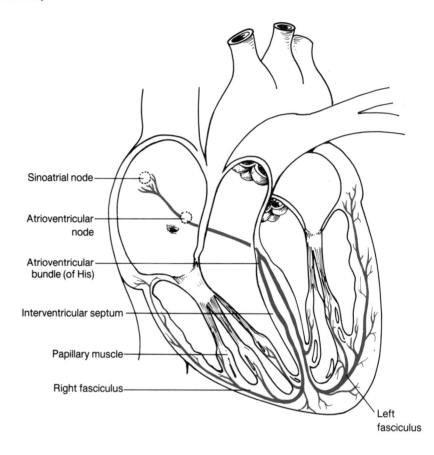

Sinoatrial node

Atrioventricular node

Atrioventricular bundle (of His)

Interventricular septum

Papillary muscle

Right fasciculus

Left fasciculus

Coronary Sinus. The opening of the coronary sinus can be seen in the medial wall of the right atrium; it is located close to the opening of the tricuspid orifice. The coronary sinus is the terminal channel into the atrium of the blood drained from the heart by the cardiac veins. The coronary sinus itself lies on the posterior surface of the heart in the coronary sulcus.

Right Atrioventricular (Tricuspid) Orifice. The tricuspid orifice is the opening between the right atrium and the right ventricle. The opening is guarded by the tricuspid valve, which will be described with the right ventricle.

Sinoatrial and Atrioventricular Nodes. It has been stated above that heart muscle can initiate and conduct impulses but it cannot transmit these impulses across the fibrous skeleton of the heart. A conducting system is required. The wall of the right atrium contains two important elements of the conducting system.

The ***sinoatrial*** (SA) ***node***, which is located in the wall of the atrium near the upper end of the crista terminalis, is a group of modified muscle cells that initiate the rhythmic heart beat (Fig. 5-10). The node is under the control of the sympathetic nervous system, which speeds up the heart beat, and the parasympathetic nervous system, which slows the heart beat. The stimulus from the SA node spreads out through the wall of the atrium and reaches the ***atrioventricular*** (AV) ***node***, which lies just superior to the opening of the coronary sinus. The AV node is the upper part of the ***interventricular bundle*** (of His) of the conducting system.

RIGHT VENTRICLE

The right ventricle has been outlined on the surface of the heart (see Fig. 5-6). The cone-like upper portion, known as the ***conus arteriosus (infundibulum)***, leads into the pulmonary artery. On the surface of the heart the right ventricle is separated from the right atrium by the ***coronary sulcus*** and from the left ventricle by the ***interventricular sulcus***, in which a major branch of the left coronary artery runs. The interior of the right ventricle shows the following features:

Wall. The wall of the right ventricle is much thicker than that of the atrium but thinner than the wall of the left ventricle (see Fig. 5-9). The muscle is raised in ridges called ***trabeculae carneae***. These trabeculae are in continuity with some prominent intraventricular projections, the ***papillary muscles***. The interventricular septum is the partition between the two ventricles; through most of its extent it is muscular, but toward its upper end it is membranous. If this membranous portion fails to develop, a patent interventricular septum results. The interventricular bundle (of His) divides into two strands at the upper border of the membranous part of the septum; one strand supplies the left ventricle and the other the right ventricle. When a patent ventricular septum exists, that is, the membranous portion of the septum is missing, the bundle runs just behind the defect, and care must be taken, when the defect is repaired, to avoid the inclusion of the bundle in the sutures.

Right Atrioventricular Valve. The tricuspid orifice is located behind the sternum at the level of the fourth and fifth intercostal spaces, to the right of the midline. The tricuspid valve has three cusps (leaves) that are attached to the ***tendinous ring*** which forms the tricuspid orifice. These three cusps completely surround the opening. The valve cusps project into the ventricle, and their serrated edges are attached to the papillary muscles by fibrous bands called the ***chordae tendineae***. It will be noted that a single prominent ridge of muscle, the ***septomarginal trabecula***, passes from the interventricular septum to the anterior papillary muscle. This has been called the moderator band for its fancied ability to prevent overdistension of the ventricle. This function is doubtful in humans because the "band" is not always complete; in some animals (oxen) it is complete. Its importance in the human heart lies in the fact that it contains elements of the conducting system from the septum to the wall of the ventricle.

The disposition of the flaps of the tricuspid valve are anterior, septal, and right-posterior.

The tricuspid valve functions as follows: when the atrium contracts (systole) and the ventricle is relaxed (diastole), blood flows through the orifice and pushes the leaves of the valve apart. When the ventricle contracts (systole), the backflow pressure of the blood brings the valve cusps together to close the orifice. The chordae tendineae prevent the valve cusps from everting into the atrium. The papillary muscles are essential as they contract in unison with the muscular wall and increase the tension on the chordae tendineae.

It should be noted that the chordae tendineae are attached to the cusps just proximal to their edges, on the septal or mural surfaces of the cusps. This leaves the surface that faces the flow of blood smooth and avoids turbulence.

Closure of the tricuspid (and mitral) valve causes the first heart sound: the "lup" of "lup-dup."

Pulmonary Valve. The pulmonary valve guards the orifice between the right ventricle and the pulmonary artery. It is located at the level of the third costal cartilage at the left side of the sternum, and in this position the second ("dup") heart sound originating from the closure of this valve is heard (see Fig. 5-6). The valve has three semilunar cusps of relatively thin and avascular fibrous tissue. At the apex of the crescentic free edge of each cusp there is a nodular thickening. When the cusps come together to prevent retrograde flow into the ventricle during its relaxation, these nodules effect complete closure of the valve at its center.

LEFT ATRIUM

The left atrium, which lies on the posterior surface of the heart, is not remarkable except that its quadrangular posterior surface receives the opening of a pulmonary vein at each corner. Sometimes there may be more than four pulmonary veins and sometimes three or less. The left auricle, similar in structure to the right auricle, hugs the base of the pulmonary artery. The left atrioventricular (mitral) orifice conducts blood from the atrium to the ventricle.

LEFT VENTRICLE

The left chamber must provide all the arterial blood flow to the systemic circulation, and to achieve this it has a very thick muscular wall. The interior is supplied with trabeculae carneae, papillary muscles, and chordae tendineae attached to the two cusps of the mitral valve.

Left Atrioventricular (Mitral) Valve. The mitral valve is located behind the sternum at the level of the fourth costal cartilage. It guards the opening between the left atrium and the left ventricle. The two cusps are arranged as an anterior (larger) leaf and a posterior (smaller) leaf. They are triangular in shape, and their bases are attached to the fibrous skeleton of the left atrioventricular orifice. The cusps act similarly to those of the tricuspid valve. The sound of their closure ("lup") is heard best at the apex beat. The reason for this has to do with the conduction of sound through the chest wall.

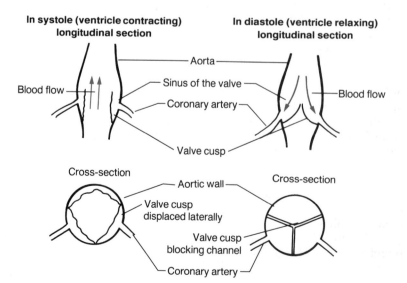

Figure 5-11.
Functioning of aortic valves. Pulmonary valves act in a similar manner.

Aortic Valve. The aortic valve is located behind the sternum at the third intercostal space, and its sound is heard best at this level at the right border of the sternum. It is the valve that guards the aortic orifice and has a structure and function similar to those of the pulmonary valve. Just distal to the aortic orifice, in the area of the valve cusps, the aorta has three bulges in its wall called the ***aortic sinuses (sinuses of Valsalva)***. The sinuses are called anterior, left, and right. The right coronary artery opens from the anterior sinus and the left coronary artery from the left sinus. There has been some disagreement concerning the naming of the aortic valves and coronary sinuses. Some anatomists and clinicians consider the positions in living persons to be right, left, and posterior, so that the left coronary artery arises from the left sinus and the right coronary artery is consigned to the right sinus. Many cardiac surgeons and radiologists overcome this problem by calling the sinus without a coronary artery the ***noncoronary sinus*** (Fig. 5-11).

SURFACE MARKINGS OF CARDIAC VALVES

The four valves can be marked on the sternum. These markings must not be interpreted too rigidly because the heart moves up and down and expands from side to side during the ***cardiac cycle*** (of contractions). The areas of auscultation (listening) of heart sounds are based on clinical observations rather than on fixed anatomical topographic dogma. The locations of heart sounds (see Fig. 5-6) are summarized below:

Pulmonary Valve: Third costal cartilage at left side of sternum
Aortic Valve: Second interspace at right edge of sternum
Tricuspid Valve: Over lower end of sternum
Mitral Valve: At cardiac apex in left fifth interspace in the
　　midclavicular line

CLINICAL NOTE

Abnormalities of the valves (congenital or acquired) fall basically into two categories: insufficiency or stenosis (narrowing). ***Valvular insufficiency*** *is the incomplete closure of a valve, allowing backflow into the proximal chamber. For example, insufficiency of the mitral valve causes (mitral) regurgitation of blood into the left atrium. This puts a strain on the left atrium, which dilates and increases the pressure in the pulmonary veins. An increase in the pulmonary (blood) pressure results, which reflects eventually on the right ventricle, causing right ventricular hypertrophy and (finally) right ventricular failure. Similarly, mitral stenosis will put stress on the left atrium. and so on.*

*　* ***Aortic insufficiency*** *or* ***aortic stenosis*** *will increase the work load of the left ventricle, leading to left ventricular hypertrophy and, again, ultimate failure.*

*　When a ventricle has exhausted its capacity to hypertrophy it will fail; thus we speak of left ventricular failure or right ventricular failure or right- (or left-) sided heart failure.*

*　The increased turbulence of blood regurgitating or flowing through a stenosed valve produces characteristic sounds or "murmurs" (bruits) that can be heard with a stethoscope.*

NERVES OF THE HEART

The nerve supply to the heart comes from the sympathetic and parasympathetic systems. Fibers from the sympathetic system and from the vagus nerve (parasympathetic) congregate at the ***cardiac plexus***, which is a dense network located below the arch of the aorta, and in front of the bifurcation of the trachea. The sympathetic fibers descend from the cervical ganglia in the neck, and there are contributions from the second to fifth thoracic ganglia. These are ***postganglionic*** fibers that have synapsed in their respective ganglia. The vagal cardiac fibers are mainly preganglionic and synapse in small ganglia located in the cardiac plexus and in some small ganglia near the sinoatrial node.

Vagal (parasympathetic) stimulation slows and weakens the heart beat; that is, it has an ***inhibitory*** effect on the SA node, whereas sympathetic stimulation causes acceleration and increased force of the heart beat.

The SA node lies at the upper end of the crista terminalis at the junction of the superior vena cava and the right atrium. From here stimuli spread over the muscles of both atria but cannot pass into the ventricles because of the fibrous skeleton that intervenes between atria and ventricles. The impulse from the atrial wall is transmitted to the AV node, which is situated just above the opening of the coronary sinus in the right atrium. From the AV node the impulse travels along the atrioventricular bundle (of His) to the posterior border of the membranous portion of the interventricular septum. Near the lower aspect of the membranous septum the bundle divides into a right and a left fasciculus to supply the muscles of the ventricles. The right fasciculus (right bundle branch) travels in the septomarginal bundle to reach the papillary muscles and the wall of the right ventricle. The left fasciculus (left bundle branch) ends in the wall and papillary muscles of the left ventricle. Impulses along these fibers of the conducting system cause contraction of the ventricular muscles (see Fig. 5-10).

A little contemplation will allow one to recognize the advantages of a nonconducting fibrous skeleton intervening between atria and ventricles. If it did not exist an impulse arising in the SA node would cause almost simultaneous contraction of all four chambers. The delay caused by transmission from the AV node to the bundle of His allows ventricular relaxation (diastole) and filling during atrial contraction (systole), and subsequently atrial relaxation (diastole) and filling during ventricular contraction (systole).

The ***pulmonary plexuses*** receive some vagal and sympathetic contributions from the cardiac plexus. Sympathetic stimulation relaxes the muscular walls of the bronchi, whereas vagal stimulation contracts these muscles and increases the secretion of mucus into the bronchi. Some fibers from the pulmonary plexus join the ***esophageal plexus*** (see below).

VESSELS OF THE HEART

A massive muscular organ such as the heart requires its own blood supply, as do, indeed, all but the smallest of blood vessels.

The coronary arteries, the origins of which have been described above, take care of the arterial supply of oxygenated blood and nutrients (Fig. 5-12). Since the terminal branches of the coronary arteries (and cardiac veins) run in the substance of the myocardium, this "feeding flow" must occur during diastole. The coronary arteries arise from the coronary sinuses just downstream from the aortic valve, which closes during ventricular diastole, thus allowing blood flow through the myocardium before the next systole occurs. The veins will be compressed and emptied during

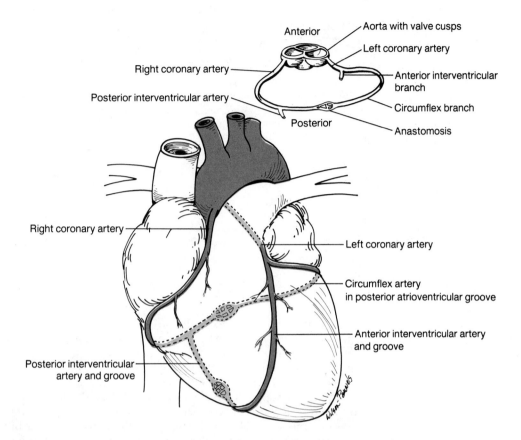

Figure 5-12.
The coronary arteries.

ventricular systole, which coincides with atrial diastole. Although venous return is mainly through the cardiac veins, which drain into the coronary sinus, there are numerous small, direct openings into each cardiac chamber: the **venae cordis minimae**. The amount of blood flowing through these direct venous openings is thought to be considerable.

Right Coronary Artery.   The right coronary artery arises from the right coronary sinus and emerges between the root of the pulmonary trunk and the right auricle; it then passes in the coronary sulcus between the right atrium and the right ventricle (atrioventricular groove). It passes to the inferior border of the heart, turns left, and continues in the atrioventricular groove to the interventricular sulcus, where it anastomoses with the left coronary artery. Before it turns to the left at the lower border of the heart it gives off a **marginal branch** to the anterior and posterior surfaces of the right ventricle.

Left Coronary Artery.   The **left coronary artery** is larger than the right. It arises from the left aortic sinus and emerges between the pulmonary trunk and the left auricle; it then turns to the left and downward in the coronary sulcus as far as the posterior end of the interventricular groove, where it anastomoses with the right coronary artery. Thus the two coronary arteries encircle the atrioventricular groove (coronary sulcus) like a crown (corona = crown).

As it emerges between the pulmonary trunk and the left auricle, the left coronary artery gives off the large and important **interventricular** branch (anterior descending artery) that descends in the anterior interven-

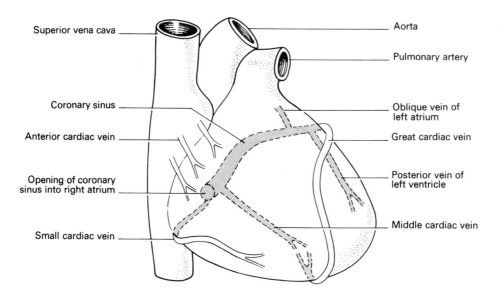

Superior vena cava
Coronary sinus
Anterior cardiac vein
Opening of coronary sinus into right atrium
Small cardiac vein

Aorta
Pulmonary artery
Oblique vein of left atrium
Great cardiac vein
Posterior vein of left ventricle
Middle cardiac vein

Figure 5-13.
Venous drainage of the heart (anterior view).

tricular groove to the apex of the heart. This branch supplies both ventricles and anastomoses with the interventricular branch of the right coronary artery in the inferior (posterior) interventricular groove. Near its commencement the left coronary artery gives off a small branch to the SA node in about 45% of cases. The anterior interventricular (anterior descending artery) artery supplies the anterior two thirds of the interventricular septum. If it becomes blocked, that part of the septum may become infarcted (lose its blood supply), and damage to the atrioventricular bundle may abolish conduction of impulses from the AV node to both ventricles or to one or the other of the ventricles. The ventricle(s) thus deprived of its conducting system will beat at its own muscular (much slower, pulse rate of 40/minute) rhythm. This is called ***heart block*** (complete or partial).

Both coronary arteries supply vessels to the wall of the ascending aorta (vasa vasorum).

Coronary Veins. The veins of the heart are called cardiac veins, and they drain by means of the ***coronary sinus*** into the right atrium (Fig. 5-13).

The ***great cardiac vein*** runs in the anterior interventricular groove, turns left in the coronary sulcus (atrioventricular groove), and joins the coronary sinus.

The ***posterior vein of the left ventricle*** runs along the left margin of the heart to join the great cardiac vein.

The ***small cardiac vein*** from the right ventricle runs in the coronary sulcus and joins the coronary sinus.

The ***middle cardiac vein*** runs in the posterior interventricular groove to join the coronary sinus.

The ***oblique vein of the left atrium*** is a remnant of the left common cardinal vein (embryology) and drains from the left atrium to the coronary sinus.

The ***anterior cardiac veins*** from the right ventricle pass directly into the right atrium.

The ***venae cordis minimae*** (Thebesian veins) were mentioned above.

CLINICAL NOTE

Insufficiency of arterial blood supply to the myocardium is a common condition. The coronary arteries become affected by the deposition of cholesterol under their endothelial inner lining, and this condition of **atherosclerosis** *can cause considerable narrowing (stenosis). If the vascular insufficiency is relative, i.e., there is sufficient flow at rest but insufficient flow when greater demands are made on the myocardium during exercise, the patient will experience pain known as* **angina pectoris**, *which can be relieved by rest.*

When complete occlusion of a coronary vessel occurs, heart muscle will die, and the angina will be constant and unrelieved until the lack of blood supply has deactivated the sensory receptors in the myocardium.

The nature of these pain receptors is not fully understood, but they are believed to be stretch receptors (and possibly other receptors).

The pain of angina pectoris is felt in the center of the chest behind the sternum, over the left anterior chest wall and often into the medial border of the left arm. On occasions the pain of angina can radiate into the neck, the jaw, or the left ear.

The neurologic pathways that transmit the sensation of pain are the sensory components of the sympathetic nervous system. It was stated in Chapter 1 that the sympathetic system of nerves contains afferent fibers that follow the course of the distribution of sympathetic fibers to the viscera but that their cell bodies are located in the posterior (sensory) root ganglia of the spinal nerves. The sympathetic nerve supply to the heart is from the second to fifth thoracic spinal segments (possibly also the first segment), and it was also stated above that some sympathetic nerve fibers to the cardiac plexus descend from the cervical ganglia (middle and inferior cervical ganglia) in the neck. Since the second to fifth thoracic spinal cord segments supply ordinary cutaneous sensory fibers (including those for pain) to the chest wall and since the second thoracic nerve carries sensation from the medial aspect of the arm to the spinal cord, it can be seen that these parts of the body have a sensory nerve supply from the same spinal cord segments that supply the heart.

Next one must realize that pain from viscera is usually interpreted imprecisely and, in its interpretation in the brain, is assigned to body wall (or limb) areas. Hence we have a situation in which pain from a viscus (the heart) is misinterpreted as coming from a somatic area. This is called referred pain, and examples of it will be encountered in other parts of the body, particularly the abdomen and the diaphragm.

The Mediastinum

The following section provides a summary of the anatomical divisions of the mediastinum. These divisions are somewhat arbitrary but help to locate structures for easy reference. Once the divisions have been considered and the contents of the divisions noted, it is preferable to describe their contained structures as whole entities rather than worry about their exact placement within mediastinal subdivisions.

SUPERIOR MEDIASTINUM

The *superior mediastinum* is that part of the thoracic content lying between the mediastinal pleurae and above a plane joining the sternal angle to the lower border of the fourth thoracic vertebra (see Fig. 5-1). It contains nerves, great vessels, esophagus, trachea, and a portion of the thymus gland, which is atrophied and not recognizable in the mature adult but prominent in infants (Figs. 6-1A and 6-1B).

The contents of the superior mediastinum may be described as three groups or layers.

Retrosternal Structures
Muscles: Sternohyoid and sternothyroid (see Chap. 27)
Thymus gland
Superior vena cava and brachiocephalic veins (right and left)

Prevertebral Structures
Esophagus
Trachea and its bifurcation at the level of the sternal angle
Thoracic duct
Left recurrent laryngeal nerve
Prevertebral muscles

Intermediate Structures
Aorta and its branches
Nerves: the vagus and the phrenic nerves. Cervical cardiac nerves from the cervical sympathetic ganglia.

POSTERIOR MEDIASTINUM

The *posterior mediastinum* is that portion of the mediastinum below the fourth thoracic vertebra and behind the fibrous pericardium. Its lower portion lies posterior to the diaphragm (see Fig. 5-1).

6

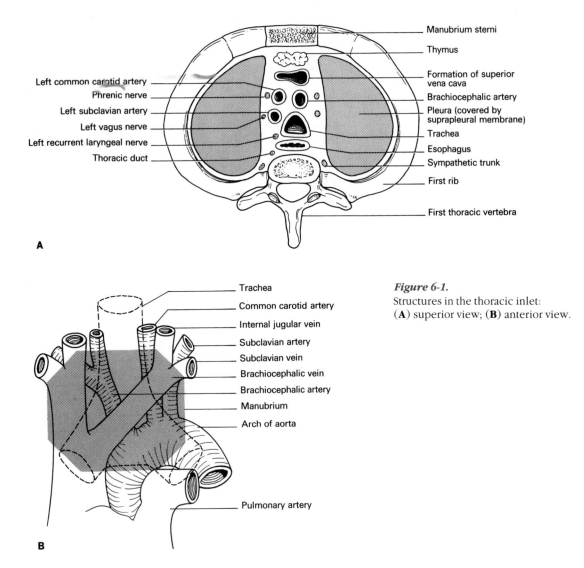

Figure 6-1.
Structures in the thoracic inlet:
(**A**) superior view; (**B**) anterior view.

The posterior mediastinum contains longitudinal and transverse structures. In general the transverse structures lie on a plane posterior to the longitudinal structures.

Main Longitudinal Structures
Descending (thoracic) aorta
Esophagus
Esophageal plexus (from vagus nerves)
Azygos and hemiazygos veins
Thoracic duct

Main Transverse Structures
Posterior intercostal arteries
Intercostal veins
The parts of the hemiazygos veins that cross the midline
The thoracic duct as it crosses from right to left

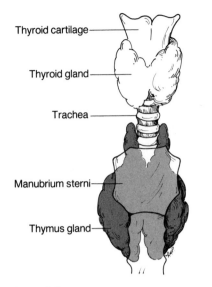

Thyroid cartilage

Thyroid gland

Trachea

Manubrium sterni

Thymus gland

Figure 6-2.
Thymus gland.

ANTERIOR MEDIASTINUM

The anterior mediastinum lies between the body of the sternum and the fibrous pericardium. Above, it is limited by the plane passing from the sternal angle to the lower border of the fourth thoracic vertebra, and its lower limit is the diaphragm (see Fig. 5-1). It contains fat, some fascia, and part of the thymus gland.

THYMUS GLAND

The thymus varies in size and appearance with age. Until the last two decades its function was completely unknown, which may account for many anatomy textbooks having given brief accounts of it in the past. The thymus is now considered to be the primary central organ of the lymphoid system. It is relatively large at birth and continues to grow into adolescence (Fig. 6-2). Subsequently it atrophies and is replaced by fatty tissue, so that it appears to be absent from the average elderly cadaver seen in the dissecting laboratory.

The thymus comprises two unequal-sized lobes, connected by areolar tissue. It lies in the superior and anterior mediastina with its narrow upper end extending into the neck, sometimes as high as the position of the thyroid gland.

The thymus has a cellular cortex and a medulla containing a reticular network of epithelial cells and lymphocytes. The lymphocytes are more densely packed in the cortex.

A more detailed description of the structure of the thymus belongs in a textbook of histology.

Among the essential functions of the thymus are lymphopoiesis, involvement in the "immune system," and possibly the production of some anticholinergic substance that interferes with neuromuscular transmission.

The Remaining Structures of the Thorax

It is time to digress from the regional fragmentation of thoracic viscera that the divisions of the mediastinum impose, and consider these structures in their contexts of anatomical continuity and function.

RESPIRATORY SYSTEM

Trachea. The trachea begins in the neck at the lower border of the cricoid cartilage. It is "horseshoe" shaped in cross-section with the posterior "open" aspect of the horseshoe covered by a membrane of connective tissue and smooth muscle (Fig. 7-1). It is approximately 11.5 cm long, and its wall is strengthened by incomplete rings of cartilage that prevent collapse of its wall during inspiration. The mucosa lining the trachea is ciliated, and the movements of the cilia are upward, helping to expel inhaled foreign particles. The mucosa is very sensitive and is responsible for a considerable part of the **cough reflex**.

At the level of the sternal angle, the trachea divides into the **right and left primary bronchi**. The point of division is called the **carina** (see Fig. 4-7). It was stated in Chapter 4 (see Fig. 4-6) that the right main bronchus is wider and leaves the trachea at a less acute angle than the left main bronchus. This accounts for the tendency for inhaled foreign matter to pass into the right bronchus more commonly than into the left bronchus.

CLINICAL NOTE

The trachea is central, in the midline in the neck, but deviates slightly to the right in the thorax. It is easily palpable in the neck, and deviation to one side or the other of the mediastinum will shift the cervical portion of the trachea to the corresponding side, a condition that can be detected by careful palpation. The air contained within its lumen is "radiolucent," and the position, shape, and size of the trachea and the carina can be seen on x-ray film.

The trachea, the primary and secondary bronchi, and the openings of the tertiary bronchi may be inspected by means of a **bronchoscope**.

Widening or distortion of the carina can be an ominous sign of invasion of this area by malignant tracheobronchial lymph nodes.

Tracheotomy is the name for a surgical procedure that makes an opening in the trachea to overcome an obstruction in the air passages at a higher level. Sometimes it is performed in patients who require prolonged

7

Figure 7-1.
Incision for tracheotomy.

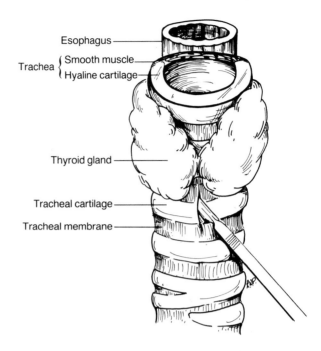

tracheobronchial suction to remove secretions (e.g., during prolonged unconsciousness) or to permit respiratory assistance by means of a ventilator. The site of the incision must be exactly in the midline of the neck, usually below the level of the isthmus of the thyroid gland (Fig. 7-1).

Right Bronchus. The right bronchus angles inferiorly and to the right and enters the lung at the level of the third costal cartilage. It is about 2.5 cm long.

Left Bronchus. The narrower left bronchus enters its lung at the level of the third intercostal space.

Many lymph nodes accompany the bronchi and trachea and congregate at the carina to form the tracheobronchial group and along the sides of the trachea to form the paratracheal group of nodes. Lymph drainage is from the pulmonary nodes and from the bronchi and trachea.

VEINS OF THE THORAX

Superior Vena Cava. The superior vena cava is formed behind the sternum at the level of the first costal cartilage by the junction of the left and right **brachiocephalic** veins. It passes downward along the right margin of the sternum to enter the right atrium at the level of the third costal cartilage.

CLINICAL NOTE

In a chest x-ray film, the superior vena cava can be seen as a shadow just lateral to the right border of the sternum (Fig. 7-2). The vena cava may be injured in blunt or penetrating thoracic trauma, and leakage of blood from it may be rather slow because of the low pressure of its contained blood. A chest x-ray film showing a widening of the mediastinal shadow

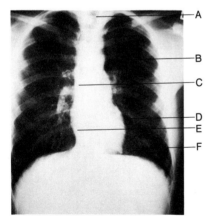

Figure 7-2.
A standard posteroanterior radiograph of the chest. Note air in trachea (*A*); aortic "knuckle" (*B*); margin of superior vena cava (*C*); margin of left ventricle (*D*); margin of right atrium (*E*); and cardiac apex (*F*).

in the region of the superior vena cava should lead to further study (by angiography) to diagnose a possible leakage of blood which, though sometimes rather slow, can be fatal.

Brachiocephalic Veins. The right brachiocephalic vein is formed behind the right sternoclavicular joint by the junction of the right subclavian and internal jugular veins.

The left brachiocephalic vein is formed behind the left sternoclavicular joint by the junction of the left internal jugular and subclavian veins. It passes to the right behind the manubrium sterni to join the right brachiocephalic vein at the commencement of the superior vena cava. In infants and young children the left brachiocephalic vein may lie above the sternal notch, that is, in the neck, and is vulnerable to injury during surgery in this region.

Inferior Vena Cava. The inferior vena cava pierces the tendinous part of the diaphragm to the right of the midline at the level of the eighth thoracic vertebra. Its adventitia is adherent to the tendinous part of the diaphragm, and this pulls apart the walls of the vessel during diaphragmatic contraction in inspiration and assists the return of venous blood to the heart.

Intercostal Veins. Most of the intercostal veins on the right and left side drain into the azygos and hemiazygos veins, respectively (see Fig. 3-10).

Azygos Vein. The azygos vein (see Figs. 3-10 and 7-3) is a continuation of the ascending lumbar vein from the right posterior abdominal wall. It pierces the right crus of the diaphragm and ascends on the posterior thoracic wall to the level of the fourth thoracic vertebra where it arches forward, superior to the root of the right lung, to enter the superior vena cava. In the cadaver it makes an impression on the mediastinal surface of the upper lobe of the right lung, and sometimes, in the living, it runs in the parenchyma of the lung, separating off a lobule of lung tissue, the azygos lobe. Most of the right intercostal veins drain into the azygos vein; because both hemiazygos veins usually drain into the azygos vein, it receives the venous blood of most of the thoracic wall.

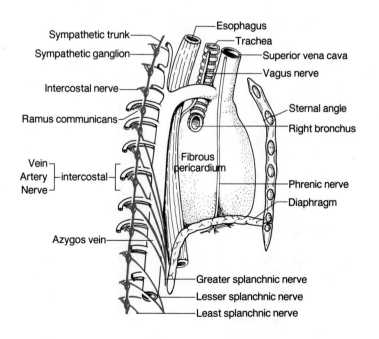

Figure 7-3.
Right surface of mediastinum with pleura removed.

Hemiazygos Veins. There are usually two hemiazygos veins (superior and inferior) on the left posterior thoracic wall (see Fig. 3-10). The inferior drains the left ascending lumbar vein and the lower intercostal spaces; the superior drains the upper intercostal spaces. At the level of the eighth thoracic vertebra both veins cross the midline to join the azygos vein. Sometimes the hemiazygos veins join each other on the left side, and then a single connection is made across the midline with the azygos vein.

ARTERIES

Aorta. The aorta arises from the left ventricle at the level of the third costal cartilage on the left of the midline. It passes upward as the ***ascending aorta*** in an arc to the middle of the manubrium sterni and then arches posteriorly, with a slight left lateral inclination, to become the ***descending aorta***. The curved, arching part is called the ***arch of the aorta***. The arch is connected to the pulmonary trunk (which divides into the pulmonary arteries in the concavity of the arch) by the ***ligamentum arteriosum***, the obliterated remnant of the ***ductus arteriosus***.

The descending aorta approaches the midline from the somewhat left-sided position of the arch and passes into the abdomen behind the median arcuate ligament of the diaphragm at the level of the 12th thoracic vertebra. In its descent through the thorax, the aorta lies on the bodies of the vertebrae and is positioned behind the esophagus, to which it gives many segmental arteries. Note that the arch of the aorta lies within the arbitrary boundaries of the superior mediastinum (Fig. 7-4).

The arch of the aorta gives off three major branches (see Fig. 6-1B), discussed below.

Brachiocephalic Artery. The brachiocephalic artery is the first major branch of the aorta. It passes upward behind the right sternoclavicular joint, where it divides into the right subclavian and the right common carotid arteries.

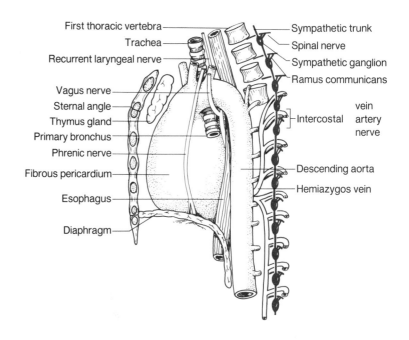

Figure 7-4.

Left surface of mediastinum with pleura removed.

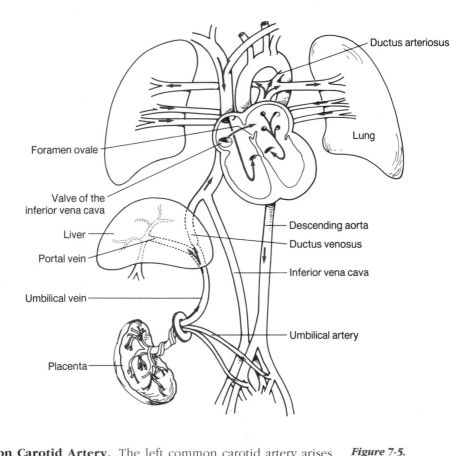

Ductus arteriosus

Lung

Foramen ovale

Valve of the
inferior vena cava

Descending aorta

Liver

Ductus venosus

Portal vein

Inferior vena cava

Umbilical vein

Umbilical artery

Placenta

Left Common Carotid Artery. The left common carotid artery arises from the highest part of the arch of the aorta and passes behind the left sternoclavicular joint to enter the neck.

Figure 7-5.
Fetal circulation.

Left Subclavian Artery. The left subclavian artery arises from the arch of the aorta just beyond the origin of the left common carotid artery. It passes behind the sternoclavicular joint into the root of the neck.

Occasionally the midpoint of the arch of the aorta gives origin to a small artery that runs vertically into the neck to supply the thyroid gland: the ***thyroidea ima artery***.

Intercostal Arteries. The lower nine pairs of intercostal arteries arise from the thoracic aorta (see Fig. 3-9). The right intercostal arteries run from the left of the midline behind the azygos vein and thoracic duct to reach their spaces. The upper two intercostal arteries are branches of the costocervical trunk that arises from the subclavian artery.

THE FETAL CIRCULATION

The circulation of blood in the fetus is quite different from the postnatal circulation. The fetus' pulmonary circulation is small in volume and is carried out against considerable resistance in the parenchyma of the unexpanded lungs. The fetus receives its gaseous exchange and nutrient services from the maternal placenta by way of the umbilical arteries and the umbilical vein (Fig. 7-5).

At birth the umbilical vessels are severed and the lungs expand, with considerable lowering of the resistance to flow of blood through them (lowering of pulmonary blood pressure).

The umbilical vein passes to the liver and continues into the portal vein, but the bulk of its blood flow is shunted into the inferior vena cava by the ***ductus venosus*** (which obliterates at birth). In the inferior vena cava the blood from the liver and from the umbilical vein mixes with blood coming from the lower limbs and lower body. This blood is carried to the right atrium, where the ***valve of the inferior vena cava*** (which can be seen as a remnant in the adult heart) directs the blood through the open foramen ovale into the left atrium, whence it passes into the left ventricle. Some of the blood from the inferior vena cava mingles with blood from the superior vena cava (draining the head and neck and upper limbs), and this stream is directed into the right ventricle. From here the blood passes into the pulmonary trunk. The lungs, being collapsed and offering greater resistance to blood flow, receive little of this blood, which instead passes through the ductus arteriosus from the pulmonary trunk (left pulmonary artery) to the arch of the aorta. Here the blood mixes with a small amount that has been pumped out by the left ventricle and passes down the aorta and through the umbilical arteries to the placenta for "refreshments." Radiographic studies have shown that the greater part of the blood that has been shunted from the right atrium to the left atrium and hence through the left ventricle to the aorta goes up through the carotid arteries to the head and neck (and brain), whereas the blood shunted from the pulmonary artery to the aorta through the ductus arteriosus goes to the lower body and back to the placenta by way of the umbilical arteries. This complex arrangement of valves and shunts and the curious hemodynamics of blood flow ensure that the most recently oxygenated blood (from the umbilical vein) reaches the head, neck, and brain areas.

At birth (or shortly thereafter) the ductus venosus, the ductus arteriosus, and the foramen ovale close and the adult form of circulation begins.

CONGENITAL ABNORMALITIES OF THE HEART AND GREAT VESSELS

If the reader has had the opportunity to study embryology he should be surprised by the fact that development of the fetus proceeds normally, most of the time. Developmental (congenital) defects occur rarely. The following is an outline of the more common congenital defects afflicting the heart and great vessels.

The heart and great vessels may be disposed so as to form a mirror image of the normal state. This is called ***dextrocardia*** or ***dextrorotation***. This may be associated with a similar rotation of abdominal viscera or be confined to the thorax.

Septal defects are relatively common, and a defect in the atrial septum owing to failure of the ***foramen secundum*** to close causes a ***patent foramen ovale***. Clinicians call this an atrial septal defect. ***Patent ductus arteriosus*** results from failure of the ductus arteriosus to close at birth, and eventually this will cause a strain on the left ventricle and an increase in the pulmonary blood pressure. Increased load on the pulmonary circulation will cause back pressure on the right ventricle and ultimately the right atrium. Thus right ventricular hypertrophy and eventual failure with right-sided heart failure may be anticipated. The condition is easily corrected by ligation of the patent ductus.

When the ductus arteriosus closes spontaneously, in the normal course of events, the obliterative process may go on to narrow the aorta at or near the ductus arteriosus, causing ***coarctation of the aorta***. In this condition the parts of the body supplied by the aorta distal to the coarctation (including the lower limbs) must derive their arterial blood supply through anastomotic channels between the subclavian arteries and their branches and the branches of the descending aorta. There will be visible and palpable enlargement of vessels around the scapula and of the intercostal vessels. The latter will cause deepening of the costal grooves and give the appearance of ***notching*** of the ribs in an x-ray film.

Congenital cardiovascular abnormalities may be classified as causing cyanosis (blue skin) or as being acyanotic. The examples given above are mostly acyanotic, but a patent ductus arteriosus can cause cyanosis when heart failure occurs.

Among the congenital abnormalities associated with cyanosis are the following:

Congenital pulmonary stenosis (stenosis = narrowing) of the pulmonary trunk by itself will cause only right ventricular hypertrophy and eventual failure. Commonly it is associated with a ventricular septal defect. The high pressure built up in the right ventricle will cause shunting of (venous) blood into the left ventricle; cyanosis will be present at birth.

The most common "blue baby" condition is the ***tetralogy of Fallot*** (Fig. 7-6). In this condition the four elements of the tetralogy are as follows:

1. pulmonary stenosis;
2. right ventricular hypertrophy resulting from pulmonary stenosis;
3. ventricular septal defect; and
4. an overriding aorta, that is, the aortic opening straddles the ventricular septal defect and receives blood from the right (hence the cyanosis) and left ventricles.

Surgical treatment for Fallot's tetralogy was first advocated by Dr. Helen Taussig, a cardiologist at Johns Hopkins Medical Center, Baltimore, and implemented by Dr. Alfred Blalock, a thoracic surgeon at the same institution. The operation consisted of anastomosing (joining) the right

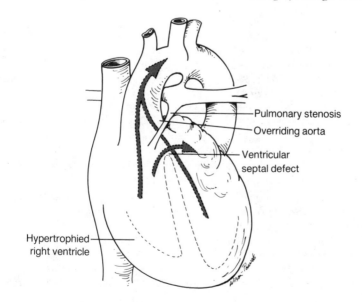

Pulmonary stenosis

Overriding aorta

Ventricular septal defect

Hypertrophied right ventricle

Figure 7-6.
Tetralogy of Fallot.

subclavian artery to the right pulmonary artery, thus bypassing the pulmonary stenosis. This procedure was carried out successfully in many young children in the 1940s and laid the foundation for more advanced cardiac surgery. With the invention of "cardiac bypass" machines a more direct attack at the septal defect and the pulmonary stenosis took the place of the ***Taussig-Blalock*** procedure.

Many other combinations of defects can occur, and some are amenable to surgical correction but are relatively rare.

THORACIC DUCT

The ***thoracic duct*** drains the lymph from most of the body (see Fig. 1-15). It begins at the cisterna chyli in the abdomen and, after passing through the aortic opening of the diaphragm, ascends in the thorax anterior to the right side of the vertebral bodies to the midthoracic region, where it gradually inclines to the left side and terminates in the neck by draining into the junction of the left subclavian and left internal jugular veins.

THE ESOPHAGUS

The esophagus is about 25 cm long and begins in the neck as the continuation of the pharynx at the level of the cricoid cartilage. It ends at the cardiac orifice of the stomach.

Its thoracic part travels through the superior mediastinum behind the trachea, is crossed by the left primary bronchus below the carina, and then continues in the posterior mediastinum, inclining somewhat to the left to pierce the diaphragm through the ***esophageal hiatus*** at the level of the tenth thoracic vertebra. In its posterior mediastinal course the descending aorta insinuates itself posterior to the esophagus. In descending order, the arch of the aorta, left bronchus and the left atrium make slight indentations in the wall of the esophagus, and these may be observed in an x-ray of the esophagus using a swallowed contrast medium, and on esophagoscopy.

CLINICAL NOTE

*The esophagus develops from the same area of the embryonic foregut as the trachea. Sometimes the separation is incomplete, and a **tracheo-esophageal fistula** may occur. In this, the communication between esophagus and trachea is usually located just above the carina, and the upper portion of the esophagus may end in the trachea or in a blind pouch (Fig. 7-7). In either event the situation exists in which the lungs may be flooded by ingested milk or by reflux of stomach acid into the trachea. This is an extremely dangerous situation for the newborn and must be corrected by surgery at an early date.*

The esophagus is a muscular tube containing striated muscle in its upper two thirds and smooth muscle in its lower third. It contracts in waves (peristalsis) and permits swallowed food to pass by this muscular action, even against gravity. Try drinking a bottle of beer while standing on your head; it can be done!

The esophagus can transmit the sensations of pain, fullness, and temperature, at least in its upper two thirds. An interesting feature of esophageal sensation is that when obstruction of the esophagus from a foreign body or a tumor occurs, the patient can frequently localize the

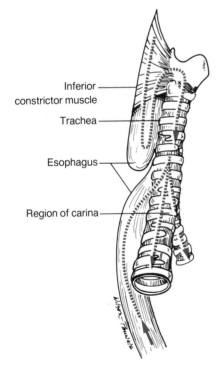

Inferior constrictor muscle

Trachea

Esophagus

Region of carina

Figure 7-7.
Tracheoesophageal fistula.

*point of obstruction and will point to a particular area over the sternum. In the thorax, its blood supply comes from multiple direct segmental branches from the aorta and its veins drain into the azygos-hemiazygos systems. The lower part of the esophagus is drained by the left gastric vein, which is a tributary of the portal vein, and here a communication exists between the portal venous system (left gastric vein) and the systemic venous system (azygos and hemiazygos veins). In the condition of **portal hypertension** caused by liver disease, the anastomosis at the lower end of the esophagus between these two venous systems may cause extreme distension of submucosal esophageal veins (**esophageal varices**) that can rupture and bleed profusely.*

LYMPHATIC DRAINAGE OF THE THORAX

Lungs. The lung tissues contain lymph vessels and lymph nodes that drain into the ***bronchopulmonary nodes*** at the roots of the lungs (see Fig. 4-7). From there efferent lymph vessels run to the ***tracheobronchial nodes*** that are clustered around the carina. These nodes drain into the paratracheal nodes, which in turn drain to the ***bronchomediastinal trunks***. The bronchomediastinal trunks also receive afferent vessels from the ***anterior mediastinal nodes*** that drain the anterior mediastinum, including the parasternal nodes (internal thoracic nodes) that lie in the anterior intercostal spaces along the internal thoracic arteries and receive afferent vessels from the medial halves of the breasts. The bronchomediastinal trunks drain into the ***thoracic duct or right lymphatic duct***.

Posterior Mediastinum. The posterior mediastinal nodes drain the posterior mediastinal structures (*e.g.*, esophagus) and the diaphragm (via the ***phrenic nodes***). The posterior mediastinal nodes drain into the ***posterior mediastinal trunks***, which in turn drain into the thoracic duct.

THE NERVES OF THE THORAX

The following are the nerves of the thorax:

1. ***phrenic nerves***;
2. ***intercostal nerves*** (described in Chap. 3);
3. ***vagus nerves*** and their plexuses and the left ***recurrent laryngeal nerve***; and
4. ***sympathetic trunks***, their ***rami communicantes***, ganglia, and splanchnic nerves.

The two phrenic nerves (right and left), which supply all the muscles of the diaphragm and receive sensory fibers from the central portion of the diaphragm, arise from the anterior primary rami of the third, fourth, and fifth cervical nerves (see Figs. 7-4 and 7-8). They enter the thorax by way of the thoracic inlet, passing posterior to the subclavian veins but anterior to the subclavian arteries. They cross from the lateral side of the inlet to the mediastinum, anterior to the internal thoracic arteries and anterior to the roots of the lungs, to reach the fibrous pericardium. They descend on its surface, between it and the mediastinal pleurae, to the diaphragm, in proximity to the pericardiacophrenic vessels.

The right phrenic nerve usually pierces the diaphragm at the inferior vena caval opening, and the left phrenic nerve pierces the diaphragm more

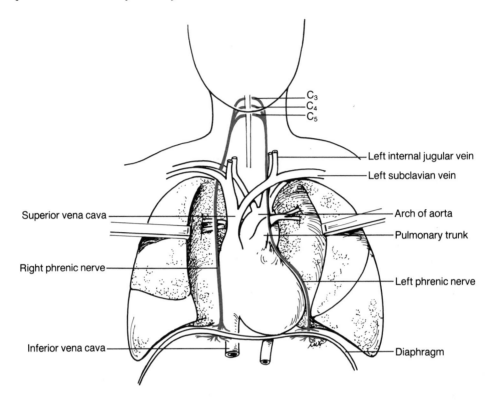

C3
C4
C5

Left internal jugular vein
Left subclavian vein
Superior vena cava
Arch of aorta
Pulmonary trunk
Right phrenic nerve
Left phrenic nerve
Inferior vena cava
Diaphragm

Figure 7-8.
Phrenic nerves.

anteriorly and laterally in its muscular portion (Fig. 7-8). Both phrenic nerves supply sensory fibers to the upper and lower surfaces of the diaphragm. On the inferior surface of the diaphragm each phrenic nerve divides into branches that supply the diaphragm and also gives a few twigs to the celiac plexus (see the Abdomen) and to the gallbladder.

An accessory phrenic nerve (usually from C5) may be present on one or both sides.

CLINICAL NOTE

The phrenic nerves are the sole motor nerves to the diaphragm. Since they arise in the neck, a spinal cord injury at or above the fourth cervical segment will cause diaphragmatic paralysis and death from respiratory failure. (The intercostal muscles will, of course, also be paralyzed because the spinal cord injury is above their origins from T1 downward.)

*The phrenic nerves give sensory fibers to the pericardium and mediastinal pleura, and pain from these areas will be referred to the shoulder tip. Both upper and lower surfaces of the diaphragm are supplied with sensory fibers from the phrenic nerves, and inflammation of either of these surfaces can be referred to the shoulder tip. More will be said about this type of **referred pain** in the section on the abdomen.*

The vagus nerves begin at the hind brain and descend through the neck and thorax into the abdomen; they have extensive distributions (which will be described in continuity in Chap. 34). They enter the thorax by passing anterior to the subclavian arteries (Fig. 7-9). In the thorax the vagus nerves contain preganglionic parasympathetic efferent ("motor") fibers that synapse with their ganglion cells in the walls of the viscera.

From these visceral ganglion cells, postganglionic fibers supply the smooth muscles and secretory cells of the viscera. There are visceral afferent (sensory) fibers in the vagus nerves, the most important of which transmit information about blood pressure (baroreceptors), the oxygen and carbon dioxide content and the pH of blood (chemoreceptors).

The right vagus nerve descends posterior to the right brachiocephalic vein and then runs posterior to the superior vena cava and lateral to the trachea. Next it passes posterior to the right primary bronchus and the root of the right lung.

The left vagus nerve descends between the left subclavian artery and the left common carotid artery, posterior to the left brachiocephalic vein. Next it crosses the lateral aspect of the arch of the aorta and passes posterior to the root of the left lung. As it reaches the inferior border (concavity) of the aortic arch it gives off the ***left recurrent laryngeal nerve***, which loops around the aortic arch to ascend into the neck in the groove between the trachea anteriorly and the esophagus posteriorly (Fig. 7-9).

Behind the roots of the lungs, both vagi join with fibers from the second, third, and fourth thoracic sympathetic ganglia to form the posterior pulmonary plexuses.

Both vagus nerves send multiple fine branches to the cardiac plexus, which lies on the surface and in the concavity of the arch of the aorta, and to pulmonary plexuses located in the hilus of each lung.

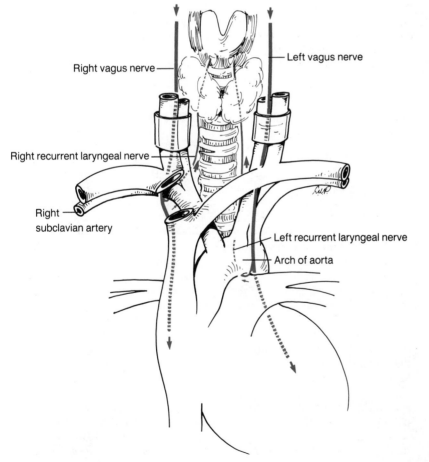

Figure 7-9.
Vagus nerves.

After contributing to the pulmonary plexi, both vagus nerves break up into several fibers that participate in the formation of the esophageal plexuses that cover the anterior and posterior aspects of the esophagus in a dense network that also receives sympathetic fibers. The left vagus nerve is distributed mainly to the anterior esophageal plexus, with the right vagus being the major contributor to the posterior esophageal plexus. The esophageal plexuses also receive fibers from the thoracic sympathetic ganglia (probably second to sixth).

At the lower end of the esophagus the esophageal network of nerve fibers consolidate to form the anterior and posterior gastric nerves. The gastric nerves leave the thorax through the esophageal opening (hiatus).

In the thorax the vagus nerves supply efferent fibers that affect the sinoatrial node (slowing of heart beat) and others that stimulate the secretion of mucus in the bronchial tree and that cause contraction of smooth muscle fibers in the walls of bronchi and bronchioles.

Vagal afferent fibers carry sensory information from ***baroreceptors*** and ***chemoreceptors*** around the great vessels to the central nervous system.

CLINICAL NOTE

The proximity of the left recurrent laryngeal nerve to the trachea, carina, and tracheobronchial lymph nodes makes it vulnerable to irritation or infiltration by mediastinal tumors (particularly carcinomata of the bronchi with secondary deposits in the lymph nodes), causing a persistent cough and later paralysis of the muscles of the left half of the larynx. This produces hoarseness of the voice. Similar left recurrent laryngeal nerve involvement can occur with gross dilatation (aneurysm) of the aortic arch.

THE SYMPATHETIC NERVOUS SYSTEM

The sympathetic nervous system comprises two trunks, one on each side of the vertebral column, which extend from the base of the skull to the lower sacral region. At each spinal nerve level there is a ***sympathetic ganglion*** that contains nerve cell bodies on which ***preganglionic*** sympathetic nerve fibers synapse (see Chap. 16). From these ganglionic neurons, ***postganglionic*** sympathetic nerve fibers are distributed to each spinal nerve. In some areas, particularly in the neck, several sympathetic ganglia are consolidated into a smaller number of larger ganglia. For example, in the neck there is not a ganglion for each cervical nerve; instead there are three larger ganglia (inferior, middle, and superior cervical ganglia) that send postganglionic sympathetic fibers to a group of nerves. Note that whatever the local "clumping" of ganglia may be, each spinal nerve receives postganglionic fibers (via gray rami communicantes) from the sympathetic trunk.

The sympathetic chains (trunks) also send off preganglionic fibers that run in a medial direction to centrally placed ganglia (*e.g.*, the prevertebral ganglia) where postganglionic fibers arise to be distributed to the viscera. Major blood vessels are surrounded by a network of postganglionic sympathetic nerve fibers. This is particularly evident around the aorta and its visceral branches, and around the internal carotid arteries that carry these fibers into the cranium. The postganglionic fibers from the sympathetic ganglia to the spinal nerves run in the ***gray rami communicantes***, and, as has been emphasized, each spinal nerve receives a gray ramus.

The ***preganglionic outflow*** of sympathetic nerve fibers from the spinal cord to the sympathetic trunks and their ganglia is confined to the thoracic segments and the first two lumbar segments of the spinal cord. This is called the ***thoracolumbar outflow***.

On the posterior thoracic wall the two sympathetic trunks can be seen and their ganglia palpated on each side of the vertebral column. The trunks rest against the heads of the ribs and lie posterior to the parietal pleura. The first thoracic ganglion blends with the lower two cervical ganglia to form the ***inferior cervical*** or ***stellate*** ganglion, but the remaining thoracic ganglia are found connected to the corresponding thoracic nerves close to their exits from the intervertebral foramina. Each thoracic (and the first two lumbar) ganglion is connected to its spinal nerve by a ***white ramus communicans*** that carries the preganglionic outflow from the lateral gray column of the spinal cord. These preganglionic fibers may synapse in their corresponding ganglia or in a ganglion at a higher or lower level. Similarly the postganglionic fibers may leave through their corresponding gray rami or pass up or down the trunk before leaving it.

The important concept is that there is only one synapse between each preganglionic and each postganglionic nerve, and this occurs at some distance from the viscera to be supplied, that is, in the ganglia of the trunks or in the prevertebral ganglia. The only exception is the medulla of the adrenal gland, which is a modified ganglion: Here, the synapse occurs in the organ itself.

The sympathetic trunks in the thorax give off preganglionic ***splanchnic nerves*** that run medially to the prevertebral plexuses. The greater splanchnic nerve carries fibers from the 5th, 6th, 7th, 8th, and 9th thoracic ganglia, the lesser splanchnic nerve from the 10th and 11th ganglia, and the least splanchnic nerve from the 12th ganglion (see Fig. 7-3). These preganglionic fibers do not synapse until they reach the ganglion cells in the prevertebral (preaortic) plexuses.

The splanchnic nerves usually reach the abdomen by piercing the crura of the diaphragm, but they may pass behind the median or medial arcuate ligaments.

The Abdomen (general description), Abdominal Wall, and Scrotum

The abdomen is that part of the body between the diaphragm and the pelvis. Its cavity extends superiorly behind the costal margins to the level of the dome of the diaphragm at the fifth intercostal space. Below, it is limited by a hypothetical plane marked by the arcuate lines of the os coxae and the promontory of the sacrum. This plane separates the abdomen from the true pelvis (pelvis minor) (Fig. 8-1).

Proper understanding of the boundaries of the abdomen and pelvis requires a knowledge of the bones involved in the formation of the abdominal walls and the walls of the pelvis.

OS COXAE (HIP BONE OR INNOMINATE BONE)

The os coxae is a composite bone consisting of three elements that are fused together in the adult: the *pubis*, the *ilium*, and the *ischium* (Fig. 8-2). The composite os coxae is a large irregular-shaped bone with a constriction in the middle and expansions above and below. The expansions are mainly for the attachments of muscles of the trunk, abdominal wall, and the lower limb. The lateral aspect of its constricted middle portion bears the cup-shaped *acetabulum* for articulation with the femur. Below the acetabulum there is a large oval hole, the *obturator foramen* (Fig. 8-2). In front, the two hip bones articulate with each other at the *pubic symphysis*, forming a secondary fibrocartilaginous joint. Behind, the two hip bones are separated by, but firmly attached to, the *sacrum*. The two hip bones form the *pelvic girdle* of the lower limbs and, in conjunction with the sacrum and coccyx, form the *pelvic ring* (see Fig. 8-1).

Ilium. The ilium is the large irregular superior part of the os coxae. Its lower part is fused with the pubis and the ischium and forms approximately the upper two fifths of the acetabulum. The upper boundary is the *iliac crest*, which terminates anteriorly in the *anterior superior iliac spine* and posteriorly in the *posterior superior iliac spine*. About 5 cm behind the anterior superior iliac spine, the crest expands into the *tubercle* of the iliac crest. The anterior spine, the posterior spine, and the tubercle are palpable landmarks and are used as reference points for anatomical and clinical purposes (see Fig. 8-2).

The blade of the ilium has an outer *gluteal surface* and an inner abdominal surface, the iliac fossa. The posterior one third of the ilium articulates with the sacrum to complete the pelvic ring.

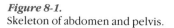

Figure 8-1.
Skeleton of abdomen and pelvis.

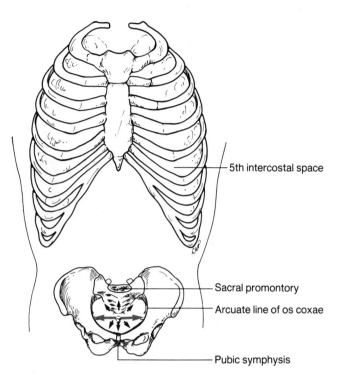

5th intercostal space

Sacral promontory

Arcuate line of os coxae

Pubic symphysis

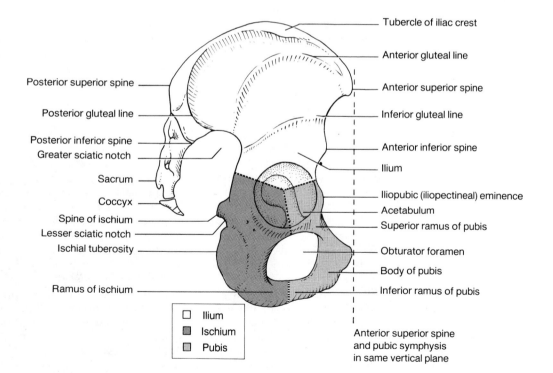

Tubercle of iliac crest

Anterior gluteal line

Posterior superior spine

Anterior superior spine

Posterior gluteal line

Inferior gluteal line

Posterior inferior spine

Anterior inferior spine

Greater sciatic notch

Ilium

Sacrum

Iliopubic (iliopectineal) eminence

Coccyx

Acetabulum

Spine of ischium

Superior ramus of pubis

Lesser sciatic notch

Ischial tuberosity

Obturator foramen

Body of pubis

Ramus of ischium

Inferior ramus of pubis

☐ Ilium
◼ Ischium
▨ Pubis

Anterior superior spine
and pubic symphysis
in same vertical plane

Figure 8-2.
Pelvic girdle (right lateral view).

Pubis. The pubis is the anterior part of the lower portion of the os coxae and meets the opposite pubis in the midline to form the pubic symphysis. It has a superior ramus, an inferior ramus, and a body. The body bears the pubic crest on its upper surface. The crest runs from the pubic symphysis medially to the ***pubic tubercle*** laterally. The superior ramus runs upward, backward, and laterally above the obturator foramen to join the ilium at the acetabulum. The anterosuperior surface of the superior ramus extends from the pubic tubercle to the ***iliopubic eminence***. It is called the pectineal surface and bears a sharp edge, the ***pectineal line*** (pecten pubis).

The inferior ramus runs downward and backward, joining with the ramus of the ischium to form the lower boundary of the obturator foramen.

Ischium. The ischium is the remaining part of the os coxae. It has a ramus and a body. The ramus completes the obturator foramen as described above, and the body bears a thick irregular bump, the ***ischial tuberosity***. This is the bony prominence of the buttock upon which the body rests in the sitting position.

The posterior aspect of the ischium presents the ***ischial spine***, and the depression below the spine is the lesser ***sciatic notch***, which, closed by ligaments, helps to form the ***lesser sciatic foramen***. Where the upper part of the body of the ischium joins the posterior part of the ilium there is a larger depression, the ***greater sciatic notch***, which, closed by ligaments, forms the ***greater sciatic foramen***.

Examination of the hip bone will show that its inner aspect, which forms parts of the abdominal and pelvic cavities, is relatively smooth, whereas the outer aspect is roughened by the attachments of muscles, their tendons and ligaments.

The following bony points of the hip bone are easily palpable in the living: the anterior superior iliac spine, the posterior superior iliac spine, the iliac crest and its tuberosity, the pubic symphysis, the pubic tubercle, and the ischial tuberosity (see Fig. 8-2). The reader should identify these points on his own body. In examining the hip bone, or the articulated pelvis, it is important to achieve correct orientation of the bones. In the upright position, the anterior superior iliac spine and the pubic symphysis lie in the same vertical plane.

THE SACRUM AND COCCYX

The sacrum and its appended coccyx form the lower part of the vertebral column and the posterior wall of the pelvis. The sacrum is a massive wedge-shaped structure that represents the fusion of five vertebral bodies. It can be seen to have anterior and posterior foramina for the passage of the anterior and posterior primary rami of the sacral nerves. The lateral aspects (lateral masses) of the base of the wedge present the ***auricular (ear-shaped) surfaces*** for articulation with the hip bones at the very strong ***sacroiliac joints***, which transmit the weight of the body to the pelvic girdle and hence to the lower limbs. The various bumps on the lateral and posterior surfaces of the sacrum represent the fused transverse and spinous processes of the five constituent vertebrae and serve as points of attachment of the massive muscles and ligaments of this area (Fig. 8-3). The prominent anterior aspect of the first sacral segment is called the ***sacral promontory***, and from this a line can be seen sweeping laterally and forward along the ilium to the iliopubic eminence. This is the ***arcuate line***, which, together with the sacral promontory, forms the upper limit of the ***true pelvis*** or pelvic inlet (see Fig. 8-1).

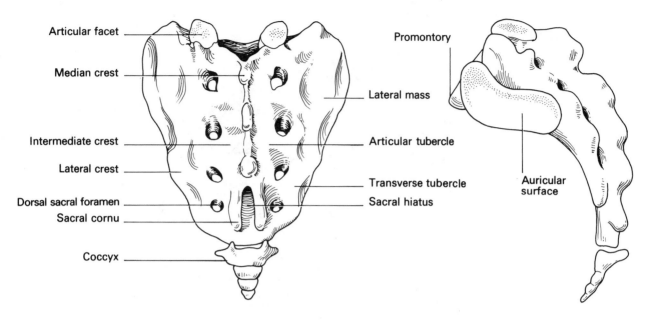

Figure 8-3.
The sacrum (posterior and lateral view).

The coccyx is the remnant of the tail and may consist of three or four fused segments. It joins the sacrum by means of a fibrous joint and has some passive mobility. The pelvic outlet is formed by the coccyx, ischial tuberosities (and attached ligaments), the bodies and rami of the ischia, and the inferior rami and bodies of the pubic bones.

LUMBAR VERTEBRAE

The five lumbar vertebrae form the lumbar spine and participate in the formation of the posterior abdominal wall. Their bodies and processes are large and their articular facets face more or less laterally and medially, thus allowing a considerable degree of flexion and extension to occur in the lumbar spine (Fig. 8-4). The lumbar spine displays a considerable curvature with its convexity forward (lordosis), and this, in combination with the prominent size of the lumbar vertebral bodies, causes a pronounced bulge of the vertebral column into the abdominal cavity so that the distance between the anterior abdominal wall and the vertebral column is only a few centimeters in a lean person. In fact the vertebral column can be palpated through the anterior abdominal wall in a thin person (Fig. 8-5).

CLINICAL NOTE

The transverse processes of the lumbar vertebrae give attachments to the strong muscles of the back and the posterior abdominal wall and, in severe lateral flexion injuries, may be fractured by the pull of their attached muscles.

REGIONS OF THE ABDOMEN

The abdomen can be divided into nine different regions by certain planes that may be outlined on the abdominal wall (Fig. 8-6). These planes mark

Figure 8-4.
Lumbar vertebra: superior and lateral views.

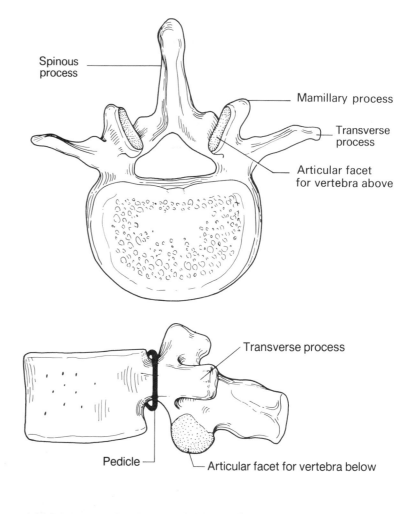

Spinous process

Mamillary process

Transverse process

Articular facet for vertebra above

Transverse process

Pedicle

Articular facet for vertebra below

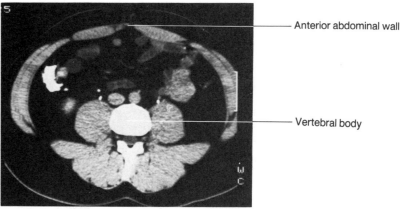

Anterior abdominal wall

Vertebral body

Figure 8-5.
CT scan (cross-section) showing bulge of vertebral column into the abdominal cavity.

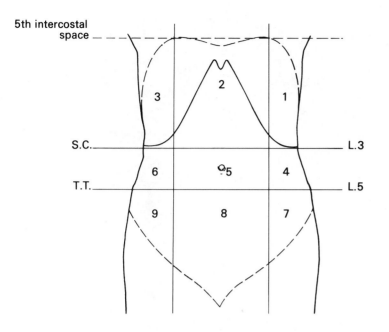

5th intercostal space

S.C.

T.T.

L.3

L.5

Figure 8-6.
Regions of the abdomen: (1) left hypochondrium; (2) epigastrium; (3) right hypochondrium; (4) left lumbar; (5) umbilical; (6) right lumbar; (7) left iliac; (8) hypogastrium; (9) right iliac. S.C. indicates subcostal plane; T.T. indicates transtubercular plane.

important landmarks in the abdomen but apply to the cadaver rather than to the living body in which the contents of the abdomen exhibit some degree of mobility and variations in location, depending on physique and posture.

The traditional anatomical planes are discussed below.

HORIZONTAL PLANES

Transpyloric Plane. Runs through the body of the first lumbar vertebra. Its anterior representation is the midpoint of a line joining the ***suprasternal notch*** to the ***pubic symphysis***.

Subcostal Plane. Runs through the body of the third lumbar vertebra and joins the lowest portion of the thoracic cage (as seen from the front) on the left to the lowest portion of the thoracic cage on the right.

Transtubercular Plane. Joins the tubercles of the iliac crests of the hip bones at the level of the fifth lumbar vertebra.

VERTICAL PLANES

The midclavicular line joins the midpoint of the clavicle with the midinguinal point. The midinguinal point is the midpoint of a line joining the anterior superior iliac spine to the pubic symphysis.

The nine regions of the abdomen are usually described using the subcostal plane and the transtubercular plane as the two horizontal references, and the two midclavicular lines as the vertical planes. With the use of these four planes, nine regions are outlined (Fig. 8-6).

CLINICAL NOTE

There is some overlap in the use of the names of the different regions as used by anatomists and clinicians. Clinically it is more usual to divide the abdomen into four quadrants (right and left upper and lower quadrants) on the basis of a horizontal plane midway between the xiphoid

Figure 8-7.
Division of the anterior abdominal wall
into quadrants.

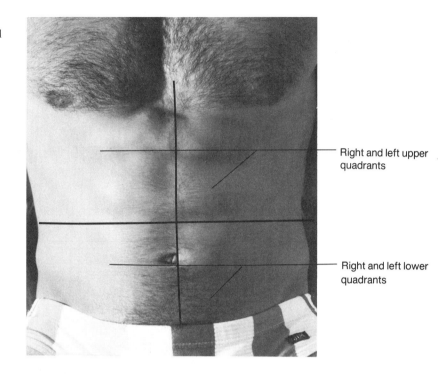

Right and left upper
quadrants

Right and left lower
quadrants

*process and the pubic symphysis and a vertical plane in the midline (Fig.
8-7). In addition, clinicians will refer to the area just below the xiphoid
as the **epigastrium** and the area above the pubis as the **suprapubic area**;
the region around the umbilicus is called the **umbilical** or **periumbilical**
area (Fig. 8-8).*

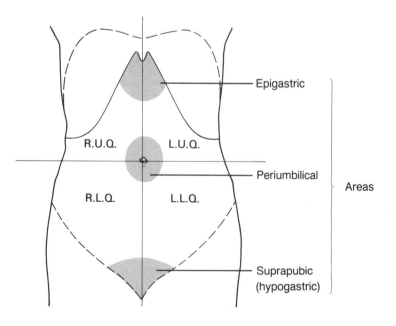

Epigastric

Periumbilical

Areas

Suprapubic
(hypogastric)

Figure 8-8.
Quadrants and areas of abdomen:
L.U.Q. = left upper quadrant; R.U.Q.
= right upper quadrant; L.L.Q. = left
lower quadrant; R.L.Q. = right lower
quadrant.

THE ANTERIOR ABDOMINAL WALL

The anterior abdominal wall is muscular; the muscles are attached to the thoracic cage, lumbar spine, ilium, and pubis. Some of these muscular attachments are indirect via fascial sheets or aponeuroses shared with other muscles (*e.g.*, those of the back).

FASCIA

The fascia of the abdominal wall is divided into superficial fascia and deep fascia.

Superficial Fascia. The superficial fascia is divided into a ***superficial layer*** that contains a variable amount of fat ***(Camper's fascia)*** and a ***deep layer*** that is largely membranous with very little fat ***(Scarpa's fascia)***.

The membranous layer of the superficial fascia is continuous with the superficial fascia of the perineum ***(Colle's fascia)***. The membranous layer of the superficial fascia is attached to the deep fascia of the thigh just below the inguinal ligament. Thus this membranous layer with its attachments to the perineal fascia and the deep fascia of the thigh forms a compartment that is open above (the membranous layer fades away above the umbilicus) but closed below. This has clinical significance, which will be referred to later (see Perineum, Chap. 15).

Deep Fascia. The deep fascia of the anterior abdominal wall is thin in places and not remarkable.

MUSCLES

There are four paired muscles in the anterior abdominal wall. Because the anterior abdominal wall extends around the flanks to the back it might be better called the "outer" abdominal wall, but convention prevails.

Rectus Abdominis

The most medial muscle is ***rectus abdominis***.

Attachments. It runs vertically from the ***crest*** and the symphysis of the pubis to the fifth, sixth, and seventh costal cartilages. It is about 7.5 cm wide and is contained within the rectus sheath (Figs. 8-9A and B). The anterior aspect of the muscle is adherent to the anterior layer of the rectus sheath by three ***tendinous intersections***: one at the umbilicus, one at the xiphoid, and one in-between (Fig. 8-9A).

The ***rectus sheath*** is a sheath formed by the aponeuroses of the three lateral muscles of the anterior abdominal wall (see Fig. 8-10). It contains the rectus abdominis muscle and the superior and inferior epigastric arteries, which anastomose with each other and lie posterior to the muscle.

The lateral border of the rectus sheath, which is easily seen in a muscular person, is called the ***linea semilunaris*** (see Fig. 8-9A). The ***linea alba*** is a tough fibrous band in which the fibers of the rectus sheaths on each side interlace. It stretches from the symphysis pubis to the xiphoid process. It is somewhat wider above than below and separates the two rectus abdominis muscles (see Figs. 8-9A and B).

Figure 8-9.
Muscles of the abdominal wall. **A.** The most superficial layer is on the left side; the next layer is on the right side. **B.** The third layer is on the right side; the deepest layer is on the left side.

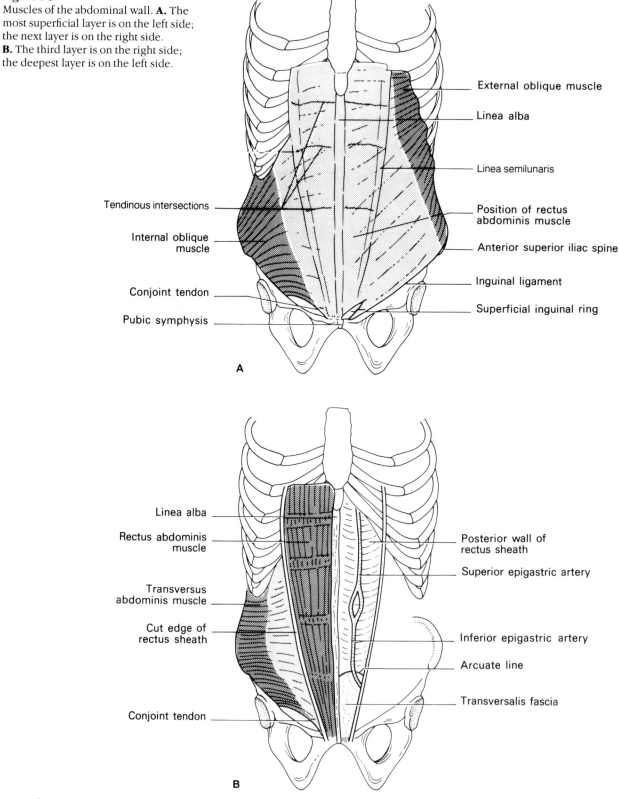

External oblique muscle

Linea alba

Linea semilunaris

Position of rectus abdominis muscle

Anterior superior iliac spine

Inguinal ligament

Superficial inguinal ring

Tendinous intersections

Internal oblique muscle

Conjoint tendon

Pubic symphysis

A

Linea alba

Rectus abdominis muscle

Transversus abdominis muscle

Cut edge of rectus sheath

Conjoint tendon

Posterior wall of rectus sheath

Superior epigastric artery

Inferior epigastric artery

Arcuate line

Transversalis fascia

B

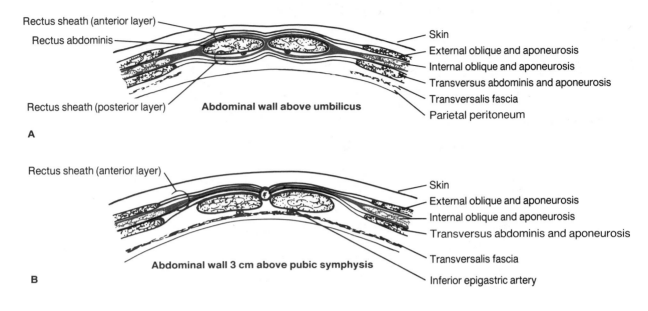

A

Rectus sheath (anterior layer)
Rectus abdominis
Rectus sheath (posterior layer)
Abdominal wall above umbilicus
Skin
External oblique and aponeurosis
Internal oblique and aponeurosis
Transversus abdominis and aponeurosis
Transversalis fascia
Parietal peritoneum

B

Rectus sheath (anterior layer)
Abdominal wall 3 cm above pubic symphysis
Skin
External oblique and aponeurosis
Internal oblique and aponeurosis
Transversus abdominis and aponeurosis
Transversalis fascia
Inferior epigastric artery

Figure 8-10.
Components of rectus sheath (cross-section).

External Oblique

The external oblique muscle is the outermost of the three "flat" muscles of the abdominal wall.

Attachments. It is attached above to the outer surfaces of the lower eight ribs (interdigitating with attachments of the serratus anterior and latissimus dorsi muscles). Its fibers pass downward, forward, and medially (see Fig. 8-9A). They become aponeurotic and blend with the aponeurosis of the internal oblique muscle to form the anterior layer of the rectus sheath. Through the latter they participate in the formation of the linea alba. The lower aponeurotic fibers fold back upon themselves to form the ***inguinal ligament***, which runs from the anterior superior iliac spine to the pubic tubercle. At its point of attachment to the pubic tubercle the inguinal ligament continues with some fibers that sweep backward and laterally to become attached to the pectineal line of the pubis. This is the ***lacunar ligament*** (or pectineal part of inguinal ligament). It is triangular and almost horizontal in erect posture. It measures about 2 cm from apex to base, and the base, which is concave, forms the medial border of the ***femoral ring***.

The lowest and most posterior fibers of the external oblique run vertically downward to be attached into iliac crest. These last described fibers give the external oblique muscle a free posterior border.

Internal Oblique Muscle

This is the middle of the three flat muscles of the abdominal wall. Its fibers run at right angles to those of the external oblique (see Fig. 8-9A).

Attachments. Below, it is attached to the lateral one half of the inguinal ligament and the anterior two thirds of the iliac crest. Behind it is firmly blended with the ***thoracolumbar fascia***. Above, the muscle is attached to the costal margin, and in front it becomes aponeurotic and splits into

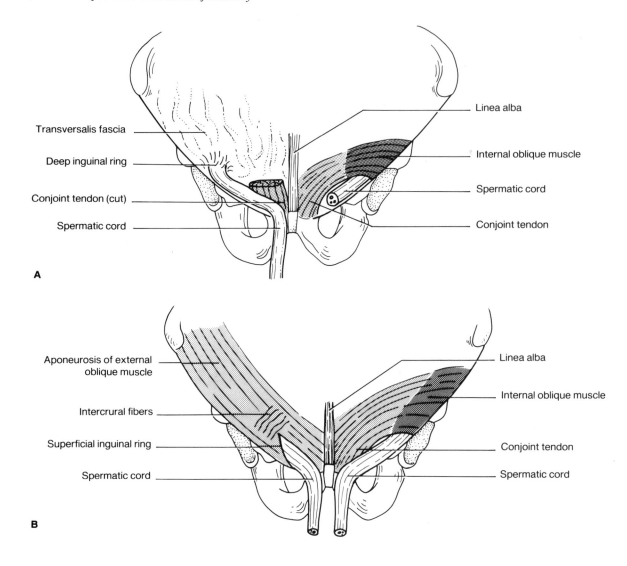

Figure 8-11.
The inguinal region: (**A**) deep;
(**B**) superficial.

two layers. The anterior layer blends with the aponeurosis of the external oblique to form the anterior layer of the rectus sheath. The posterior layer blends with the aponeurosis of the transversus abdominis to form a layer that becomes the posterior lamella of the rectus sheath above but, at a point midway between the pubis and the umbilicus, blends with its own anterior layer and with the aponeurosis of the transversus abdominis to be continuous with the anterior layer of the rectus sheath. It can be seen that the rectus sheath has no posterior layer below this line of blending, the arcuate line (see Fig. 8-10).

The lowest fibers of the internal oblique, those that arise from the inguinal ligament, blend with the lower fibers of the transversus abdominis muscle and arch over the spermatic cord to lie posterior to it (or the round ligament) and to fuse with the lower part of the lateral border of the rectus sheath. This attachment, which also extends onto the pectineal line of the pubis, is called the **_conjoint tendon_**. The conjoint tendon and the muscles converging upon it play an important part in the anatomy and integrity of the inguinal canal (see below), but careful examination will show that the conjoint tendon is, indeed, the lower part of the rectus sheath, attached to the pubic crest, pubic tubercle, and the pectineal line (see Figs. 8-11A and B).

Transversus Abdominis Muscle

This is the innermost of the three flat muscles of the abdominal wall. It is attached below to the lateral one third of the inguinal ligament, the iliac crest, and the conjoint tendon, posteriorly to the thoracolumbar fascia, above to the inner surfaces of the lower six ribs (interdigitating with the diaphragm) and anteriorly to the linea alba. The fibers of the transversus abdominis run in a horizontal direction (see Fig. 8-9B).

Pyramidalis

This is a small triangular muscle (which may be absent) on the anterior aspect of the rectus abdominis. It is attached to the body of the pubis below and the linea alba above.

Nerve Supply

The external oblique muscle is supplied by the anterior primary rami of the lower six thoracic nerves (*i.e.*, the lower six intercostal nerves). The internal oblique, the transversus abdominis, and the recti are supplied by the anterior primary rami of the lower six thoracic nerves and the *first lumbar nerve*. The *ilioinguinal nerve* and the *iliohypogastric nerves* are branches of the anterior primary ramus of the first lumbar nerve. The iliohypogastric nerve supplies the internal oblique and transversus abdominis muscles near the groin, as well as the skin over the groin. The ilioinguinal nerve pierces the internal oblique muscle at the level of its lower fibers and then lies in the inguinal canal below the spermatic cord (round ligament in the female) and emerges through the external ring of the inguinal canal to supply the skin on the upper medial aspect of the thigh and the root of the penis.

The lower six thoracic nerves run between the internal oblique and transversalis muscles, give off lateral cutaneous branches to the abdominal skin, and enter the rectus sheath to supply the rectus muscle before terminating in the skin of the abdominal wall near the midline.

The cutaneous nerve supply to the abdominal wall is provided in oblique strips (with considerable overlap, as was seen in the thorax) by successive thoracic (intercostal) nerves and the first lumbar nerve; for example, the level of the epigastrium is supplied by the seventh and the umbilical region by the tenth thoracic nerve. Not only do the lower six thoracic nerves and the first lumbar nerve supply the skin and muscles of the abdominal wall, but also they supply the parietal peritoneum lining the interior of the abdominal wall, just as the intercostal nerves supply the parietal pleura.

Consequently abdominal disease irritating the parietal peritoneum will cause localized pain along these segments and will also cause reflex spasm of the abdominal muscles of the region. This palpable spasm of the muscles is called ''guarding'' or splinting by clinicians.

CLINICAL NOTE

Incisions through the muscles of the anterior abdominal wall for the purpose of gaining access to the abdominal cavity must be planned with proper regard for the innervation of the abdominal muscles and the positions and directions of the nerves of the abdominal wall.

Abdominal Incisions

The following are some of the more frequently used abdominal incisions.

Midline Incision cuts vertically through the linea alba, which is a relatively bloodless zone and avoids the nerves of the abdominal wall.

Paramedian Incision is placed vertically about 2 cm from the midline and cuts through the anterior rectus sheath. Lateral displacement of the rectus muscle avoids damaging its nerve and blood supply; the posterior rectus sheath and peritoneum are incised to gain access to the abdominal cavity.

Transverse Incisions provide good access, and because the nerves of the abdominal wall assume a nearly horizontal position as they approach the midline, the rectus muscle can be cut horizontally without threat to its nerve supply.

Subcostal Incisions provide good access to the gallbladder on the right and the spleen or proximal stomach on the left. They are kept a least 2.5 cm below the costal margin to avoid the seventh thoracic nerve, which hugs that margin, and cut across the fibers of the oblique, transversus, and rectus muscles.

Muscle Splitting (Gridiron) Incisions avoid the cutting of muscle fibers and rely on the ability to separate muscle fibers in the lines of their directions. If kept relatively small they provide adequate access and avoid tearing or stretching of nerves. The most common example of a muscle splitting incision is the *McBurney* incision for appendectomy. The external oblique aponeurosis and the internal oblique and transversus abdominis muscles are split along their fibers. Little sewing is needed on closing because the muscle fibers will fall back into their natural positions.

Actions of Abdominal Muscles

It is difficult to conceive a situation in which abdominal muscles would act individually. Generally they act together to maintain or increase intraabdominal pressure in activities such as emptying the bladder or rectum, parturition, coughing, vomiting, or deep expiration. Acting together the two recti help in flexion of the lumbar spine, and the "flat" muscles can act together to produce lateral flexion or twisting movements.

TRANSVERSALIS FASCIA

The transversalis fascia is the fascia on the deep surface of the transversus abdominis muscle and the posterior wall of the rectus sheath. It forms the only posterior wall of the rectus sheath below the arcuate line. In living persons, the transversalis fascia is more obvious in the lower parts of the abdominal wall and can be seen as a definite layer in lower abdominal incisions and in surgery of the inguinal canal (hernia surgery). It plays some part in maintaining the integrity of the inguinal canal.

A layer of ***extraperitoneal fatty tissue*** separates the transversalis fascia from the parietal peritoneum.

The ***parietal peritoneum*** (see Fig. 8-10) is a connective tissue membrane lined by mesothelial cells, which give it a smooth moist surface.

THE UMBILICUS

The umbilicus is the scarred remnant of the attachment of the ***umbilical cord*** to the abdominal wall. Usually this scar tissue contracts to form the familiar dimple. It is located in the linea alba, and, if scarring is incomplete, a defect will persist in infants producing an ***umbilical hernia***. In many young people this defect will heal during early childhood, but sometimes its persistence may require surgical repair.

UMBILICAL LIGAMENTS

The inner (posterior) aspect of the lower part of the anterior abdominal wall shows a median ridge and paired medial and lateral ridges. These are called ***umbilical ligaments***, and the first two represent the remnants of certain embryonic structures. The ***median umbilical ligament*** runs from the apex of the urinary bladder to the umbilicus and represents the obliterated remnant of the ***urachus***. Rarely, it may fail to obliterate and present a small but continuous leakage of urine at the umbilicus. Sometimes an isolated segment may persist and become a urachal cyst. These lesions are easy to repair or excise.

The ***paired medial umbilical ligaments***, one on each side of the midline, represent the remnants of the obliterated ***umbilical arteries*** and run from the internal iliac arteries in the pelvis to the umbilicus. The paired ***lateral umbilical ligaments*** are the ridges raised by the existing and functioning inferior epigastric arteries, which are branches of the external iliac arteries. They run upward behind the rectus abdominis muscles and enter the rectus sheaths at the arcuate lines. At a higher level they anastomose with the ***superior epigastric*** branches of the internal thoracic arteries.

INGUINAL REGION

The inguinal region merits attention because it is the most common site for hernias in both sexes. (Hernias are more common in males.) The region is an area of potential weakness in the abdominal wall because it has to allow the passage of the ***spermatic cord*** in the male and the ***round ligament*** of the uterus in the female. Understanding of the inguinal region requires some comments on the migration of the testis (Fig. 8-12).

MIGRATION OF TESTIS

The embryonic testis is formed on the posterior abdominal wall deep to the transversalis fascia and peritoneum. As the fetus grows the testicle migrates to the inguinal region and then passes through the abdominal wall to reach the scrotum, where it is usually found at birth or shortly thereafter. In its course it follows the path already laid out for it by the processus vaginalis and the gubernaculum (Fig. 8-12A).

The ***processus vaginalis*** is an outpouching of the peritoneum that occurs at about the 12th week of fetal life (Fig. 8-12B). The processus passes through the transversalis fascia, under the arching lower fibers of the internal oblique and transversus abdominis muscles and their conjoint tendon, and through the aponeurosis of the external oblique to reach the scrotum. In so doing, the processus vaginalis takes with it a covering of each layer through which it passes. These layers will later cover the ***ductus deferens***. The downward migrating testis makes its move at about the seventh month of fetal life but travels behind the processus vaginalis (Fig. 8-12C). Thus the ductus deferens and the testis have the same coverings as the processus vaginalis, all the way into the scrotum. When the testis reaches the scrotum, the proximal part of the processus vaginalis usually obliterates but its distal part in the scrotum persists to form the ***tunica vaginalis*** of the testis (Fig. 8-12D).

The ovary of the female never leaves the abdominal cavity, but a homologue of the gubernaculum, the round ligament of the uterus, passes

Figure 8-12.
Congenital inguinal hernia. (After G. Reid)

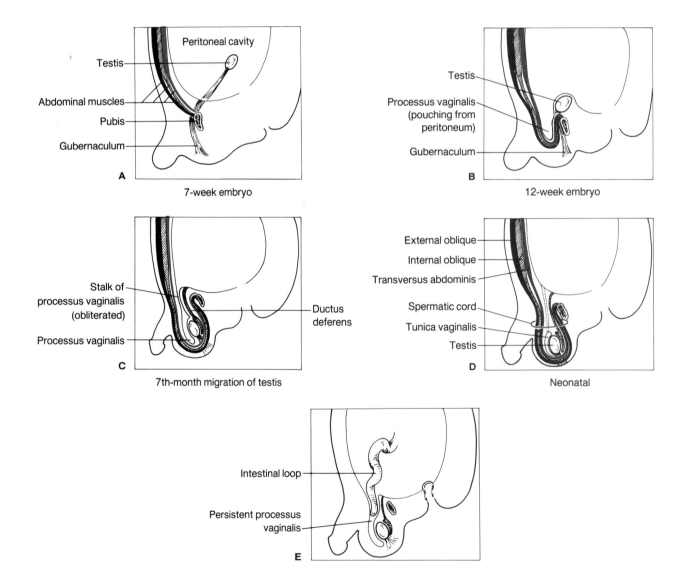

through the inguinal canal behind the processus vaginalis to end in the labium majus.

If the proximal part of the processus vaginalis fails to obliterate (at birth), the stage is set for the existence of a congenital inguinal hernia (in either sex) (Fig. 8-12E).

THE INGUINAL CANAL

The oblique channel in the lower abdominal wall through which the testis, with its ductus deferens and its blood, lymph, and nerve supply, has traveled is the inguinal canal. It is about 4 cm long and lies parallel to the *inguinal ligament*, which stretches from the pubic tubercle to the anterior superior iliac spine (Fig. 8-13).

The lateral or upper end of the inguinal canal is the point at which the processus vaginalis and the testis passed through the transversalis fascia and is called the *internal ring* (of the inguinal canal). The internal ring lies just above the inguinal ligament and can be identified best from the abdominal aspect of the anterior abdominal wall, lying just lateral to the origin of the inferior epigastric artery from the external iliac artery (Figs. 8-14 and 8-15). From the exterior its position can be estimated at being just above the inguinal ligament in line with the pulsation of the femoral artery.

The floor of the canal is the J-shaped lower margin of the inguinal ligament. The roof of the canal is made of the arched fibers of the internal oblique and transversus muscles as they sweep medially to become the conjoint tendon. The anterior wall is the external oblique aponeurosis and

Figure 8-13.
The inguinal, lacunar, and pectineal ligaments (right side). The pelvis is tipped posteriorly to show the lacunar ligament.

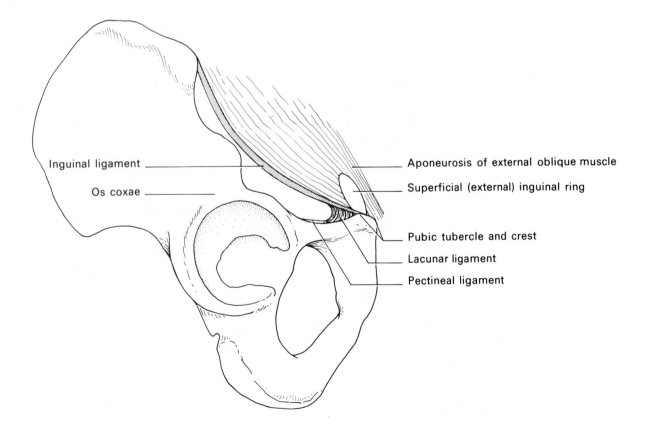

Inguinal ligament

Os coxae

Aponeurosis of external oblique muscle

Superficial (external) inguinal ring

Pubic tubercle and crest

Lacunar ligament

Pectineal ligament

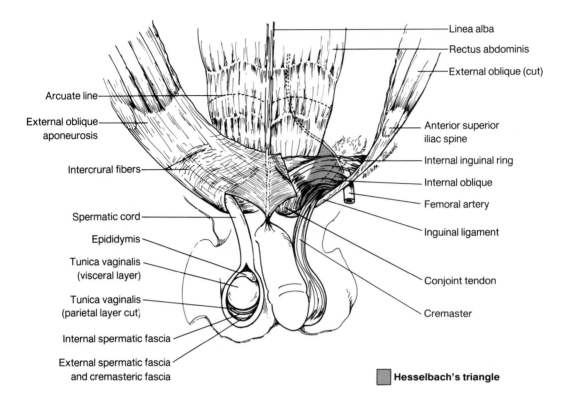

Hesselbach's triangle

Figure 8-14.
Inguinal region and coverings of the ductus deferens.

the posterior wall is the transversalis fascia (Figs. 8-14 and 8-15). Near the internal ring the anterior wall is reinforced by the arched fibers of the internal oblique and transversus abdominis muscles before they sweep upward and posteriorly to form the roof of the canal. Posterior to the external ring the posterior wall is reinforced by the conjoint tendon as it is about to become the lower part of the lateral border of the rectus sheath.

The ***external ring*** of the inguinal canal is in the aponeurosis of the external oblique muscle. It is located just above the inguinal ligament and just lateral to the pubic tubercle. It has a lateral crus (or limb) attached to the pubic tubercle and a medial crus that is continuous with the rectus sheath. The opening of the external ring can be palpated in thin persons. Splitting of the external oblique aponeurosis at the external ring is pre-

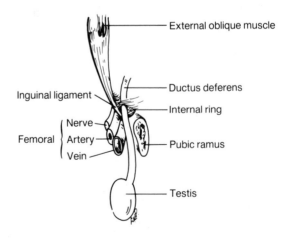

Figure 8-15.
The inguinal ligament (medial view).

vented by some obliquely running fibers in the aponeurosis just lateral to the external ring (see Fig. 8-14 and Intercrural Fibers).

The inguinal canal transmits the **spermatic cord** and the ilioinguinal nerve in males and the round ligament and the ilioinguinal nerve in females.

The main constituents of the spermatic cord are the **ductus deferens** of the testis, the **testicular artery**, a **pampiniform plexus** of spermatic veins (from the testis), the testicular lymphatics, and the genital branch of the genitofemoral nerve (a branch of the lumbar plexus). Also contained in the spermatic cord are the cremasteric artery (from the inferior epigastric artery), the artery of the ductus deferens (from the inferior vesical artery), and sympathetic nerves accompanying these arteries. Note that the common clinical name for the ductus deferens is the vas deferens—hence the term **vasectomy**.

As the spermatic cord travels through the inguinal canal it receives a covering from each layer through which it passes. It picks up the membrane-like **internal spermatic fascia** from the transversalis fascia at the internal ring. Next it obtains a mantle of looped muscle fibers from the internal oblique, the cremaster muscle, and cremasteric fascia. Finally, it receives a covering from the external oblique at the external ring, the **external spermatic fascia** (Fig. 8-16).

The **cremaster muscle** consists of a series of loops that surround the cord and are supplied by the genital branch of the genitofemoral nerve. In cold weather the cremaster muscle contracts and hitches the testis toward the abdominal wall. In some mammals it will retract the testis inside the abdominal wall, presumably to maintain thermostasis.

Cremasteric Reflex. The cremasteric reflex is elicited by stroking the skin of the upper part of the medial aspect of the thigh. If the genitofemoral nerve and its spinal cord segment are intact, the testicle will retract upward.

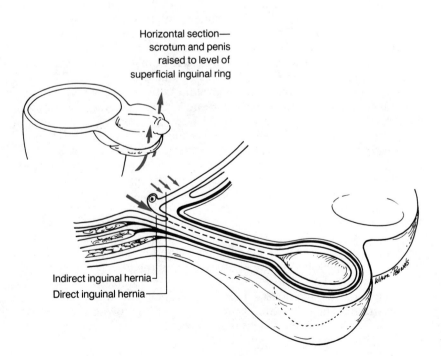

Horizontal section—
scrotum and penis
raised to level of
superficial inguinal ring

Indirect inguinal hernia
Direct inguinal hernia

Figure 8-16.
Direct and indirect inguinal hernias (penis removed).

Inguinal Hernia

A *hernia* is the protrusion, usually as the result of pressure, of a structure through a barrier limiting an anatomical compartment. If the hernia is through the abdominal wall the herniated structure is usually covered by peritoneum, which forms the *hernia sac*. The most common form of hernia in both sexes is an inguinal hernia. It is caused by a combination of abdominal pressure and a weak point in the wall of the inguinal canal. The parietal peritoneum bulges into the inguinal canal and abdominal content, often omentum or bowel, follows the peritoneum, and lies within the peritoneal hernia sac. If the hernia enlarges sufficiently to pass through the external ring, it will find little resistance to impede its further passage into the scrotum or labium majus.

Factors Tending to Prevent Inguinal Hernia. The inguinal canal presents an area of weakness in the abdominal wall, but its structure is such as to prevent herniation. The obliquity of the inguinal canal makes protrusion rather difficult because increased intra-abdominal pressure, such as results from heavy lifting, will push the posterior muscular wall against the anterior muscular wall and tend to reduce the capacity of the canal.

The *internal oblique* fibers passing in front of the internal ring and the insertion of the conjoint tendon behind the external ring will contract during abdominal exertion and produce a valve-like effect on the inguinal canal and its two openings. Thus the external ring is protected by the pressure of the conjoint tendon and the rectus abdominis muscle pressing against the external oblique aponeurosis, and the internal ring is protected by the arched fibers of origin (from the inguinal ligament) of the internal oblique muscle.

There are two types of inguinal hernias (Fig. 8-16), discussed below.

Indirect Inguinal Hernia. This is called indirect because there is no direct break through the walls of the canal. It is caused by a *persistent patent processus vaginalis* that allows abdominal content to slide along its open portion at the internal ring and to progress into the canal throughout the length of its patency. These indirect hernias, which may be regarded as congenital, may manifest themselves at any age from infancy (when persistent crying may be a factor that increases intra-abdominal pressure) to adulthood (when heavy lifting may contribute). In this type of hernia there is no initial breakdown in the muscular and fascial structures in the inguinal canal, but with time these tissues will become attenuated as a result of the persistent pressure of the hernia on surrounding structures. If the patent processus vaginalis communicates with the *tunica vaginalis* of the testis there will be little resistance to the progression of the hernia content into the scrotum. Even if there is no communication with the tunica vaginalis, persistence of the hernia will eventually stretch the processus vaginalis (the hernia sac) into the scrotum.

At this stage it will be a useful anatomical exercise to enumerate the covering layers of an indirect inguinal hernia. This can be done in a didactic manner beginning at the peritoneum of the hernia sac or, in a more interesting way, by considering the layers the surgeon will encounter in her operation. Either approach requires an understanding of the anatomy of the region. The anatomist's approach will begin at the internal ring, and he will remember that the processus vaginalis is covered by internal spermatic fascia (transversalis fascia). The next layer will be the cremasteric fascia and cremaster muscle. After emerging from the external ring, the final layer will be external spermatic fascia (from the external oblique aponeurosis).

The surgeon will view matters from the opposite approach. First she will cut through the skin from the external ring to a point well lateral to the position of the internal ring. Next she will incise the superficial layer of the superficial fascia, followed by the cutting of the deep (membranous) layer of the superficial fascia. Next the external oblique aponeurosis is split along its fibers laterally from the external ring. The first structure to be looked for and protected will be the ilioinguinal nerve, which will lie anterior and inferior to the spermatic cord. The cremaster muscle and the cremasteric fascia are then incised, and the hernia sac (which is usually empty because of the muscle relaxing effect of the anesthetic) will be visible but still covered by a thin layer of tissue, the internal spermatic fascia. When the latter has been incised the hernia sac can be freed, tied off at the internal ring, and the stump permitted to retract into the abdominal cavity. This is all that may be needed, other than the closure of the previously incised layers, in infants or young persons. When the hernia is of long standing and has produced stretching and attenuation of the walls of the canal, some kind of muscle repair is required, but that is beyond the scope of this textbook.

Direct Inguinal Hernia.   This type of hernia does not pass through the internal ring. Instead it forces its way through the transversalis fascia medial to the internal ring and then pushes its way into the layers of the spermatic cord from behind. Again, from the surgeon's view, after cutting skin and superficial fascia and opening the external oblique, the surgeon will encounter cremaster muscle and cremasteric fascia before finding the hernia sac. Next the sac may be covered by a thin layer of transversalis fascia, but usually the hernia has broken through the transversalis fascia. Occasionally a direct hernia may be more medially placed and stretch or break through the conjoint tendon.

A direct hernia always passes through the lower part of the **inguinal triangle (Hesselbach's triangle)** (see Fig. 8-14). This triangle may be seen from the abdominal aspect of the anterior abdominal wall and is the triangular area the base of which is the inguinal ligament and the two sides are the inferior epigastric artery laterally and the lateral border of the rectus abdominis muscle medially.

CLINICAL NOTE

Surgeons have, in the past, emphasized teaching their students how to diagnose the difference between a direct and an indirect hernia. (This has been a favorite examination topic.) However, this test is not infallible, and the final diagnosis is often made only at the time of surgery, when an indirect hernia will reveal itself by coming through the internal ring, that is, lateral to the inferior epigastric artery, whereas a direct hernia will be placed more medially.

The basic clinical approach is to reduce the hernia back into the abdominal cavity by gentle pressure and then to apply pressure with a finger over the internal ring (1 cm above the uppermost palpable point of pulsation of the femoral artery). If the patient is then instructed to produce a hard cough, the finger pressure will sometimes prevent the emergence of an indirect hernia. If it is a direct hernia it will pop out medial to the point of digital pressure.

SCROTUM

The scrotum, testis, and spermatic cord are best studied at this time, since the last two originate as contents of the abdomen and therefore relate to the abdomen. The penis and vulva are considered to be parts of the perineum and are described in Chapter 15.

The scrotum consists of skin and superficial fascia and a layer of smooth, involuntary muscle, the dartos, within the superficial fascia.

CLINICAL NOTE

*The **dartos** muscle, when contracted, gives the scrotal skin a wrinkled appearance in cold weather. The subcutaneous tissue of the scrotum is very thin, and a small accumulation of tissue fluid in it, such as the edema of an insect bite, can produce quite alarming-looking swellings of the scrotum, particularly in young children.*

*The superficial fascia of the scrotum is continuous with the membranous layer of the superficial fascia of the abdominal wall (Scarpa's fascia) (Fig. 8-17). It provides the scrotum with a septum dividing the scrotal sack into two compartments, one for each testis. Posteriorly this layer of fascia is continuous with the membranous layer of superficial fascia of the perineum (Colle's fascia), which is attached to the inferior aspect of the body of each pubic bone and to the conjoint (inferior) rami of the ischium and pubis as far back as the center of the perineum, where it becomes attached to posterior edge of the **perineal membrane**.*

*The **spermatic cord** passes deep to the superficial fascia from the testis to reach the external ring and hence the inguinal canal.*

Tunica Vaginalis. The tunica vaginalis is the distal remnant of the processus vaginalis. Therefore it is a sac of peritoneal origin that envelops the testis in two layers. Its ***parietal*** layer lines the wall of the scrotum and its ***visceral*** layer covers the testis and epididymis (see Fig. 8-14).

CLINICAL NOTE

*An accumulation of fluid, from inflammation or trauma, for example, may fill the potential space between the visceral and parietal layers of the tunica vaginalis. The resulting swelling in the scrotum may become quite large. This condition is called a **hydrocele**. A hydrocele can be distin-*

Figure 8-17.
Superficial fascia of the perineal region.

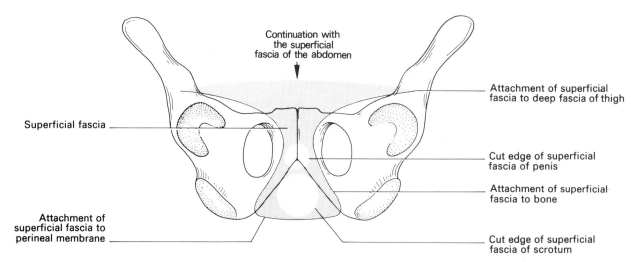

Continuation with the superficial fascia of the abdomen

Attachment of superficial fascia to deep fascia of thigh

Superficial fascia

Cut edge of superficial fascia of penis

Attachment of superficial fascia to bone

Attachment of superficial fascia to perineal membrane

Cut edge of superficial fascia of scrotum

guished from a large inguinal hernia extending into the scrotum by the fact that the swelling of the hydrocele does not extend above the neck of the scrotum. A hernial swelling in the scrotum can be felt to continue upward and laterally into the inguinal canal.

TESTIS AND SPERMATIC CORD

The testis is covered by the tunica vaginalis in all but its posteromedial portion. Deep to the visceral layer of the tunica vaginalis the testis is surrounded by a dense white layer of connective tissue, the ***tunica albuginea*** (Fig. 8-18). The testis consists of a large number of ***seminiferous tubules*** that produce the sperms. Between the seminiferous tubules are the ***inter-***

Figure 8-18.
Right testis (transverse section). The shaded area is a potential space between the parietal and visceral layers of the tunica vaginalis.

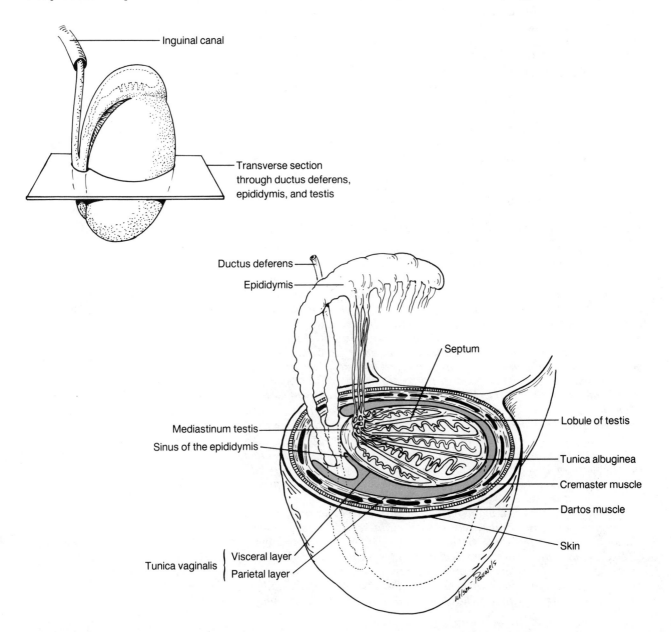

stitial cells that produce the male hormone *testosterone*. The seminiferous tubules are divided into groups *(lobules)* by *septa* of connective tissue that pass from the tunica albuginea to the posterior part of the testis. This posterior part is the *mediastinum testis*. The seminiferous tubules form a network *(rete testis)* in the mediastinum testis. *Efferent ductules* pass from the rete to the upper end *(caput)* of the epididymis (Fig. 8-18).

EPIDIDYMIS

The epididymis is found on the posterior surface of the testis. It is a coiled tubular structure that receives the efferent ductules at its head (caput). The *body* of the epididymis lies against the body of the testis, partially separated from it by the *sinus of the epididymis*. This sinus is lined by the visceral layer of the tunica vaginalis. The lower part of the epididymis is the *tail*, and it is attached to the testis. Here the coiled tube of the epididymis becomes continuous with the *ductus deferens*. The testis "hangs" in the scrotum from the ductus deferens (vas deferens). Mature sperms are stored in the epididymis until they are ejaculated.

    The spermatic cord was described with the inguinal canal. Remember that it contains the ductus deferens, which is a firm cord-like structure that can be recognized easily by palpation and the nerves, blood vessels, and *lymph vessels* of the testis. These testicular lymphatics drain to the *aortic* group of lymph nodes along the posterior abdominal wall in the region of origin of the testicular artery. Hence a tumor of the testis may produce a large abdominal mass if it has spread to the aortic nodes. The lymphatics of the *skin* of the scrotum drain to the *inguinal* lymph nodes in the upper part of the thigh.

CLINICAL NOTE: UNDESCENDED TESTIS

The testis develops on the posterior abdominal wall in the region of the kidney and subsequently "migrates" to the scrotum (see Fig. 8-12A). In fact its caudal end becomes attached to the ventral embryonic abdominal wall quite early, and as the body grows the testis maintains this attachment to the ventral abdominal wall by a piece of fibromuscular connective tissue, the **gubernaculum**. *It may be argued that the body grows away from the attachment of the gubernaculum rather than that the testis migrates. The gubernaculum is present early in the embryonic development of the testis and has four slips that are attached to parts of the body wall that subsequently become located in the scrotum, the primary location; the perineum; the root of the penis; and the upper femoral triangle near the termination of the great saphenous vein. If the testis fails to descend it may remain lodged anywhere from the abdominal cavity to the inguinal canal or in the subcutaneous tissue around the external ring. If it appears lodged in the inguinal canal or at the external ring it can often be brought into the scrotum by surgical means. This should be done before the child is 3 or 4 years of age because spermatogenesis will not occur in the undescended or ectopic testis.*

    Maldescent occurs if the testis follows one of the other attachments of the gubernaculum, and rarely such a maldescended testis can be found in the perineum, at the root of the penis or in the femoral triangle.

    Care must be taken before declaring a testis undescended. A testis may appear lodged at the external ring because of contraction of the cremaster muscle in a cold or fractious child and yet be perfectly lodged in the scrotum at other times. The clue to diagnosis is the appearance of the "empty" scrotum. If the testis normally resides in the scrotum but is rather mobile, the scrotum will have a normally developed yet empty appearance. If the testis has never descended to its normal position, the scrotum will be underdeveloped or may even be absent on that side.

APPENDICES OF THE TESTIS

One or two appendices may be found on the testis. One occurs at its upper pole, and another is sometimes found at the head of the epididymis. These are small cyst-like structures representing embryonic remnants.

The Peritoneum and Viscera

Before one can understand the abdominal viscera and their locations, it is necessary to study the peritoneum and the peritoneal cavity.

PERITONEUM

The peritoneum consists of a single layer of mesothelial cells resting on a thin layer of connective tissue. It has a blood supply, nerve supply, and lymphatic drainage. The configuration of the peritoneum follows the principle of a serous membrane, as outlined in Chapter 4, that is, it has a parietal layer and a visceral layer. The two layers are separated by a thin film of fluid, and the peritoneal cavity is more of a potential space than an actual cavity in the intact living body. This may not be very apparent in cadavers or in the living body after the abdominal wall and peritoneum have been exposed surgically, but the principle applies.

Parietal Peritoneum. The parietal peritoneum lines the abdominal wall but does not surround a viscus.

Visceral Peritoneum. The visceral peritoneum is that portion of the peritoneum surrounding a viscus.

Peritoneal Cavity. This is the potential space between the parietal peritoneum and the visceral peritoneum. The peritoneal cavity may be compared to the interior of a deflated balloon. If the viscera are compared to structures pushed into the outside of the balloon, it will be noted that each of these structures is covered by two layers of balloon, and the two layers of balloon are in internal surface-to-surface contact with one another (Fig. 9-1). If a balloon has been blown up and deflated a few times, its interior is made moist by condensation from the breath; the two layers will slide easily on each other. The two layers of the peritoneum slide easily on each other. If one now pushes a fist in from one side, one can consider the fist to be the equivalent of an abdominal viscus (*e.g.,* small bowel). The layer of the balloon in contact with the fist is the ***visceral layer***, and the outer layer is the ***parietal layer*** of peritoneum (Fig. 9-1). Note that at the wrist the two layers are continuous, that is, the visceral layer becomes continuous with the parietal layer.

MESENTERY

A mesentery (Fig. 9-2) is a double layer of peritoneum connecting a viscus to the abdominal wall. If one uses the analogy of the balloon and pushes

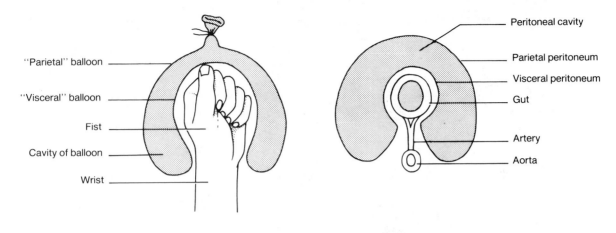

in an object to invaginate one wall, the object will be covered by "visceral peritoneum," and the portion of the balloon as it passes down on either side of the fingers supporting the object is similar to a mesentery.

Note: A viscus covered with, but not surrounded by, peritoneum (*e.g.*, kidney) is ***retroperitoneal*** (Fig. 9-2).

OMENTUM

An omentum (Fig. 9-3) is a double layer of peritoneum attached to the stomach. By definition the ***lesser omentum*** attaches to the ***lesser curvature*** of the stomach and the ***greater omentum*** attaches to the ***greater curvature*** of the stomach.

LIGAMENT

In some situations the layers of peritoneum running from one viscus to another or from a viscus to the body wall are called ***peritoneal ligaments***. They are given special names because they have special importance.

RECESS

In certain areas folds of peritoneum form culs-de-sac, which are blind sacs with a single opening to the rest of the peritoneal cavity. These peritoneal recesses have some clinical importance that will be discussed later.

Figure 9-1.
Comparison of the peritoneal cavity to a balloon.

Figure 9-2.
Transverse section (schematic) of the peritoneal cavity.

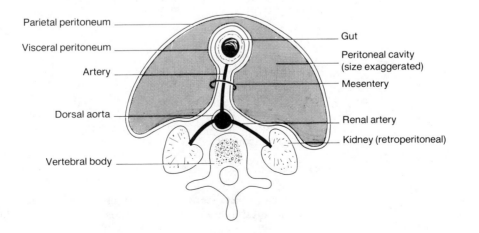

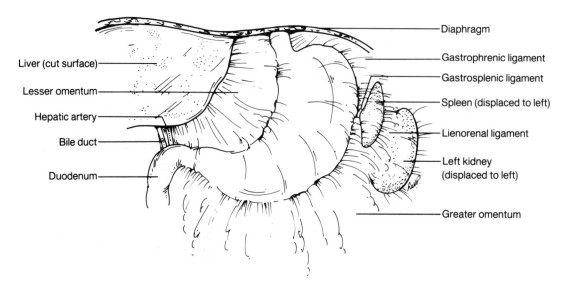

Figure 9-3.
Anterior view of the stomach and its peritoneal attachments. Note that the left kidney and spleen have been displaced to the left.

THE ABDOMINAL CAVITY

The abdominal cavity (as opposed to the peritoneal cavity) is that portion of the body within the abdominal wall.

PELVIS

The pelvis comprises two parts:

1. ***Pelvis Major (false pelvis)***: that part of the pelvis which is found above the brim of the true pelvis (see Fig. 8-1). In other words it is that part of the lower abdominal cavity flanked by the walls of the iliac fossae.
2. ***Pelvis Minor (true pelvis)***: lies below the brim of the pelvis (made up of the sacral promontory and the two arcuate lines of the hip bones).

ABDOMINAL VISCERA AND THEIR ATTACHMENTS

The following will be a brief preview of the abdominal viscera and their disposition within the abdominal cavity (Fig. 9-4).

LIVER

The liver is situated in the right upper quadrant of the abdomen below the diaphragm and extends downward as far as the costal margin. It is connected to the diaphragm by the ***triangular ligaments*** (right and left) and to the anterior abdominal wall by the ***falciform ligament***. The latter is a remnant of the embryonic ***ventral mesentery***. All of the above mentioned ligaments are composed of peritoneum.

ESOPHAGUS

The esophagus pierces the diaphragm just to the left of the midline at the level of the eighth thoracic vertebra. It has a course approximately 2.5 cm

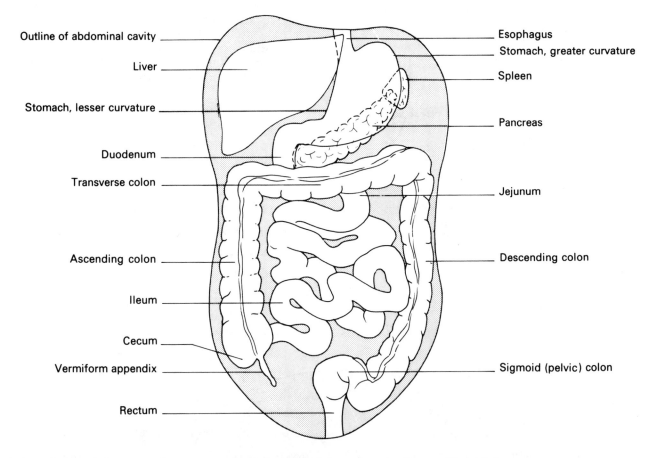

Outline of abdominal cavity
Liver
Stomach, lesser curvature
Duodenum
Transverse colon
Ascending colon
Ileum
Cecum
Vermiform appendix
Rectum

Esophagus
Stomach, greater curvature
Spleen
Pancreas
Jejunum
Descending colon
Sigmoid (pelvic) colon

long in the abdomen and terminates at the cardia of the stomach. The esophagus is covered by peritoneum on its anterior and lateral aspects.

Figure 9-4.
Anterior view of general plan of the viscera of the abdominal cavity.

STOMACH

The esophagus enters the stomach at the ***cardiac orifice***. The stomach is a large distensible J-shaped structure lying in the left upper quadrant of the abdominal cavity. Its anterior and posterior surfaces are covered by visceral peritoneum. Its lesser and greater curvatures (upper and lower borders) are connected to other viscera by mesentery-like structures, the omenta.

Lesser Omentum. The lesser omentum connects the lesser curvature of the stomach to the liver. The lesser omentum has a free edge on the right, and here the ***bile duct, hepatic artery,*** and ***portal vein*** are situated (between the two layers of the lesser omentum) (see Figs. 9-3 and 9-5B).

Gastrophrenic Ligament. Passing from the upper part of the greater curvature of the stomach to the diaphragm is a double layer of peritoneum, the gastrophrenic ligament (see Figs. 9-3 and 9-5).

Gastrosplenic and Lienorenal Ligaments. These ligaments run together as a unit from the greater curvature of the stomach to the posterior abdominal wall in the region of the left kidney. The spleen lies between these two ligaments, which, in fact form a mesentery for the spleen (see Figs. 9-3 and 9-5). The gastrosplenic ligament is a layer of visceral perito-

Figure 9-5.
Lesser sac of peritoneal cavity.
A. Anterior view of stomach, liver, and spleen. Lines indicate levels of cross-sections. **B.** At level of liver. **C.** At level of epiploic foramen.

neum running from the greater curvature to the spleen; it surrounds the spleen and is then reflected onto the left kidney to form the **_lienorenal ligament_**. As it reaches the kidney (which is a retroperitoneal structure), this same layer of peritoneum becomes continuous with the **_parietal_** peritoneum of the posterior abdominal wall (Fig. 9-5).

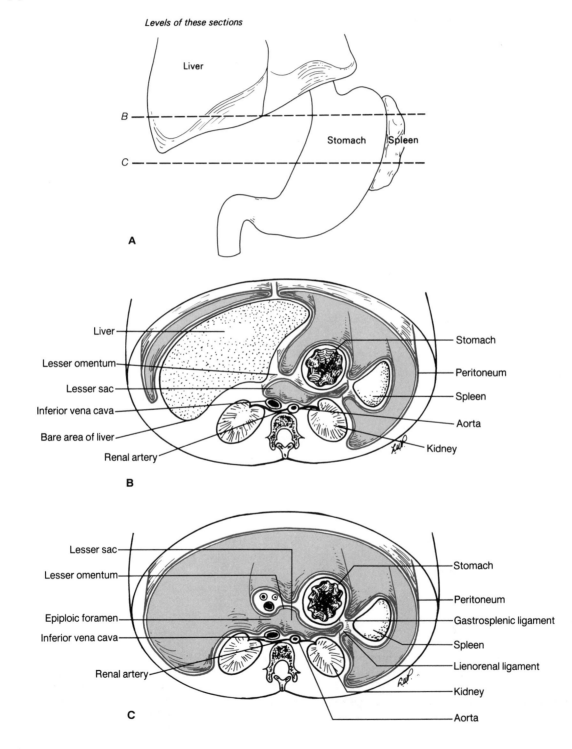

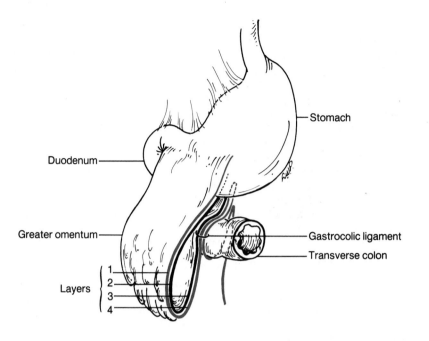

Figure 9-6.
Greater omentum cut to show layers that form it.

Greater Omentum.   The greater omentum (Figs. 9-6 and 9-7) runs inferiorly from the greater curvature of the stomach and then loops back upon itself to attach to the transverse colon, which runs across the abdomen just below the stomach. The greater omentum may be long enough to reach into the pelvis.

The greater omentum consists of four layers of peritoneum in the embryo. The peritoneum in front of the stomach and the peritoneum from the posterior wall of the stomach (layers 1 and 2) descend to the inferior

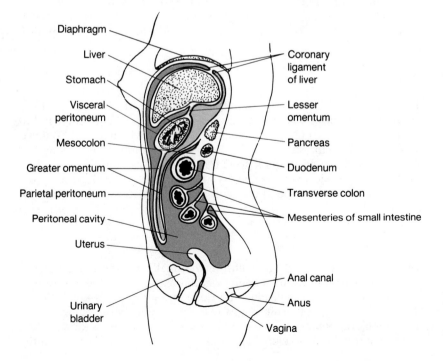

Figure 9-7.
Sagittal section of the female abdominopelvic cavity showing the peritoneum, the omenta, and the mesenteries.

edge of the greater omentum. There they turn upward and form layers three and four. These two posterior layers return to the posterior abdominal wall, where layer three ascends to become the posterior parietal peritoneum of the lesser sac. Layer four passes downward and forward to surround the transverse colon and becomes the ***transverse mesocolon***. Subsequently the posterior layer of the greater omentum fuses with the anterior (superior) layer of the transverse mesocolon so that it "appears" that the transverse colon is surrounded by the greater omentum. Usually the second and third layers fuse so that the appearance is one of a two-layered omentum. During gastric surgery it is necessary to separate the transverse mesocolon from the anterior two layers of the greater omentum, and then the cleavage between layers two and three can be dissected. The term ***gastrocolic ligament*** is used to describe the tissue joining the transverse colon to the stomach. It is composed of the four (fused) layers that have been described.

CLINICAL NOTE

Peritonitis generates a sticky fluid containing fibrin that tends to "glue" the greater omentum to inflamed areas; this limits the spread of infection within the peritoneal cavity.

DUODENUM

The duodenum is the first part of the small intestine. It is 25 cm long, and its name refers to the duodecimal (count by twelves) system because earlier clinicians considered its length to be the same as the width of 12 fingers. German anatomists refer to it as the "zwölffingerdarm" (the 12-finger gut).

The duodenum starts at the distal (pyloric) end of the stomach and has a C shape. It descends from the pylorus and then crosses the midline to become continuous with the jejunum to the left of the midline. Only the first part of the duodenum has a mesentery, the lesser omentum, which it shares with the stomach. The remainder of the duodenum is retroperitoneal (Fig. 9-7).

JEJUNUM

The jejunum and the next part of the small intestine (the ileum) pass from the ***duodenojejunal*** junction to the right iliac fossa. The jejunum and ileum possess a mesentery from which the jejunum, which forms about two fifths of the small intestine, and the ileum are suspended freely. This gives the 500 to 600 cm small gut a good deal of mobility, and loops of small bowel often reside in the pelvis. In the usual anatomical position the greater omentum is draped between the small intestine and the anterior abdominal wall.

ILEUM

The ileum forms the distal three fifths of the small intestine and terminates at the ***ileocecal*** junction to become continuous with the large intestine.

The composite small intestine (jejunum + ileum) is much longer than the attachment of its mesentery to the posterior abdominal wall. This attachment, the ***root*** of the mesentery, runs from the left of the vertebral column at the duodenojejunal junction downward and to the right of the vertebral column to the right iliac fossa and is only about 15 cm long. Thus it is obvious that the mesentery must be shaped like a fan as it spreads from its 15 cm attachment to the posterior abdominal wall to its attachment to the 6-meter-long small intestine.

CECUM

The cecum is a sack-like extension of the large intestine located in the right iliac fossa. The ileum joins the cecum at the ***ileocecal valve*** (see Fig. 10-14). The cecum does not have a mesentery, but there is a recess of the peritoneal cavity, the ***retrocecal recess***, posterior to its lower part (see Fig. 9-10).

VERMIFORM APPENDIX

The ***vermiform*** (worm-like) ***appendix*** (see Fig. 10-14), known as the "appendix" for short, is a narrow outpouching of the lower end of the cecum. It has its own small mesentery ***(mesoappendix)*** connecting it to the mesentery of the small intestine (see Fig. 9-10).

ASCENDING COLON

The ascending colon is continuous with the cecum and passes upward toward the liver on the right side of the posterior abdominal wall; it is retroperitoneal. At the ***right colic (hepatic) flexure*** it becomes continuous with the transverse colon.

TRANSVERSE COLON

The transverse colon, which does have a mesentery ***(the transverse mesocolon)***, passes across the abdomen from the ***right colic flexure*** to the ***left colic (splenic) flexure*** to continue as the descending colon. Because it has a mesentery, the transverse colon is variable in position. It may be at the level of the transpyloric plane or its midpoint may droop into the pelvis. The transverse colon is easily identifiable by the greater omentum which is attached to it.

DESCENDING COLON

The descending colon starts at the left colic flexure and is a retroperitoneal structure that descends along the left posterior abdominal wall to the left iliac fossa.

SIGMOID COLON

The sigmoid colon (sigmoid = S-shaped), also known as the ***pelvic*** colon, is that portion of the large intestine between the descending colon and the rectum. It has a V-shaped mesentery, with two arms, one horizontal and the other vertical.

RECTUM

The rectum has no mesentery; it is partly covered by peritoneum in its upper two thirds. The lower third of the rectum passes through the muscles that form the pelvic floor and becomes continuous with the anal canal.

ANAL CANAL

The anal canal is the terminal part of the alimentary tract and will be considered with the pelvis and perineum.

KIDNEYS AND SUPRARENAL GLANDS

The two kidneys (see Fig. 11-2) lie on each side of the vertebral column about 5 cm from the midline. Each kidney extends approximately from the level of the 12th thoracic vertebra to the level of the third lumbar vertebra. The bulge produced by the two kidneys can be seen and palpated through the parietal peritoneum of the posterior abdominal wall. Note that the posterior abdominal wall slopes backward and laterally from the vertebral column, so that each kidney faces anteriorly and laterally. Immediately above each kidney, and forming a cap on the kidney, is a **suprarenal (adrenal) gland**. The kidneys and suprarenal glands are retroperitoneal. The **ureters** run downward from each kidney to enter the bladder in the pelvis. The ureters are retroperitoneal.

CONTENTS OF THE PELVIS

The chief contents of the true pelvis (pelvis minor) are as follows.

Rectum. The rectum is covered by peritoneum on its upper two thirds. The upper third has peritoneum in front and at each side; the middle third is covered by peritoneum in front only, and the lower third of the rectum has no peritoneal covering. The rectum lies in the posterior part of the pelvis and follows the curve of the sacrum.

Urinary Bladder. The urinary bladder is found in the anterior part of the pelvis. It is extraperitoneal, but its superior surface is covered by peritoneum.

Female Genital Organs. The female genital organs consist of the **uterus, ovaries, uterine (Fallopian) tubes,** and **broad ligaments**. They lie in the female pelvis between the rectum and the bladder.

Male Genital Organs. The male genital organs are not covered by peritoneum except for the tips of the **seminal vesicles**. They will be considered later.

SPECIAL FEATURES OF THE PERITONEAL CAVITY

LESSER SAC OR OMENTAL BURSA

The stomach, the lesser omentum, and the anterior two layers (two of four) of the greater omentum form the anterior wall of the **lesser sac** or **omental bursa**. The lesser sac is a completely enclosed recess of the peritoneal cavity (see Fig. 9-5B) that communicates with the rest of the peritoneal cavity (greater sac) only behind the free edge of the lesser omentum at the upper border of the duodenum (see Figs. 9-5B and 9-7). This opening is called the **epiploic foramen** or foramen of Winslow (Fig. 9-8).

GREATER SAC

The greater sac of the peritoneum constitutes all of the peritoneal cavity that has not been described as lesser sac.

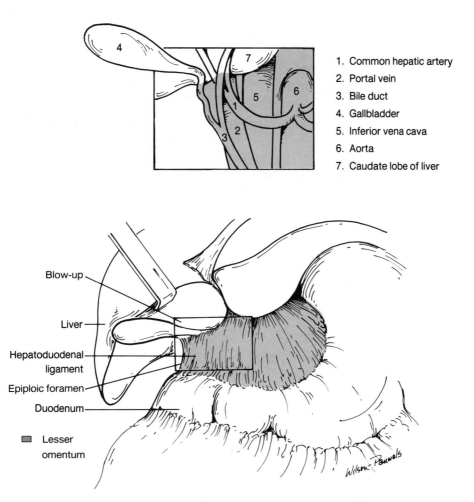

1. Common hepatic artery
2. Portal vein
3. Bile duct
4. Gallbladder
5. Inferior vena cava
6. Aorta
7. Caudate lobe of liver

Figure 9-8.
Right border and contents of lesser omentum.

PARACOLIC GUTTERS

Lateral to the ascending colon and the descending colon, respectively, the peritoneum can be seen to form a gutter as it is reflected from these portions of the colon to the posterior abdominal wall (Fig. 9-9). Since the living colon contains feces and gas, the *paracolic gutters* may be quite deep. Collection of fluid from the upper abdomen, such as gastric content escaping into the peritoneal cavity from a perforated ulcer, reaches the pelvis and lower abdomen by way of the paracolic gutters.

HEPATORENAL RECESS (MORRISON'S POUCH)

The *hepatorenal recess* (pouch of Morrison) is the lowest point in the peritoneal cavity when the patient is lying flat on his back; it is a recess of the greater sac just to the right of the epiploic foramen. It is bounded medially by the right kidney and above by the right lobe of the liver. Fluid from this recess will drain into the rest of the peritoneal cavity via the right paracolic gutter (Fig. 9-9).

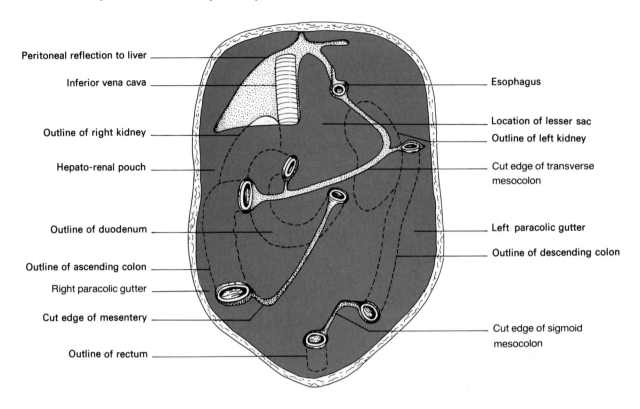

Peritoneal reflection to liver

Inferior vena cava

Esophagus

Outline of right kidney

Location of lesser sac

Outline of left kidney

Hepato-renal pouch

Cut edge of transverse mesocolon

Outline of duodenum

Left paracolic gutter

Outline of descending colon

Outline of ascending colon

Right paracolic gutter

Cut edge of mesentery

Outline of rectum

Cut edge of sigmoid mesocolon

Figure 9-9.
Peritoneal reflections. The dashed lines represent retroperitoneal viscera left in the abdominal cavity but covered by peritoneum. The colored area represents the parietal peritoneum. The solid lines represent cut edges of the peritoneum where viscera have been removed from the abdominal cavity.

CLINICALLY IMPORTANT FEATURES OF THE PERITONEAL CAVITY

PERITONEAL RECESSES

A peritoneal recess is usually found where an organ that was retroperitoneal acquires a mesentery, or where one that has a mesentery becomes retroperitoneal. The commonly found ones are around the duodenojejunal junction and the ileocecal junction (Fig. 9-10). The clinical importance

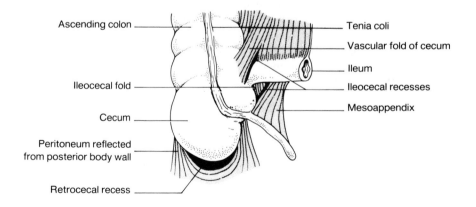

Ascending colon

Tenia coli

Vascular fold of cecum

Ileum

Ileocecal fold

Ileocecal recesses

Mesoappendix

Cecum

Peritoneum reflected from posterior body wall

Retrocecal recess

Figure 9-10.
Peritoneal folds and recesses around the ileocecal junction.

of these recesses is that they may form the sites of ***internal*** hernias. That means that a loop of gut (usually small bowel) may be trapped in one of the recesses and become obstructed at the edges of the recess. Appendectomy may become difficult if the vermiform appendix is located in the ***retrocecal recess***.

There is little need for the reader to learn all the names and locations of these recesses because, in practice, internal hernias are quite rare. The ***superior paraduodenal*** recess should be mentioned because its superolateral margin contains the ***inferior mesenteric vein***. This vein must be protected during surgery for the relief of a ***strangulated*** internal hernia at this site.

EPIPLOIC FORAMEN (FORAMEN OF WINSLOW)

The epiploic foramen (see Figs. 9-5C and 9-8) lies immediately superior to the first part of the ***duodenum***, the peritoneal covering of which forms the inferior boundary of the foramen. The posterior wall of the foramen is the peritoneum (posterior abdominal wall parietal peritoneum), which covers the ***inferior vena cava***. This peritoneum runs from the duodenum to the liver. The superior boundary of the foramen is the peritoneum covering the ***caudate process*** of the liver. The anterior boundary of the foramen is the ***(right) free edge of the lesser omentum***, which contains the ***bile duct***, the ***hepatic artery***, and the ***portal vein***. The first and last structures extend from the porta hepatis to the posterior aspect of the pancreas (Fig. 9-8).

LESSER SAC OF PERITONEUM (OMENTAL BURSA)

The lesser sac may now be delimited. It is entered through the epiploic foramen (which will admit the tip of a finger). If the examining finger then passes superiorly, it runs toward the diaphragm posterior to the caudate lobe of the liver, which bulges into the lesser sac. The lesser sac is limited to the right by the reflection of the peritoneum from the inferior vena cava to the liver. The superior boundary is the reflection of peritoneum from the diaphragm to the lesser omentum. It is limited anteriorly by the lesser omentum and the reflection of its two peritoneal layers around the anterior and posterior aspects of the stomach, and the second layer (of four) of the greater omentum. The second and third layers of the greater omentum are usually fused (although they can be separated by careful dissection), so that for practical purposes the lesser sac ends at, or near, the greater curvature of the stomach. The posterior wall of the lesser sac is the posterior abdominal wall covered by parietal peritoneum; the left boundary is formed by the gastrophrenic and lienorenal ligaments and by the left border of the greater omentum.

CLINICAL NOTE

Pathologic processes involving structures that surround the lesser sac can cause accumulation of fluid in the lesser sac. This fluid may trickle out of the epiploic foramen into the remainder of the peritoneal cavity, or it can be trapped in the lesser sac, if the epiploic foramen becomes sealed by inflammatory reaction.

Examples would be acute inflammation of the pancreas (acute pancreatitis) that can produce an abscess or a pseudocyst, the walls of which will be the boundaries of the lesser sac; and perforation of a posterior wall gastric ulcer.

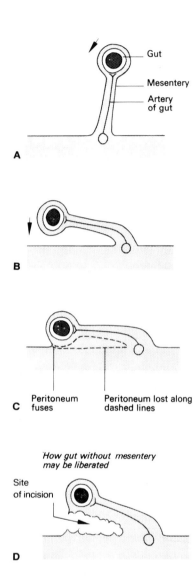

Figure 9-11.
How a portion of gut loses its mesentery, and how this may be used to advantage in surgical procedures.

SUBPHRENIC ABSCESSES

Clinicians recognize a number of potential spaces below the diaphragm in which inflammatory fluids can accumulate and become sealed off to form ***subphrenic abscesses***.

The right and left ***subphrenic spaces*** lie between the diaphragm and the respective lobes of the liver. They are separated by the falciform ligament.

The right ***subhepatic*** space is the ***hepatorenal recess*** (Morrison's pouch), and the left subhepatic space is the lesser sac.

The locations of these spaces are basically posterior, and subphrenic abscesses are drained through posterior incisions, usually through the bed of the 12th rib, taking care not to injure the costodiaphragmatic reflection of the pleura. Not surprisingly, the presence of a subphrenic abscess will frequently irritate the diaphragmatic pleura and cause an effusion of fluid into the pleural space. This can be detected on a radiograph.

SURGICAL IMPORTANCE OF MESENTERIES AND THEIR EMBRYOLOGY

Various structures lose their mesenteries during embryologic development. The duodenum, ascending colon, and descending colon, which originally had mesenteries but lost them by fusion with the parietal peritoneum, may be separated from the posterior abdominal wall by cutting along the attachment of the original visceral peritoneum to the parietal peritoneum (Fig. 9-11). This mobilization is easily accomplished, and the surgeon can reflect the duodenum or ascending colon without difficulty.

In any operation in which it is necessary to anastomose (join) the jejunum to the stomach, the jejunum must be brought up in front of the transverse colon or passed through a surgically created opening in the transverse mesocolon.

Adhesions. Adhesions in the abdomen are the fixation to each other of two or more structures originally covered by peritoneum. Actually an adhesion is nothing more than fibrous scar tissue, the result of the biological process of repair after injury or inflammation. Following intra-abdominal inflammation, trauma, or surgery, adhesions between viscera or between viscera and abdominal wall can be expected. Indeed it is the capacity to form adhesions that allows two cut ends of intestine to heal to each other long after the sutures or staples used to join them have cut through the tissues into which they were inserted.

Adhesions may cause kinking and obstruction of loops of bowel, but there is considerable variation in the number and density of adhesions found in the postsurgical or postinflammatory peritoneal cavity.

The Digestive Tract

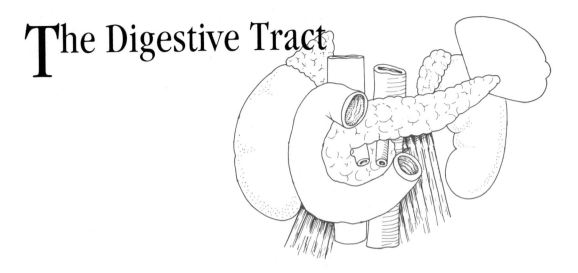

The digestive tract extends from the mouth to the anus. It is a channel which, although it passes through the body, communicates with the exterior at both ends, and its lumen remains continuous with the exterior. A swallowed coin will pass unchanged (except for some color changes) from mouth to anus. Contents of the digestive tract become internal only after they, or their digestive products, have been absorbed through the intestinal wall.

The digestive tract is a muscular tube lined by epithelium and possesses several very complicated outgrowths, for example, liver and pancreas, into which food does not penetrate. The purpose of the digestive tract is to transform food and drink into absorbable breakdown products and to absorb these products. It also serves as an excretory channel for unabsorbable ingested matter and the excretions of its appendages, such as certain constituents of bile from the liver.

The plan of the tube is basically simple (Fig. 10-1). The lumen is lined by **mucosa** that consists of a layer of epithelial cells, some connective tissue (lamina propria), and a thin layer of smooth muscle, the muscularis mucosae. The mucosa varies in appearance and function in different components of the digestive tract. Outside the mucosa is a layer of connective tissue, the **submucosa** with its blood vessels, nerves, and lymphatics. Outside this is the muscular wall of the tube arranged in two layers of **smooth** muscle: an inner circular layer and an outer longitudinal layer. The outermost covering is peritoneum, which may be replaced by adventitia (connective tissue) in places.

Some of the epithelial cells are modified to produce various secretions (*e.g.*, mucus and enzymes). These cells may occur singly or in glands (*e.g.*, liver or pancreas).

ESOPHAGUS, STOMACH, FIRST PART OF THE DUODENUM, AND SPLEEN

The spleen is not part of the digestive tract, but its intimate relationship and proximity to the stomach makes its description appropriate in this chapter.

ABDOMINAL ESOPHAGUS

The esophagus was described in Chapter 7. Its abdominal course is quite short (2.5 cm). It enters the abdomen through the **esophageal hiatus** of the diaphragm and enters the stomach at the **cardia**.

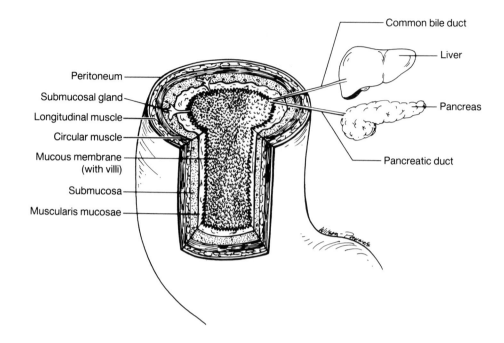

Figure 10-1.
Schematic representation of transverse section of duodenum.

STOMACH

The stomach is the first major part of the digestive tract (alimentary canal) below the diaphragm. It is a large J-shaped organ (Fig. 10-2).

Cardia. The cardia of the stomach is at the point of entry of the esophagus. At the cardia the stratified squamous epithelium of the esophagus changes abruptly into the columnar secretory epithelium of the stomach.

Fundus. The fundus of the stomach is the highest portion of the stomach; it is a pouch above and to the left of the cardia. The fundus rests against the diaphragm and usually contains a large bubble of air that can be seen on an x-ray.

Lesser Curvature. The lesser curvature is the right upper border of the stomach. It extends from the cardia to the pylorus. The **lesser omentum** is attached to it.

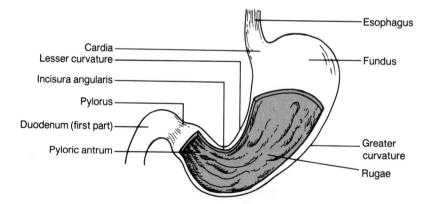

Figure 10-2.
The stomach (anterior view).

Greater Curvature. The greater curvature passes from the fundus to the pylorus and constitutes the left and lower border of the stomach. It gives attachments to the ***gastrophrenic*** and ***gastrosplenic*** ligaments as well as to the ***greater omentum***.

Pyloric Antrum. The pyloric antrum is the part of the stomach leading to the narrow ***pyloric canal***.

Pylorus. The pylorus is the outlet of the stomach. It contains a thick layer of circular muscle that gives the pylorus a sphincteric function. The pylorus is easily identified during surgery. Its muscular sphincter is palpable, and its surface is marked by a constant small vertically running vein (the vein of Mayo). The pylorus connects the stomach to the duodenum. The functional dimensions of the pylorus (as seen radiologically after the ingestion of a barium meal) appear to be 1 cm in diameter and 1 cm in length.

Incisura Angularis. The incisura angularis (angular notch) is at the lower end of the lesser curvature and marks the junction of the body of the stomach with the pyloric antrum. The incisura angularis is a feature of the embalmed stomach and a radiologic feature, caused by muscle contraction. It is not particularly evident in the living during surgery. An operation called ***antrectomy*** is used as part of the treatment of peptic ulcers, and the antrum is usually considered to be the distal third of the stomach.

Body of Stomach. The body of the stomach is that part joining the cardia and fundus on the left to the pyloric antrum on the right.

Rugae. Longitudinal folds in the mucous membrane of the stomach can be demonstrated radiologically. These folds are called rugae, and the arrangement of the mucosa into these folds permits distension of the mucosal lining together with the muscular wall of the stomach after the ingestion of food (see Figs. 10-2 and 10-6).

Orientation of the Stomach

The stomach is a large and rather mobile structure. In a thin individual the greater curvature may extend down to the brim of the pelvis. The locations of the cardia and the pylorus are relatively constant with the cardia lying about 4 cm to the left of the midline behind the seventh costal cartilage. The pylorus lies in the ***transpyloric plane*** (L1) about 2.5 cm to the right of the midline.

Structure of the Stomach

Like the remainder of the digestive tract, the muscles of the stomach are basically disposed in an outer longitudinal layer and an inner circular layer, interspersed by some obliquely running fibers. The circular arrangement is very evident at the pylorus, where the thick ring-like pyloric sphincter can be felt. Although no similar circular arrangement is macroscopically or microscopically demonstrable at the cardia, there is a ***functional sphincter*** at the cardia that can be demonstrated by manometry (measurement of intraluminal pressure) through a balloon introduced into the esophagus and by radiology. The ***cardiac sphincter*** is in the lower esophagus or at the gastroesophageal junction. The obliquely running muscle fibers of the stomach loop like an inverted U around the cardia, and the two limbs of the U run along the anterior and posterior aspects

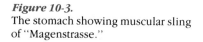

Figure 10-3.
The stomach showing muscular sling
of "Magenstrasse."

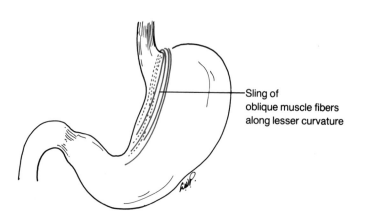

Sling of
oblique muscle fibers
along lesser curvature

of the lesser curvature (Fig. 10-3). This is believed to create a "highway"
for gastric content along the lesser curvature (German anatomists call it
the Magenstrasse). This arrangement may account for the fact that ***gastric
peptic ulcers*** occur most commonly along the lesser curvature. A gastric
ulcer seen on the greater curvature of the stomach is treated with great
suspicion and must be considered to be malignant until proven otherwise.

The mucosal rugae are absent or relatively sparse at the lesser curvature. Without going into physiological details it may be stated that the gastric juice contains mucus, hydrochloric acid, and digestive enzymes.

Relations of the Stomach

Figure 10-4.
The bed of the stomach. Where organs
have not been depicted, the stomach
rests against the diaphragm or the
muscles of the posterior abdominal
wall.

The ***bed of the stomach*** (Fig. 10-4) is that portion of the posterior abdominal wall against which the stomach rests in the recumbent position in the
cadaver. In living persons, mobility and distensibility of the stomach cause
it to cover a wider area.

The ***diaphragm*** is in contact with the esophagus, the cardia, and part
of the fundus.

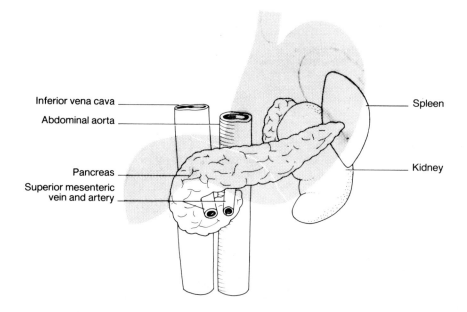

Inferior vena cava

Abdominal aorta

Spleen

Pancreas

Kidney

Superior mesenteric
vein and artery

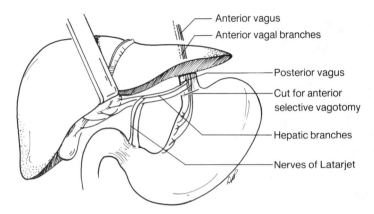

Anterior vagus
Anterior vagal branches
Posterior vagus
Cut for anterior
selective vagotomy
Hepatic branches
Nerves of Latarjet

Figure 10-5.
Selective vagotomy.

The **spleen** is in contact with the fundus.

The **left kidney** and **left suprarenal gland** are in contact with the upper part of the body of the stomach.

The major posterior relationship of the stomach is the **pancreas**.

The **transverse colon** and **transverse mesocolon** are related to the posterior aspect of the greater curvature.

The **anterior surface** of the stomach is related to the diaphragm, the left lobe of the liver, and the anterior abdominal wall.

Nerve Supply of the Stomach

Postganglionic sympathetic efferent fibers from the celiac plexus are distributed with the arteries to the stomach. The **efferent secretomotor** fibers to the stomach come from the vagus nerves (Fig. 10-5). The anterior and posterior vagi enter the abdomen through the esophageal hiatus. The anterior vagal fibers are the conglomerations of the anterior esophageal plexus and originate mainly from the left vagus nerve. The posterior vagus nerve is reconstituted from the posterior esophageal plexus and is mostly of right vagus nerve origin. The posterior vagus nerve is larger and is easily palpable as a cord-like structure. It gives off a large branch to the celiac plexus before running along the posterior aspect of the lesser curvature and gives off branches to the anterior and posterior aspects of the stomach.

The anterior vagus nerve usually branches at, or above, the esophageal hiatus and enters the abdomen in close contact with the wall of the esophagus, behind its anterior covering of visceral peritoneum. This nerve, or each of its three or four subdivisions, gives off a branch to the liver. These **hepatic** branches of the anterior vagus nerve run in the lesser omentum. The remainder of the anterior vagus nerve(s) runs down the anterior aspect of the lesser curvature. These anterior and posterior gastric nerves supply the smooth muscle of the stomach and influence the production of the acid. Near the pylorus both anterior and posterior **gastric nerves** give off branches (which resemble crows' feet) to the pylorus. These latter fibers are known as the nerves of **Latarjet**.

CLINICAL NOTE

*The cause of **duodenal** ulcer is held to be excessive secretion of gastric acid particularly during the **nervous phase** of gastric secretion. Since parasympathetic stimuli are carried by the vagus nerves (the synapse between preganglionic and postganglionic elements being in the **myenteric***

Figure 10-6.
Barium meal and follow-through
showing stomach and rugae (*a*);
gallbladder (*b*); duodenum (*c*); and
loops of small intestine (*d*).

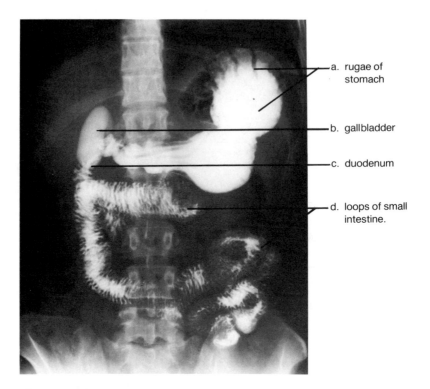

a. rugae of
   stomach

b. gallbladder

c. duodenum

d. loops of small
   intestine.

*plexus in the stomach wall), cutting the vagus nerves (**vagotomy**) can
be a very effective surgical treatment for duodenal ulcer. Total vagotomy
(high in the esophageal hiatus) has undesirable side effects owing to the
denervation of the small intestine, liver, and gallbladder. **Selective vagotomy** with preservation of the celiac branch of the posterior vagus nerve
and preservation of the hepatic branches of the anterior vagus nerve(s)
is preferred. This procedure has the side effect of reducing gastric motility
and slowing gastric emptying. **Highly selective vagotomy** cuts the anterior and posterior gastric nerves and leaves the nerves of Latarjet intact
(Fig. 10-5).*

*Radiologic examination of the stomach is performed by means of an
ingested contrast agent (barium meal) (see Fig. 10-6).*

*Direct inspection of the gastric interior has been made efficient by
the invention of the **fiberoptic gastroscope**, the flexibility of which permits inspection of the gastric and duodenal lumens.*

FIRST PART OF DUODENUM

The pylorus continues into the first part of the duodenum that has a mesentery, the right margin of the lesser omentum. This part of the duodenum
is horizontal and about 5 cm long; then the duodenum turns inferiorly
to form the second, ***descending part***. The first part of the duodenum is
separated from the inferior vena cava by the structures in the right margin
of the lesser omentum (the ***common bile duct, portal vein,*** and a branch
of the hepatic artery, the ***gastroduodenal artery***).

Radiologically the first part of the duodenum has the appearance of
a smooth triangular structure, the ***duodenal cap*** (Fig. 10-6).

SPLEEN

The spleen developed in the dorsal mesogastrium (Fig. 10-7). It is attached to the stomach by the ***gastrosplenic (gastrolienal) ligament*** and to the left kidney by the ***lienorenal ligament***. It is covered by peritoneum except at the ***hilus***, where the splenic artery enters and the splenic vein leaves.

The spleen is 12 to 15 cm long and 5 to 8 cm wide. Usually it fits comfortably into the examining hand. It lies parallel to the 9th, 10th, and 11th ribs with its long axis along the 10th rib. The anterior tip (lower pole) of the spleen does not usually extend anteriorly beyond the midaxillary line. A spleen that is palpable below the costal margin is usually pathologically enlarged.

The outer surface of the spleen is convex and adapts to the shape of the diaphragm, which separates it from the lower ribs; the anterior border is notched. The inner surface is pyramidal in shape and accommodates the tip of the pancreas.

Posteriorly the spleen rests against the left kidney, anteriorly it is related to the fundus of the stomach and inferiorly it is in contact with the left (splenic) colic flexure.

CLINICAL NOTE

*The spleen is often injured in blunt trauma to the upper abdomen or lower thorax. If ruptured, it will bleed profusely because its substance is soft and pulpy and its capsule is thin. These properties of its parenchyma and capsule make repair of the lacerated spleen difficult, but repair is preferable to **splenectomy** (removal of spleen) since the spleen plays an important role in the immune system, especially in children and youths.*

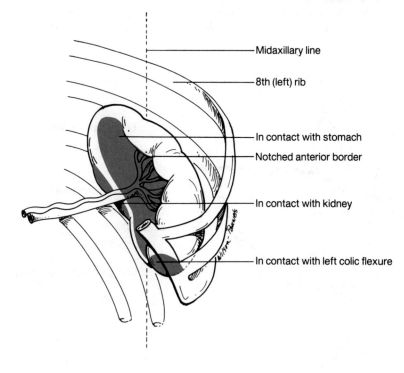

- Midaxillary line
- 8th (left) rib
- In contact with stomach
- Notched anterior border
- In contact with kidney
- In contact with left colic flexure

Figure 10-7.
Spleen.

Occasionally accessory spleens are present; they are to be found in the peritoneal ligaments of the spleen or in the greater omentum. (These structures are of posterior mesogastrium origin.)

Splenectomy may be necessary for certain diseases of the blood, and care should be taken to remove any accessory spleens.

Arterial Supply of Stomach, Spleen, and Liver

These three structures are grouped together because they receive their arterial blood supply from the same source, the ***celiac artery*** (celiac trunk or celiac axis). The celiac artery is a short 1 to 2 cm trunk arising from the center of the anterior aspect of the aorta just below the median arcuate ligament at the level of the 12th thoracic vertebra. It divides into three terminal branches: the ***left gastric artery***, the ***hepatic artery***, and the ***splenic artery*** (Fig. 10-8).

At its origin the celiac artery is surrounded by a dense plexus of sympathetic nerve cells and fibers, the ***celiac plexus***. Each branch of the celiac artery is accompanied by postganglionic sympathetic nerve fibers from this plexus.

Left Gastric Artery. The left gastric artery passes upward and to the left behind the posterior wall of the lesser sac. It reaches the lesser curvature at the cardia and continues along the lesser curvature to anastomose with the right gastric artery near the pyloric antrum. At the cardia the left gastric artery gives off an esophageal branch that ascends into the thorax through the esophageal hiatus.

The Common Hepatic Artery (Hepatic Artery). The common hepatic artery, more commonly called the hepatic artery, arises from the celiac artery and passes downward and to the right behind the posterior peritoneal wall of the lesser sac. The hepatic artery reaches the duodenum below the epiploic foramen and gives off the large gastroduodenal artery and the smaller right gastric artery. The hepatic artery then enters between the two

Figure 10-8.
Arterial supply of stomach, spleen, and first part of duodenum.

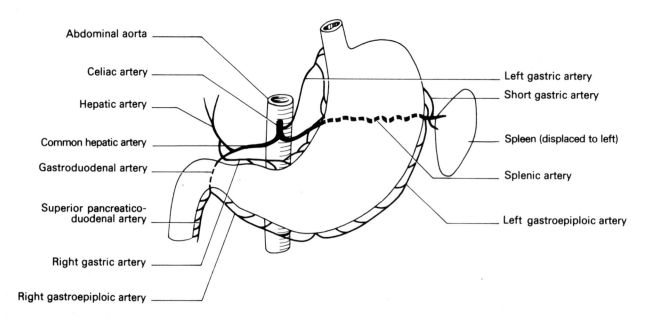

layers of peritoneum that form the right margin of the lesser omentum and passes upward to the porta hepatis to supply the liver. In the lesser omentum it lies to the left of the ***bile duct*** and in front of the ***portal vein***.

Below the porta hepatis the hepatic artery divides into right and left hepatic arteries to the right and left lobes of the liver. The right hepatic artery gives off a branch, the ***cystic artery***, to the ***gallbladder***. The cystic artery usually runs behind the bile duct or right hepatic duct to reach the gallbladder, but there is considerable variation in its course and origin. This must be borne in mind during gallbladder surgery so that the right hepatic artery is not divided by mistake for the cystic artery.

The ***right gastric artery*** passes along the lesser curvature to anastomose with the left gastric artery. The right and left gastric arteries give off multiple branches at the lesser curvature to supply the upper third to one half of the stomach.

Gastroduodenal Artery. This artery runs inferiorly, posterior to the first part of the duodenum, between the duodenum and the neck of the pancreas, making a groove in the pancreas. Below the first part of the duodenum it divides into two branches:

1. The ***superior pancreaticoduodenal artery*** runs in the curvature of the duodenum, where the head of the pancreas is lodged, and anastomoses with the ***inferior pancreaticoduodenal artery***, a branch of the ***superior mesenteric artery***.
2. The other branch of the gastroduodenal artery is the ***right gastroepiploic artery***, which runs to the left along the greater curvature of the stomach at the attachment of the greater omentum to anastomose with the ***left gastroepiploic artery***, a branch of the splenic artery.

Splenic artery. The splenic artery is the third branch of the celiac artery and follows a very tortuous course behind the peritoneum of the lesser sac along the upper border of the pancreas to enter the spleen at the ***splenic hilus***. The splenic artery supplies the body of the pancreas, and at the hilus of the spleen it gives off four to six ***short gastric arteries*** that run in the ***gastrosplenic (gastrolienal) ligament*** to supply the fundus of the stomach and left part of the greater curvature.

The ***left gastroepiploic artery*** arises from the splenic artery just proximal to the hilus of the spleen and runs to the right along the greater curvature of the stomach to anastomose with the ***right gastroepiploic artery***. It supplies the greater curvature of the stomach and the greater omentum.

The splenic artery usually divides into several branches just before entering the spleen.

CLINICAL NOTE

*Duodenal ulcers are most common in the first 2 cm of the duodenum. If the ulcer occurs on the anterior wall and is neglected to the point where it perforates the wall of the duodenum, gastric and duodenal content will be spilled into the peritoneal cavity and cause **peritonitis**. If the ulcer is located in the posterior wall, erosion into the pancreas, sometimes with disastrous hemorrhage from the gastroduodenal artery or one of its branches, is more likely. Remember that anterior duodenal ulcers are prone to perforate and that posterior duodenal ulcers are born to bleed.*

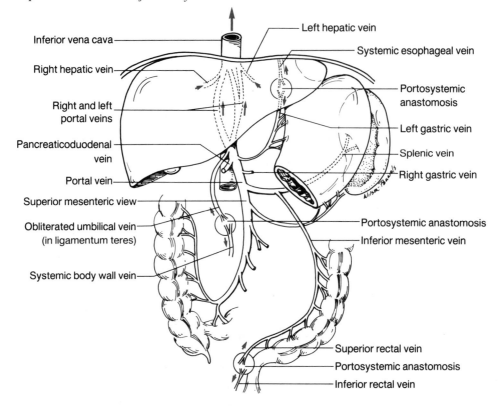

Inferior vena cava

Right hepatic vein

Right and left portal veins

Pancreaticoduodenal vein

Portal vein

Superior mesenteric view

Obliterated umbilical vein (in ligamentum teres)

Systemic body wall vein

Left hepatic vein

Systemic esophageal vein

Portosystemic anastomosis

Left gastric vein

Splenic vein

Right gastric vein

Portosystemic anastomosis

Inferior mesenteric vein

Superior rectal vein

Portosystemic anastomosis

Inferior rectal vein

Figure 10-9.
Portosystemic anastomoses.

Gastric ulcers are also prone to perforate. An anterior ulcer will perforate into the greater sac, and a posterior gastric ulcer will perforate into the lesser sac.

Venous Drainage of Stomach and Spleen

A new concept, that of a ***portal venous system***, will be introduced: a portal venous system is a system of veins that drains a specific area of the body first to a specific organ. Then the blood from this organ drains into the general (systemic) venous circulation. The hepatic ***portal system of veins*** is the major example. Another example is the portal circulation of the pituitary gland.

The ***portal vein*** is formed by the junction of the ***splenic*** and the ***superior mesenteric veins*** behind the neck of the pancreas (Fig. 10-9). The splenic vein emerges from the hilus of the spleen and runs behind the body of the pancreas.

The ***superior mesenteric vein*** drains the ***small intestine, ascending colon,*** and the right two thirds of the ***transverse colon*** and receives tributaries from the greater curvature of the stomach.

More will be said later about the portal venous system and its clinical importance (see p. 146).

The ***splenic vein*** drains the ***spleen***, portions of the ***stomach***, and the ***descending and sigmoid colons*** and ***upper rectum***. The last three structures drain into the splenic vein by way of the ***inferior mesenteric vein***, which joins the splenic vein behind the body of the pancreas.

Portal Vein This very large vein passes behind the first part of the duodenum and into the (right) free edge of the lesser omentum, with the bile duct and hepatic artery in front of it, to reach the porta hepatis, where it divides into right and left divisions to reach the liver tissue. It drains blood from the gastrointestinal regions mentioned above to the liver, where the absorbed products of digestion are deposited. The venous blood of the portal vein filters through the liver lobules and then becomes collected by the tributaries of the three hepatic veins that join the inferior vena cava.

INTESTINES

SMALL INTESTINE

The small intestine comprises the ***duodenum, jejunum, and ileum***. The duodenum, a largely retroperitoneal structure, will be considered later (see p. 140) because its retroperitoneal position makes it inaccessible to examination until the rest of the small intestine and the transverse colon have been removed.

Structure

The small bowel has a mucosa, submucosal connective tissue, an inner circular layer of smooth muscle, and an outer longitudinal layer of smooth muscle. The mucosa is thrown into folds called ***plicae circulares***, and these are much more prominent in the jejunum (Fig. 10-10). The ileum has a somewhat thinner wall, less prominent mucosal folds, and conspicuous patches of lymphoid tissue in its submucosal layer ***(Peyer's patches)***. There is no distinct demarcation between jejunum and ileum. The jejunum comprises the upper two fifths of the small bowel, and the total length of the small intestine (when freed of its mesentery and laid out) averages 6.5 meters in the adult.

   The jejunum can be distinguished from the ileum by its thicker wall and by its mesentery, which is less fat laden and thus demonstrates the typical ***arcade*** formation of the branches of the superior mesenteric artery more clearly (Fig. 10-10). These arterial arcades are more numerous in the

Figure 10-10.
Typical sections of jejunum and ileum indicating classically described differences between these portions of the small bowel.

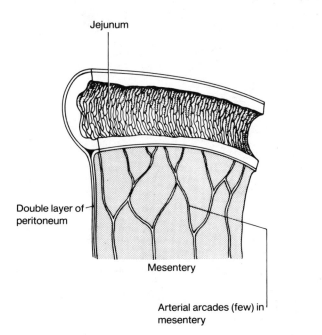

Jejunum

Double layer of peritoneum

Mesentery

Arterial arcades (few) in mesentery

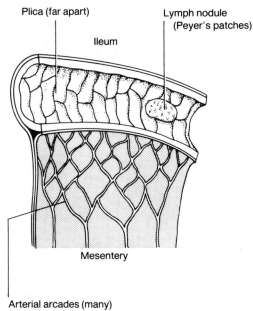

Plica (far apart)

Lymph nodule (Peyer's patches)

Ileum

Mesentery

Arterial arcades (many)

ileum, and there may be no arcades in the first two loops of jejunum. Note that the arteries of the small intestine enter the bowel at its attachment to the mesentery, the ***mesenteric border*** of the intestine.

CLINICAL NOTE

Radiologic examination of the small intestine can be carried out by means of the ingestion of barium sulphate or other contrast material. The plicae circulares give some clue on an x-ray film to differentiating jejunum from ileum, but the overlap of multiple, ubiquitous loops makes interpretation difficult.

*The small intestine absorbs breakdown products of digestion; major disease or surgical removal of small intestine may cause **malabsorption**. When extensive surgery on the small intestine is necessary, the problem of malabsorption has to be remembered.*

Meckel's Diverticulum. Meckel's diverticulum is a remnant of the vitelline duct of the embryo. When present, in about 2% of subjects, it is usually 2 inches long and occurs 2 feet from the ileocecal valve. It may be patent to the umbilicus and cause fecal discharge at the umbilicus, or it may be connected to the umbilicus by a fibrous band (Fig. 10-11).

CLINICAL NOTE

*This fibrous band may trap a loop of intestine between itself and the anterior abdominal wall and create an **internal hernia**. The diverticulum may become inflamed and mimic the symptoms and signs of acute inflammation of the vermiform appendix.*

*Occasionally **ectopic gastric mucosa** may occur in a Meckel's diverticulum. This ectopic mucosa can produce gastric secretions with the occurrence of a **peptic ulcer** in or near the diverticulum. Such ulceration may cause gastrointestinal bleeding.*

Mesentery. The mesentery of the small intestine has a base about 15 cm long but fans out to a span of about 6.5 meters long at the intestinal attachment. From base (at the posterior abdominal wall) to its intestinal attachment the mesentery has a depth of about 20 to 25 cm.

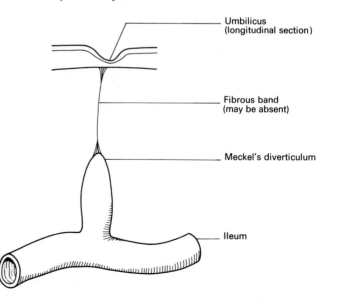

Umbilicus
(longitudinal section)

Fibrous band
(may be absent)

Meckel's diverticulum

Ileum

Figure 10-11.
Meckel's diverticulum.

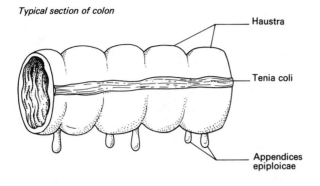

Typical section of colon

Haustra

Tenia coli

Appendices epiploicae

Figure 10-12.
A section of colon indicating its characteristic features.

LARGE INTESTINE

The cecum, colon, and rectum constitute the large intestine. The muscular walls of the large intestine are an inner circular layer and an outer longitudinal layer. The longitudinal muscle fibers are confined to three bands, the ***teniae coli*** (Fig. 10-12). The teniae coli begin at the base of the vermiform appendix and are somewhat shorter than the length of the large intestine so that the intestine becomes bunched up like a concertina. Between the teniae coli the wall of the large intestine bulges outward as pouches called ***haustra***. The colon also exhibits ***appendices epiploicae***, which are small sacs of peritoneum-covered fat hanging from the surface of the large intestine. The teniae coli are replaced by a full layer of longitudinal muscle in the appendix. The teniae are first seen on the cecum, and they coalesce to reconstitute a full layer of longitudinal muscle in the rectum.

Colon

The colon comprises four named parts (Fig. 10-13):

1. The ***ascending colon*** is similar in structure to the rest of the colon but usually does not have a mesentery. At the ***right colic (hepatic) flexure*** it is continuous with the ***transverse colon***.
2. The ***transverse colon*** runs from the right colic flexure across the abdomen to the ***left colic (splenic) flexure***. It has a long mesentery that allows the transverse colon considerable vertical displacement, and the central part of the transverse colon may be found anywhere from the transpyloric plane to the pelvis. It should be remembered that the greater omentum hangs from the transverse colon and enjoys similar mobility.
3. The ***descending colon*** is similar to the ascending colon in lacking a mesentery. It begins at the left colic flexure and ends in continuity with the sigmoid colon.
4. The ***sigmoid colon*** has a V-shaped mesentery and is folded upon itself like the letter S (hence its name).

Cecum

The cecum is a blind sac at the lower end of the ascending colon. The ileum empties into it at the ***ileocecal valve*** (Fig. 10-14). The two lips of the valve are disposed in the horizontal plane and are surrounded by the circular muscle of the ileum. The ileocecal valve usually prevents reflux of cecal content into the terminal ileum.

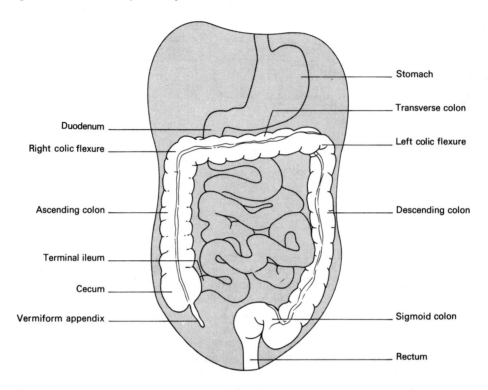

Duodenum

Right colic flexure

Ascending colon

Terminal ileum

Cecum

Vermiform appendix

Stomach

Transverse colon

Left colic flexure

Descending colon

Sigmoid colon

Rectum

Figure 10-13.
Anterior view of the large bowel in the abdominal cavity.

Vermiform Appendix. The vermiform appendix, which is usually referred to as the appendix, is a narrow part of the large bowel. It hangs from the apex of the cecum and varies in length from 5 cm to 15 cm or more. Its lumen is narrow (about the diameter of an ordinary pencil). The submucosa of the appendix contains much ***lymphoid tissue***. In children the amount of lymphoid tissue is relatively large. In children and the elderly, the wall of the appendix is quite thin. The location of the appendix may be found by following the teniae coli of the cecum to the base of the appendix. The appendix is free to move because it has a small mesentery,

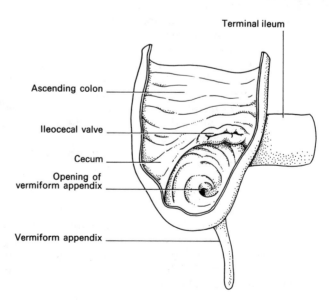

Terminal ileum

Ascending colon

Ileocecal valve

Cecum

Opening of vermiform appendix

Vermiform appendix

Figure 10-14.
Anterolateral view of the interior of the cecum with the anterior wall removed.

the ***mesoappendix***. This mobility accounts for the variability of location of the appendix. In about 70% of people it will be found behind the cecum ***(retrocecal)***; in 25% it will hang over the pelvic brim ***(pelvic appendix)***; and the remaining 5% of positions are in front of the ileum ***(preileal)*** or behind the terminal ileum ***(retroileal)***.

Mesenteric Folds of the Ileocecal Area. Three peritoneal folds are found in the ileocecal area:

1. The ***mesoappendix*** is the mesentery of the vermiform appendix. It runs from the mesentery of the ileum to the appendix, and its free edge contains the ***appendicular artery***, a branch of the ***ileocolic artery***. The artery is accompanied by the vein of the same name.
2. The ***ileocecal fold*** (bloodless fold of Treves) is a fold of peritoneum joining the ileum, cecum, and the base of the appendix. It may be mistaken for the mesoappendix. It is not always "bloodless."
3. The ***vascular fold of the cecum*** is a fold of peritoneum joining the mesentery of the small intestine to the cecum at its junction with the ascending colon; it carries a branch of the ileocolic artery.

The above-mentioned folds allow the formation of two peritoneal recesses: one posterior to the vascular fold and the other posterior to the bloodless fold. These recesses are potential sites for internal hernia formation. These hernias are rare. At times there is a gap in the mesentery of the terminal ileum; internal herniation can occur through this mesenteric defect. One of us (JAL) has operated on three such cases but has never seen an internal hernia in the paracecal fossae (recesses around the cecum mentioned above).

CLINICAL NOTE

*The narrow lumen of the appendix, narrowed even more when the abundant submucosal lymphoid tissue is inflamed by infection (often of viral origin), paves the way for **acute appendicitis**, which probably starts with obstruction of the appendix. In children and the elderly, the thin wall of the appendix facilitates early rupture. The appendicular artery is the only blood supply of the appendix, and its position in the mesoappendix is close to the wall of the appendix. When appendicitis occurs the artery will frequently become **thrombosed** (i.e., the blood in it will clot with obstruction of the artery). This leads to **gangrene** of the appendix and makes it even more prone to rupture.*

*Like the remainder of the intestinal tract the appendix has an autonomic afferent (sensory) nerve supply (from the sympathetic system), and when the appendix becomes inflamed and obstructed its muscular wall contracts vigorously, causing ill-defined **colicky pain**, felt in the midabdominal (umbilical) area. This is an example of referred pain. The ileocecal area of the gut is supplied with sympathetic fibers from the tenth thoracic spinal cord segment, and the tenth thoracic somatic nerve supplies the skin of the umbilicus. More will be said about referred pain in Chapter 12.*

*When the inflammation has reached the serosal (peritoneal) covering of the appendix the nearby **parietal peritoneum** will be irritated, and then the clinical picture will change from vague midabdominal cramps to severe pain in the right lower quadrant of the abdomen, aggravated* *by movement and the pressure of the physician's examining hand. When the appendix is in the pelvic position, the rectum or bladder may be irritated, causing spurious diarrhea or bladder symptoms.*

*The right lower quadrant tenderness will be accompanied by reflex contraction of the abdominal muscles of the region; clinicians call this phenomenon **guarding or splinting**.*

Rectum

The rectum begins where the sigmoid colon loses its mesentery. The teniae coli coalesce to form a complete layer of longitudinal smooth muscle. The upper third of the rectum has peritoneum in front and on either side but not behind. The middle third of the rectum is covered by peritoneum only on its anterior surface, and the lower third has no peritoneal covering. The rectum follows the hollow of the sacrum and, belying its name (rectum = straight), has three flexures (or bends)—two with their concavities to the left and one with its concavity to the right. The rectal mucosa opposite these flexures is elevated into crescentic folds (the valves of Houston). The rectum ends at the ***anal canal***.

Arterial Supply of the Intestines

Superior Mesenteric Artery.   The ***superior mesenteric artery*** (Fig. 10-15) is the artery of the embryonic ***midgut***; the celiac artery (see above) is the artery of the foregut. The superior mesenteric artery supplies the distal duodenum, the small intestine, and the large intestine to the junction of the middle and distal thirds of the transverse colon. It arises from the abdominal aorta about 1 cm below the origin of the celiac artery, at the level of the first lumbar vertebra. At its origin it lies posterior to the pancreas and then passes between the body and uncinate process of the pancreas, crossing the third part of the duodenum to enter the mesentery of the small intestine. The main trunk of the artery follows the root of the mesentery to the right iliac fossa. The superior mesenteric artery and its branches are surrounded by autonomic nerve fibers derived from the superior mesenteric plexus. The superior mesenteric artery has many named branches that arise in the following order (variations are not infrequent):

Inferior Pancreaticoduodenal Artery.   This is the first branch. It runs in the groove between the head of the pancreas and the lower part of the duodenum, supplying both anterior and posterior branches to these structures before anastomosing with the superior pancreaticoduodenal artery.

Middle Colic Artery.   The middle colic artery arises just after the inferior pancreaticoduodenal artery. It runs downward and forward in the transverse mesocolon and supplies the transverse colon, anastomosing with the upper left colic artery at the junction of the middle and distal thirds of the transverse colon. On the right the middle colic artery anastomoses with the right colic artery near the hepatic flexure of the colon.

Right Colic Artery.   The right colic artery arises from the arch of the superior mesenteric artery in the root of the small bowel mesentery. It passes to the right across the posterior abdominal wall to supply the ascending colon and to anastomose with the middle colic artery above and the ileocolic artery below.

Ileocolic Artery.   This is the terminal branch of the superior mesenteric artery, being its continuation. It runs toward the ileocecal junction, giving

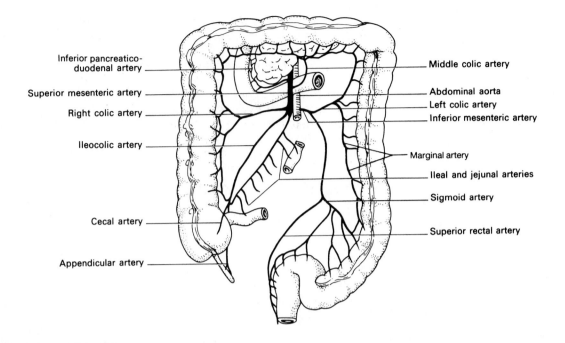

Inferior pancreatico-duodenal artery
Superior mesenteric artery
Right colic artery
Ileocolic artery
Cecal artery
Appendicular artery

Middle colic artery
Abdominal aorta
Left colic artery
Inferior mesenteric artery
Marginal artery
Ileal and jejunal arteries
Sigmoid artery
Superior rectal artery

off a recurrent ileal artery to the mesenteric border of the terminal ileum, the ***appendicular artery***, and cecal branches to the cecum. The terminal part of the ileocolic artery ascends along the medial border of the ascending colon to anastomose with the right colic artery.

Jejunal and Ileal Arteries. These vessels arise from the left convexity of the superior mesenteric artery at intervals along its course in the root of the small bowel mesentery. As they pass to the small intestine they form ***arterial arcades*** (see Fig. 10-10) from which the terminal branches to the bowel originate. These arcades ensure alternate pathways for blood to reach any part of the small intestine.

Inferior Mesenteric Artery. The inferior mesenteric artery is the artery of the hindgut; it arises from the aorta about 5 cm below the origin of the superior mesenteric artery at the level of the third lumbar vertebra and behind the third part of the duodenum. It passes to the left and downward across the posterior abdominal wall to terminate at the ***superior rectal artery***. It gives off the ***left colic artery***, which divides into ascending and descending branches to supply the descending colon and distal transverse colon, anastomosing with the middle colic artery and the ***sigmoid artery***.

The sigmoid arteries (there may be two or three sigmoid arteries) are also branches of the inferior mesenteric artery; they supply the sigmoid colon and anastomose with the descending branch of the left colic artery above and the superior rectal artery below.

The ***marginal artery*** (artery of Drummond) runs about 1 cm from the mesenteric border of the colon. It may extend along the whole length of the colon but is most apparent from the middle of the transverse colon to the lower sigmoid colon. It is of importance in maintaining the blood supply to the colon after surgical resection of part of the colon or rectum.

CLINICAL NOTE

The existence of the marginal artery enables the surgeon to ligate the inferior mesenteric artery at its origin and remove it together with its sur-

Figure 10-15.
Arterial supply to small and large intestines.

rounding lymph nodes during surgery for sigmoid colon and rectal cancer. This high ligation is desirable in order that the highest lymph nodes
draining the area be removed.

The end arteries to the colon enter around the periphery of the bowel
instead of entering at the mesenteric border, as they do in the small intestine. There are three points of entry between the teniae coli. These entry
points are potential weak spots in the colonic wall, and, because of the
high pressure of gas and feces within the colon, outpouching of mucosa
may occur in these areas. This condition is called **diverticulosis**. If the
diverticula become infected, the condition of **diverticulitis** results.

DUODENUM AND PANCREAS

The duodenum is that part of the small intestine that lies between the
pylorus and the jejunum. It is about 25 cm long and lies in a C-shaped
curve on the posterior abdominal wall surrounding the head of the pancreas (Fig. 10-16). The duodenum is divided into four parts for descriptive
purposes. It must be recognized that because of the C-shaped configuration, the two ends are only 5 cm apart. Originally the duodenum had a
dorsal and a ventral mesentery, but the dorsal mesentery fused with the
posterior parietal peritoneum so that all but the first 2.5 cm of the first part
are **retroperitoneal**. The first 2.5 cm of the duodenum are suspended in
the **lesser omentum**, and it will be remembered that this part of the duodenum forms the inferior boundary of the epiploic foramen to the lesser
sac of the peritoneum. The lesser omentum is the remnant of the ventral
mesentery of the duodenum and stomach.

Superior (First) Part of Duodenum. The first part of the duodenum is
about 5 cm long and runs posteriorly and to the right from the pylorus
toward the neck of the gallbladder.

Figure 10-16.
The pancreas and duodenum *in situ.*
The numbers on the duodenum
represent the following portions:
(1) superior, (2) descending,
(3) horizontal, (4) ascending.

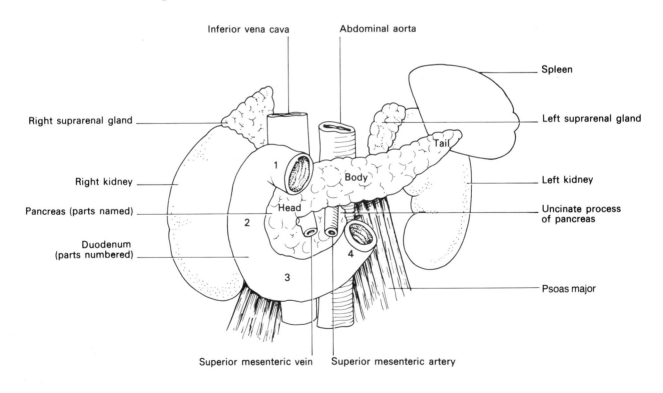

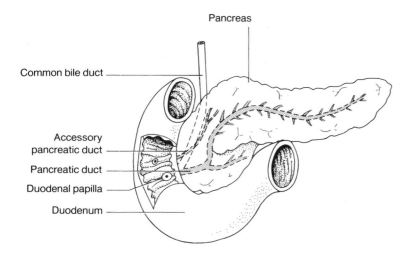

Pancreas

Common bile duct

Accessory
pancreatic duct

Pancreatic duct

Duodenal papilla

Duodenum

Figure 10-17.
The ducts of the pancreas and their
relation to the bile duct and
duodenum.

Its relations include, posteriorly, the inferior vena cava, right suprarenal gland (a retroperitoneal structure), common bile duct, portal vein,
and the gastroduodenal artery; anteriorly, the peritoneum, liver, and gallbladder; superiorly, the neck of the gallbladder and the epiploic foramen;
inferiorly, the pancreas.

Descending (Second) Part of Duodenum.   The descending part of the
duodenum is 10 cm long and is retroperitoneal; it passes almost directly
downward, parallel to and to the right of the inferior vena cava.

Its relations include posteriorly, the right kidney, the pelvis of the
right kidney, the right renal vessels, and the psoas major muscle; anteriorly, the transverse mesocolon, transverse colon, mesentery, and some
small intestine; and medially, the pancreas, common bile duct, and the
pancreatic duct(s) (Fig. 10-16). These last two structures approach the duodenum from the medial side and pass behind it (Fig. 10-17) to open into
its posteromedial wall. The point of entry of the common bile duct and
pancreatic duct into the duodenum marks the junction of the ***embryonic
foregut*** with the ***embryonic midgut***.

Horizontal (Third) Part of Duodenum.   The horizontal part of the duodenum is about 10 cm long and passes horizontally and to the left across
the posterior abdominal wall, to the left of the midline; it is retroperitoneal. Its relations include, posteriorly, the psoas major muscle, the right
ureter, inferior vena cava, and the aorta; anteriorly, the small bowel mesentery, loops of small intestine, and the superior mesenteric vessels; and
superiorly, the origin of the superior mesenteric artery and the pancreas.

Ascending (Fourth) Part of Duodenum.   The last part of the duodenum
is only 2.5 cm long and passes upward and to the left to the level of the
second lumbar vertebra.

Its relations include the left psoas major muscle posteriorly and the
pancreas medially.

Junction of the Duodenum and Bile and Pancreatic Ducts

The ***common bile duct*** enters the second part of the duodenum on its
posteromedial aspect at about its midpoint. The ***pancreatic duct*** enters
at the same spot. These two ducts usually enter a common chamber, the
ampulla (of Vater) (Figs. 10-20 and 10-22). The common bile duct con-

tains a sphincter of circular smooth muscle above its entry into the ampulla; the pancreatic duct lacks such a sphincter.

The location of the ampulla is marked on the internal wall of the duodenum by the **duodenal papilla**. The apex of the papilla contains the opening of the ampulla. This opening is surrounded by another layer of circular smooth muscle, the **sphincter of the ampulla (sphincter of Oddi)**.

PANCREAS

The pancreas (Fig. 10-17) is a long (25 cm) organ that lies retroperitoneally across the posterior abdominal wall. Its head is cradled in the C-shaped concavity of the duodenum, and its tail touches the spleen at the splenic hilus. It is an organ that comprises many small lobules connected by tough connective tissue. It has both an **exocrine** and an **endocrine** function. The exocrine cells (the majority of the cells) secrete the pancreatic juices that contain digestive enzymes, including **amylase**, for the breakdown of carbohydrates into **glucose** and a powerful fat-digesting enzyme, **lipase**.

Endocrine cells are found in the connective tissue between the lobules, and these **islet cells** (cells of the islets of Langerhans) produce **insulin**, an important hormone concerned with carbohydrate storage, mobilization, and metabolism.

Head of Pancreas.   The head of the pancreas is the right extremity of the gland and rests in the curve of the duodenum. From its left lower margin the **uncinate process** passes to the left behind the superior mesenteric vessels.

Posteriorly the head rests on the inferior vena cava, the right renal vessels, and the left renal vein. The uncinate process rests against the abdominal aorta posteriorly. The **common bile duct** and the **portal vein** groove the upper part of head of the pancreas posteriorly. The common bile duct may lie posterior to the head or may be embedded in its substance.

Neck of Pancreas.   The **neck** of the pancreas is below and behind the pylorus. It is covered by peritoneum and lies in front of the junction of the superior mesenteric vein with the splenic vein, that is, the beginning of the portal vein. The confluence of the superior mesenteric and splenic veins makes a groove on the posterior aspect of the neck of the pancreas.

CLINICAL NOTE

*Because the common bile duct rests behind, or in the substance of, the head of the pancreas, a carcinoma of the pancreas can cause obstruction of drainage of bile from the liver, resulting in **jaundice**.*

Body of Pancreas.   The body of the pancreas is covered anteriorly by peritoneum and is separated from the body of the stomach by the lesser sac of peritoneum. It provides attachment for the transverse mesocolon. Its posterior relations include the aorta, superior mesenteric artery, left kidney, splenic vein, and left suprarenal gland.

Tail of Pancreas.   The tail of the pancreas is the prolongation of the body into the lienorenal ligament. The tip of the tail touches the hilus of the spleen, and care must be taken not to injure this tail during the surgical operation of splenectomy.

Duct of Pancreas

The duct of the pancreas (duct of Wirsung) is shaped like a lopsided Y (see Fig. 10-17). The long arm of the Y stretches the full length of the body and tail of the pancreas, and the short arm stretches through the head and into the uncinate process. The two arms join in the substance of the head and pass parallel to the common bile duct to the second part of the duodenum. An *accessory pancreatic duct* (duct of Santorini) is sometimes present, and it drains the superior part of the head of the pancreas and opens separately into the duodenum above the duodenal papilla.

CLINICAL NOTE

Because the bile duct and the pancreatic duct have a common opening into the duodenum, an obstruction in the ampulla by a gallstone or by spasm of the muscle of the sphincter of the ampulla, secondary to duodenal inflammation, may cause a reflux of bile into the pancreas. This can produce the pathologic condition of acute pancreatitis in which the pancreatic parenchyma is inflamed to the extent that trypsin, amylase, and lipase are released into the surrounding structures and into the bloodstream. The release of trypsin aggravates the condition because it literally digests the pancreatic tissue. The release of lipase will break down fat in the peritoneal covering of the pancreas; these released fatty acids will undergo a process of saponification that causes white plaques of soapy material to be deposited on the peritoneal coverings of the pancreas. The release of amylase into the bloodstream can be measured by the blood's **serum amylase** *content. This test is an aid in diagnosing acute pancreatitis.*

Severe inflammation or trauma to the pancreas will set up extensive peritonitis with massive outpouring of fluid into the peritoneal cavity. The pancreas is a major structure behind the posterior wall of the lesser sac, and a large collection of fluid may pour into the lesser sac. If the epiploic foramen becomes plugged by inflammatory exudate a **pseudocyst** *will develop in the lesser sac. These cysts may become quite large, be palpable, and displace the stomach upward and the transverse colon downward, a condition that can be demonstrated by means of contrast radiography (barium meal and barium enema).*

Blood Supply of Pancreas and Duodenum

The body of the pancreas gets much of its arterial supply from the splenic artery that runs along its upper border. The *superior pancreaticoduodenal* artery from the gastroduodenal artery and the *inferior pancreaticoduodenal* artery from the superior mesenteric artery supply the head and neck region of the pancreas.

LIVER, GALLBLADDER, AND BILE DUCTS

LIVER

The liver (Fig. 10-18) is the largest gland in the body, weighing about 1 to 2.5 kg, and represents about 2.5% of body weight. It receives the breakdown products of digestion from the small intestine by way of the portal vein. These products are metabolized and stored in the liver. Additional functions include the following:

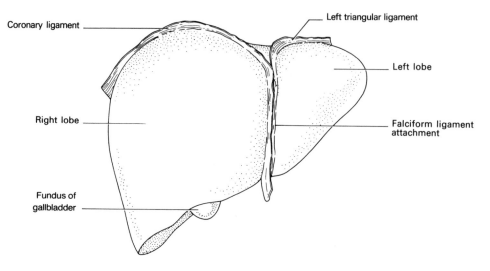

Figure 10-18.
The liver (anterior view).

1. Excretion of the products of the breakdown of hemoglobin in the form of bile pigments;
2. Secretion of bile, which is essential for the digestion and absorption of fats.

Anatomy textbooks describe many depressions and angles, but these are mostly the results of embalming. The liver in living persons is a relatively soft and resilient mass that does not exhibit all the contours found in the embalmed cadaver.

Shape and Position. The liver is roughly triangular (or wedge) shaped. The bulk of it lies in the *right hypochondrium* separated from the pleura and lungs by the diaphragm.

Surface Marking. The liver is hidden by the lower thoracic cage. Its superior surface rises as high as the fifth intercostal space and is convex upward. Its left extremity reaches to the left midclavicular line and its right extremity to the costal margin in the midaxillary line. The right side descends close to the costal margin, and the inferior margin follows approximately along the costal margin to reach the left extremity of the superior surface.

CLINICAL NOTE

*The lower border of the right lobe and middle portions of the liver can be felt in persons with a wide subcostal angle. The edge should be smooth, not hard or tender. In pathologic enlargement caused by heart failure, cirrhosis, or malignancy, the edge may descend considerably and be hard, irregular, or tender, depending on the pathologic cause for the **hepatomegaly** (enlarged liver).*

Blunt or penetrating trauma to the lower thorax (especially on the right side) may injure the liver with (at times) considerable bleeding and fragmentation of liver parenchyma.

Surfaces of the Liver. The upper diaphragmatic surface of the liver is smooth. Its posterior surface is deeply grooved to accommodate the inferior vena cava. The visceral (inferior) surface (Fig. 10-19) is concave and is directed inferiorly and posteriorly. The **gallbladder** and the **porta hepatis** are located on this surface. The porta hepatis is the region where vessels and bile ducts enter or leave the liver.

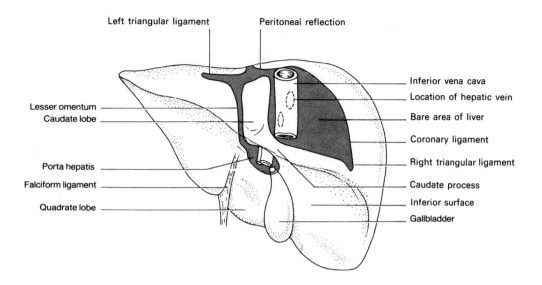

Lobes of the Liver. The liver is divided anatomically into right and left lobes with the ***falciform ligament*** (see Fig. 10-18) demarcating the division. The ***quadrate lobe*** is that part to the right of the falciform ligament and to the left of the gallbladder. The ***caudate lobe*** is a small lobe on the posterior aspect of the liver projecting into the lesser sac (see Fig. 10-19). The right and left halves of the liver are functionally separate, and the division is determined by the blood supply of the "half." Each receives its own arterial blood supply (right or left branch of hepatic artery) and has its own portal venous supply and systemic venous drainage. Similarly the right and left hepatic (bile) ducts correspond to the distribution of the blood supply. This last description follows recognized surgical anatomical lines developed in the study of liver resection, and there appears to be little overlap between these segments or halves. The surgical anatomical plane of division runs approximately from the fossa for the gallbladder in front to the emergence of the hepatic veins posteriorly.

In subsequent clinical studies the reader will encounter further segmentation of the liver based on ***ultrasonography***. It is possible to demonstrate the echo of the liver parenchyma with a beam of ultrasound. Ultrasonographers find the three hepatic veins easy to identify and use the middle hepatic vein to distinguish left from right and the right and left hepatic veins to provide further subdivision of the right and left segments.

Peritoneal Reflections Around the Liver

The peritoneal reflections of the liver will be found to be continuous with each other, and the present account will start on the anterosuperior aspect of the viscus.

Falciform Ligament. The falciform ligament connects the liver with the anterior abdominal wall and the diaphragm. The falciform ligament is a double layer of peritoneum; above, the right layer passes to the right to form the anterior leaf of the ***coronary ligament***, and the left leaf passes to the left to form the anterior layer of the left ***triangular ligament***. The free edge of the falciform ligament contains the ***ligamentum teres***, the remnant of the ***umbilical vein*** of the fetus. (The lumen of this vein persists, although no blood passes through it. The ligamentum teres can be exposed and opened for purposes of blood transfusion, for example.)

Figure 10-19.
The liver (posterior view showing the visceral surface). Peritoneal reflections are indicated by the solid line surrounding the darkly stippled area.

Coronary Ligament. The coronary ligament receives its name because it surrounds the bare area of the liver like a crown. The bare area is that part of the liver which is not covered by peritoneum; it contains the groove for the inferior vena cava and is the site of emergence of the hepatic veins (see Fig. 10-19.)

Right Triangular Ligament. This is the right extremity of the coronary ligament.

Left Triangular Ligament. The left layer of the falciform ligament passes to the left over the superior surface of the left lobe of the liver and at the left extremity of the liver folds back upon itself to cover the inferior aspect of the left lobe of the liver. Thus the left triangular ligament is formed.

Lesser Omentum. The posterior leaf of the left triangular ligament becomes continuous with the lesser omentum, which connects the liver to the lesser curvature of the stomach (see Fig. 9-7). The attachment of the lesser omentum to the liver follows the course of the ***ductus venosus*** (ligamentum venosum of post fetal life), which in the fetus joined the umbilical vein to the inferior vena cava at the porta hepatis.

Blood Supply and Venous Drainage of the Liver

The blood supply to the liver is from two sources: the ***portal vein*** and the ***hepatic artery***. All venous drainage is via the ***hepatic veins***.

Portal Vein. The portal vein is formed behind the neck of the pancreas by the junction of the splenic vein with the superior mesenteric vein (see Fig. 10-9). The splenic vein receives drainage from the inferior mesenteric vein, and the superior mesenteric vein receives venous drainage from all the territories supplied by the superior mesenteric artery . The left and right gastric veins and the superior pancreaticoduodenal veins drain directly into the portal vein. Thus the portal vein receives the venous blood of the intestinal tract from the stomach to the rectum as well as from the spleen and pancreas.

After its formation the portal vein runs behind the first part of the duodenum and ascends in the free margin of the lesser omentum to the porta hepatis, where it divides into right and left branches to supply the liver.

CLINICAL NOTE

*The alert reader will have noticed that the above description of the portal vein did not claim that it drained all the blood from the intestinal tract. The left gastric vein, via its esophageal tributary, and the superior rectal vein, via its anastomosis with the **inferior rectal veins**, communicate with the **systemic venous circulation** at the lower end of the esophagus and in the anal canal (**portosystemic anastomoses**; see Fig. 10-9). (A less important systemic anastomosis is that of the "obliterated" umbilical vein with the systemic veins of the abdominal wall.)*

*If obstruction to flow occurs in the portal system (e.g., in severe cirrhosis of the liver or from extrahepatic thrombosis of the portal vein), the increased pressure in the portal system will engorge the anastomotic channels and produce such conditions as **esophageal varices** or **hemorrhoids**. (Note that although esophageal varices are nearly always caused* *by **portal hypertension**, hemorrhoids have much more common and less exotic causes.)*

Rectal Veins. The ***superior rectal vein*** drains by way of the inferior mesenteric vein into the portal system. The ***middle*** and ***inferior rectal veins*** are part of the ***systemic venous system*** and drain into the inferior vena cava by way of the internal iliac and pudendal veins. The anastomosis with the portal system occurs in the ***anal canal***.

Esophageal Veins. The veins of the esophagus drain via the thoracic veins into the systemic system. The veins of the cardia of the stomach drain into the portal system. At the lower end of the esophagus the systemic and portal system anastomose via the veins in the lower esophagus.

Hepatic Veins. The hepatic veins are very short but important veins that drain the liver parenchyma into the inferior vena cava (see Fig. 10-9). There are usually three hepatic veins, but two of them may join in the liver substance so that only two veins will be apparent on the liver surface. These veins are very short but are quite large. It should be noted that all of the blood that enters the liver through the hepatic artery and the portal vein drains through the hepatic veins to the inferior vena cava.

Hepatic Artery. The hepatic artery is a branch of the celiac artery; it passes in the free edge of the lesser omentum to the porta hepatis. The artery is the only source of fully oxygenated blood to the liver, and if one of its two terminal branches is ligated the portion of the liver supplied by it will probably die. The right and left divisions of the hepatic artery each supply approximately one half of the liver substance.

Cystic Artery. The cystic artery (see Fig. 10-20) is the arterial supply to the gallbladder. It usually arises from the right hepatic artery and first runs posterior to the hepatic duct and then just superior to the cystic duct to reach the gallbladder.

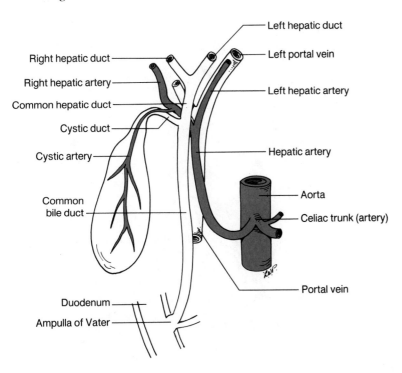

Figure 10-20.
Biliary apparatus, hepatic artery, portal vein.

Anomalies of Arterial Supply. The arterial supply of the liver and of the gallbladder is very variable. The following is a list of some of the variations that may be encountered.

1. The left hepatic artery may arise from the left gastric artery.
2. The whole hepatic artery may arise from the superior mesenteric artery.
3. The hepatic artery may arise from the left gastric artery.
4. The hepatic artery may arise from the splenic artery.
5. The cystic artery may arise from the gastroduodenal artery.
6. The right hepatic artery may be a branch of the gastroduodenal artery.

CLINICAL NOTE

The surgeon must be aware of these and other variations. It is best to be sure that what looks like cystic artery is indeed the blood supply of the gallbladder and not a variation that also supplies the liver. Ligation of the right hepatic artery by mistake for the cystic artery is one common hazard.

BILIARY APPARATUS

The bile is secreted from the cells of the liver into bile canaliculi, which join to form ductules which in turn unite to form intrahepatic bile ducts. These coalesce, and a major right and left hepatic duct emerge from each lobe.

The **right hepatic duct** drains approximately the right half of the liver, and the **left hepatic duct** drains approximately the left half. These areas of drainage correspond to the areas supplied by the right and left hepatic arteries and right and left divisions of the portal vein.

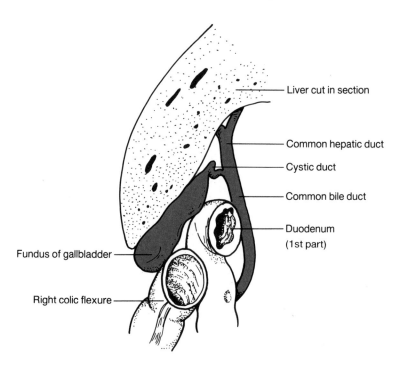

Liver cut in section

Common hepatic duct

Cystic duct

Common bile duct

Duodenum (1st part)

Fundus of gallbladder

Right colic flexure

Figure 10-21.
Lateral view of biliary apparatus showing relationships to duodenum and colon.

Figure 10-22.
Confluence of common bile and pancreatic ducts at ampulla of Vater.

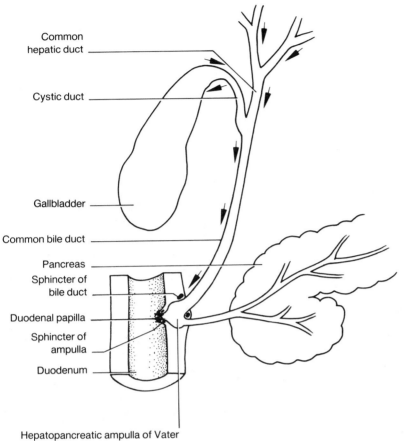

Common hepatic duct

Cystic duct

Gallbladder

Common bile duct

Pancreas

Sphincter of bile duct

Duodenal papilla

Sphincter of ampulla

Duodenum

Hepatopancreatic ampulla of Vater

The ***common hepatic duct*** is formed by the union of the right and left hepatic ducts in, or just below, the porta hepatis. The ***cystic duct***, carrying bile to and from the gallbladder, joins the common hepatic duct at a rather variable level. Below the point of junction with the cystic duct the common hepatic duct is usually called the ***common bile duct***.

Bile Duct (Common Bile Duct). The common bile duct is formed by the junction of the cystic duct and the common hepatic duct (see Figs. 10-21 and 10-22). It passes in the free edge of the lesser omentum in front of the ***portal vein*** and to the right of the ***hepatic artery***. These three structures are often referred to as the ***portal triad***. The common bile duct passes posterior to the first part of the duodenum and the head of the pancreas (or in its substance) and enters the posteromedial aspect of the second part of the duodenum at the duodenal papilla.

CLINICAL NOTE

*Anything obstructing the free passage of bile down the biliary tree may cause **jaundice**. The most common cause of such obstruction is a **gall-stone** that has escaped from the gallbladder and has come to reside in the common bile duct; such **calculi** may be multiple.*

The only remedy for a stone in the common bile duct used to be surgery. With the advent of the flexible fiberoptic gastroscope it is sometimes possible to cut the sphincter of the duodenal papilla with an instru-

*ment passed down the channel of the gastroscope. If successful this proce-
dure may permit the spontaneous passage of gallstones from the common
bile duct.*

GALLBLADDER

The gallbladder stores and concentrates bile. It lies on the visceral (inferior) surface of the right lobe of the liver and has a capacity of approximately 25 ml. The gallbladder is suspended from the liver so that its ***fundus*** is the lowest part and usually projects beyond the liver edge (see Fig.
10-18). The fundus of the gallbladder is located approximately at the ninth
costal cartilage in the midclavicular line. The ***body*** of the gallbladder projects upward on the under surface of the liver toward the porta hepatis and
is continuous with the ***neck*** of the gallbladder. The ***cystic duct*** is a narrow
structure of variable length that joins the neck of the gallbladder to the
common hepatic duct (see Fig. 10-20).

The gallbladder is covered on its inferior surface by peritoneum; occasionally it may have a mesentery. Rarely the gallbladder may be completely embedded in the substance of the liver. The superior surface of
the gallbladder is in contact with the liver, and small arteries and veins
commonly pass from the liver substance to the gallbladder. Small bile
channels may, on occasion, drain from the liver directly into the hepatic
surface of the gallbladder.

The fundus is usually in contact with the anterior abdominal wall just
lateral to the linea semilunaris; the body is in contact with the transverse
colon, and the neck of the gallbladder is in proximity to the first part of the
duodenum (see Fig. 10-21).

The gallbladder is a storage reservoir for bile. When there is no demand for bile from the intestinal tract (bile aids the digestion of fats), the
sphincter of the bile duct will be closed. The liver continuously secretes
bile, which accumulates in the gallbladder where the volume of the bile
is concentrated considerably by the absorption of water. When a demand
for bile occurs, the gallbladder contracts and empties bile into the bile
duct and the sphincter of the bile duct relaxes, allowing bile to enter the
duodenum.

CLINICAL NOTE

*There seem to be no detectable problems with digestion if the gallbladder
has to be removed. The biliary tree takes over the storage (and possible
water absorption) functions, and the sphincter of the common bile duct
regulates the release of bile into the duodenum (see Fig. 10-22).*

*There are some variations in the termination of the cystic duct; they
are rare but surgically important. The cystic duct may join the right (or
left) hepatic duct; the cystic duct may be absent and the gallbladder connected directly to the bile duct; the cystic duct may be long and lie parallel to or even be incorporated in the wall of the common bile duct; and
the gallbladder may be doubled.*

Gallbladder pain *(biliary colic) is believed to be caused by violent
contractions of the (smooth) muscle wall of the viscus. This would occur
if a gallstone blocks the cystic duct. The pain is very severe and may be
intermittent, rising to a crescendo and then abating for a short time before recurring. This pattern is typical of* ***colicky pain*** *arising in a hollow
viscus. The pain of* ***biliary colic*** *typically is felt in the epigastrium and
referred also to the right upper quadrant of the abdomen and possibly to*

*the right scapular region. This is an example of **referred pain**, the impulses traveling to the seventh or eighth thoracic spinal cord segments via the sympathetic visceral afferent nerves. The locations of referral are explained by the fact that T7 and T8 supply the skin of the epigastric and scapular regions.*

Occasionally biliary colic will be referred to both upper quadrants or sometimes to the left upper quadrant and the left scapular region. If we realize that the gallbladder is a midline structure that finds secondary residence on the right side, these aberrant symptoms become self-explanatory (i.e., right-sided pain is caused by a preponderance of right- over left-sided nerve supply, whereas the reverse applies in the case of left-sided pain). When, as a result of obstruction to the cystic duct, the gallbladder becomes inflamed it will irritate the parietal peritoneum of the right upper quadrant, and, at this stage, there will be constant right upper quadrant pain, tenderness, and muscle guarding in this area, even if the original colic was felt on the left side.

LYMPHATIC DRAINAGE OF THE INTESTINAL TRACT

The lymphatic drainage of the intestinal tract is of great physiological and clinical importance. Lymph vessel plexuses exist throughout the submucosal layers of the intestinal tract (Fig. 10-23). These plexuses drain into lymph vessels that accompany the arterial supply. In the small intestine these lymph vessels carry absorbed fats and the larger molecular breakdown products of the digestion of fats. This fat-laden lymph, which is called **chyle**, will eventually reach the **cisterna chyli** from which it will be transported by means of the **thoracic duct** to the venous circulation. In

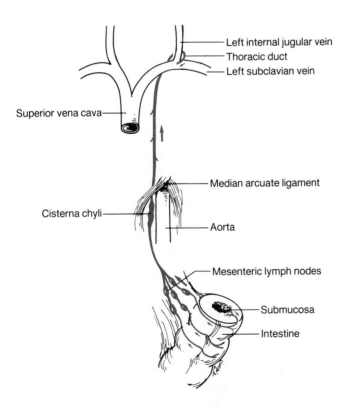

Figure 10-23.
Lymphatic drainage of gut and cisterna chyli into the thoracic duct.

living persons the chyle-filled lymphatics are often easily visualized in the small bowel mesentery.

Small lymph nodes are found where arteries enter the gut. These nodes drain into larger nodes along the major blood vessels. These nodes can be seen aligned along the vessels, particularly in the mesentery. Ultimately these nodes drain into nodes grouped around the origins of the major arteries. There are ***celiac, superior mesenteric,*** and ***inferior mesenteric*** lymph nodes. These specifically named groups of nodes drain into the ***aortic*** or ***para-aortic*** nodes around the aorta. The lymph nodes of the stomach communicate with, and partially drain into, mediastinal lymph nodes. The lymph from the rectum and upper half of the anal canal drains into the inferior mesenteric nodes, but the ***lower half of the anal canal*** drains into the ***inguinal lymph nodes***.

The Kidneys and Suprarenal Glands

The kidneys are critically important in maintaining water, electrolyte, and acid–base homeostasis. They also eliminate from the body certain waste products, and function as an endocrine organ to produce ***erythropoietin***, which stimulates the production of red blood cells.

The kidneys lie on the posterior abdominal wall on either side of the vertebral column with their long axes sloping downward and slightly laterally. Because of the prominence of the vertebral column bulging into the abdominal cavity, the anterior surfaces of the kidneys face laterally as much as they do anteriorly (Fig. 11-1). The upper parts of the kidneys lie under the protection of the lower part of the thoracic cage.

KIDNEYS

Each kidney is bean shaped (Fig. 11-2) and approximately 10 cm long, 5 cm wide, and 2.5 cm thick. The kidneys lie in the retroperitoneal space and are embedded in a layer of fat, the ***perirenal (perinephric) fat***, or fascia. The posterior surfaces of the kidneys are separated from the vertebral column by the psoas major muscles. Each kidney has an upper pole and a lower pole, an anterior surface and a posterior surface, and a lateral border and a medial border. Their upper poles are in contact with the diaphragm posteriorly.

The upper pole of each kidney is at the level of the 11th or 12th thoracic vertebral body with the lower pole extending down to the body of the third lumbar vertebra. The bulk of the liver causes the right kidney to lie at a somewhat lower level than the left kidney. This makes the lower pole of a normal-sized right kidney often palpable in a thin individual. In general the hilus of the kidney lies at the transpyloric plane. The upper pole lies approximately 4 cm and the lower pole approximately 6 cm from the midline.

Each kidney has certain important relationships, discussed below.

Anterior Surface. The anterior surface of the kidney is smooth and is related as follows: The ***right kidney*** is related to the right suprarenal gland, liver (separated by peritoneal cavity), second part of the duodenum, and the right colic flexure. The suprarenal, duodenal, and colic areas are not covered by peritoneum. The ***left kidney*** is related to the left suprarenal gland, stomach, spleen, jejunum, pancreas, and descending colon. The areas in contact with the suprarenal gland, pancreas, and descending colon are not covered by peritoneum.

11

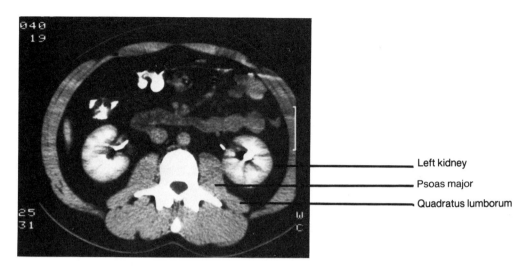

Figure 11-1.
CT scan (cross-section) of abdomen showing position of kidneys.

Figure 11-2.
Kidneys, ureters, and suprarenal glands.

Posterior Surface. Posteriorly, each kidney lies on muscle. These muscles are, above, the diaphragm, medially the psoas major, and laterally the quadratus lumborum and sometimes the transversus abdominis. The reader is reminded that the upper part of the kidney, lying on the diaphragm, is separated only by the diaphragm from the ***costodiaphragmatic recess*** of the pleura.

Lateral Border. The lateral border is smooth and featureless.

Medial Border. The medial border of the kidney contains the sinus, which is a deep slit through which pass the renal artery, renal vein, and

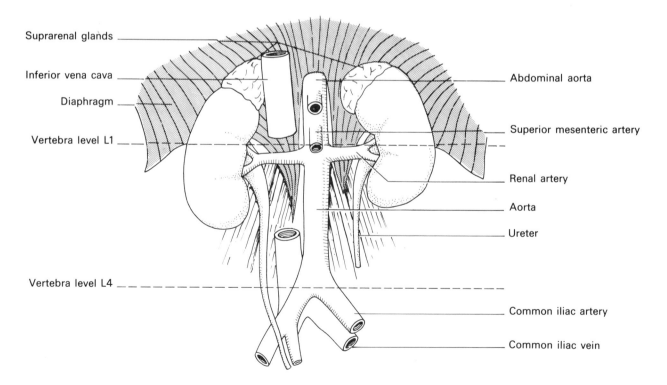

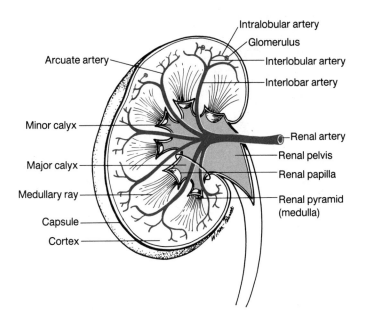

Arcuate artery
Intralobular artery
Glomerulus
Interlobular artery
Interlobar artery
Minor calyx
Renal artery
Renal pelvis
Major calyx
Renal papilla
Medullary ray
Renal pyramid (medulla)
Capsule
Cortex

Figure 11-3.
Arterial supply of right kidney (coronal section).

ureter. From anterior to posterior the order of these structures is vein, artery, ureter. Above the sinus the medial border is in contact with the suprarenal gland. Below the sinus the medial border is smooth and the ureter descends close to it.

CLINICAL NOTE

Surgical approach to the kidney is usually through an incision in the loin. The 12th rib may be resected, and care must be taken not to injure the pleura.

STRUCTURE OF THE KIDNEY

If a coronal section of the kidney is inspected it will be seen (Fig. 11-3) that the soft tissue is surrounded by a firm, fibrous **capsule**. The **renal sinus**, a depression on the medial border of the kidney, is normally lined by capsule and contains vessels and the upper expanded end of the ureter, the **renal pelvis**. The renal pelvis is so shaped that there is an upper and a lower division called the upper and lower **major calyces**. Each major **calyx** branches into several **minor calyces**. Urine reaches the minor calyces through the tips of the **renal papillae** and then flows through the major calyces to the **renal pelvis** into the ureter. The ureter carries the urine to the **urinary bladder**. This is not a passive flow; it is an active procession aided by peristaltic contractions of the (smooth) muscular walls of the calyces, renal pelvis, and ureter. These peristaltic contractions are regulated and initiated by specialized muscle tissue located in the subepithelial layers of the minor calyces; this activity resembles that of the **pacemaker** of the heart.

The kidney tissue proper can be divided into two parts, **cortex** and **medulla**. The cortex of the kidney is pale, interspersed with darker strands called **medullary rays**. These medullary rays run through the cortical tissue to the **renal pyramids**, which are composed of medullary tissue. Between the pyramids run **renal columns**, which are cortical tissue. The **glo-**

Figure 11-4.
A nephron and its collecting system.

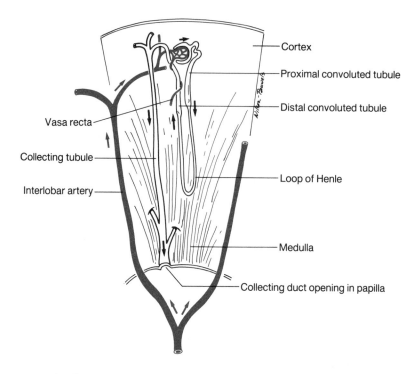

meruli, the filtering units of the kidney, are mainly located in the cortical tissue outside the pyramids. Other portions of the ***nephron*** (the functional unit of the kidney) are located in the cortex or medulla (Fig. 11-4).

The ***collecting tubules*** originate in the medullary rays and get larger as they approach the apex of the pyramid. At the apex of the pyramid the tubules empty into the minor calyx. The tip of a pyramid entering a minor calyx is called a ***papilla***. Each minor calyx receives up to three renal papillae.

CLINICAL NOTE:

*The ureters, renal pelves, and major and minor calyces can be outlined radiographically in two ways. A **retrograde pyelogram** is produced by passing a catheter through a **cystoscope** into the ureteral opening in the bladder. The injection of a radiopaque (iodine bound) dye that can be excreted and concentrated by the kidney will produce an **intravenous pyelogram** (Fig. 11-5A).*

Doubling of the renal pelvis is not very rare. When this condition occurs, the major calyces will also be doubled and there may be two ureters with separate bladder openings or the double ureters may fuse to form a single structure before draining into the bladder (Fig. 11-5B).

***Ectopic** ureters may occur and may open into the urethra or vulva and, being beyond the control of the bladder sphincter, cause incontinence of urine.*

BLOOD SUPPLY OF KIDNEY

The ***renal arteries*** are major branches of the abdominal aorta, and the renal veins drain into the inferior vena cava. These vessels will be described in greater detail in the section dealing with the posterior abdominal wall.

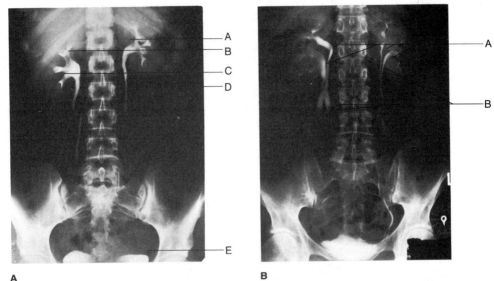

A   B

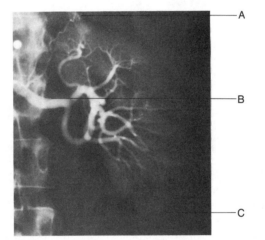

C

Each renal artery enters the sinus of the kidney and separates into branches that divide into interlobar arteries before becoming the arcuate arteries, which pass between the cortex and the medulla. The arcuate arteries branch extensively to ultimately become the afferent arterioles of the glomeruli. The kidneys may be divided into segments based on the arterial supply.

CLINICAL NOTE

*Not infrequently the renal artery will divide into branches before entering the renal sinus or there may be a double renal artery or an accessory renal artery arising from the aorta or internal iliac artery. Any of these variations may cause an abnormally placed artery to cross the **uretero-pelvic junction** and kink the ureter or interfere with the normal generation or transmission of the ureteral peristaltic impulses. A distension of the renal pelvis may result from this structural or functional obstruction. This condition is called **hydronephrosis**.*

Figure 11-5.
A. Normal intravenous pyelogram outlining components of the urinary tracts and bladder: pelvis of left kidney (*A*); major calyx of right kidney (*B*); minor calyx of right kidney (*C*); left ureter (*D*); urinary bladder (*E*).
B. Intravenous pyelogram showing bilateral doubling of renal pelves (*A*); ureters (*B*). **C.** Selective angiogram of the left kidney: upper pole (*A*); left renal artery (*B*); and lower pole (*C*).

The divisions of the renal artery, whether they occur outside the renal sinus or within it, are end arteries, that is, they do not anastomose with each other so that division or ligation of one of them will result in the death of the corresponding segment of the kidney.

*The kidneys develop in the embryonic pelvic area and migrate upward. If migration fails, a pelvic kidney, receiving its arterial supply from the internal iliac artery, occurs. Both kidneys may remain in the pelvis and be joined at their lower poles. This is the so-called **borseshoe kidney**.*

It is possible to demonstrate the arterial blood supply of the kidneys by means of radiopaque contrast material injected into a renal artery through a catheter introduced into the femoral artery (Fig. 11-5C).

The veins of the kidneys correspond to the arteries and come together to form the renal veins, which drain into the inferior vena cava.

LYMPHATIC DRAINAGE OF THE KIDNEYS

The lymphatics from the kidneys follow the renal vessels and drain into the **aortic nodes** located at the branching of the renal arteries from the abdominal aorta.

NERVE SUPPLY OF KIDNEYS

The kidneys are supplied by a plexus of autonomic nerves that pass from the aortic plexus along and surrounding the renal arteries to the renal parenchyma.

RENAL FASCIA AND FAT

The kidney, surrounded by its capsule, lies in an envelope of fat known as the **perirenal fat** (perinephric fat) (Fig. 11-6). Surrounding the perirenal fat is a layer of connective tissue, the **renal fascia**. The upper and lateral parts of the renal fascial envelopes are closed and in continuity with the anterior and posterior layers. Inferiorly and medially the fascia is open, and the fascia passes medially in front of and behind the aorta. Think of two pillow cases positioned with the open ends placed inferiorly (one surrounding each kidney) and connected to each other by a sleeve from

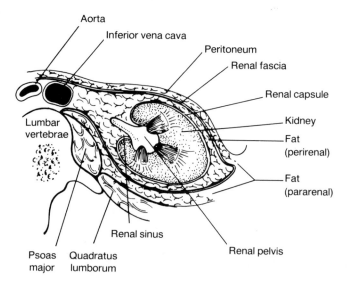

Figure 11-6.
Transverse section of the kidney and renal fascia.

hilus to hilus. These two layers are almost impossible to demonstrate in the dissecting laboratory, but radiologic studies using air injection can show this communication across the posterior abdominal wall. Inferiorly the posterior layer fuses with the transversalis fascia and anteriorly it just fades away. The ***pararenal fat*** lies outside the renal fascia but is covered by peritoneum.

CLINICAL NOTE

*The perirenal fat can become infected by a lesion in the kidney or by blood-borne bacteria. A **perirenal (perinephric) abscess** may result. The perirenal fat is less radiopaque than the kidney or psoas major muscle. As a result the outline of the kidney and the lateral border of the psoas major muscle can often be demonstrated on an x-ray film without the use of contrast media.*

*In a thin person the kidney may slip downward out of the renal fascia. This is called a **floating kidney**. Supposedly this can cause kinking and obstruction of the ureter. It is very doubtful that this kinking ever occurs in a floating kidney.*

URETERS

Each ureter is about 25 cm long; about half of its length resides in the abdomen and the other half lies in the pelvis. Only the abdominal portion will be considered here; the pelvic part of the ureter will be described later (see Chap. 14).

The ureter is a tube with a thick muscular wall, made up of an outer covering of fibrous connective tissue that is continuous with the renal capsule, and an outer layer of ***circular muscle*** and an inner layer of ***longitudinal muscle***. These muscle layers are continuous with the muscles of the pelvis and the calyces of the kidney. The submucosa is loose, and the mucosa (with transitional epithelium) is thrown into about six folds, which allow for considerable distension of the ureter.

Each ureter runs downward along the medial border of the psoas major muscle near the tips of the transverse processes of the lumbar vertebra. The ureter is, of course, retroperitoneal, but it adheres to the posterior aspect of the peritoneum and tends to be lifted up with the peritoneum when the latter is elevated or dissected. This close peritoneal relationship makes it very vulnerable during surgery, and the wise surgeon will always identify and protect the ureter, particularly during colonic or pelvic surgery.

Both ureters leave the abdomen by crossing the brim of the pelvis just anterior to the division of the ***common iliac arteries*** into their two terminal branches (see Fig. 11-2).

The right ureter lies posterior to the duodenum, the right colic vessels, the mesentery, and the terminal ileum. The left ureter lies posterior to the left colic vessels, the sigmoid colon, and the sigmoid mesocolon.

CLINICAL NOTE

The ureter propels urine to the bladder by peristaltic contractions that are initiated in the smooth muscle of the calyces. The ureter has an abundant nerve supply from the aortic plexus of autonomic nerves. If a small foreign object, for example, a stone or blood clot, gets into the ureter, the peristaltic waves will increase in force to the extent of causing the severe, inter-

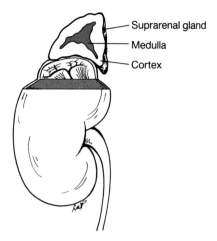

Figure 11-7.
Right suprarenal gland.

mittent pain of *ureteral (renal) colic*. This pain is mediated by sympathetic afferent fibers that supply the ureter in a segmental manner, and the pain will migrate from the loin to the lower abdominal quadrant, the groin, testicle (or vulva), and medial aspect of upper thigh; this progression of pain is one of *referred pain* to the spinal cord segments that supply the skin and body wall as well as to the ureter. These segments may be considered to be from about T9 to L1.

Renal stones (calculi) may obstruct the ureter at its three narrowest points: the pelviureteral junction, the brim of the pelvis, and the ureterovesical junction. Obstruction will cause distension of the proximal ureter, renal pelvis, and calyces and result in *hydroureter* and *hydronephrosis*.

The ureters receive their arterial blood supply from the renal arteries, from segmental branches from the lumbar arteries, and from the arteries to the bladder. All these vessels anastomose with each other, but the link between the vesicular branches and the lumbar branches is weak and cannot be relied upon. During surgery it is advisable to dissect only short segments of ureter from its surrounding connective tissue for fear of disturbing its blood supply.

SUPRARENAL GLANDS

The two suprarenal (adrenal) glands are situated like caps on the upper poles of the two kidneys (see Figs. 11-2 and 11-7).

Right Suprarenal Gland. The right suprarenal gland is pyramidal in shape and projects posterior to the inferior vena cava and an adjacent part of the liver. It may extend enough distally to lie behind the duodenum. Posteriorly it rests on the diaphragm. The *hilus* of the right gland is on its anterior surface, and from it a short vein leaves to enter the inferior cava directly. This vein is important in the operation of *adrenalectomy* because it and its attachment to the inferior vena cava are easily torn.

The arterial supply to the right suprarenal gland is from the renal artery, directly from the aorta and from the right phrenic artery (the first branch of the abdominal aorta).

Left Suprarenal Gland. The left gland is semilunar in shape and about 3 to 5 cm long. It is related anteriorly to the stomach and pancreas and posteriorly to the diaphragm. Its hilus is anterior, and the left suprarenal vein joins the left renal vein. The arterial blood supply is similar to that of the right suprarenal gland.

STRUCTURE OF A SUPRARENAL GLAND

Each suprarenal gland consists of a cortex and a medulla (Fig. 11-7). The *adrenal cortex* secretes certain important steroids controlling salt, water, and carbohydrate metabolism. It also secretes both male and female sex hormones.

The *suprarenal medulla* is a large modified sympathetic nervous system ganglion. It develops from the same *neural crest* tissue as do the sympathetic ganglia. The cells of the medulla secrete epinephrine (adrenalin) and norepinephrine. The innervation of the medulla is by *preganglionic* fibers of the sympathetic system. The cells of the medulla themselves are the end organs, and synapse occurs right in the medulla.

The Posterior Abdominal Wall

The posterior abdominal wall primarily consists of muscles attached to the vertebral column, os coxae, and ribs. It includes fascia, vessels, and nerves.

MUSCLES AND FASCIA OF THE POSTERIOR ABDOMINAL WALL

The major muscles of the posterior abdominal wall are all covered on their internal surfaces by important deep fascia.

MUSCLES

There are three important paired muscles in the posterior abdominal wall: psoas major, quadratus lumborum, and iliacus.

Psoas Major Muscle

The psoas major muscle (Fig. 12-1) is a cylindrical muscle that is a major flexor of the hip joint.

Proximal Attachments. To the bodies of the 12th thoracic and all five lumbar vertebrae and to arches of fascia that bridge between the bone of the vertebral bodies but do not attach to the intervertebral disks.

Distal Attachment. The lower end of the psoas major muscle ends in a tough tendon that merges with the tendon of the iliacus and, passing behind the inguinal ligament, attaches to the lesser trochanter of the femur.

Nerve Supply. The nerve supply to psoas major is by direct branches from the anterior primary rami of L1, L2, and L3.

Action. A major flexor of the hip joint and causes anterior and lateral flexion of the lumbar spine.

Psoas Minor Muscle

This is a thin strip of muscle, absent in 40% of bodies. It has a very prominent and long shiny tendon that attaches to the pectineal line and iliopubic eminence of the os coxae.

Quadratus Lumborum

The quadratus lumborum (Fig. 12-1) is a thin, flat muscle of the flank.

12

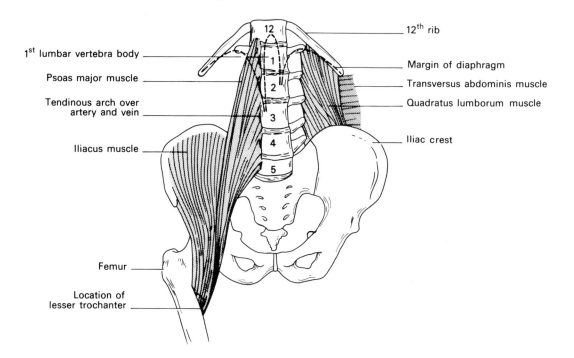

1st lumbar vertebra body

Psoas major muscle

Tendinous arch over artery and vein

Iliacus muscle

Femur

Location of lesser trochanter

12th rib

Margin of diaphragm

Transversus abdominis muscle

Quadratus lumborum muscle

Iliac crest

Figure 12-1.
Muscles of the posterior abdominal wall.

Distal Attachments. To the posterior part of the iliac crest, to the iliolumbar ligament, and to the fifth lumbar transverse process.

Proximal Attachments. To the remaining lumbar transverse processes and to the 12th rib. To reach its attachment to the 12th rib the quadratus lumborum muscle passes behind the lateral arcuate ligament of the diaphragm.

Nerve Supply. Its nerve supply is from the anterior primary rami of T12 to L4.

Action. When active the quadratus lumborum fixes the 12th rib and produces lateral flexion of the lumbar spine.

Transversus Abdominis

This muscle was considered with the anterior abdominal wall.

FASCIA

The fascia of the posterior abdominal wall has some clinical importance.

Fascia over Iliopsoas. The psoas major and the iliacus muscles and their fascia are usually considered together because the two muscles blend in the iliac fossa and share a common lower attachment. The fascia lying anterior to the psoas major muscle passes from the bodies of the vertebrae over the psoas major to blend with the fascia covering the anterior aspect of the quadratus lumborum muscle. Superiorly the fascia is thickened to form the medial arcuate ligament of the diaphragm. The fascia in front of the iliacus muscle is very dense. It attaches to the iliac crest and the brim of the pelvis and is continuous with the transversalis fascia.

CLINICAL NOTE

Tuberculosis of the spine may be considered by the reader to be a rarity, but it is still seen in underdeveloped areas of the world. A tuberculous abscess occurring in the lumbar spine tracks downward in the substance of the psoas muscle, confined by its fascia, and may present as a swelling in the groin in the femoral triangle.

Fascia over the Quadratus Lumborum. The fascia of the quadratus lumborum is attached to the transverse processes of the lumbar vertebrae, iliac crest, and 12th rib and is continuous with the transversalis fascia (Fig. 12-1). The fascia on the anterior and posterior surfaces of the quadratus lumborum muscle helps to form the ***thoracolumbar fascia***, which provides attachment for other muscles.

NERVES OF THE POSTERIOR ABDOMINAL WALL

The nerves of the posterior abdominal wall are divisible into ***somatic nerves*** and ***nerves of the autonomic nervous system***.

Autonomic Nervous System

The autonomic nervous system is made up of the sympathetic and parasympathetic systems. The basic principle of the sympathetic system was discussed in Chapter 7. Chapter 16 describes the ***autonomic nervous system***.

Abdominal Sympathetic Trunks. Following the pattern outlined in Chapter 7, the sympathetic trunks (Figs. 12-2A and 12-2B) consists of two chains of four to five lumbar ganglia, one chain on each side of the vertebral column. The trunks enter the abdomen behind the medial arcuate ligaments of the diaphragm and run downward in the groove between the vertebral column and the psoas major muscle. The right trunk is usually located behind the inferior vena cava, and on both sides the lumbar vessels pass behind the trunk.

Passing posterior to the common iliac vessels the trunks enter the pelvis (Figs. 12-2A and 12-2B), run on the anterior surface of the sacrum, and join each other in the midline opposite the last piece of the sacrum to form the ***ganglion impar***.

Although the ganglia of the abdominal and pelvic parts of the sympathetic trunk are connected to the segmental somatic nerves by gray rami communicantes, only the upper two lumbar ganglia have ***white rami*** (Figs. 12-2A and 12-2B).

Medial branches from the sympathetic lumbar ganglia are known as ***lumbar splanchnic nerves***. These medial branches pass to the ganglia and plexuses around the major branches of the aorta to synapse with cells in these ganglia (celiac ganglion, aortorenal ganglion, superior mesenteric ganglion, inferior mesenteric ganglion). The postganglionic fibers from these ganglia (plexi) accompany the branches of the abdominal aorta and constitute the dense network around these arteries that the student finds difficult to remove when he tries to display these aortic branches. Other major (sympathetic) contributors to the aortic plexi are the ***splanchnic nerves*** (greater, lesser, and least), which were described earlier as passing from the thoracic sympathetic trunks, through the crura of the diaphragm, to reach the abdomen.

These aortic plexuses also receive parasympathetic fibers from the vagus nerve. The parasympathetic nerves do not synapse in the aortic ganglia but make their final neuronal contacts in the walls of the viscera for which they are destined.

Figure 12-2.
A. Abdominopelvic sympathetic nervous system showing paravertebral and prevertebral ganglia. **B.** (*facing page*) Vagus and pelvic splanchnic (parasympathetic) nerves. Arrows indicate distribution to parasympathetic ganglia located close to or within the walls of the target organ.

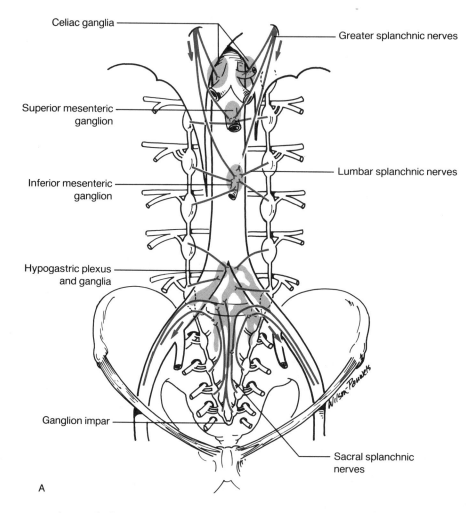

Celiac ganglia

Greater splanchnic nerves

Superior mesenteric ganglion

Lumbar splanchnic nerves

Inferior mesenteric ganglion

Hypogastric plexus and ganglia

Ganglion impar

Sacral splanchnic nerves

A

The medially running lumbar splanchnic fibers from the lumbar sympathetic trunks form the **hypogastric plexus**, which is in continuity with the aortic plexus. The hypogastric plexus is located in front of the last lumbar vertebral body and the promontory of the sacrum. From the hypogastric plexus two **hypogastric nerves** pass into the pelvis and give dense networks of fibers to the common iliac arteries, the internal iliac arteries, the external iliac arteries and their branches. In addition, medial branches from the sacral ganglia contribute **sacral sympathetic splanchnics** to the inferior hypogastric plexuses. This is how the pelvic viscera receive their sympathetic nerve supply.

Abdominal and Pelvic Parasympathetic Nerve Supply. The parasympathetic nerve supply to the abdominal viscera is divided into a cranial component supplied by the **vagus nerves** and a sacral component derived from the **pelvic splanchnic nerves** branching from the anterior primary rami of sacral nerves 2, 3, and 4. The vagus nerves supply the foregut and midgut structures as far caudally as the left colic flexure. The presynaptic vagal fibers enter the plexuses around the major blood vessels to the gut and travel with the blood vessels to the end organ where they synapse in

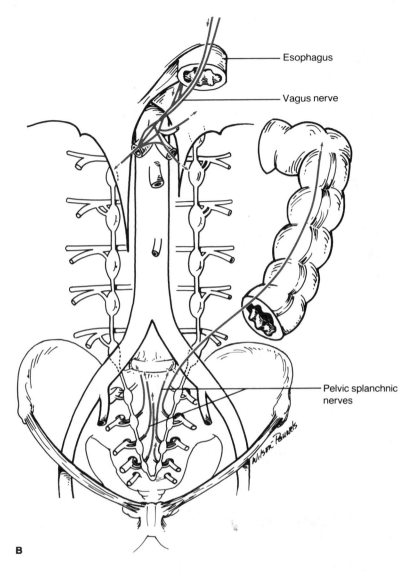

Esophagus

Vagus nerve

Pelvic splanchnic nerves

B

minute ganglia close to, or in the walls of, the organs they supply. From the left colic flexure, the hindgut is supplied by pelvic splanchnic nerves from S2, S3, and S4. The pelvic splanchnics (parasympathetics) supply all of the intestinal tract that receives blood from the inferior mesenteric artery. The pelvic splanchnic (parasympathetic) fibers to the descending and sigmoid colons first join with the inferior hypogastric nerves but leave the plexuses to travel independently until reaching the vasa recti of the gut tube, where they then travel with perivascular plexuses to reach the descending and sigmoid colons. The remainder of the abdominal and pelvic viscera receive their parasympathetic fibers through periarterial plexuses around branches of the internal iliac arteries.

Somatic Nervous System

The somatic nervous system of the abdomen consists almost entirely of the **lumbar plexus**, a complex arrangement of the fibers of the anterior primary rami of the lumbar nerves, formed for the most part in the substance of the psoas major muscle.

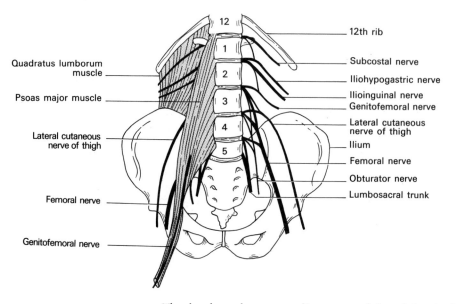

Figure 12-3.
Lumbar plexus. The left psoas major muscle has been removed.

The lumbar plexus supplies parts of the abdominal wall and the lower extremity. It combines with the **sacral plexus** to form the **lumbosacral plexus**.

The lowest thoracic nerve (T12) forms the **subcostal nerve**, which passes behind the lateral arcuate ligament of the diaphragm, runs across the quadratus lumborum muscle to reach the transversus abdominis muscle, which it pierces, and supplies the abdominal wall between the umbilicus and the pubic symphysis (Fig. 12-3).

Lumbar Plexus. The anterior primary rami of lumbar nerves 1, 2, 3, and 4 take part in the formation of the lumbar plexus (Fig. 12-3). A few direct twigs pass to the psoas and quadratus lumborum muscles. All of the other branches are named.

The plexus is found in the psoas major muscle, which must be dissected piecemeal to reveal the nerves and their plexiform arrangements.

The **iliohypogastric nerve** (L1) passes from the lateral border of the psoas muscle over the posterior abdominal wall to reach the transversus abdominis. It travels through the abdominal wall supplying muscle and skin along a band that ends above the external (superficial) ring of the inguinal canal.

The **ilioinguinal nerve** (L1) follows a similar course at a slightly lower level. It enters the inguinal canal near the internal ring and travels through the canal, lying superficial to and below the spermatic cord, to exit the canal at the superficial (external) ring. It supplies the skin over the superomedial part of the thigh, the scrotum, and the root of the penis; in the female it supplies an area of the vulva. On the analogy of an intercostal nerve, the ilioinguinal nerve may be considered the collateral branch of L1. It varies in size in inverse proportion to that of the iliohypogastric nerve.

CLINICAL NOTE

The ilioinguinal nerve may be injured during hernia repair. If cut, an area of anesthesia will result. This area corresponds to the area of skin that can be felt by a hand introduced into a trouser pocket. If not cut but involved in postoperative scar tissue, the traumatized ilioinguinal nerve will produce a burning sensation over a similar area.

The ***genitofemoral nerve*** (L1 and L2) emerges from the ***anterior*** surface of the psoas major muscle and proceeds inferiorly until it splits above the inguinal ligament into a femoral and a genital branch. The femoral branch descends behind the inguinal ligament in front of the femoral artery and supplies a small triangle of skin over the upper femoral triangle. The genital branch enters the internal ring of the inguinal canal and supplies the cremaster muscle and the labium majus.

The ***lateral cutaneous nerve of thigh*** (L2 and L3) emerges from the lateral border of psoas major and runs across the iliacus muscle to the lateral end of the inguinal ligament. Here it passes behind or through the inguinal ligament to supply the skin of the lateral aspect of the thigh.

The ***femoral nerve*** (L2, L3, and L4) runs in the groove between the psoas major and iliacus muscles, behind the inguinal ligament, to enter the thigh just lateral to the femoral artery. Its subsequent course will be described with the lower limb.

The ***obturator nerve*** (L2, L3, and L4) appears on the ***medial*** aspect of the psoas major muscle at the brim of the pelvis and will be discussed later. Some of its territory may be supplied by an ***accessory obturator nerve***.

It will have been observed that the femoral and obturator nerves come from the same lumbar anterior primary rami. Anterior primary rami to the limbs divide into ventral (anterior) and dorsal (posterior) divisions for the supply of the morphologic ventral (flexor) and dorsal (extensor) aspects of the limbs. The obturator nerve represents the ventral division and the femoral nerve the dorsal division.

The ***lumbosacral trunk*** (L4 and L5). This large trunk is formed by the remainder of the anterior primary ramus of the fourth lumbar nerve and the anterior primary ramus of the fifth lumbar nerve. It passes along the medial border of the psoas major muscle over the ala of the sacrum and is a major component of the important ***sacral plexus***.

ABDOMINAL AND VISCERAL PAIN

The interpretation of pain felt in the abdomen requires consideration of the nerve supply of abdominal ***viscera*** and of the abdominal wall. It must be remembered that the term ***abdominal wall*** refers not only to the obvious anterior and posterior limits, but also to the diaphragm and the floor of the pelvis. Although the pelvic portion of the abdominal cavity has not yet been described in this book, we shall consider the whole question of abdominal pain and abdominal sensations in this segment of the book. The reader is encouraged to refer back to this section, from time to time.

A clear distinction must be made between pain arising in viscera and pain from the abdominal wall. This requires an understanding of the neurologic pathways involved in the transmission of visceral afferent impulses, as distinct from somatic afferent pathways.

SOMATIC PAIN

Sensation from the ***parietal peritoneum*** is carried in the segmental somatic sensory nerves of the region to their neurons in the posterior root ganglia of the spinal nerves. The central processes of these posterior root ganglion cells make connections at different levels of the central nervous system, and many of them reach levels of the brain where these sensory impulses can be analyzed and correctly interpreted as to the nature of the sensation, (*e.g.*, touch, temperature, or pain) and to the location on the body wall from which the impulse originates.

The interpretation of the site of origin of such impulses may be misleading when different parts of the body receive their sensory nerve supply from the same spinal cord segments. This misinterpretation of the site of origin of pain has already been mentioned in Chapter 3 on the thorax, with particular reference to the diaphragm and the lower intercostal nerves. It will be summarized and reiterated in the following paragraph.

The ***diaphragm*** receives some of its sensory nerve supply from the ***phrenic nerve*** that arises from the third, fourth, and fifth cervical segments of the spinal cord (see Fig. 7-8). The diaphragm is covered by ***parietal pleura*** above and by ***parietal peritoneum*** below. Painful stimuli from the central portion of the diaphragm will be misinterpreted by the brain as coming from the area of skin supplied by these cervical segments and will be felt in the tip of the shoulder and the root of the neck. This is ***referred pain***. The lower six intercostal nerves supply the lower parietal costal pleura and peripheral diaphragmatic pleura as well as the abdominal wall. Thus pain arising in the lower thoracic wall or peripheral diaphragmatic pleura can be misinterpreted as coming from the abdominal wall. An example is the occurrence of ***abdominal pain*** with abdominal muscle guarding (reflex contraction of abdominal muscles) in, for instance, pneumonia affecting the peripheral part of the diaphragmatic pleura or lower thoracic parietal pleura. This is another example of referred pain.

In general pain from the parietal peritoneum of the anterior, lateral, and posterior abdominal wall is localized quite accurately. Further examples appear below.

VISCERAL PAIN

The sensation of visceral pain is transmitted by ***visceral afferent*** fibers running with the distribution of the ***sympathetic system***. The neurons of these visceral afferent fibers are located in the posterior root ganglia of the spinal nerves from the same spinal cord segments that give rise to the ***preganglionic*** sympathetic outflow, that is, T1 to L2. Visceral pain will be interpreted as coming from somatic areas supplied by these spinal cord segments. Cardiac pain being referred to the anterior chest wall and inner aspect of the left arm has already been discussed (see Chap. 5).

The abdominal viscera can tolerate many insults, but they do not like being stretched, obstructed, or distended. During a surgical abdominal procedure carried out on a conscious patient under local anesthetic, the stomach, duodenum, or intestines can be cut, burned, or crushed without pain being felt. However, distension of the viscera or pulling on the mesenteries will be painful.

The basic visceral abdominal pain is ***colic***. Colic is believed to arise in stretch receptors in the muscular walls of the viscera and occurs during violent contractions of these muscular walls. These violent contractions may occur in infections such as gastroenteritis in which the inflamed, irritable gut tries to empty itself, because of mechanical obstructions such as a stone in the ureter, or during the course of physiological phenomena such as uterine contractions during childbirth or menstruation.

There are degrees of colic, and it is difficult to draw a line of definition between a mild bellyache from dietary indiscretion and the severe pain of a bowel obstruction. A typical colic consists of waves of pain that increase in severity and may cause the victim to double over. The wave reaches a crescendo and then partially abates before the next wave hits.

The part of the abdomen in which the colic will be felt depends on the viscus involved. ***Foregut derivatives*** such as stomach and gallbladder will produce pain that is located in the ***epigastrium***. ***Midgut*** areas such as small intestine and proximal large intestine (including the vermiform appendix) produce pain centered around the ***umbilicus***, and ***hindgut*** pain (*e.g.*, sigmoid colon) will be located in the ***hypogastrium***. Although severe, the localization of visceral colic from the gut is rather indefinite.

Gallbladder colic (biliary colic) is an interesting example of visceral pain. The pain is extremely severe and is felt in the epigastrium, retrosternal region, right hypochondrium, and right ***scapular area***. The explanation is that all these areas are supplied by the seventh (eighth) thoracic nerve(s). When the gallbladder becomes inflamed it will irritate the anterior abdominal wall parietal peritoneum, and constant pain and tenderness will occur in the right upper quadrant.

The pain of ***appendicitis*** begins as appendicular colic felt vaguely in the umbilical region, then it localizes in the right lower quadrant of the abdomen, with tenderness and muscle guarding when the parietal peritoneum of the area becomes inflamed. Many appendices lie in the retrocecal position, and in these cases the posterior parietal peritoneum will become inflamed and the patient will resent stretching of the psoas major muscle when the physician tries hyperextension of the hip joint; indeed, the patient will often lie in bed with the right hip flexed.

KIDNEY AND URETERAL PAIN

Renal pain occurs when the kidney is inflamed or distended. It is felt in the loin, and this location corresponds to the somatic nerve supply of the area from T10 (and T11). It may also be felt in the front of the abdomen at the same level. The pain of ***ureteral colic*** such as occurs during the passage of a kidney stone down the ureter is very severe and follows the somatic representation of the visceral nerve supply of the ureter. Beginning in the loin it moves to the flank, to the lower quadrant of the abdomen, and then into the groin, penis, labium majus, and even occasionally to the medial aspect of the upper thigh (T10–L1).

Pain from traction on the mesentery that may occur when a loop of intestine becomes twisted will be referred to the back, as will the pain of a posteriorly penetrating peptic ulcer.

UTERINE PAIN

Uterine pain, for example, labor pains or menstrual cramps, are felt in the hypogastrium and posteriorly in the lumbar area at and below the level of the iliac crest, corresponding to the nerve supply of the uterus from the hypogastric nerves whose spinal cord segments are T10–L1. The reason that the posteriorly felt pain from the uterus is at this low level is that the ***cutaneous branches*** of the ***posterior primary rami*** of the lower thoracic and lumbar nerves run downward and supply body wall structures (including skin) at a level much below the emergence of the corresponding spinal nerve from the vertebral canal. Dysmenorrhea (excessively painful menstrual cramps) can be relieved by cutting the hypogastric nerves (presacral neurectomy), but pain from stretching of the cervix will be unaffected by this procedure.

Pain from the cervix is transmitted by the (parasympathetic) pelvic splanchnic nerves (S2, S3, and S4). The cervix cannot be dilated without pain, but it can be cauterized without pain. Pain from the cervix uteri is generally felt in the perineum and lower sacral area.

The urinary bladder receives afferent fibers from both the hypogastric nerves and from the pelvic splanchnic nerves. Lesions affecting the T11–L2 spinal nerves abolish the pain of an overdistended bladder, but cutting of the hypogastric nerves does not relieve the pain of bladder cancer.

So far we have assigned visceral pain and sensations mainly to the afferent fibers in the sympathetic system. There is no doubt that the parasympathetic system (vagus nerve and sacral parasympathetic nerves) contains many afferent fibers that have their neurons located in the ganglia of the vagus nerve and in the posterior root ganglia of the appropriate sacral nerves. Many of these afferents are probably concerned with the transmission of afferent impulses related to physiological states that never reach consciousness (*e.g.*, chemoreceptors and baroreceptors). The vague sensations that do reach consciousness and may be mediated by parasympathetic afferents include hunger, gastric fullness, nausea, and rectal distension. In addition (as mentioned above), it is probable that pain from the cervix uteri and urinary bladder is transmitted by the pelvic splanchnic nerves.

Segmental Sympathetic Supplies
Head and neck T1–T5
Upper limb T2–T5
Lower limb T10–L2
Heart T1–T5
Bronchopulmonary T2–T4
Esophagus (caudal part) T5–T6
Stomach T6–T10
Small intestine T9–T10
Large gut to splenic flexure T11–L1
Distal colon and rectum L1–L2
Liver and gallbladder T7–T9
Spleen T6–T10
Pancreas T6–T10
Kidney T10–L1 (L2)
Ureter T11–L2
Adrenal (suprarenal) T8–L1
Testis and ovary T10–T11
Bladder T11–L2
Prostate and prostatic urethra T11–L1
Uterus T12–L1
Uterine tube T10–L1

ARTERIES OF POSTERIOR ABDOMINAL WALL

The arteries of the posterior abdominal wall arise from the descending (abdominal) aorta.

AORTA

The aorta passes through (actually behind) the diaphragm under cover of the median arcuate ligament at the level of the 12th thoracic vertebra, in the midline. It descends in the abdomen in front of the vertebral column, diverting a little to the left, and at the level of the fourth lumbar vertebra the aorta divides into two *common iliac arteries*, which pass along the brim of the pelvis. Halfway to the inguinal ligament (Fig. 12-4), anterior to the sacroiliac joint, the common iliac artery divides into the *internal* and *external iliac arteries*. The external iliac artery continues along the brim of the pelvis and passes behind the midpoint of the inguinal ligament to become the *femoral artery*.

The aorta passes posterior to the lesser sac, pancreas, left renal vein, the third part of the duodenum, the root of the mesentery, and the *aortic plexus of nerves*. It lies anterior to the vertebrae and intervertebral discs.

Some of the branches of the aorta have already been studied; they are summarized below.

Unpaired Visceral Branches
Celiac artery
Superior mesenteric artery
Inferior mesenteric artery

Paired Visceral Branches
Renal Arteries. The distribution of these vessels to the kidneys has already been described. It remains to be said that the renal arteries are large and they arise from the aorta at the level of the first lumbar vertebra, the right being a little higher. The

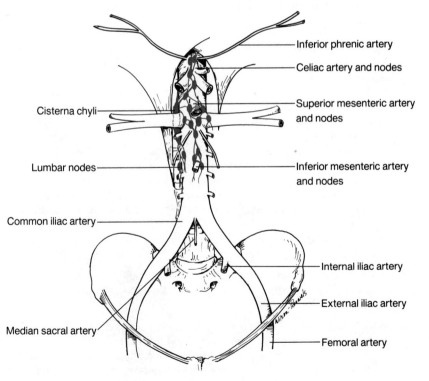

Inferior phrenic artery
Celiac artery and nodes
Superior mesenteric artery and nodes
Cisterna chyli
Inferior mesenteric artery and nodes
Lumbar nodes
Common iliac artery
Internal iliac artery
External iliac artery
Median sacral artery
Femoral artery

Figure 12-4.
Abdominal aorta and lymph nodes.

renal arteries run across the posterior abdominal wall at right angles from the aorta, crossing the crura of the diaphragm. The right renal artery will pass behind the inferior vena cava, the right renal vein, the head of the pancreas, and the second part of the duodenum. The left renal artery runs behind the left renal vein, the body of the pancreas, and the splenic and inferior mesenteric veins. The renal arteries give origin to some of the suprarenal arteries and to the ureteral arteries.

Testicular (Ovarian) Arteries. The testicular (ovarian) arteries arise from the abdominal aorta just below the origins of the renal arteries. They pass into the pelvis and will be described therein later.

Paired Somatic Branches. These arteries supply the diaphragm and the body wall. The first branches are the ***phrenic arteries***, which arise just below the diaphragm and supply that structure. Four ***lumbar arteries*** run laterally across the posterior abdominal wall behind the sympathetic trunks and behind the inferior vena cava. They pass deep to the arched fibers of origin of the psoas major muscle and supply the quadratus lumborum and the deeper muscles of the back.

Unpaired Somatic Branch. The ***median sacral artery*** is the terminal and vestigial continuation of the aorta beyond its bifurcation. It passes down the center of the sacrum and fades away at the coccyx.

VEINS OF THE POSTERIOR ABDOMINAL WALL

The two ***external iliac veins*** begin behind the inguinal ligament as continuations of the two femoral veins. They pass along the pelvic brim medial to the external iliac arteries and join with the ***internal iliac*** veins to form the ***common iliac*** veins. At the level of the fifth lumbar vertebra, to the right of the origins of the common iliac arteries, the common iliac veins unite to form the ***inferior vena cava***.

The inferior vena cava (Fig. 12-5) ascends on the vertebral bodies, grooves the posterior surface of the liver, and at the level of the eighth thoracic vertebra passes through the central tendon of the diaphragm and the pericardium to enter the right atrium of the heart. It is covered by peritoneum throughout most of its course. It passes behind the root of the mesentery, duodenum, head of pancreas, epiploic foramen, common bile duct, hepatic artery, portal vein, and liver. Posteriorly it lies on the vertebral column, the right psoas major muscle, and the right crus of the diaphragm. The right sympathetic trunk may be overlapped by the right border of the inferior vena cava.

The inferior vena cava receives tributaries from the viscera and from the body wall.

Visceral Veins. These are the renal, hepatic, right testicular (ovarian), and right suprarenal veins. The left testicular (ovarian) and suprarenal veins enter the left renal vein.

Somatic Veins. Other tributaries of the inferior vena cava drain the body wall. The ***inferior phrenic*** veins drain the abdominal surface of the diaphragm; the ***lumbar*** veins drain the posterior abdominal wall and are joined to each other by paired ***ascending lumbar*** veins, which parallel

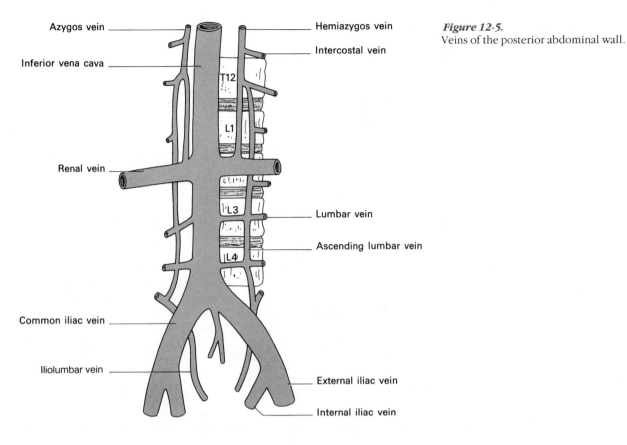

Azygos vein

Inferior vena cava

Renal vein

Common iliac vein

Iliolumbar vein

T12

L1

L3

L4

Hemiazygos vein

Intercostal vein

Lumbar vein

Ascending lumbar vein

External iliac vein

Internal iliac vein

Figure 12-5.
Veins of the posterior abdominal wall.

the inferior vena cava and lie anterior to the transverse processes of the lumbar vertebrae.

At its upper end each ascending lumbar vein is joined by a subcostal vein. The upward continuation behind the diaphragm forms the ***azygos*** vein on the right and the ***hemiazygos*** vein on the left. The ascending lumbar vein may be duplicated or interrupted along its length into several shorter segments.

ALTERNATIVE VENOUS PATHWAYS IN THE ABDOMEN

It may be necessary to ligate the inferior vena cava or it may become blocked by pathologic conditions. When this happens the blood has two main alternative pathways of return to the heart.

1. Via the ascending lumbar, hemiazygos, and azygos veins. The azygos vein empties into the superior vena cava.
2. The ***vertebral plexus of veins***, passing in the spinal canal and bodies of the vertebrae, anastomoses with the lumbar and intercostal veins, giving a second alternate pathway.

CLINICAL NOTE

*The abundance and large size of the veins of the vertebral plexus of veins is thought to be responsible for the spread of cancer cells from the lower abdomen, notably the **prostate gland**, to the vertebral bodies.*

LYMPHATICS OF THE POSTERIOR ABDOMINAL WALL

The abdominal aorta is surrounded by multiple lymph nodes that appear to form a continuous chain. At different levels these nodes receive the lymph from certain territories, and different names are applied to these groups according to their field of drainage (see Fig. 12-4).

1. ***Celiac Nodes.*** Drain the stomach, duodenum, lower esophagus, gallbladder, liver, and head of the pancreas.

2. ***Superior Mesenteric Nodes.*** Receive lymph from the entire territory supplied by the superior mesenteric artery.

3. ***Inferior Mesenteric Nodes.*** Receive lymph from the areas of supply of the inferior mesenteric artery and have some surgical significance: the aim of cancer surgery is to remove the primary tumor and its lymphatic drainage. Rarely is this possible in the abdomen without infringing on the blood supply of other viscera.

    In surgery for cancer of the rectum or lower colon it is possible to remove the inferior mesenteric nodes right up to the aorta because the ***marginal artery of Drummond*** of the colon permits division of the inferior mesenteric artery at its origin without depriving the remaining colon of its arterial blood supply.

4. The ***lumbar group of lymph nodes*** are clustered around the origins of the renal arteries and the lower aorta and also receive lymph drainage from the ***testes*** and ***ovaries***. These nodes can be removed surgically to treat renal or testicular cancer, and this dissection is often carried into the thorax in an attempt to remove, at a higher level, the glands into which the renal group drain.

The ***cisterna chyli***, the sac-like lower extremity of the ***thoracic duct***, lies between the aorta and the right crus of the diaphragm at the level of L2. It is about 5 cm long and receives drainage from the ***lumbar*** and ***intestinal lymph trunks*** of both sides. The lumbar lymph trunks drain the lumbar and iliac lymph nodes which, in turn, receive lymph from the lower part of the body and the lower limbs, the kidneys, testes, ovaries, and pelvic viscera.

The ***intestinal lymph*** trunks receive drainage from the celiac, superior mesenteric, and inferior mesenteric nodes.

The Wall of the Pelvis

The wall of the true pelvis primarily consists of muscles on a framework of bones. The bones involved are the sacrum, coccyx, pubis, ischium, and the lower part of the ilium.

The superior opening of the true pelvis, the **pelvic inlet**, is formed posteriorly by the promontory and alae of the sacrum, and on each side and in front by the arcuate line of the ilium, the pectineal line of the pubis, and the crest of the pubis (Fig. 13-1).

JOINTS OF THE PELVIS

The joints involved in the formation of the pelvis are the lumbosacral, sacrococcygeal, and sacroiliac joints and the pubic symphysis (Fig. 13-1).

Lumbosacral Joint. The lumbosacral joint is the joint between the fifth lumbar and the first sacral vertebrae. It has all the usual ligaments of a spinal joint complex plus the tough **lumbosacral ligament** running from the transverse process of the fifth lumbar vertebra to the ala of the sacrum.

CLINICAL NOTE

The fifth lumbar vertebra may be fused to the sacrum. This condition is called **sacralization**. *The spine, laminae, and inferior articular processes of the fifth lumbar vertebra may be fused as a unit but remain separated from the rest of the pedicle and body of the vertebra by fibrocartilaginous tissue; this is believed to be a congenital abnormality and is of no significance unless separation of the bony units occurs, allowing the body and superior articular facets of the fifth lumbar vertebra to slide forward; this condition is called* **spondylolysis** *and can cause a visible back deformity and pain (Fig. 13-2).*

Sacrococcygeal Joint. The sacrococcygeal joint includes a small intervertebral disc and some unimportant ligaments. The joint usually exhibits a considerable range of passive flexion and extension movements.

Sacroiliac Joint. The sacroiliac joint is between the auricular surface of the sacrum and the auricular surface of the ilium on each side (Fig. 13-3). It is a synovial joint and is surrounded by some strong ligaments:

    Anterior sacroiliac ligament is relatively thin and covers the anterior aspect of the joint.

13

Figure 13-1.
Bony pelvis and pelvic inlet (true pelvis).

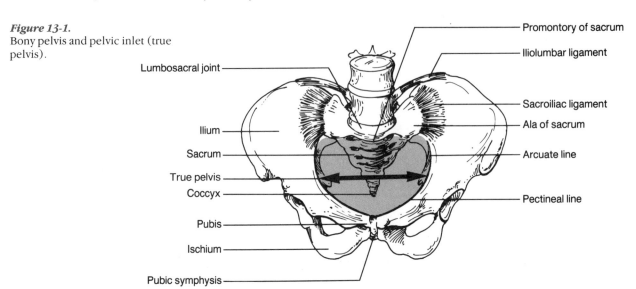

Promontory of sacrum

Iliolumbar ligament

Lumbosacral joint

Sacroiliac ligament

Ala of sacrum

Ilium

Arcuate line

Sacrum

True pelvis

Coccyx

Pectineal line

Pubis

Ischium

Pubic symphysis

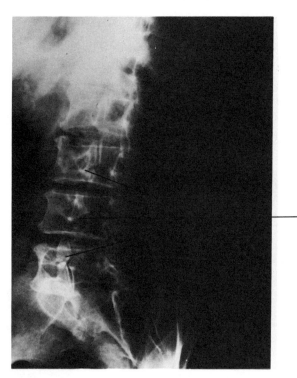

A

—A

—B

B

Figure 13-2.
A. Radiograph (oblique view) showing spondylolysis. Note the defects indicated at *A.* **B.** Radiograph (lateral view) spondylolisthesis showing defect (*A*) and anterior displacement of body of 5th lumbar vertebra on sacrum (*B*). (Radiographs courtesy of Dr. R. Pototschnik, Toronto Western Hospital).

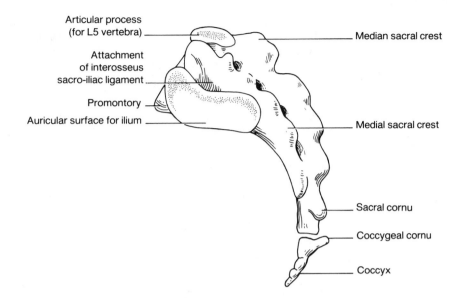

Articular process
(for L5 vertebra) _____

Attachment
of interosseus
sacro-iliac ligament _____

Promontory _____

Auricular surface for ilium _____

_____ Median sacral crest

_____ Medial sacral crest

_____ Sacral cornu

_____ Coccygeal cornu

_____ Coccyx

Figure 13-3.
Sacrum (left lateral view).

Posterior sacroiliac ligament is also rather thin and extends from the ilium to the lateral sacral crest.

The *interosseous sacroiliac ligament* is a very thick and strong ligament that connects the rough area behind the auricular surface of the ilium to the ala of the sacrum (Fig. 13-1).

The *iliolumbar ligament* runs from the transverse process of the fifth lumbar vertebra to the iliac crest.

The *sacrotuberous ligament* and the *sacrospinous ligament* are not ligaments of the sacroiliac joint but are in proximity and play a part in the formation of the pelvis and its outlet (see Fig. 15-1).

The sacroiliac joint is covered posteriorly by the erector spinae and gluteus maximus muscles. Anterior to the joint lie the psoas major and iliacus muscles. It is crossed on its pelvic surface by the lumbosacral trunk, piriformis muscle, and blood vessels.

The sacroiliac joint allows a slight amount of rotation; the degree of mobility is limited by the very dense and tough interosseous ligament. However, in the terminal weeks of pregnancy the ligaments become softer and looser, and considerable rotation of the sacrum enables the lower end of the sacrum to tip posteriorly to allow passage of the fetal head during childbirth.

Pubic Symphysis. The pubic symphysis is a secondary cartilaginous joint. There is a fibrocartilaginous disc between the two pubic bones, and the joint is supported by strong anterior and inferior ligaments. The posterior and superior ligaments are relatively weak. Although there is frequently a cavity in the fibrocartilaginous discs (more common in females) the joint is *not* a synovial joint.

MUSCLES OF THE PELVIS

Obturator Internus Muscle. The obturator internus muscle (see Fig. 15-6) is attached to the inner aspect of the os coxae and the obturator membrane. On the os coxae it is attached to the pubis, pubic rami, ischial ramus, and body of the ischium as well as the obturator membrane. The fibers of the muscle pass posteriorly through the lesser sciatic foramen, making a right angle bend around the body of the ischium to be attached below to the greater trochanter of the femur.

The two obturator internus muscles form a substantial part of the lateral walls of the pelvis.

Obturator Membrane. This fibrous sheet covers the pelvic aspect of the obturator foramen. Its superolateral corner is deficient to allow the obturator nerve and vessels to pass through the upper and lateral aspect of the obturator foramen.

Piriformis Muscle. The piriformis (see Fig. 19-3) forms a useful landmark in the gluteal region (see p. 244). It is attached proximally to the anterior (pelvic) surfaces of the second, third, and fourth segments of the sacrum. It passes out of the pelvis into the gluteal region through the greater sciatic foramen and is attached distally to the tip of the greater trochanter of the femur. Its nerve supply is from L5, *S1*, and S2. When the hip joint is extended the piriformis acts as a lateral rotator of the thigh. When the hip joint is flexed the piriformis is an abductor of the thigh.

Levator Ani Muscle. The levator ani muscle originates from fascia covering the obturator internus muscle along a line drawn from the back of the body of the pubis to the ischial spine (see Fig. 15-3). This origin is deficient anteriorly, leaving a gap between the two levatores ani. The prostate gland or the vagina is found in this gap, surrounded by the medial margins of the two levatores ani muscles. The fibers of the levator ani muscle pass medially and inferiorly to meet each other in a midline raphe, in much the same manner as the walls of a funnel. Parts of the levator ani can be named according to their origin from different parts of the os coxae, but only the ***puborectalis*** deserves special mention (see Pelvic Diaphragm below). The nerve supply to the levator ani is from the third and fourth sacral nerves.

Coccygeus Muscle. The posterior portion of the sheet of muscle that forms the levator ani runs from the coccyx to the ischial spine. (If we had a tail it would wag it.)

STRUCTURES OF THE PELVIC DIAPHRAGM

The ***pelvic diaphragm*** consists of the two levatores ani muscles and the two coccygeus muscles.

The muscles of the pelvic diaphragm thus originate from a line drawn from the body of the pubis to the ischial spine, one on each side. Below, the muscle fibers of this funnel-shaped diaphragm are attached to the following structures:

1. Each side of the ***prostate*** or ***vagina***.
2. The ***central perineal tendon*** (perineal body), which is a fibrous ***junction*** located at the midpoint of the perineum.
3. Each side wall of the ***anal canal***.
4. The ***anococcygeal ligament*** (anococcygeal raphe), which is a fibrous band located between the anus and the tip of the coccyx.
5. The ***coccyx***.

The chief function of the pelvic diaphragm is to form the floor of the pelvis. This function is particularly important in man, who was so unwise as to assume the biped erect posture. Like all muscle the pelvic diaphragm is covered by fascia on both its upper and its lower surfaces, and these fascial layers have some considerable importance, as will be described later.

Puborectalis. The puborectalis forms a sling of muscle that runs from the body of the pubis around the posterior aspect of the anorectal junction in such a way that it pulls the anorectum forward so that the rectum forms an angle with the anal canal (see Fig. 14-3B). This angle takes much of the weight of the fecal mass and relieves the anal sphincter of much of the pressure of accumulated feces in the rectum.

NERVES OF THE PELVIS

The *lumbosacral trunk* (L4, L5) passes inferiorly in front of the sacroiliac joint to join the sacral plexus. The rest of the pelvic nerves come from the sacral and coccygeal nerves.

The *sacral plexus* (Fig. 13-4), a major nerve plexus, is formed from the *anterior* and *posterior* divisions of the *anterior primary rami* of L4, L5, S1, S2, and S3. The plexus lies in the pelvis anterior to the piriformis muscle and posterior to the internal iliac vessels and ureter; it divides into several major nerves:

1. *Sciatic Nerve.* The sciatic nerve leaves the pelvis to enter the gluteal region through the greater sciatic notch (Fig. 13-4). This nerve is the major nerve of the posterior aspect of the thigh and for the whole of the leg. It will divide into the *tibial nerve* and the *common peroneal nerve*. The nerve fibers destined for the tibial nerve arise from the *anterior divisions* of L4, L5, S1, S2, and S3. The fibers destined for the common peroneal nerve come from the *posterior divisions* of L4, L5, S1, and S2.

Figure 13-4.
Sacral plexus.

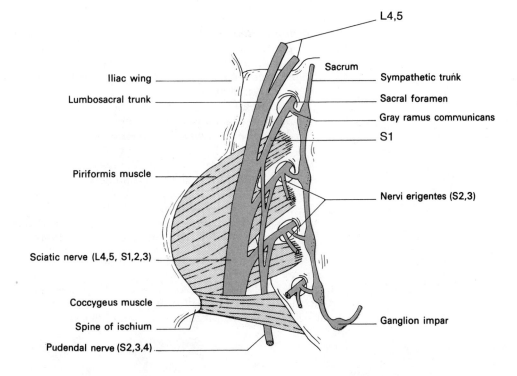

Iliac wing

Lumbosacral trunk

Piriformis muscle

Sciatic nerve (L4,5, S1,2,3)

Coccygeus muscle

Spine of ischium

Pudendal nerve (S2,3,4)

L4,5

Sacrum

Sympathetic trunk

Sacral foramen

Gray ramus communicans

S1

Nervi erigentes (S2,3)

Ganglion impar

Figure 13-5.
Major branches of the sacral plexus.

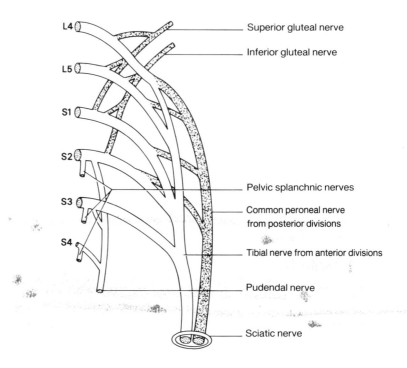

2. **Pudendal Nerve**. The pudendal nerve (Figs. 13-4 and 13-5) will be encountered in the description of the perineum (see p. 206). It arises from the **anterior divisions** of S2, S3, and S4.
3. **Superior Gluteal Nerve**. This important nerve (Fig. 13-5) of the gluteal region comes from the **posterior divisions** of L4, L5, and S1. It leaves the pelvis through the greater sciatic foramen to supply the gluteus medius, gluteus minimus, and tensor fascia latae muscles.
4. **Inferior Gluteal Nerve**. The inferior gluteal nerve arises from the **posterior** divisions of L5, S1, and S2. It passes through the greater sciatic notch to supply the gluteus maximus muscle.
5. **Other branches** of the sacral plexus include the nerve to piriformis, posterior divisions of S1, S2, and S3; branches to the levator ani and external anal sphincter, ventral divisions of S3 and S4; the nerve to quadratus femoris, anterior divisions of L4, L5, and S1; the **nerve to obturator internus**, anterior divisions of L5, S1, and S2, which leaves the pelvis through the greater sciatic notch and reenters through the lesser sciatic notch, lying medial to the pudendal vessels and nerve as it crosses the ischial spine; and the posterior cutaneous nerve of the thigh (S1, S2, and S3), to be described with the lower limb.

Many of the above-mentioned nerves have cutaneous branches, which will be detailed in the description of the lower limb.

Obturator Nerve. The obturator nerve comes from the lumbar plexus, arising from the **anterior** divisions of L3, L4, and L5. It emerges from the medial border of the psoas major muscle and runs across the brim of the pelvis to the anterolateral wall of the pelvis. It then passes through the obturator foramen to supply the adductor muscles and some of the skin of the thigh.

The **coccygeal plexus** is a small network of fibers from the last sacral and coccygeal nerves. It supplies the coccygeus muscle and a small area of skin in the coccygeal area.

AUTONOMIC NERVOUS SYSTEM IN THE PELVIS

The **sympathetic components** consist of the sacral portions of the sympathetic trunks and gray rami communicantes passing from the sympathetic ganglia to the spinal nerves. The hypogastric nerves and branches from the hypogastric plexus surround arteries supplying pelvic viscera.

The two **sympathetic trunks** pass anterior to the body of the sacrum medial to the anterior sacral foramina (see Fig. 13-4) to unite at their termination in the **ganglion impar** on the anterior surface of the fourth segment of the sacrum. A variable number of ganglia occur on the sacral part of the sympathetic chain. The sacral ganglia supply gray rami to the sacral nerves and **sacral splanchnic branches** to the hypogastric plexus (see Fig. 12-2A).

The **parasympathetic nerves** of the pelvis are the **pelvic splanchnic nerves** (nervi erigentes), which arise from the second, third, and fourth sacral segments of the spinal cord. The pelvic splanchnic nerves carry the preganglionic sacral parasympathetic outflow to the pelvic viscera. They synapse in the walls of the viscera (see Fig. 12-2B).

Parasympathetic nerves are distributed to all the pelvic viscera, and three groups of pelvic splanchnic nerves are worthy of more detailed description.

1. The parasympathetic fibers to the bladder control bladder function, stimulating contraction of the detrusor muscle and relaxation of the internal sphincter urethrae.
2. The parasympathetic nerves to the **distal transverse colon, descending colon, and sigmoid colon** have a long oblique retroperitoneal path to reach their destinations.
3. Branches from the pelvic splanchnic nerves supply preganglionic efferent fibers to the male and female internal sex organs and to the erectile tissues of the external genitalia in both sexes.

ARTERIES OF THE PELVIS

Four different arteries enter the pelvis; the first is a single vessel, and the next three are bilateral structures:

1. **Median Sacral Artery.** This small artery is the continuation of the abdominal aorta into the pelvis. It runs anterior to the body of the sacrum, covered by a layer of pelvic fascia (Waldeyer's fascia) (see Fig. 12-4).
   Note: Waldeyer's fascia is of surgical significance in resections of the rectum using the perineal approach.
2. **Superior Rectal Artery.** The superior rectal artery is a branch of the inferior mesenteric artery. It supplies the rectum as far as the anal canal (see p. 139).
3. **Ovarian Artery.** The ovarian artery (like the testicular artery) is a branch of the abdominal aorta. It arises just below the renal arteries to descend into the pelvis medial to the ureter. In the pelvis it is crossed by the terminal portion of the ureter.

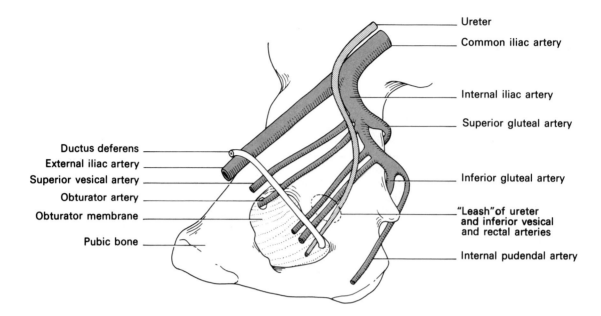

Ureter
Common iliac artery
Internal iliac artery
Superior gluteal artery
Inferior gluteal artery
"Leash" of ureter
and inferior vesical
and rectal arteries
Internal pudendal artery

Ductus deferens
External iliac artery
Superior vesical artery
Obturator artery
Obturator membrane
Pubic bone

Figure 13-6.
Right lateral wall of the pelvis in the male. The term "leash" for the bundle including ureter and vessels was suggested by J. C. B. Grant.

4. **Internal Iliac Artery**. This is one of the two terminal divisions of the common iliac artery and provides the major blood supply to the viscera of the pelvis, as well as supplying the musculoskeletal part of the pelvis and the gluteal region of the lower limb.

   The internal iliac artery arises from the division of the common iliac artery at the level of the sacroiliac joint and passes inferiorly, and somewhat posteriorly, on the wall of the pelvis (Fig. 13-6). The ureter maintains an important anteroinferior relationship to the internal iliac artery.

BRANCHES OF THE INTERNAL ILIAC ARTERY (Fig. 13-6)

1. The **obturator artery** runs inferior to the obturator nerve to leave the pelvis through the obturator foramen and supply muscles on the medial aspect of the thigh. A pubic branch anastomoses with the inferior epigastric branch of the external iliac artery; this may be quite a large anastomosis, the **abnormal obturator artery**. This artery may pass lateral or medial to the femoral ring. If it lies on the medial aspect of the femoral ring it may be in danger of being cut if the lacunar ligament (see p. 95) has to be cut to relieve a strangulated femoral hernia.
2. The **umbilical artery** runs superior to the obturator nerve and ends as the medial umbilical ligament (obliterated part of the fetal umbilical artery). Before its termination the umbilical artery gives off the **superior vesical artery**, which supplies the urinary bladder.
3. The **inferior vesical artery** occurs only in the male. It passes to the base of the bladder and supplies the seminal vesicles, prostate, and ductus deferens.
4. The **vaginal artery** is the female homologue of the inferior vesical artery. It supplies the base of the bladder and the vagina.

5. The ***uterine artery*** crosses the floor of the pelvis in the root of the broad ligament and passes superior to the lateral fornix of the vagina to reach the uterus (see Fig. 14-12). It is separated from the vagina by the ureter; in this situation the ureter is in danger of being cut or ligated during a hysterectomy. The uterine artery then passes superiorly along the lateral aspect of the uterus and ends in the upper part of the broad ligament by anastomosing with a branch of the ovarian artery.

6. The ***middle rectal artery*** passes in the ***lateral ligament*** of the rectum to the inferior part of the rectum and the prostate (or vagina).

    Note: The inferior vesical (or vaginal) artery, the middle rectal artery, the uterine artery, and the ureter all run across the floor of the pelvis in a so-called ***leash***. The region of the base of the bladder, the rectum, and the vagina (or prostate) are served by vessels from this leash rather than by individual arteries to specific viscera.

7. The ***internal pudendal artery*** passes inferior to the piriformis muscle, out through the greater sciatic notch, over the ischial spine (or sacrospinous ligament), and then through the lesser sciatic notch to enter the perineum.

8. The ***superior*** and ***inferior*** gluteal arteries pass out of the pelvis through the greater sciatic foramen to supply the gluteal muscles.

VEINS OF THE PELVIS

The ***internal iliac vein*** joins the ***external iliac vein*** to form the ***common iliac vein***, which in turn joins with its counterpart of the opposite side to form the ***inferior vena cava*** at the level of the fifth lumbar vertebra (see Fig. 12-5). The various branches of the internal iliac artery are accompanied by venous plexuses, which come together to form the internal iliac vein. These plexuses are the rectal, vesical, uterine, prostatic, and vaginal plexuses. All of these plexuses may intercommunicate. These venous plexuses also communicate with the vertebral plexus of veins around the bodies of the vertebrae. This communication may provide direct channels for the spread of malignant cells from pelvic viscera to the vertebral bodies, spinal cord, and brain.

LYMPHATIC VESSELS OF THE PELVIS

The lymph vessels of the pelvis accompany the arteries, and there are chains of external, internal, and common iliac nodes as well as a few sacral nodes. The common iliac nodes drain into the lumbar nodes (see p. 14 and Fig. 1-15).

The Pelvis: Differences Between Male and Female Bony Pelvis

The following list is meant for reference, not memory. It attempts to point out the salient differences between the pelves of the two sexes. There are many sex-determined differences between most bones of the skeleton, depending on such variables as age of fusion of epiphyses, body size, and muscularity. The female pelvis differs from the male pelvis in a functional sense. The female pelvis requires certain parameters of shape and size to permit the full-term fetus to negotiate the ***birth canal*** in a safe, easy, and mechanically effective manner. When all is said and done, the important question is simply, "Can a baby's head get through this pelvis (birth canal)?"

Many alterations in the form of the pelvis are possible, and females frequently have a type of pelvis that is rather like that of a male. This is an ***android*** pelvis. The more typically female pelvis is called ***gynecoid***. If the size or shape of the female pelvis is inadequate for the natural birth of a normal-sized fetus, a state of ***disproportion*** is said to exist. An android pelvis may be adequate for parturition, whereas a small gynecoid pelvis can cause difficulties.

14

Male (Fig. 14-1A)
The pelvis is a long section of a short cone.
The upper part of the sacrum is curved.
The pubic arch is less than a right angle.
Pubic symphysis is relatively deep.
The ischial spines turn inward.
The greater sciatic notch is narrow.
The inlet is heart shaped.
The margins of the pubic rami are everted by the crura (roots) of the penis.
Distance between pubic tubercles is relatively less.

Female (Fig. 14-1B)
The pelvis is a short section of a long cone.
The sacrum is shorter and the promontory less pronounced; the lower portion of the sacrum is more curved.
The pubic arch forms nearly a right angle. Pubic symphysis is less deep.
Ischial spines turn inward less than in the male.
Greater sciatic notch is wider.
Inlet is larger and more circular.
Pelvis is less rugged.
Pubic rami are less everted.

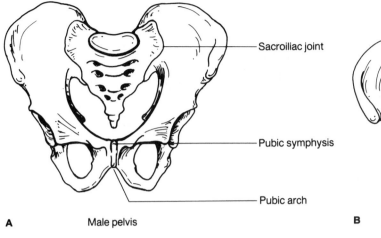

Sacroiliac joint

Pubic symphysis

Pubic arch

A Male pelvis

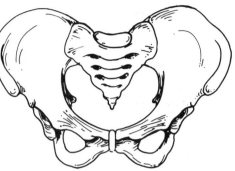

B Female pelvis

Figure 14-1.
Male and female pelves.

Ischial tuberosities are not everrted.
Ilia are less vertical.
Iliac fossae are deep.
There is more curve to the iliac crest.
Anterior superior spines are closer together.

Distance between pubic tubercles is relatively greater.
The ischial tuberosities are everted.
Ilia are more vertical.
Iliac fossae are shallow.
Iliac crest is less curved, as seen from above.
Anterior superior spines are further apart.

PELVIC VISCERA

The term *pelvic viscera* is confined to structures that are rooted in the pelvis, as opposed to structures that may move in and out of the pelvis, such as small intestine and omentum (Figs. 14-2A and 14-2B).

The *floor* of the pelvis consists of the two levatores ani muscles, which were described on page 178. The floor is deficient anteriorly where the vagina (or the prostate) is surrounded by the medial margins of the levatores ani. The pelvis contains considerable accumulations of loose are-olar tissue, which acts as the padding that holds the viscera in position. Some of the pelvic viscera are covered by peritoneum.

The *pelvic colon, rectum, urinary bladder, ureters, blood vessels, lymphatics,* and *nerves of the pelvis* are basically common to both sexes. The pelvic viscera common to both sexes will be described with the male pelvis.

The specific male viscera are the *prostate gland, prostatic urethra, ductus deferens, seminal vesicles,* and the *ejaculatory ducts*.

The special female structures are the *uterus, round ligament, broad ligament, ovaries, uterine tubes, vagina,* and the *female urethra* (see Figs. 14-2B, 14-13).

Pelvic Colon

The pelvic colon is the continuation of the descending colon, and the name change occurs at the brim of the pelvis. The pelvic colon has a V-

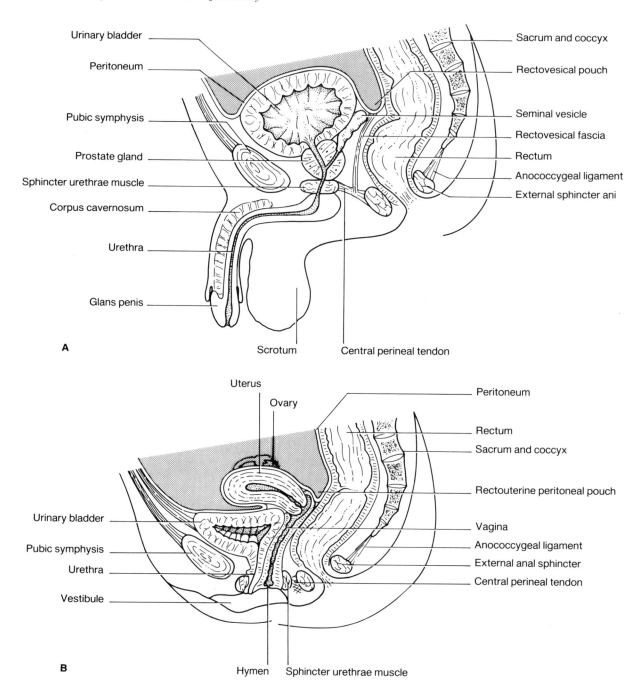

Figure 14-2.
Median section of the pelvis. **A.** Male
(with bladder partially filled).
B. Female (with bladder empty).

shaped mesentery, the ***pelvic (sigmoid) mesocolon***. This mesentery ends
at the level of the third segment of the sacrum, and at this point the pelvic
colon becomes the rectum.

Rectum

The ***rectum*** (Fig. 14-3A) is about 12 cm long, beginning at the third segment of the sacrum, following the curve of the sacrum and coccyx to about
2 cm below the tip of the coccyx, where it turns posteriorly to become the

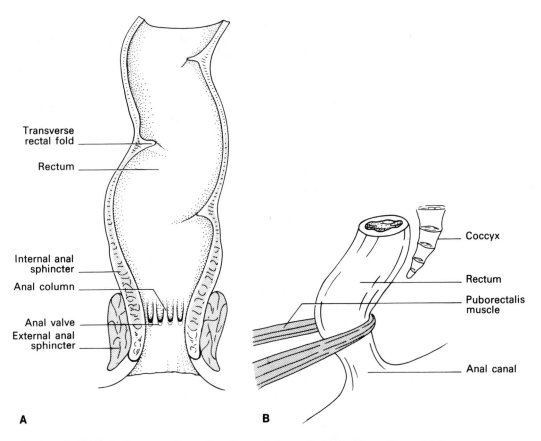

A   **B**

Figure 14-3.
Rectum and anal canal. **A.** Coronal section. **B.** Lateral view showing puborectalis muscles.

anal canal. Note that this is the area where the sling of the ***puborectalis*** (Fig. 14-3B) is located.

The rectum has three gentle ***flexures***; the upper and lower are concave to the left, and the middle one is concave to the right. The upper third of the rectum is covered by peritoneum in front and on each side; the middle third has peritoneum only in front; and the lower third has no peritoneal covering.

Arterial Supply.   The ***superior rectal artery*** (the terminal branch of the inferior mesenteric artery) supplies the lower end of the sigmoid colon and the upper part of the rectum.

The ***middle rectal artery*** is a small branch of the internal iliac artery and runs in a condensation of pelvic connective tissue, the ***lateral rectal ligament***, to supply the middle and lower parts of the rectum and upper anal canal.

CLINICAL NOTE

In excision of the rectum the lateral rectal ligaments have to be divided. They bleed very little, and a special effort to ligate the middle rectal artery may not be needed.

Rectovesical Pouch (Rectouterine Pouch in the Female).   The rectovesical pouch (see Figs. 14-2A and 14-2B) is that part of the peritoneal cavity where the peritoneum is reflected from the junction of the middle and third parts of the rectum onto the bladder in the male (or vagina and uterus in the female).

Interior of the Rectum. At the three flexures of the rectum there are folds of mucous membrane. These are the ***transverse rectal folds (valves of Houston)*** (see Fig. 14-3A).

CLINICAL NOTE

The lumen of the rectum and lower sigmoid colon can be inspected by a rigid instrument, the **sigmoidoscope**. *The transverse rectal folds may make passage of the instrument difficult. The invention of the fiberoptic flexible sigmoidoscope and colonoscope has made direct inspection of the rectal and colonic lumina much more efficient. These are instruments for specialists. The family physician must be capable of inspecting the anal canal and lower rectum by means of a* **proctoscope**; *he must remember the posterior angulation of the anal canal on the rectum in order to carry out this examination safely and with minimal discomfort to the patient.*

Anal Canal

The anal canal is the terminal portion of the digestive tract. It begins below the lower dilated end of the rectum (the ***rectal ampulla***) and is about 4 cm long in the male adult. The anterior wall is slightly shorter than the posterior wall. In the empty state its lumen presents a triradiate slit. Posteriorly it is in contact with the anococcygeal ligament (raphe) (see Fig. 15-4). Its sides are surrounded and padded by the fatty tissue of the ischiorectal fossa (see Fig. 15-5).

The upper half of the anal canal is lined by stratified columnar epithelium and is sensitive only to pressure and distension; the lower half is lined by stratified squamous epithelium (without hair follicles, sweat glands, or sebaceous glands), which is supplied by the inferior rectal nerve (S2 and S3) and the perineal branch of S4; it is ***exquisitely sensitive*** to ***touch*** and ***pain***.

The mucosa of the upper half is plum colored in living persons owing to the submucosal plexus of veins, which are tributaries of the superior rectal veins. The mucosa of the upper half of the anal canal has six to ten folds, the ***anal columns***. The lower ends of the anal columns are joined by small transverse folds of mucous membrane called ***anal valves***. The line of attachment of the anal valves is called the ***pectinate line***. The

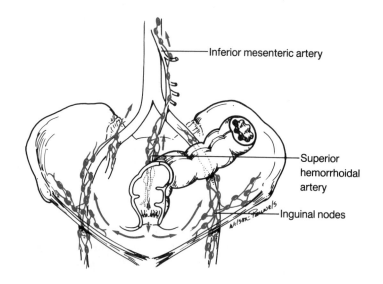

Inferior mesenteric artery

Superior hemorrhoidal artery

Inguinal nodes

Figure 14-4.
Lymphatic drainage of rectum, anal canal, and anus.

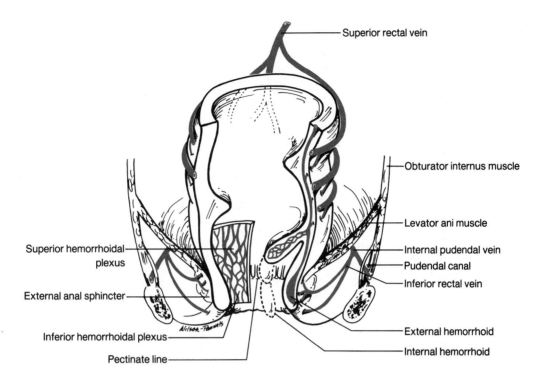

Superior rectal vein

Obturator internus muscle

Levator ani muscle

Internal pudendal vein

Pudendal canal

Inferior rectal vein

Superior hemorrhoidal plexus

External anal sphincter

Inferior hemorrhoidal plexus

Pectinate line

External hemorrhoid

Internal hemorrhoid

Figure 14-5.
Venous drainage of the rectum and anus showing internal and external hemorroids.

spaces formed by the lower ends of the anal columns and the anal valves are called ***anal sinuses***. The mucosa of the anal canal above the pectinate line contains mucus secreting glands, which are particularly numerous in the anal sinuses and which may penetrate beyond the submucosa into and through the internal sphincter muscle layer of the anal canal. These glands may be the site of abscess formation.

The muscles of the anal canal consist of the external sphincter, which is made of striated muscle and is continuous with the fibers of levator ani above, and the internal sphincter, which is a thickened circular layer of smooth muscle fibers continuous with the circular muscle layer of the rectum. For details of the anal musculature, see Chapter 15.

CLINICAL NOTE: LYMPHATIC DRAINAGE *(Fig. 14-4)*

*The upper half of the anal canal drains into lymph vessels that run along the superior rectal vessels and ultimately drain into the inferior mesenteric nodes. The lower half of the anal canal drains into the **inguinal nodes**.*

*   ***Hemorrhoids** (Fig. 14-5). The superior rectal artery divides into terminal branches to supply the anal canal (one branch corresponding to each anal column), and there are three particularly prominent arteries, located on the left side, the right posterior side, and the right anterior side of the anal canal. These arteries are accompanied by tributaries of the superior rectal vein (superior hemorrhoidal vein). The disposition of the veins resembles that of the arteries. These superior rectal veins anastomose with the inferior rectal veins draining into the internal pudendal veins. Thus, here we have a **portosystemic anastomosis**. In portal hypertension there may be distension of the superior hemorrhoidal veins, a condition referred to as **hemorrhoids**, or **piles**. However, by far the most common cause of hemorrhoids is **repetitive straining during defecation**.*

This is usually caused by persistent constipation but may also be the result of repeated bouts of diarrhea or of inflammation of the anorectum, called **proctitis**. *The superior hemorrhoidal veins form a vascular network within a connective tissue stroma in the submucosal layer of the anal canal. In the three positions already mentioned these networks are most prominent and form so-called cushions that can be seen in the normal anal canal on proctoscopy and that probably aid in the complete closure of the anal canal.*

Inspection and palpation of the anal margin in living persons will reveal an outer ridge-like ring, the lower border of the external sphincter, then a narrow groove (the intersphincteric groove), and next another ring, the lower border of the internal sphincter. The pedicles of internal hemorrhoids are within the inner ring; external hemorrhoids are in the perianal space.

The **perianal space** *is the area between the perianal skin (below), the lower border of the internal sphincter (above), the lower anal mucosa (medially), and the most lateral of the fibromuscular strands (continuous with the longitudinal muscle of the rectum) which intersperse the subcutaneous part of the external sphincter and are inserted into the perianal skin.*

Passage of a hard stool may tear one of the anal valves and anal skin below it. This is similar to a "hangnail"; this condition is called an **anal fissure** *and can be extremely painful. The tear exposes the internal sphincter, which goes into painful spasm. The condition is usually easily cured by simple division of the internal sphincter.*

The anal glands may become infected and produce abscesses. If the abscess is low lying it will be a perianal abscess. If a gland at a higher level is infected the pus may spread into the ischiorectal fossa and produce an ischiorectal abscess.

Rectal Examination. *"The physician who does not put his finger in the rectum will eventually put his foot in it"* is an old medical saying. Apart from the mouth and pharynx, the anal canal and lower rectum are the only parts of the digestive tract accessible to digital examination.

When it is realized that some 75% of cancers of the large bowel occur in the part of the rectum that can be palpated, the truth of the above statement becomes apparent.

By rectal examination one can palpate the lumen of the anal canal and lower rectum as well as structures surrounding the rectum. The sling of the puborectalis can be felt as a firm ridge at the anorectal junction posteriorly (see Fig. 14-3B). The following extraneous structures can be palpated: the prostate gland, the bulb of the penis, the cervix of the uterus, the base of the bladder (if the bladder is full), the sacrum and coccyx, the ischial tuberosities, the conjoint rami of the pubis and ischium, and any structures of sufficient firmness lying in the rectovesical or rectovaginal pouch (**pouch of Douglas**).

Urinary Bladder

There is no difference between a male and a female bladder. The differences occur in the arrangement of the urethras and in relationships to other pelvic structures.

The male urinary bladder rests against the pubis separated from it by the **retropubic space** (cave of Retzius) (see Fig. 14-2A). Its superior aspect and the upper part of its base (posterior aspect) are covered by perito-

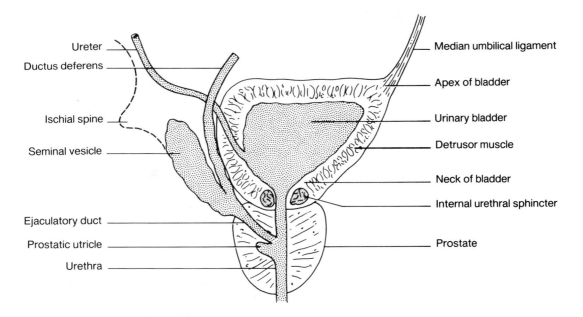

Ureter

Ductus deferens

Ischial spine

Seminal vesicle

Ejaculatory duct

Prostatic utricle

Urethra

Median umbilical ligament

Apex of bladder

Urinary bladder

Detrusor muscle

Neck of bladder

Internal urethral sphincter

Prostate

Figure 14-6.
Schematic sagittal section of the urinary bladder and internal genitalia in males.

neum, and in its empty state the bladder lies totally within the pelvis. Note that the bladder of the infant and young child is nearly an abdominal viscus because of the smaller size of the pelvis. As the adult bladder fills it lifts the peritoneum off the anterior abdominal wall, and in the full state the anterior wall of the bladder is in contact with the lower part of the abdominal wall without the peritoneum intervening. It is possible to drain the full, obstructed bladder by means of a ***suprapubic catheter*** without entering the peritoneal cavity.

The bladder is shaped like a triangular pyramid with the ***base*** facing posteriorly (Fig. 14-6). Two of its walls face anterolaterally, and these are in contact with endopelvic fascia covering the levator ani muscles. The ***apex*** of the bladder points against the pubic symphysis (see Figs. 14-2A and 14-2B). The ***neck*** of the bladder is the inferior part and rests on the ***prostate gland*** (see Fig. 14-2A). The lumina of the bladder and urethra become continuous at the ***bladder neck***. The remnant of the ***urachus***, the ***median umbilical ligament***, passes from the apex of the bladder to the umbilicus.

Wall of the Bladder. The wall of the bladder is composed of smooth muscle, the ***detrusor muscle***. The muscle fibers are arranged in many layers running in different directions but converge on the bladder neck to form the ***internal urethral sphincter*** (Fig. 14-6).

Interior of the Bladder. The interior of the bladder is lined by transitional epithelium on a lax submucosa. This causes the interior of the empty bladder to appear wrinkled, but as the bladder fills these folds are smoothed out.

The ***trigone*** (Fig. 14-7) is found at the base of the bladder and has a totally different appearance. The mucosa is not wrinkled and is paler in color than that of the remaining bladder wall. The trigone is triangular in shape with the apex of the triangle being pointed toward the urethral opening at the bladder neck. The other two corners show the openings of the ureters. The ureters enter the bladder obliquely, and this obliquity creates a valve effect that prevents reflux of urine into the ureters. Connecting the two ureteral orifices, and forming the base of the trigone, is the

Figure 14-7.
The trigone of the bladder.

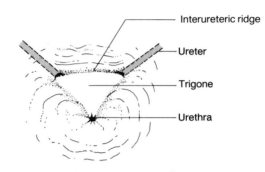

interureteric ridge

Ureter

Trigone

Urethra

interureteric ridge—the ridge formed by a thickening in the underlying muscle.

The ***internal urethral orifice*** lies immediately superior to the prostate gland and is surrounded by the musculature of the bladder neck, the ***sphincter vesicae*** (internal urethral sphincter).

Micturition (Urination). Basically the act of micturition is one of contraction of the detrusor muscle with simultaneous relaxation of the sphincter of the bladder (internal urethral sphincter) and relaxation of the external urethral sphincter. Changes in bladder position and urethral alignment occur with probable relaxation of the levatores ani muscles, allowing the neck of the bladder to descend and the base of the bladder to assume a funnel shape. This change in configuration allows some urine to enter the proximal urethra, which sets up a strong reflex that causes the detrusor to contract to its fullest ability with simultaneous relaxation of the external urethral sphincter. At the end of micturition the levatores ani contract and everything returns to its original configuration.

CLINICAL NOTE

*Inflammation of the bladder mucosa causes a desire to micturate frequently, and the act may be quite painful. Certain chronic inflammations of the bladder will cause thickening of the wall so that the bladder loses capacity because its walls cannot expand. This will cause further **frequency of micturition**. Obstructions at the bladder neck, whether from prostatic enlargement or other causes, will slow the urinary stream and may cause **retention of urine**. If the obstruction is incomplete the bladder never empties completely, and frequent micturition may occur when the fullness reaches a maximum, only the **overflow** amount being expelled.*

Arteries of the Bladder. The arterial blood supply to the bladder is from two sources: the superior vesical artery from the umbilical artery, and the ***inferior vesical artery*** from the internal iliac artery (see Fig. 13-6).

Nerve Supply of the Bladder. Fibers of the pelvic splanchnic nerves (parasympathetic) (see Fig. 12-2B) synapse in the wall of the bladder and are motor nerves to the detrusor muscle and inhibitory to the internal sphincter.

The ***sympathetic*** nerve supply to the bladder is derived from the lower two thoracic and the upper two lumbar nerves by means of the ***hypogastric*** nerves. They probably inhibit detrusor contraction and promote muscle tone at the internal sphincter. This has been disputed, and it may be that these sympathetic fibers are merely vasomotor to the bladder.

Sensations of pain and of bladder fullness are transmitted by both parasympathetic and sympathetic fibers, with the parasympathetic path-

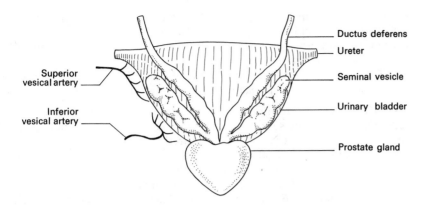

Figure 14-8.
Posterior view of base of the bladder
and adjacent organs.

ways predominating as far as pain is concerned. The pain of an invading
cancer of the bladder cannot be relieved by division of the hypogastric
nerves.

CLINICAL NOTE

*The distended bladder may be injured by lower abdominal trauma or by
fractures of the pelvis. The rupture is usually extraperitoneal with extravasation of urine into the pelvic and anterior abdominal wall connective
tissue. If the tear involves the superior surface of the bladder, urine will
escape into the peritoneal cavity.*

*Obstruction to the outflow of urine (e.g., from prostatic enlarge-
ment) can cause hypertrophy of the detrusor muscle, which then presents
fasciculi of thickened muscle running in all directions. This is called a
trabeculated bladder; the bladder mucosa may pouch out between the
trabeculae, forming one or more diverticula which, having no muscle
wall of their own, will fail to empty during micturition.*

*Chronic increase in intravesical pressure may also cause back pres-
sure in the ureters, which may distend and pass the pressure on to the
renal pelves, resulting in **hydroureter** and **hydronephrosis**.*

*The interior of the bladder may be inspected by means of a **cystoscope**, and the ureters can be catheterized (e.g., for the injection of radiopaque dyes) through the cystoscope.*

Prostate Gland

The prostate gland (Figs. 14-8 and 14-9) is about the size of a walnut and
is somewhat cone-shaped, with the **apex** pointing downward and resting
on the (external) sphincter urethrae muscle. The **base** of the cone rests

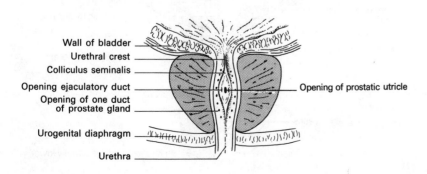

Figure 14-9.
Coronal section of the prostatic urethra
showing its posterior wall.

against the bladder neck. The prostate lies in its own sheath of endopelvic fascia and is surrounded by a ***prostatic plexus*** of veins. With this fascial covering and its venous plexus the prostate is cradled between the medial edges of the levator ani muscles.

The prostate is composed of glandular tissue and fibromuscular tissue. The secretions of the prostatic glands are added to the seminal fluid.

Prostatic Urethra. The prostatic urethra begins at the trigone of the bladder and runs to the superior layer of deep fascia of the sphincter urethrae muscle. The interior of the prostatic urethra is lined by mucous membrane and exhibits certain striking features. These are best seen in a coronal section (Fig. 14-9). The ***urethral crest*** is a central vertical ridge on the posterior wall of the prostatic urethra. Its central part is expanded to form the ***colliculus seminalis*** (verumontanum). The ***prostatic utricle***, which may be the homologue of the uterus and upper vagina, is seen as a vertical slit in the center of the colliculus seminalis. The openings of the ***ejaculatory ducts*** are found on the colliculus as small slits on each side of the opening of the prostatic utricle. The ***prostatic sinuses*** are grooves on either side of the urethral crest. The ducts of the multiple prostatic glands drain into these prostatic sinuses.

Ductus Deferens (Vas Deferens) (Fig. 14-10). The ductus deferens is the male duct carrying sperms from the testis to the ejaculatory duct. It passes from the testis through the inguinal canal and enters the abdomen (retroperitoneally) through the internal ring. It crosses the brim of the pelvis and then crosses in front of and above the ureter on its way toward the prostate gland. As it approaches the prostate it passes above the upper pole of the ***seminal vesicle*** and then medial to the body of the seminal vesicle to join with the duct of the seminal vesicle to form the ***ejaculatory duct***. Opposite the base of the bladder the ductus deferens exhibits a dilatation called the ***ampulla***.

Seminal Vesicle (see Fig. 14-10). The seminal vesicle is about 5 cm long and 1 cm wide and is attached to the base of the bladder by a fascial sheath. It lies on the posterior surface of the bladder and extends upward and laterally. Its duct, located at its lower end, joins with the ductus deferens

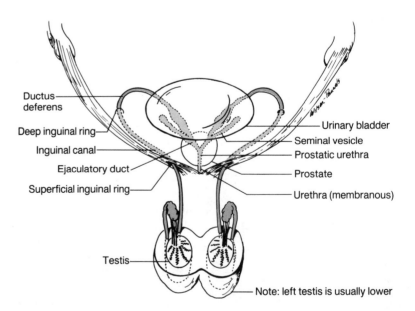

Figure 14-10.
Course of the ductus deferens.

to form the ejaculatory duct. The seminal vesicle is a long, coiled, tube-like glandular structure that secretes part of the seminal fluid. Its name is a misnomer: it does not store sperm. The upper pole of the seminal vesicle, which reaches up to the level of the ischial spine, can sometimes be felt on rectal examination.

Ejaculatory Duct. The ejaculatory duct is formed by the junction of the ductus deferens and the duct of the seminal vesicle at the base of the bladder. It enters the prostate and opens into the prostatic urethra lateral to the prostatic utricle (see Fig. 14-10).

Rectovesical Fascia

The *rectovesical (Denonvilliers') fascia* is a fascial septum that passes from the perineal body to the peritoneum of the floor of the rectovesical pouch; it is the fused remnant of the peritoneal layers of an original, much deeper rectovesical pouch (see Fig. 14-2A). It lies between the anal canal and rectum and the prostate gland and seminal vesicles. The rectovesical fascia is an important landmark in surgical removal of the rectum and anal canal; it provides a plane of dissection which allows separation of the original fused layers of peritoneum without injuring the anorectum or the capsule of the prostate and its venous plexus.

Ureter

In males, the ureter crosses the external iliac artery at its origin (see Fig. 13-6). It runs anteroinferior to the internal iliac vessels to the region of the ischial spine, where it curves forward, superior to the levator ani muscle, to reach the posterosuperior angle of the bladder. In the pelvis, as in the abdomen, the ureter is extraperitoneal, and the only structure that passes between the ureter and the pelvic peritoneum is the terminal end of the ductus deferens.

PELVIC VISCERA: FEMALE (see Fig. 14-2B)

The urinary bladder and the ureter are similar to those of the male. The bladder is a triangular pyramid with its apex at the upper level of the pubic symphysis. The base or posterior wall is separated from the rectum by the cervix of the uterus and the upper part of the vagina. There being no prostate gland in the female, the urethra leaves the neck of the bladder to pass into the fascial tissue covering the anterior wall of the vagina.

The female urethra is only 3 to 4 cm long. Its upper end is surrounded by a few fibers of the sphincter urethrae muscle. The rest of the urethra is embedded in the anterior wall of the vagina and empties into the vestibule. The female urethra is short and straight and easily catheterized.

The internal female genitalia consist of the uterus, vagina, uterine tubes, ovaries, and the ligaments associated with these structures.

Vagina

The *vagina* is the lower part of the female genital tract (see Figs. 14-2B and 14-11) and consists of a muscular tube about 7.5 cm long. It opens into the vestibule inferiorly. It lies between the bladder in front and the rectum behind and is embraced by the medial margins of the levatores ani.

The *cervix* (neck) of the uterus projects into the anterior wall of the vagina near its upper extremity. The cervix protruding into the vaginal

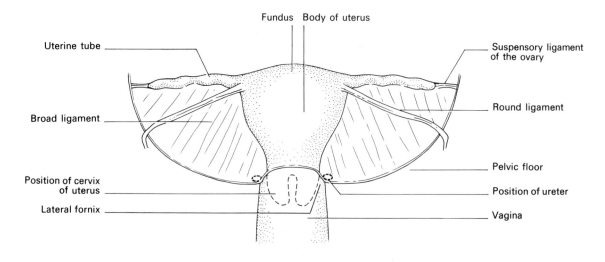

Fundus Body of uterus

Uterine tube

Suspensory ligament
of the ovary

Broad ligament

Round ligament

Pelvic floor

Position of cervix
of uterus

Position of ureter

Lateral fornix

Vagina

Figure 14-11.
Anteroinferior view of the uterus and
broad ligament.

cavity forms recesses anterior, posterior, and lateral to the cervix; these
four recesses are called *fornices* (singular, fornix; fornix=arch).

The upper limit of the posterior wall of the vagina (that part of the
vaginal wall forming the posterior fornix) is covered by peritoneum, which
is then reflected onto the anterior surface of the rectum. The pouch so
formed is the *rectouterine pouch or pouch of Douglas*.

CLINICAL NOTE

*It is possible to pass an instrument through the posterior fornix into the
peritoneal cavity to drain fluid (e.g., pus) or to visualize the pelvis
through a* **colposcope**. *This procedure is called* **posterior colpotomy (or
colposcopy)**; *if a surgical instrument* **accidentally** *penetrates a vaginal
fornix in the course of a vaginal operation it is called* **malpractice**.

Relations of the Vagina. Anteriorly the vagina is related to the bladder
and the urethra. Laterally the upper margin of the vagina is attached to the
base of the broad ligament; posteriorly the vagina is in contact with the
rectum, separated from it only by a fascial septum.

Hymen. The hymen is a fold of skin across the lower margin of the vaginal opening. After the start of sexual activity only a few nodules on the
vaginal wall remain as the remnants of the hymen. These are called the
carunculae hymenales and are of no practical significance.

CLINICAL NOTE

*The hymen may be sufficiently complete so as to require surgical incision
before intercourse can be accomplished. Rarely the hymen will be a complete septum, and when the pubescent female menstruates for the first
time there will be no exit for the bloody discharge. Accumulation of this
material in the vagina and uterus (through several menstrual cycles) will
eventually cause pressure symptoms in the lower abdomen. Micturition
may be impaired because of the upward displacement of the bladder and
stretching of the urethra which this vaginal distension can cause. This
clinical condition is called* **hematocolpos**, *caused by an* **imperforate
hymen**.

Uterus

The uterus (Figs. 14-11 and 14-12) is approximately 7 cm long, 7 cm wide, and 2 cm thick (from back to front). It consists of four parts: cervix, isthmus, body, and fundus. The body is angled forward on the cervix (***anteflexed***), and the uterus as a whole is tipped forward from the vertical (***anteverted***) (see Fig. 14-2B). The superolateral parts of the uterus (where the fundus meets the body) are continuous with the ***uterine (fallopian) tubes***.

The ***cervix*** is the lower end of the uterus and projects into the vagina through the upper end of its anterior wall. The highest, rounded part of the uterus is the ***fundus***. The fundus of the nonpregnant uterus lies above the apex of the bladder behind the upper part of the pubic symphysis and is not palpable by abdominal examination. In pregnancy the fundus will rise as high as the epigastrium.

Peritoneal Covering. The peritoneum covers the uterus anteriorly and posteriorly except for the intravaginal part of the cervix. It reflects forward onto the bladder, and posteriorly at the level of the posterior fornix it reflects onto the rectum to form the rectouterine pouch. Laterally the peritoneum forms folds that extend from the uterus to the side walls of the pelvis. These folds are known as ***broad ligaments*** and contain the uterine tubes in their upper borders. At the base of the broad ligament the two peritoneal layers are separated by connective tissue, the ***parametrium***.

The wall of the uterus comprises three distinct layers:

1. ***Serous***. The serous coat consists of peritoneum and some connective tissue.
2. ***Muscular***. The muscular coat is called the ***myometrium*** and is a thick layer of smooth muscle. The density of this layer is so great that the uterine cavity is a mere slit in the resting state of the uterus. The myometrium proliferates enormously during

Figure 14-12.
Ovary and uterus (posterior view).

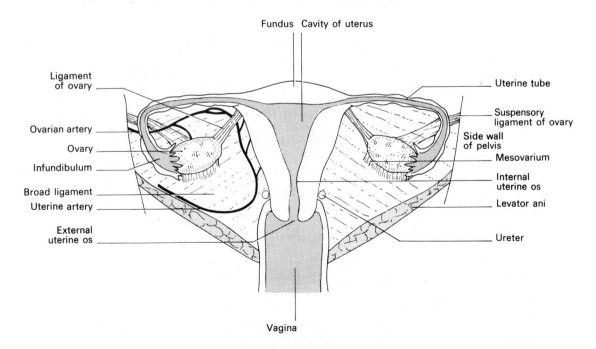

pregnancy so that it can accommodate the fetus and fetal membranes and their contents.

3. ***Mucosa***. The mucosa of the uterus is called the **endometrium**; it proliferates during the menstrual cycle and is then sloughed off to be discharged during menstruation.

Cavity. The cavity of the uterus (Fig. 14-12) is really a potential cavity since the endometrium converts it to a mere slit. It is relatively broad at the two horns, where it is continuous with the lumina of the uterine tubes. Below, the cavity is continuous with the ***internal uterine os*** at the upper end of the cervix. The cervix empties into the vagina at the ***external uterine os***.

Uterine Arteries. The uterine arteries (Fig. 14-12) are important branches of the internal iliac arteries which cross the floor of the pelvis around the lateral fornices of the vagina to reach the uterus. They run up the wall of the uterus and anastomose with branches of the ovarian arteries. On their way to the bladder the ***ureters*** cross ***inferior*** to the uterine arteries and in proximity to them and the lateral fornices. The ureters are at great risk during surgical procedures on the uterus and ovaries, and the prudent surgeon will always identify them before ligating or cutting any tissue in this region of the pelvis (Fig. 14-12).

Uterine Tubes

The uterine tubes (see Figs. 14-11 and 14-12) are about 10 cm long and run from each lateral horn (cornu) of the uterus to the lateral wall of the pelvis in the upper border of the broad ligament. Just before reaching the lateral pelvic wall they turn back upon themselves to open into the peritoneal cavity. The open end of the uterine tube (the ***ostium***) surrounds a portion of the ovary. ***Note***: the ostium of the uterine tube is the only communication between the ***peritoneal cavity*** and the outside environment. The part of the broad ligament attached to the uterine tubes is called the ***mesosalpinx***. There are four parts to the uterine tube.

1. ***Uterine (intramural)***. The uterine part is the short portion of the tube within the wall of the uterus.
2. ***Isthmus***. The isthmus is a narrow part of the tube about 2.5 cm long, lateral to the uterine wall.
3. ***Ampulla***. The ampulla is the wider part of the tube extending from the isthmus to the infundibulum.
4. ***Infundibulum***. The infundibulum is the funnel-shaped end of the uterine tube. It has ragged edges or ***fimbria***. The fimbria contain continuations of the smooth muscle walls of the tubes and tend to "clutch" the ovary as they lurk in wait for the discharge of an ovum at ovulation.

At ovulation the ovum is picked up by the infundibulum and carried along the tube to the uterus. Fertilization by a sperm usually occurs in the uterine tube, but the fertilized ovum is not embedded until it reaches the uterine cavity.

CLINICAL NOTE

*If transit of the fertilized ovum down the uterine tube is delayed, the ovum can become embedded in the uterine tube. This is called a **tubal pregnancy** or **ectopic pregnancy**. Division of the ovum will occur, and the*

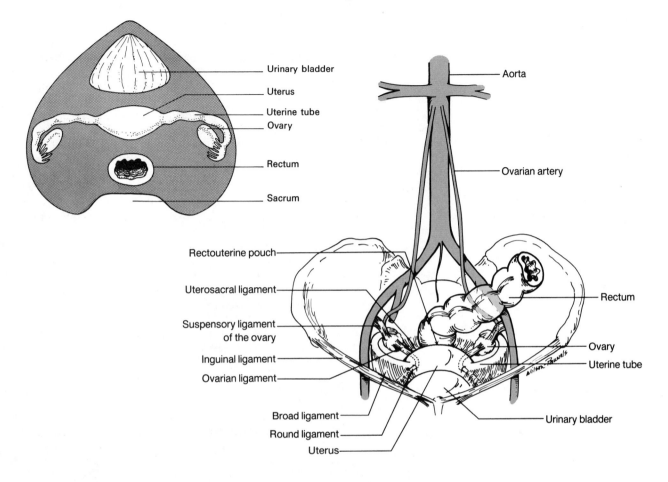

Urinary bladder
Uterus
Uterine tube
Ovary
Rectum
Sacrum

Aorta
Ovarian artery

Rectouterine pouch
Uterosacral ligament
Suspensory ligament of the ovary
Inguinal ligament
Ovarian ligament
Broad ligament
Round ligament
Uterus

Rectum
Ovary
Uterine tube
Urinary bladder

Figure 14-13.
Anterior view of pelvic viscera (female). Inset shows a superior view of pelvic viscera.

embryo and its highly vascular membranes will enlarge within the tube until the tube ruptures with a great deal of pain and blood loss. (This usually occurs around the 6th week of pregnancy.)

Female sterilization can be performed by ligation of the uterine tubes. This can be done operatively by opening the abdominal cavity or **endoscopically** *by inserting a* **laparoscope** *through a small subumbilical incision and crushing or burning the tube with an electrical current or by applying occluding rings. If subsequent pregnancy is desired the tubes can be reconstructed by microsurgery with varying degrees of success.*

Ovary

The ovary (Fig. 14-12) is about 3 cm long, 2 cm wide, and 1 cm thick. It is located on the posterosuperior surface of the broad ligament, but, because of its suspension from the broad ligament by the **mesovarium**, the ovary enjoys some mobility and variation of location in living persons. Because of the configuration of the broad ligament the ovary is usually found near the pelvic wall (Fig. 14-13).

The ovary presents an irregular surface with scars where ova have previously been shed into the peritoneal cavity. The surface of the ovary is not covered by peritoneum. The ovary and the interior of the infundibulum of the uterine tube are the only two structures in the peritoneal cavity that are not covered by peritoneum. At **ovulation** (approximately every 28 days) an ovum is shed into the peritoneal cavity and picked up by the infundibulum to be passed along the uterine tube to the uterine cavity.

The ovary is attached to the broad ligament by the ***mesovarium***, a fold of peritoneum forming a small mesentery for the ovary.

From the lateral pelvic wall, a thickening of fibrous tissue, the ***suspensory ligament*** of the ovary (infundibulopelvic ligament), runs to the infundibulum and the ovary. The ovarian artery and vein are located in the suspensory ligament. Like the testicular artery, the ovarian artery is a branch of the abdominal aorta arising just below the renal artery at the level of the second lumbar vertebra. The ovarian artery runs downward retroperitoneally to cross the external iliac artery at the brim of the pelvis and then enters the suspensory ligament (Fig. 14-13). The ovarian artery anastomoses with the uterine artery in the broad ligament.

The ***hilus*** of the ovary is the point at which the ovarian vessels enter and exit.

The ***ligament of the ovary*** is a band of fibrous tissue that runs in the broad ligament connecting the uterine pole of the ovary to the lateral aspect of the uterus. The ligament, if dissected out carefully, can be traced through the wall of the uterus to its anterior surface, where it becomes continuous with the ***round ligament*** and passes in the broad ligament to the internal ring of the inguinal canal. After passage through the inguinal canal the round ligament is attached to the subcutaneous tissue of the labium majus. The ligament of the ovary and the round ligament are the homologues of the ***gubernaculum*** of the testis.

Supports of the Uterus

The normal position of the uterus is maintained by certain structures and by the normal configuration of the other pelvic organs.

The uterus is normally found in the ***anteverted*** and ***anteflexed*** position. This means that it is angled forward on the upper end of the vagina and that it exhibits a forward angulation at the junction of body and neck (cervix). ***Retroversion*** and ***retroflexion*** may occur and can cause certain clinical problems.

Broad Ligament. The broad ligament (see Figs. 14-11, 14-12, and 14-13) is a fold of peritoneum attaching the lateral margins of the uterus to the lateral walls of the pelvis and to the pelvic floor. The upper free edges of the broad ligament contain the uterine tubes. The broad ligament also contains the ligament of the ovary and the round ligament. There is some doubt about the role of the broad ligament in maintaining the normal position of the uterus; it is more of a peritoneal reflection, that is, the mesentery of the uterus and tubes. The broad ligament gives attachment on its posterosuperior surface to the mesovarium.

Uterosacral Ligaments. Passing from the sacrum to the cervix, deep to the peritoneum but superior to the levatores ani, is a condensation of the pelvic fascia, the uterosacral ligament. The two uterosacral ligaments hold the cervix in its normal position relative to the sacrum and relative to the midline (Fig. 14-13).

Lateral Cervical Ligaments. The lateral cervical (cardinal) ligaments are condensations of pelvic fascia running from the lateral pelvic walls to the cervix.

Peritoneal Pouches in the Female Pelvis

The pelvic peritoneum passes from the rectum anteriorly over the posterior fornix of the vagina and then onto the body of the uterus, over the fundus of the uterus to its anterior surface. It is then reflected onto the posterior wall of the bladder, over the superior surface of the bladder to the anterior abdominal wall. The uterus is separated from the rectum by the **rectouterine pouch** (of Douglas) (see Fig. 14-2B).

Lymphatic Drainage of the Female Internal Genitalia

The lymphatic drainage of the female reproductive organs is of great clinical importance because cancer of the cervix is common and cancer of the body of the uterus is relatively common.

1. The ovaries, the fundus of the uterus, and the uterine tubes drain to the **aortic nodes** at the origins of the ovarian arteries. A few lymphatics will accompany the round ligament and drain into the inguinal nodes.
2. The body of the uterus drains into lymph vessels located in the broad ligament, and these vessels drain into the internal iliac nodes.
3. The cervix drains in three directions: laterally to the external iliac nodes, posterolaterally along the uterine vessels to the internal iliac nodes, and posteriorly along the uterosacral ligaments to the sacral nodes.
4. The lymphatics of the upper two thirds of the vagina follow the lymphatics of the cervix, but the lower one third drains into the inguinal nodes.

CLINICAL NOTE: VAGINAL EXAMINATION

Vaginal examination gives the clinician an excellent opportunity not only to examine the vagina and cervix but to palpate many structures in the female pelvis. Through the anterior vaginal wall, tumors of the bladder may be palpable, and similar problems in the rectum may be felt through the posterior vaginal wall. Of course, adequate examination of the female pelvis requires both vaginal and rectal examination. The size, location, and texture of the cervix are easily palpable, and the cervix and interior of the vagina can be inspected by means of a speculum. The normal-sized uterus can be felt by bimanual palpation, that is, two fingers in the vagina and the other hand on the lower abdominal wall. Between them, they can determine the size, texture, and orientation of the uterus. The ovaries are palpable through the lateral fornices, and any abnormal mass in the pelvis is usually readily accessible to bimanual palpation.

The Perineum

The **perineum**, which corresponds to the outlet of the pelvis, is that region bounded by the pubic symphysis, pubic rami, ischial rami and tuberosities, sacrotuberous ligament, and coccyx. It contains the anus and, in the male, the root of the scrotum and penis or, in the female, the vulva.

The bony structures of the pelvic outlet have already been studied. The **sacrotuberous ligament** has a broad attachment to the sacrum above. It passes laterally and inferiorly to attach as a thick but narrow band to the ischial tuberosity. The **sacrospinous ligament** stretches from the lateral aspect of the sacrum to the spine of the ischium. Together the sacrospinous and sacrotuberous ligaments convert the greater and lesser sciatic notches of the os coxae into the **greater** and **lesser sciatic foramina** (Fig. 15-1).

The perineum is diamond shaped, and a transverse line through its center divides it into the anterior **urogenital** and posterior **anal triangles**. The transverse line dividing the two perineal triangles joins the midpoints of the two ischial tuberosities.

PELVIC DIAPHRAGM

The **pelvic diaphragm** closes the pelvic outlet in much the same way that a funnel might close it (Fig. 15-2). The formation of this diaphragm by **levatores ani** and **coccygeus** muscles was described in Chapter 13.

External Anal Sphincter. The external sphincter of the anus is formed by voluntary muscle in continuity with the lower fibers of the levator ani muscles. The sphincter muscle fibers are arranged in a circular fashion surrounding the anal canal and its orifice (Fig. 15-3). It is conventional to describe the external sphincter as having three parts.

1. The **subcutaneous** part of the sphincter is a scattered collection of muscle fibers lying in the **perianal space**. The continuity of the muscle ring of the subcutaneous part of the anal sphincter is interrupted by fibromuscular prolongations of the longitudinal coat of smooth muscle of the rectum and anus inserted into the perianal skin. These insertions, which contain many elastic fibers, give the perianal skin its puckered appearance.
2. The **superficial** part of the external sphincter is a fairly dense and truly circular layer of muscle that finds attachment to the perineal body in front and the anococcygeal raphe behind.

15

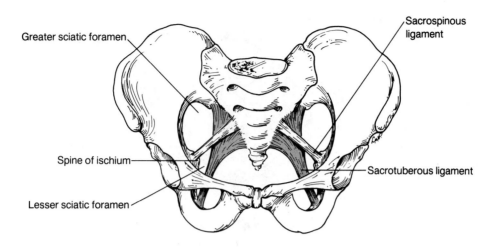

3. The **deep** part of the external sphincter is also circular in disposition but is more in continuity with the levatores ani muscles. The external sphincter is responsible for voluntary fecal continence and is supplied by **S2** and **S3** through the **inferior rectal** branch of the **pudendal nerve** and by the **perineal branch** of **S4**. The levatores ani muscles actually elevate the anal canal over a fecal mass. There is also an internal sphincter of smooth muscle fibers that is under sympathetic nervous control.

The anterior fibers of levatores ani muscles surround the vagina and actually form a **sphincter vaginae**. In the male these fibers support the **prostate gland** and are called the **levatores prostatae**.

UROGENITAL DIAPHRAGM

The **urogenital diaphragm** runs from the rami of the pubis and ischium (conjoint rami) of one side to the conjoint rami of the other side. In other

Figure 15-1.
Ligaments forming the foramina of the pelvis.

Figure 15-2.
The pelvic diaphragm (posterior) as represented by a funnel.

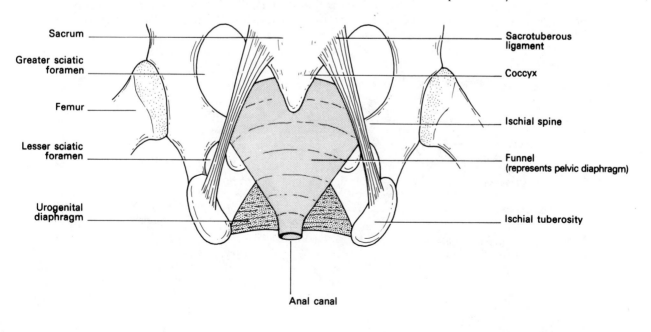

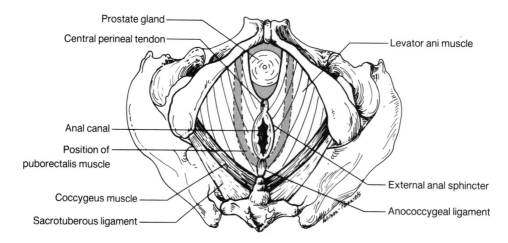

Figure 15-3.
Inferior view of pelvic diaphragm.

words, it closes the anterior portion of the pelvic outlet much as a piece of cardboard would close it if the cardboard was placed from side to side across the opening (Figs. 15-4 and 19-5).

Basically the urogenital diaphragm is formed by the ***sphincter urethrae*** muscle, which is attached to the conjoint rami. The sphincter urethrae encircles the urethra in males and encircles the vaginal outlet in females. In females a few fibers of sphincter urethrae surround the urethra.

Deep Fascia. The superior surface of the urogenital diaphragm is covered by deep fascia which is continuous with the pelvic fascia. The deep fascia on the inferior surface of the urogenital diaphragm is especially dense and forms the ***perineal membrane***. It is continuous with the superior fascia of the urogenital diaphragm anteriorly and posteriorly. The perineal membrane is attached to the conjoint rami on each side and forms the base for the attachments of the greater parts of the penis or clitoris. The perineal membrane is pierced by the ***urethra*** and, in females, by the ***vagina***.

Deep Pouch. The sphincter urethrae muscle is covered by deep fascia above and below (perineal membrane); this arrangement has been called the deep pouch of the perineum. The deep pouch contains vessels and nerves that will be described later.

Figure 15-4.
Pelvic diaphragm overlaid by urogenital diaphragm.

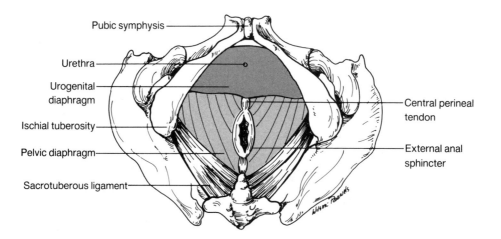

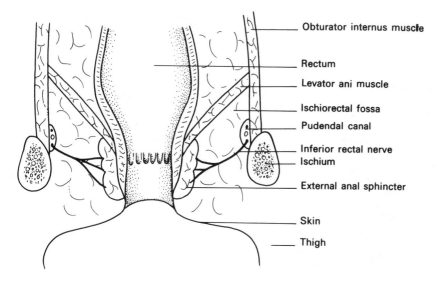

Obturator internus muscle

Rectum

Levator ani muscle

Ischiorectal fossa

Pudendal canal

Inferior rectal nerve

Ischium

External anal sphincter

Skin

Thigh

Figure 15-5.
Coronal section of ischiorectal fossa.

ISCHIORECTAL FOSSA

The ***ischiorectal fossa*** (Fig. 15-5) is a space filled with fat and fascia lying between the levator ani (medially) and the side wall of the pelvis (laterally). It is limited inferiorly by skin. The ischiorectal fossa can be understood only if it is remembered that the levator ani is shaped like a funnel. The ischiorectal fossa will then be seen to be wide inferiorly but narrowing superiorly and ending at the point where the levator ani is attached to the fascia covering the obturator internus muscle.

The ischiorectal fossa continues anteriorly, superior to the urogenital diaphragm. This is known as the ***anterior recess*** of the ischiorectal fossa.

The posterior boundary of the ischiorectal fossa is the sacrotuberous ligament. The anterior boundary of the inferior part of the ischiorectal fossa is the posterior edge of the urogenital diaphragm, but the anterior recess passes superior to this diaphragm.

Contents of the Ischiorectal Fossa. The ischiorectal fossa (Fig. 15-5) contains various structures. The largest of these is a pad of stringy fat that is mobile enough to be compressed by a fecal mass passing through the anal canal. It fills the ischiorectal fossa. It might be instructive for the reader to realize that collections of fat, wherever they may be found, perform a similar function: that of filling spaces and allowing movements of adjacent structures.

The ***inferior rectal nerve*** comes from the pudendal canal (see below) and runs anteromedially and superficially across the ischiorectal fossa to supply the ***external anal sphincter***. The ***inferior rectal vessels*** pass from the internal pudendal vessels, in the pudendal canal, to the anal canal. The ***scrotal nerves and vessels*** (labial in females) pass from the pudendal canal anteriorly and inferiorly to reach their destination (Fig. 15-6).

CLINICAL NOTE

The ischiorectal fossa may become the site for abscess formation. The infection usually spreads to the fossa from the anal glands. The poor vascu-

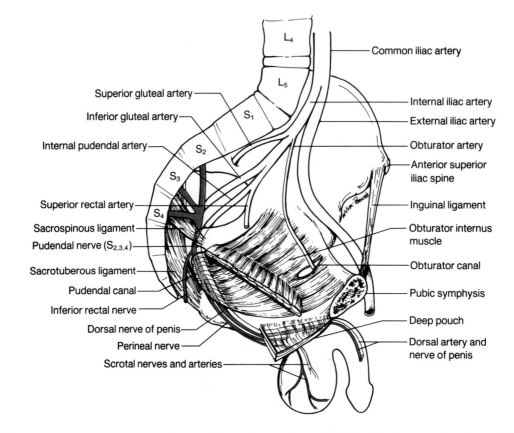

Figure 15-6.
Medial surface of left wall of pelvis showing contents of the pudendal canal and the deep pouch.

larity of fatty tissue allows the infection to spread quite rapidly, and the result is an extremely painful, bulging mass. The bulge may be seen from the exterior and felt from the interior (by rectal examination). Incision of an ischiorectal abscess must be carried out with care to avoid damage to the inferior rectal nerve.

PUDENDAL CANAL (ALCOCK'S CANAL)

The **pudendal canal** contains an important group of vessels and nerves (Fig. 15-6). The **pudendal nerve** and the **internal pudendal artery** and **vein** leave the pelvis on the surface of the obturator internus muscle, covered by its fascia. They leave the pelvis above the origin of levator ani and pass through the greater sciatic foramen, across the spine of the ischium and sacrospinous ligament, and enter the perineum through the lesser sciatic foramen. They still lie on the surface of obturator internus and deep to its fascia, but they are now **below** the origin of levator ani. Indeed, they are in the ischiorectal fossa, the medial wall of which is the anus and levator ani. The lateral wall of the ischiorectal fossa, which contains the pudendal canal, is made up of the ischial tuberosity and the conjoint rami of the pubis and ischium, and these are covered by the obturator internus muscle.

The pudendal nerve (S2, S3, and S4), the principal nerve of the perineum, splits part way along the pudendal canal into the **dorsal nerve** of the **penis** and the **perineal nerve**. These two nerves run forward on either side of the internal pudendal artery. The perineal nerve gives off the **scrotal (or labial)** branches and then continues to supply the muscles of the urogenital diaphragm.

The ***dorsal nerve of the penis (or clitoris)*** runs through the deep pouch to act as a sensory nerve to its named end organ.

Inferior Rectal Vein. This vein passes from the inferior part of the anal canal to drain into the internal pudendal vein in the pudendal canal. The inferior rectal vein anastomoses in the submucosal layer of the anal canal with the ***superior rectal vein***, which is a tributary of the ***portal venous system***. The inferior rectal vein (and the middle rectal vein, which is of little clinical importance) drains into the ***systemic venous system***. In the condition of ***portal hypertension*** this anastomosis between inferior rectal vein (systemic) and superior rectal vein (portal) can cause enlargements of the anastomosing veins, producing varicosities known as hemorrhoids. ***Note***: Do not think that hemorrhoids are particularly common in portal hypertension nor that this constitutes a ***common*** cause for hemorrhoids. The ***true*** nature and the structure of hemorrhoids were described with the detailed structure of the anal canal (see pp. 188–189).

Caution. The reader is advised to proceed no further with a description of the perineum until he has understood the structure and disposition of the pelvic and urogenital diaphragms and the ischiorectal fossa. This little bit of anatomy is important and, often, little understood by family physicians. Yet it is one of the foundations of the highly specialized fields of obstetrics and gynecology, proctology, and urology.

PERINEAL POUCHES

The ***deep perineal pouch*** has been described as that portion of the perineum that is enclosed by the deep fascia covering the superior and inferior aspects of the sphincter urethrae muscle (Figs. 15-7A and 15-12).

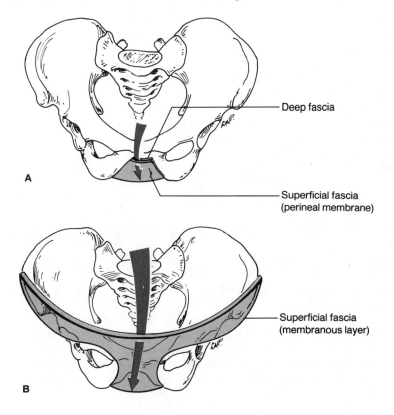

— Deep fascia

— Superficial fascia (perineal membrane)

— Superficial fascia (membranous layer)

Figure 15-7.
Arrows indicate **(A)** deep perineal pouch and **(B)** superficial perineal pouch.

The **superficial perineal pouch** is located between the superficial fascia of the perineum (Colle's fascia) and the **perineal membrane**. It will be recalled that the superficial fascia is attached to the conjoint rami (of pubis and ischium) and to the posterior edge of the perineal membrane. This means that the superficial pouch is limited laterally by the junction of superficial fascia and bone, and posteriorly by the junction of superficial fascia and perineal membrane. The superficial pouch is open anterior to the body of the pubic bone, where the superficial fascia is continuous with the superficial fascia of the abdominal, scrotal, and penile walls (Scarpa's fascia) (Fig. 15-7B).

The contents of the superficial pouch are not generally considered to include the body of the penis or the scrotum. However, if one remembers that the superficial fascia in this area continues into the scrotum and along the penis one will see that in fact these two structures could be considered to be contents of the superficial pouch.

MALE PERINEUM

The male perineum is easier to understand than the female and will be considered first. Many of the male structures have female counterparts, and an understanding of the perineal anatomy in the male will lead to relatively simple understanding of the female perineum.

PENIS

The penis has a **root** and a **body**. The root has its origin in the superficial perineal pouch (Figs. 15-8A, 15-8B, and 15-8C).

The body of the penis comprises two **corpora cavernosa** and one **corpus spongiosum**. The two corpora cavernosa lie side by side on the dorsum of the penis, and the corpus spongiosum, traversed lengthwise by the **urethra**, lies inferior (ventral) to the two corpora cavernosa (Fig. 15-8A). These three structures form the body of the penis. The root of the penis consists of two **crura** continuous with the corpora cavernosa and a **bulb** continuous with the corpus spongiosum.

The root and body of the penis comprise mainly **erectile tissue**, which is a specialized tissue consisting primarily of blood vessels. The arterial components are the **helicine** arteries, which are coiled arteries in the erectile tissue. Under parasympathetic stimulation from the **nervi erigentes** (**S2** and **S3**), which pass from the pelvic hypogastric plexus alongside the internal pudendal arteries, the helicine arteries dilate. These dilated arteries press on the veins and prevent the return of venous blood. This causes engorgement of the erectile tissue. This engorgement within the tough fibrous coverings of the penile erectile tissue results in erection of the penis. The sympathetic fibers also follow the blood vessels and cause the helicine arteries to contract, relieving the pressure on the veins of the penis and allowing the penis to revert to a state of flaccidity.

Root. The root of the penis consists of the bulb and two crura and their associated muscles.

The **crus** of the penis is really a tube of dense connective tissue filled with erectile tissue. Each crus is continuous with a corpus cavernosum of the body of the penis. Each crus is attached to a conjoint ramus of the pubis and ischium, and at the anterior end of the inferior ramus of the pubis the two crura come together at the base of the body of the penis and thereafter

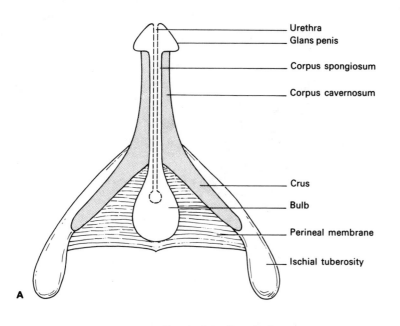

Urethra
Glans penis
Corpus spongiosum
Corpus cavernosum
Crus
Bulb
Perineal membrane
Ischial tuberosity

A

Figure 15-8.
Root and body of penis: **(A)** without muscles; **(B)** cross-section of body of penis; **(C)** with muscles.

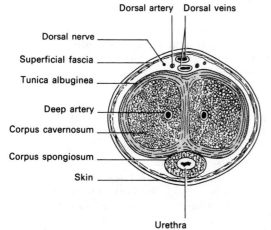

Dorsal artery   Dorsal veins
Dorsal nerve
Superficial fascia
Tunica albuginea
Deep artery
Corpus cavernosum
Corpus spongiosum
Skin
Urethra

B

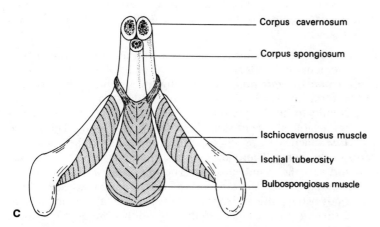

Corpus cavernosum
Corpus spongiosum
Ischiocavernosus muscle
Ischial tuberosity
Bulbospongiosus muscle

C

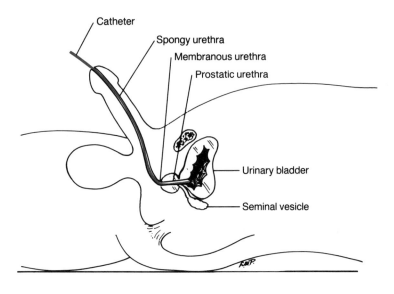

Catheter
Spongy urethra
Membranous urethra
Prostatic urethra
Urinary bladder
Seminal vesicle

Figure 15-9.
Catheterization of male urethra and relevant anatomy.

parallel each other as corpora cavernosa within the body of the penis (Fig. 15-8A).

The ***bulb*** of the penis also consists of a tube of connective tissue filled with erectile tissue. It is attached to the perineal membrane. The urethra pierces the perineal membrane, enters the bulb of the penis, and runs through the corpus spongiosum to the tip of the penis. Thus the urethra is surrounded by the erectile tissue of the corpus spongiosum. That portion of the urethra contained within the bulb and the corpus spongiosum is called the ***spongy*** (penile) urethra.

The (striated) ***muscles*** of the root of the penis surround both the crura and the bulb (Fig. 15-8C). The muscle surrounding each crus is called the ***ischiocavernosus*** and that surrounding the bulb, the ***bulbospongiosus***. By their contractions these muscles increase the pressure within the erectile tissue and cause further erection of the penis. The bulbospongiosus can effect pressure on the urethra and helps in the ejaculation of semen and expulsion of urine from the urethra.

Body. The body of the penis consists of the two corpora cavernosa and the corpus spongiosum, surrounded by deep and superficial fascia (Fig. 15-8C).

1. The ***corpus cavernosum*** (paired) is a continuation of the crus of the penis and ends at the base of the glans of the penis.
2. The ***corpus spongiosum***, surrounding the urethra, passes on the ventral (inferior) side of the penis, and its expanded distal extremity forms the ***glans*** of the penis. The glans is covered by a retractable sleeve of skin, the ***prepuce*** (foreskin). The ***frenulum*** attaches the prepuce to the undersurface of the penis.
3. ***Veins*** of the penis. There are two major dorsal veins of the penis, one superficial and one deep. The superficial vein passes to the external pudendal vein, and the deep vein passes behind the body of the pubis and in front of the perineal membrane, where a small gap exists between the anterior margin of the perineal membrane and the body of the pubis, to the plexus of veins around the prostate gland (prostatic venous plexus).

4. **Dorsal** and **deep arteries** of the penis. These two paired arteries are branches of the internal pudendal artery. They supply the erectile tissue.

5. **Dorsal nerve** (paired) of penis. A branch of the pudendal nerve, the dorsal nerve of the penis, passes in the deep pouch and then out its anterior end to reach the dorsum of the penis. It runs along the dorsal length of the penis and is a sensory nerve (see Fig. 15-6).

6. **Suspensory ligament** of the penis. The suspensory ligament is a condensation of superficial fascia passing from the pubis and lower linea alba of the rectus sheath to the dorsal surface of the penis.

The heavy condensations of connective tissue around the corpora cavernosa, corpus spongiosum, crura, and bulb collectively are spoken of as the **tunica albuginea** (see Fig. 15-8B).

DEEP POUCH OF PERINEUM

The deep pouch of the perineum is that portion enclosed by the **deep** fascia covering the superior and inferior aspects of the sphincter urethrae (see Figs. 15-7A and 15-12). It contains the following structures:

1. Sphincter urethrae muscle.
2. Membranous urethra (male).
3. Urethra and lower end of vagina (female).
4. Internal pudendal vessels.
5. Dorsal nerves of penis (or clitoris).
6. Bulbourethral glands (male).

Urethra. The urethra runs from the bladder to the tip of the penis (Fig. 15-9); it is divided into three parts:

1. **Prostatic urethra**, which runs in the pelvis from the neck of the urinary bladder, through the prostate gland to the upper surface of the deep pouch.

2. **Membranous urethra**, which pierces the upper fascial surface of the deep pouch and leaves through the inferior fascia of the deep pouch (perineal membrane) to become continuous with the spongy urethra.

3. **Spongy urethra**, which runs in the bulb and the corpus spongiosum.

This division of the urethra into three named parts is not just an exercise of anatomical semantics but has practical applications:

CLINICAL NOTE

*When the penis is dependent in its flaccid state, the urethra assumes the shape of the letter S lying on its side. The narrowest part of the urethra is at the **external meatus** at the tip of the penis. The widest part is the spongy urethra in the bulb. The lumen narrows in the membranous portion, and the membranous portion is relatively rigid in its position. The prostatic urethra is wider and more distensible. If a catheter has to be inserted into the male bladder, any sized catheter that will fit into the external meatus will fit the rest of the lumen (Fig. 15-9). The catheter must be flexible to negotiate the right angle bend at the transition from the spongy portion to the membranous portion, or, if a rigid instrument such as a cystoscope* *is used, it must have an angle near its tip to negotiate this bend. Just proximal to the external meatus at the tip of the penis the urethra contains an oval dilatation, the **fossa navicularis**, which may contain a valve-like flap of mucosa on its superior aspect. A rigid instrument should be introduced with its tip downward to avoid being snagged in this valve.*

Bulbourethral Glands. These are two glands, situated in the deep pouch, the ducts of which pierce the perineal membrane to empty their secretions into the spongy (bulbous) part of the urethra.

Internal Pudendal Artery. This artery passes from the pudendal canal into the deep pouch (see Fig. 15-6). While in the pudendal canal, and before entering the pouch, it gives off scrotal branches. Within the deep pouch the artery gives off branches to the root of the penis.

Perineal Nerve. The perineal nerve is one of two terminal branches of the pudendal nerve. It gives off scrotal nerves and then passes into the deep pouch to supply the sphincter urethrae muscle (see Fig. 15-6).

Dorsal Nerve of the Penis. This sensory nerve is the other terminal branch of the pudendal nerve. From the deep pouch it passes along the dorsal surface of the penis (see Fig. 15-6).

FEMALE PERINEUM

The female perineum is most easily understood as a variant of the male perineum. Figures 15-10 and 15-11 may help the reader to understand the difference. The main features of difference follow:

The **vagina**, which pierces the urogenital diaphragm.
The **urethra**, which is in the anterior wall of the vagina.
The **clitoris**, which does not surround the urethra.

SUPERFICIAL POUCH (FEMALE)

The **clitoris** (Figs. 15-10 and 15-11) is the homologue of the penis and consists of two corpora cavernosa, similar to those of the male. The **glans**

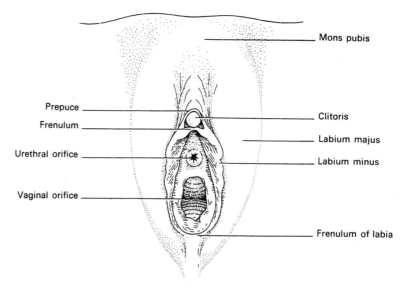

Figure 15-10.
The vulva.

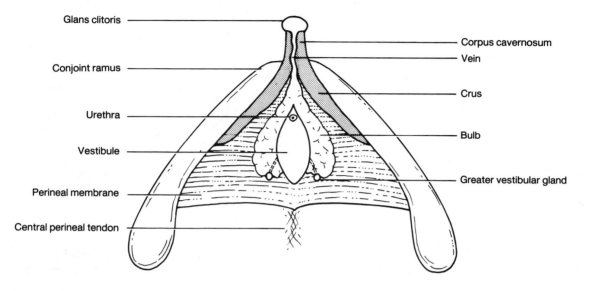

Glans clitoris

Conjoint ramus

Urethra

Vestibule

Perineal membrane

Central perineal tendon

Corpus cavernosum

Vein

Crus

Bulb

Greater vestibular gland

Figure 15-11.
The clitoris.

clitoris is connected to the bulb by only a few veins; there is no corpus spongiosum and no urethra in the clitoris. The crura are composed of erectile tissue but are smaller than those in the male.

Bulb.   The bulb of the vestibule is split by the vagina so that the bulb forms two masses of erectile tissue, one on each side of the vaginal opening. The bulb is connected to the glans by a few veins (Fig. 15-11).

Greater Vestibular Gland.   The female has, at the posterior border of the bulb on either side, a gland that drains by ducts into the vagina. (These greater vestibular glands are the homologues of the male bulbourethral glands [Fig. 15-11].)

CLINICAL NOTE

The greater vestibular glands may become infected or cystic: **Bartholin's cyst or abscess**.

The external genitalia may be anesthetized by infiltration of the region close to the pudendal nerve with local anesthetic. (This can be particularly useful during childbirth.) The needle is introduced through the vagina into the area of the ischial spine that can be palpated by a finger placed in the vagina. This procedure is called a "pudendal nerve block."

DEEP PERINEAL POUCH (FEMALE)

The deep perineal pouch of the female contains basically the same structures as that of the male except for the absence of the bulbourethral glands. Otherwise the constituents are as follows (Fig. 15-12):

> Sphincter urethrae muscle.
> Internal pudendal vessels.
> Dorsal nerve of clitoris.
> Branches of the perineal nerve to the sphincter urethrae.
> The parts of the vagina and the urethra traversing the deep pouch.

The sphincter urethrae muscle of the female surrounds the urethra ***and*** the vagina. The part of the sphincter urethrae surrounding the urethra is not very bulky, and the female bladder depends for its continence on

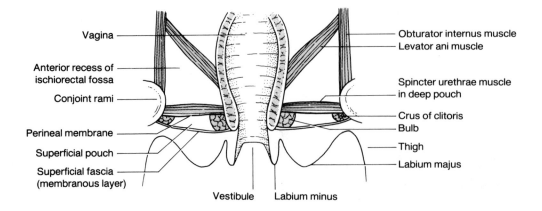

Vagina

Anterior recess of ischiorectal fossa

Conjoint rami

Perineal membrane

Superficial pouch

Superficial fascia (membranous layer)

Obturator internus muscle

Levator ani muscle

Spincter urethrae muscle in deep pouch

Crus of clitoris

Bulb

Thigh

Labium majus

Vestibule   Labium minus

Figure 15-12.
Coronal section of the vagina showing its relationship to the pelvic and urogenital organs.

the intimate relationship between vagina and urethra. This is achieved by a fibromuscular band that is attached to the back of the body of the pubis and to the conjoint pubic rami, on each side. This arrangement creates an optimum angle for continence between the urethra and the bladder neck. Any disruption of this fibromuscular band will allow prolapse of the anterior wall of the vagina with its attached urethra (***cystocele***) and alter the ***urethrovesical angle***, resulting in urinary incontinence. Such prolapse is not uncommon after repeated stretching of these tissues following multiple childbirths.

EXTERNAL GENITALIA (FEMALE)

The following names are attached to the different parts of the female genitalia, which, as a group, are called the ***vulva*** (see Fig. 15-10).

Vagina. That part of the genital system between the opening of the vulva and the uterus is called the vagina. The orifice of the virginal vagina is partially closed by the ***hymen***.

Hymen. This thin incomplete membrane is just superior to the vaginal opening.

Urethral Opening. The urethral opening is found between the clitoris and the opening of the vagina.

Vestibule. This is the area between the labia minora. It includes the vaginal and urethral openings.

Mons Pubis. The mons pubis is a low, rounded eminence produced by the presence of a pad of fat anterior to the pubis. The mons is continuous with the labia majora.

Labia Majora. The two labia majora are fat-filled skin folds on either side of the ***pudendal cleft***. The ***round ligaments*** of the uterus terminate in the subcutaneous tissue of the labia majora. Each labium majus is the homologue of one half of the (unfused) scrotum (see Fig. 15-10).

Labia Minora. The labia minora are two hairless folds of skin between the labia majora and the vestibule.

Clitoris. The clitoris is the homologue of the penis and is found at the anterior end of the vestibule. The labia minora divide to enclose the clitoris. The portion of the labia minora that passes anterior to the clitoris is the ***prepuce*** and that which passes posterior to the clitoris is the ***frenulum***.

Frenulum of Labia (Fourchette). Posteriorly the labia minora blend with the labia majora to form a transverse ridge called the ***frenulum*** of the labia. Do not confuse it with the frenulum of the clitoris.

CLINICAL NOTE: EPISIOTOMY

At normal delivery the head of the baby may be too large for the vestibular opening. This could result in a tear in the perineum. Such a tear may rip into the anus or rectum. To avoid this spontaneous tear the obstetrician may elect to make an incision into the vestibule. This cut is called an **episiotomy** *and is usually made in the fourchette a little lateral to the midline to avoid the anus. An episiotomy has to be repaired after delivery, and if the birth was ''natural'', that is, without anesthetic, a pudendal nerve block of local anesthetic agent may be very suitable for the performance and repair of the episiotomy.*

The Autonomic Nervous System

REVIEW

The autonomic nervous system has been described with a certain amount of fragmentation in previous sections of this book to preserve the essential regional description of the human body that the authors consider to be the most practical and logical progression for the reader. A systemic approach is to be found in the more comprehensive textbooks on anatomy.

The autonomic nervous system is divided into sympathetic and parasympathetic components and requires a considerable amount of understanding. Although it is said that the sympathetic system is used in emergencies (flight or fight mechanism) and the parasympathetic system is used by the body at rest, this is *not* necessarily always true. For instance, a student who is frightened by an upcoming examination has obvious evidence of both sympathetic and parasympathetic stimulation; his hands are sweaty (sympathetic), his bowels are churning (parasympathetic), and his heart pounds (sympathetic).

FUNCTION

The function of the autonomic nervous system may be illustrated by the male student who wakes slowly and quietly one morning from a deep, refreshing sleep and whose first functions are those of the ***parasympathetic nervous system***.

1. His heart beats slowly (vagus nerves).
2. He may awake with an erection (pelvic splanchnic nerves).
3. He turns on the light, causing his pupils to constrict (oculomotor nerve light reflex).
4. The bright light causes his eyes to water (greater petrosal nerve).
5. He looks at his watch and his eyes accommodate for close vision (oculomotor nerve).
6. He empties his bladder (pelvic splanchnic nerves).
7. He smells bacon cooking and he salivates (lesser petrosal nerves and chorda tympani).
8. His gastric and intestinal juices flow (vagus nerves).
9. Peristalsis begins in his stomach (vagus nerves).
10. Peristalsis reaches his colon and his bowels move (vagus and pelvic splanchnic nerves).

11. The internal anal sphincter relaxes to permit defecation (pelvic splanchnic nerves).
12. He has intercourse with his wife and ejaculates (pelvic splanchnic nerves). His wife's orgasm is a function of her pelvic splanchnic (and somatic) nerves.

The opposite actions to the foregoing are performed by the **sympathetic system**.

1. He leaves the house on a cold winter morning and his cheeks blanch.
2. He has to dodge heavy traffic and his palms begin to sweat.
3. He is almost hit by a car, causing the hairs on the back of his neck to stand up.
4. He jumps out of the way and his heart pounds.
5. His blood pressure rises (general sympathetic vasoconstrictor and cardiac responses).
6. He reaches the university, almost late, runs and gasps for breath (bronchial dilatation).

PARASYMPATHETIC (CRANIOSACRAL) OUTFLOW

The **parasympathetic fibers** come from either the brain (cranial nerves III, VII, IX, and X) or from the sacral nerves (S2, S3, and S4). The latter are called **pelvic splanchnic nerves**. All of these nerves have at least some parasympathetic fibers and have effects, as noted above, on the structures that they supply. The pelvic splanchnic nerves supply the bowel below the left colic flexure as well as the pelvic and perineal organs. The parasympathetic system, like the sympathetic system, has **one** synapse with a secondary neuron outside the central nervous system. These synapses are in the walls of the target organs (thoracic, abdominal, pelvic, and perineal viscera) or in ganglia located close to the target organs (ciliary, pterygopalatine, submandibular, and otic ganglia) (Fig. 12-2B).

SYMPATHETIC (THORACOLUMBAR) OUTFLOW

The thoracolumbar outflow is the **preganglionic** sympathetic outflow (Fig. 16-1). The **preganglionic fibers** originate in the intermediolateral column of the gray matter of the spinal cord from the **first thoracic** to the **second lumbar** spinal cord segments, inclusively. The fibers leave by way of the anterior column of gray matter in the anterior roots of the spinal nerve and enter the corresponding ganglia of the **sympathetic trunk** through the **white rami communicantes**. The gray rami communicantes carry **postganglionic fibers** from the sympathetic trunk to the spinal nerves, where they are distributed by the anterior and posterior primary rami.

These gray rami are found at **all** levels from **C1** to **S5**, but they simply carry **postganglionic fibers** (ones that have already synapsed) and do not represent direct connections with the spinal cord (Fig. 16-1). **All** preganglionic fibers to the head and neck region leave the spinal cord at **T1** and enter the sympathetic trunks by white rami to ascend in the sympathetic trunks in the neck and synapse in cervical ganglia.

Splanchnic fibers are **preganglionic fibers** that do not synapse in ganglia of the sympathetic trunk but run medially to synapse in **prevertebral ganglia** (e.g., celiac, superior mesenteric [see Fig. 12-2A]). As a rule,

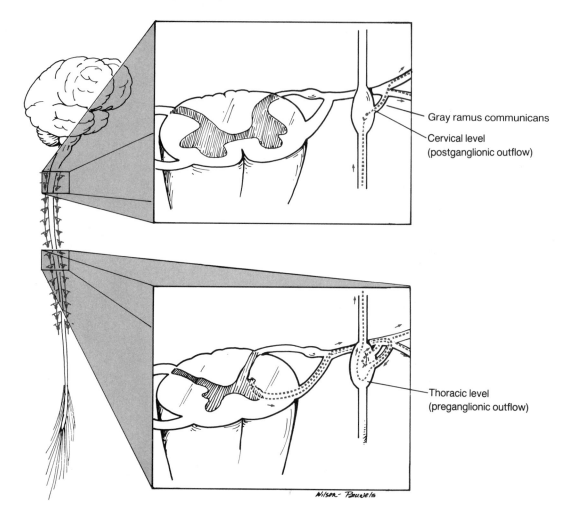

Figure 16-1.
Sympathetic (thoracolumbar) outflow.

the sympathetic fibers synapse in ganglia that are situated at a distance from the end organ. There are exceptions to this rule, the most notable being the medulla of the suprarenal gland, which is a modified ganglion; its cells are modified nerve cells, and preganglionic fibers synapse directly with them.

To summarize, there is **one** synapse between preganglionic fibers and end organs. These synapses occur in the ganglia of the sympathetic trunk, in prevertebral ganglia, such as the celiac ganglion, **or** in the medulla of the suprarenal gland.

SYMPATHETIC TRUNK

The sympathetic trunk runs from the level of the first cervical vertebra to the level of the last segment of the sacrum, where it joins with the trunk of the opposite side at the **ganglion impar** (Fig. 16-2). Morphologically there is a ganglion for each vertebral level, but in some areas one or more ganglia become fused into single, larger ganglia. The most obvious of these fused ganglia are the **superior cervical ganglion**, which lies at the level of the second to third cervical vertebrae and represents the fusion of the upper four cervical ganglia, the **middle cervical ganglion**, which lies

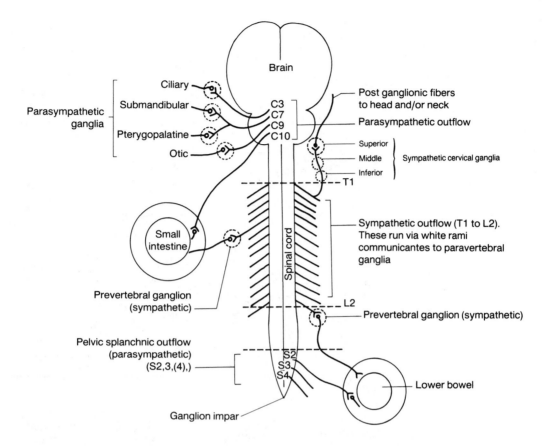

at the level of the sixth cervical vertebra and represents the fusion of the fifth, sixth, and (seventh) cervical ganglia, and the ***inferior cervical ganglion*** (stellate ganglion), which is found anterior to the neck of the first rib and represents the fusion of the (seventh) and eighth cervical ganglia and the first thoracic ganglion (Fig. 16-2). In the thorax there is usually one ganglion to each vertebral body, and in the abdomen and pelvis there is often some fusion of ganglia.

The neurons that synapse in these ganglia may do so in a variety of ways. The simplest example is one in which the preganglionic fibers of a thoracic segment synapse at its numerically equivalent ganglion and the postganglionic fibers leave at that level. Other possibilities are the ascent or descent of preganglionic fibers to synapse at different levels and the emergence of postganglionic fibers at different levels (Fig. 16-3).

Figure 16-2.
Sympathetic and parasympathetic systems: schematic anterior view indicating craniosacral and thoracolumbar outflows. Note that in general the parasympathetic fibers synapse in the wall of the viscus while the sympathetic synapse is in a prevertebral ganglion.

AUTONOMIC PLEXUSES

The autonomic supply to viscera of the abdomen, pelvis, and thorax and the sympathetic supply to structures of the head and neck usually travel in plexuses along arteries. In the head and neck the parasympathetic supply usually travels with cranial nerves, and, in this area, the synapses of the parasympathetic nervous system are in ganglia located a short distance from their end organ (*e.g.*, otic and submandibular ganglia) (see Fig. 16-2).

In the abdomen there is a major autonomic plexus along the abdominal aorta (the ***aortic plexus***) that receives preganglionic fibers from the

Figure 16-3.
Left sympathetic trunk.

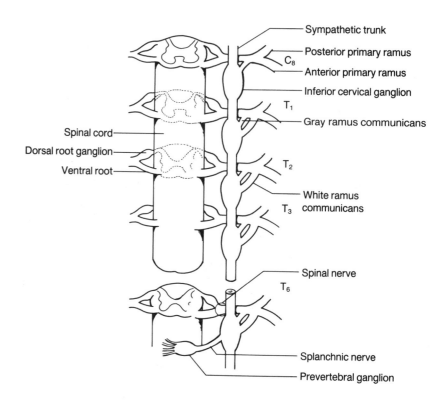

splanchnic nerves, including the lumbar splanchnic nerves (from L1 and L2 ganglia). Its parasympathetic fibers come from the vagus nerve and, at a lower level, from the pelvic splanchnic nerves (see Fig. 12-2). This plexus sends extensions along the branches of the aorta, and each of these plexuses takes its name from the artery with which its fibers will travel (*e.g.,* **celiac plexus, superior mesenteric plexus, inferior mesenteric plexus, aorticorenal plexus**).

The **aortic plexus** continues into the pelvis as the **hypogastric plexus**. From this plexus, nerves supply the uterus and its appendages, and the other pelvic structures.

The hypogastric plexus contains sympathetic and parasympathetic fibers. The latter come from S2, S3, and S4 and are the **pelvic splanchnic nerves** (nervi erigentes) (see Fig. 12-2B).

VISCERAL AFFERENT (SENSORY) SYSTEM

There is some uncertainty about the precise pathways by which visceral sensation is transmitted to the central nervous system. Visceral afferent impulses can be considered under several headings:

1. Afferent impulses that do ***not*** reach consciousness. Afferent impulses traveling along sympathetic and parasympathetic fibers carry information about the state of tone or contraction in smooth muscle, blood pressure (baroreceptors), the chemical makeup of the blood (chemoreceptors), the degree of fullness of hollow viscera, and more. Baroreceptor and chemoreceptor impulses probably travel through the parasympathetic system (vagus and glossopharyngeal nerves, *e.g.;* see Chap. 34), as do

those carrying information about the status of hollow viscera (vagus and pelvic splanchnic nerves). These unconscious activities are of a reflex nature and probably occur at the brain stem or spinal cord level.

2. Afferent impulses of which we are vaguely conscious. These include such sensations as hunger, nausea, and fullness of rectum and bladder. They are appreciated consciously, localized with varying degrees of accuracy, and probably belong to the afferent pathways of the parasympathetic system.

3. Painful, consciously appreciated impulses from viscera. These are believed to travel in nerves of the sympathetic system. The fibers follow a distribution similar to that of postganglionic sympathetic fibers, and their neurons are located in the posterior root ganglia of the spinal nerves. Localization of the sites of origin of these painful impulses is poorly interpreted in the brain, and such pain is usually referred to somatic segments which send their sensory impulses to the same spinal cord segments. This arrangement would account for the localization of foregut pain in the epigastrium, midgut pain in the umbilical region, cardiac pain in the chest wall and left arm, and the many other types of visceral pain that were described in the regional sections of this book.

The Lower Limb: General

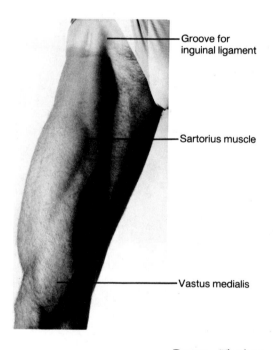

Groove for
inguinal ligament

Sartorius muscle

Vastus medialis

The lower limb is comparable to the upper limb despite the fact that they have rotated from the embryonic position through 90 degrees in opposite directions. The embryonic center of each limb is taken as the axial line. The ***preaxial side*** of the upper extremity is the radial side and the ***postaxial side***, the ulnar side.

In the lower limb the ***preaxial side*** is medial. Therefore, the cutaneous innervation of the medial side of the lower limb comes from a higher level than that for the lateral side (Fig. 17-1).

Because of this, the thumb and great toe, which are homologous, will lie on opposite sides; the knee and the elbow, which are homologous, flex in opposite directions. The fibula corresponds to the ulna and the tibia corresponds to the radius.

LOWER LIMB AS A WEIGHT BEARER

The lower extremity has been modified to bear the weight of the body and to provide mobility. This requires strong bones and stable yet mobile joints:

Bones. The bones are heavy and strong with well-marked areas for muscular and ligamentous attachments.

Joints. The strength of the joints is produced by good bony configurations or strong ligaments, or both.

Muscles. The muscles are powerful and less capable of delicate movements than those of the upper limb.

Nerves. Muscles that produce delicate movements require more nerve fibers per muscle than do muscles that produce coarse movements. Thus the muscles of the lower limb have a lower nerve/muscle fiber ratio than do the muscles of the upper limb; this comparison is particularly applicable to the hand versus the foot. Similarly there are relatively fewer sensory nerve fibers in the lower limb, accounting for the less well-defined appreciation of tactile sensation in the lower limb (Fig. 17-1).

17

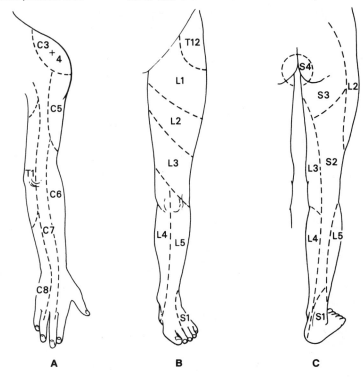

Upper limb-posterior view | Lower limb-anterior view | Lower limb-posterior view

A   **B**   **C**

Figure 17-1.
Segmental cutaneous innervation of lower extremity (**B** and **C**) compared with similar distribution in upper extremity (**A**).

Blood Vessels. The greater bulk and work load of the lower limb require large arteries and veins.

Modification of the Foot. The feet must take the impact of the weight of the whole body at every step. They must sustain the body's weight during prolonged standing and must be adaptable to standing and walking on uneven surfaces. Thus the foot is more differentiated and functionally specialized than the hand. The basic change in the foot is that it has resilient arches that absorb shock, adjust to uneven ground, and can act as levers to transfer the body weight from heel to toes in pushing off to take a step.

Notes on Terminology. In the description of the limbs and in subsequent chapters, muscle attachments will be described as upper and lower, or ***proximal*** and ***distal***, replacing the older terms of "origin and insertion." This is done to comply with modern usage and to make the reader aware of the function of parts of the body as the weight-bearing machinery of the human biped.

When we describe the innervation of muscles the appropriate nerve will be named, and the spinal cord segments responsible for that specific innervation will be given in parentheses, for example, "femoral nerve (L2, L3, and L4)."

The lower limb is divided into named parts: ***thigh*** (upper part of lower limb above the knee), ***gluteal region*** (region of the buttock), ***leg*** (the part of the limb below the knee and above the ankle), and ***foot***.

The Thigh: Anterior and Medial Aspects

BONES OF THE PELVIS AND THIGH

The **pelvic girdle** is a strong ring of bone that articulates with the vertebral column and the lower limbs. Unlike the **pectoral girdle** it is both a weight bearer and part of a body cavity (the pelvis). It sacrifices mobility for stability and strength.

PELVIC GIRDLE

The pelvic girdle consists of the two **os coxae**. Each os coxae is formed by three bones: the **ilium**, the **ischium**, and the **pubis** (Fig. 18-1A).

The os coxae was described in some detail in Chapter 8, and the following account will summarize its salient features, as they apply to its function in the pelvic girdle.

Bony Landmarks (Figs. 18-1A and 18-1B). In order to understand the anatomy of the thigh it is necessary to be able to identify the following landmarks:

Pectineal line	Iliac crest and tubercle
Pubic rami	Anterior superior iliac spine
Pubic crest and tubercle	Anterior inferior iliac spine
Ischial spine and tuberosity	Posterior superior iliac spine
Ischial ramus	Posterior inferior iliac spine
Obturator foramen	Greater sciatic notch
Acetabulum	Lesser sciatic notch

Femur. The femur is the massive bone of the thigh; it is the homologue of the humerus of the arm. The following landmarks should be identified (Fig. 18-2):

Head and fovea	Pectineal line
Neck	Linea aspera
Greater trochanter	Medial condyle
Lesser trochanter	Lateral condyle
Intertrochanteric line (anterior)	Medial and lateral epicondyles
Intertrochanteric crest (posterior)	Intercondylar fossa
Quadrate tubercle	Supracondylar lines (medial and lateral)
Gluteal tuberosity	Adductor tubercle

18

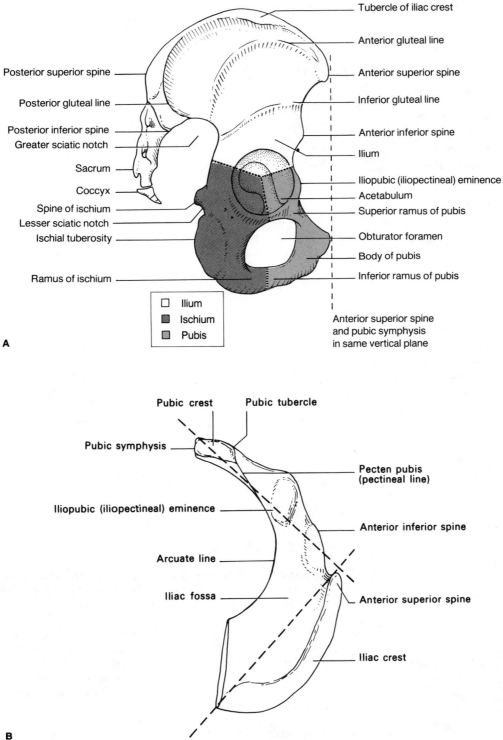

Tubercle of iliac crest

Anterior gluteal line

Posterior superior spine

Anterior superior spine

Posterior gluteal line

Inferior gluteal line

Posterior inferior spine

Anterior inferior spine

Greater sciatic notch

Ilium

Sacrum

Iliopubic (iliopectineal) eminence

Coccyx

Acetabulum

Spine of ischium

Superior ramus of pubis

Lesser sciatic notch

Obturator foramen

Ischial tuberosity

Body of pubis

Ramus of ischium

Inferior ramus of pubis

☐ Ilium
■ Ischium
▨ Pubis

Anterior superior spine
and pubic symphysis
in same vertical plane

A

Pubic crest   Pubic tubercle

Pubic symphysis

Pecten pubis
(pectineal line)

Iliopubic (iliopectineal) eminence

Anterior inferior spine

Arcuate line

Iliac fossa

Anterior superior spine

Iliac crest

B

Figure 18-1.
A. Pelvic girdle (right lateral view).
B. Right os coxae (superior view).

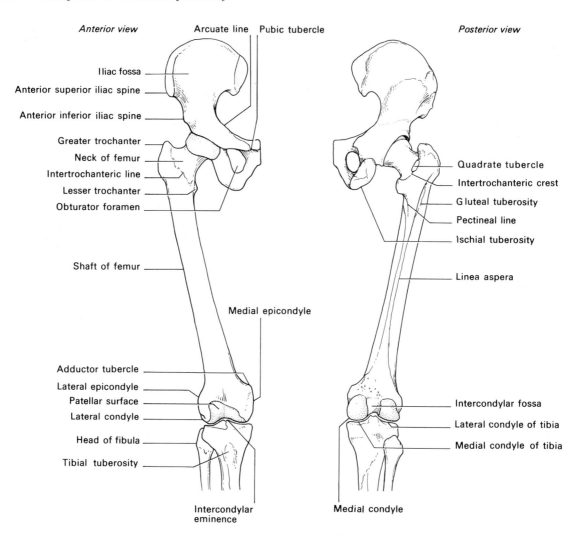

Anterior view   Arcuate line   Pubic tubercle   *Posterior view*

Iliac fossa

Anterior superior iliac spine

Anterior inferior iliac spine

Greater trochanter   Quadrate tubercle
Neck of femur   Intertrochanteric crest
Intertrochanteric line   Gluteal tuberosity
Lesser trochanter   Pectineal line
Obturator foramen   Ischial tuberosity

Shaft of femur   Linea aspera

Medial epicondyle

Adductor tubercle
Lateral epicondyle
Patellar surface   Intercondylar fossa
Lateral condyle   Lateral condyle of tibia
Head of fibula   Medial condyle of tibia
Tibial tuberosity

Intercondylar   Medial condyle
eminence

Figure 18-2.
Bones of right hip, thigh, and knee.

INGUINAL LIGAMENT

The ***inguinal ligament*** is the lower thickened edge of the ***aponeurosis*** of the external oblique muscle of the abdomen. It runs from the anterior superior iliac spine to the pubic tubercle and forms an identifiable surface anatomy feature (Fig. 18-3).

The ***lacunar ligament*** is a part of the medial end of the inguinal ligament and consists of fibers that pass posteriorly to the pectineal line of the pubis, leaving a sharp, curved, free lateral border that forms the medial margin of the femoral ring. The continuation laterally of these fibers along the pectineal line of the pubis forms the ***pectineal ligament*** (Cooper's ligament) of the pubis.

CLINICAL NOTE

The pectineal ligament is used surgically in the repair of inguinal hernias. It blends with the periosteum of the pectineal line and with the fascia covering the pectineus muscle.

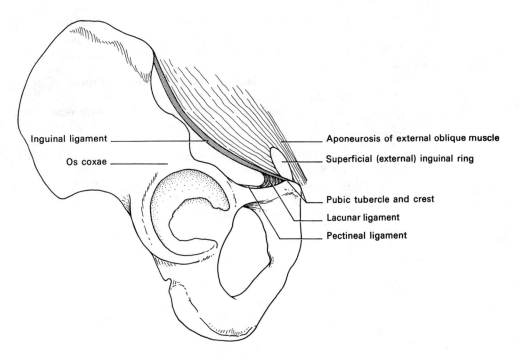

Inguinal ligament

Os coxae

Aponeurosis of external oblique muscle

Superficial (external) inguinal ring

Pubic tubercle and crest

Lacunar ligament

Pectineal ligament

SUPERFICIAL FASCIA

The superficial fascia of the abdomen is attached to the deep fascia of the thigh just below the inguinal ligament. The superficial fascia of the thigh contains some fat and certain important structures.

CUTANEOUS NERVES

Certain cutaneous nerves supply the skin of the anterior, medial, and lateral aspects of the thigh (see Fig. 12-3).

Ilioinguinal Nerve (L1) and **Genitofemoral Nerve** (L1 and L2). These nerves supply the skin of the thigh immediately below the medial half of the inguinal ligament.

Lateral Cutaneous Nerve of the Thigh (L2 and L3). The lateral surface of the thigh is supplied by this nerve. It splits into anterior and posterior branches so that it supplies not only the lateral side of the thigh, but also the lateral portions of the anterior and posterior surfaces.

CLINICAL NOTE

*The lateral cutaneous nerve of the thigh travels from the iliac fossa of the abdomen into the thigh either behind or **through** the substance of inguinal ligament. This point is very close to the medial aspect of the anterior superior iliac spine. The nerve may become trapped behind or in the inguinal ligament. This may cause pain referred to the area of distribution of the nerve, and eventually loss of sensation in this area may result. This condition, called **meralgia paresthetica,** may require surgical relief of the trapped nerve.*

Figure 18-3.
The inguinal, lacunar, and pectineal ligaments (*right side*). The pelvis is tipped backward to show the lacunar ligament.

Branches of the Femoral Nerve. The femoral nerve supplies muscles and skin on the anterior aspect of the thigh. It is usually described as having anterior and posterior divisions, and the cutaneous nerves come from its anterior division.

The ***intermediate cutaneous nerve of thigh*** supplies the anterior aspect of the thigh as far as the knee.

The ***medial cutaneous nerve of thigh*** becomes cutaneous about halfway down the thigh and supplies both the anterior and posterior aspects of the medial side of the thigh.

The ***saphenous nerve*** is the largest cutaneous branch of the femoral nerve. It travels down the thigh close to the femoral artery and does not pierce the deep fascia to become cutaneous until it reaches the lower part of the knee. Then it runs down the medial aspect of the leg, just posterior to the ***great saphenous vein***, to supply the medial aspect of the leg, ankle, and foot.

CLINICAL NOTE

The proximity of the saphenous nerve to the great saphenous vein makes it vulnerable to damage during varicose vein surgery. A permanent loss of sensation may result if it is "stripped" in the course of a vein-stripping operation.

CUTANEOUS VEINS

Great (Long) Saphenous Vein. The great and small saphenous veins are two major vessels that drain the subcutaneous tissue of the lower limb. The great saphenous vein starts at the medial end of the dorsal venous arch of the foot and runs ***anterior*** to the medial malleolus of the ankle, following the medial border of the tibia to the medial aspect of the knee joint (about a hand's breadth posterior to the medial edge of the patella). From here it continues upward on the medial aspect of the thigh to the ***saphenous opening*** in the deep fascia of the thigh, about 2.5 cm below and lateral to the pubic tubercle. At the saphenous opening the great saphenous vein enters the femoral vein (Fig. 18-4).

CLINICAL NOTE

The great saphenous vein has to carry a long column of blood from the ankle to the groin. To facilitate the antegrade flow of blood and to prevent retrograde flow it is supplied with multiple valves that permit flow only in the upward direction. If these valves become incompetent (due to causes that will be described later) the condition of varicose veins will occur in the long saphenous vein and its tributaries.

*The great saphenous vein is frequently used, particularly in infants, for the insertion of an intravenous line if no other peripheral veins are found suitable. In "cutting down" on this vein at the ankle, it is important to remember that the vein runs **anterior** to the medial malleolus and that the saphenous nerve is closely applied to its posterior aspect.*

The great saphenous vein is an eminently suitable vessel for use in arterial bypass surgery. Cardiovascular surgeons will dissect out its whole length and may use it to bypass major arteries, such as the femoral artery or coronary arteries.

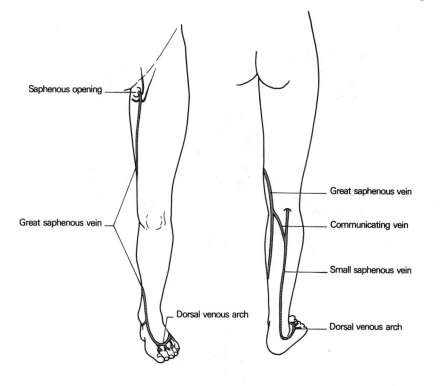

Figure 18-4.
The saphenous veins. **Left.** Anterior
view, left lower extremity. **Right.**
Posterior view, right lower extremity.

DEEP FASCIA

The deep fascia of the thigh is called the ***fascia lata*** (lata = wide or broad)
(Fig. 18-5). It provides a dense envelope for the thigh and prevents the
bulging outward of thigh muscles during their contraction. Thus it en-
hances the effectiveness of contraction of these muscles. The fascia lata
also gives attachment to some of the thigh muscles, and its lateral aspect,
from the tubercle of the iliac crest to the lateral aspect of the tibia, is especially thickened to form the ***iliotibial tract*** (Fig. 18-5). The iliotibial tract
gives attachment to the gluteus maximus and tensor fasciae latae muscles,
two important muscles in providing stability for the hip joint (see below).

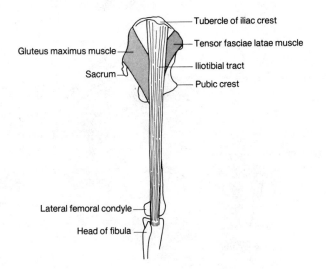

Figure 18-5.
Iliotibial tract and muscles inserting
into it (lateral view of right lower
extremity).

The deep fascia of the front of the thigh is deficient just below the inguinal ligament to allow the passage of the great saphenous vein. This **saphenous opening** has sharp lateral, superior, and inferior margins and the opening is closed by the **cribriform fascia** (a plug of fat that transmits three small tributaries of the great saphenous vein). The femoral vein lies deep to the saphenous opening.

THIGH (ANTERIOR ASPECT)

The muscles of the thigh are large and powerful and consist of three groups separated by **intermuscular septa** of deep fascia (Fig. 18-6). Intermuscular septa are extensions of the enveloping deep fascia of an area, passing from the deep fascia to ridges on the underlying bone. They separate groups of muscles and play an important role in providing large areas for attachment of muscles. They divide the anatomical part into well-defined, relatively unyielding compartments. Each compartment has its own group of muscles and neurovascular bundle.

The muscles grouped on the anterior aspect of the thigh are, in general, extensors of the knee joint (some also flex the hip joint) and are supplied by the **femoral nerve**.

The medial muscles of the thigh are **adductors** of the hip joint and are supplied by the **obturator nerve**.

The muscles on the posterior aspect of the thigh are extensors of the hip and flexors of the knee joint and are supplied by the **sciatic nerve**. The erect body is balanced on the lower limb by these heavy muscles acting like guy wires. In the lower limb it is particularly inappropriate to talk

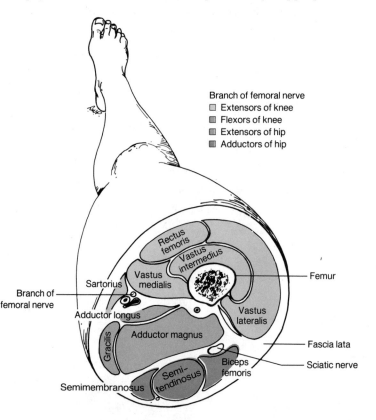

Branch of femoral nerve
☐ Extensors of knee
☐ Flexors of knee
☐ Extensors of hip
☐ Adductors of hip

Figure 18-6.
Cross-section of right thigh.

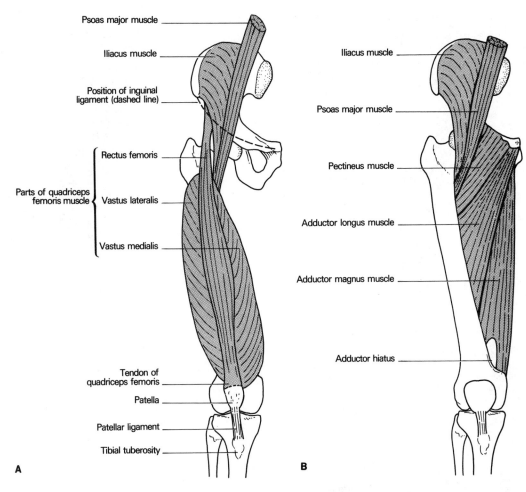

Psoas major muscle

Iliacus muscle

Position of inguinal ligament (dashed line)

Parts of quadriceps femoris muscle {
Rectus femoris

Vastus lateralis

Vastus medialis
}

Tendon of quadriceps femoris

Patella

Patellar ligament

Tibial tuberosity

A

Iliacus muscle

Psoas major muscle

Pectineus muscle

Adductor longus muscle

Adductor magnus muscle

Adductor hiatus

B

Figure 18-7.
Muscles of the right thigh, anterior view. **A.** Extensor group. **B.** Adductor group.

of origins and insertions of muscles because the action of the muscle will depend on the posture of the body. For example the gluteal muscles may extend the lower limb at the hip joint, but, just as frequently, they will act from their lower attachments on the hip joint to provide balance for the standing body or to retrieve the whole body from a bent over or squatting position. For these reasons the terms "origin" and "insertion" have been replaced by "attachments" whenever possible in this book.

MUSCLES OF THE THIGH (ANTERIOR ASPECT)

Iliopsoas

This powerful muscle is really a composite structure made up of the fusion of the *iliacus* and *psoas* muscles near their lower attachments (Fig. 18-7). The proximal attachments were described in Chapter 12. The distal attachment is to the lesser trochanter of the femur. The iliopsoas passes deep to the inguinal ligament from the abdomen to its lower attachment.

Nerve Supply. Psoas from lumbar nerves (*L1, L2,* and L3), iliacus from the femoral nerve (*L2* and L3).

Action. The iliopsoas is a strong flexor of the hip joint. In the standing position it steadies the hip joint and prevents the body from tipping backward at the hip joint.

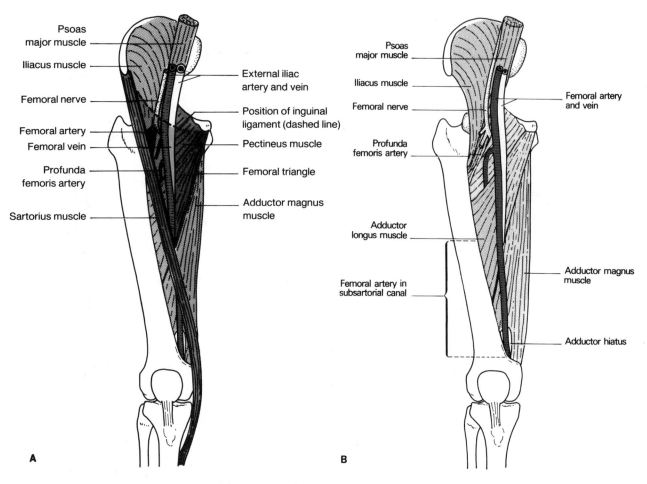

A

B

Figure 18-8.
Nerves and vessels of the front of the right thigh. **A.** The femoral triangle. **B.** Femoral artery and vein (full length).

Tensor Fasciae Latae

Proximal Attachment. The anterior part of the iliac crest.

Distal Attachment. The iliotibial tract (see Fig. 18-5).

Nerve Supply. Superior gluteal nerve (L4 and L5).

Action. This muscle can abduct and flex the hip joint, but in the upright posture it steadies the body on the thigh.

Sartorius

The sartorius is a long narrow strip of muscle that stretches from the anterior superior iliac spine to the upper end of the medial surface of the shaft of the tibia (Fig. 18-8).

Proximal Attachment. The anterior superior iliac spine.

Distal Attachment. Proximal end of the medial surface of the shaft of the tibia.

Nerve Supply. Femoral nerve (L2 and L3).

Action. Weak flexion, abduction, and lateral rotation of the thigh.

The sartorius covers the femoral vessels through most of their course in the ***adductor canal***; the muscle forms the roof of the adductor (subsartorial) canal.

Quadriceps Femoris

This muscle comprises four parts: the ***rectus femoris, vastus lateralis, vastus medialis,*** and ***vastus intermedius*** (see Fig. 18-7A).

Attachments. The rectus femoris is attached above to the anterior inferior iliac spine and to an area of ilium above the acetabulum (reflected head of rectus femoris). The three vasti are attached to the shaft of the femur and the medial and lateral intermuscular septa. The fibers of the four heads of quadriceps femoris come together in the ***quadriceps tendon***, which attaches to and surrounds the patella before continuing as the ***ligamentum patellae*** and attaching to the tuberosity of the tibia.

Nerve Supply. The quadriceps femoris is supplied by the femoral nerve (L2, ***L3***, and ***L4***).

Action. Acting through the ligamentum patellae and the patella (which is a ***sesamoid bone*** that changes the direction of pull of a muscle), the quadriceps femoris extends the knee joint and, through the action of rectus femoris, flexes the hip.

CLINICAL NOTE

*The ligamentum patellae is struck to elicit the **patellar reflex** (knee-jerk). The powerful quadriceps muscle is a strong stabilizer of the knee joint. Following injury to the knee joint the quadriceps will waste (lose bulk from inactivity) rapidly, particularly vastus medialis. Considerable physiotherapy is required to regain substance and power of the quadriceps after knee injury or surgery.*

Pectineus

The pectineus is a short, flat member of the adductor group of muscles (see Fig. 18-7B).

Proximal Attachment. Pectineal line of the os coxae.

Distal Attachment. Pectineal line of the femur.

Nerve Supply. From the femoral nerve (***L2*** and L3).

Action. Adduction and flexion of the hip.

Adductor Longus

The adductor longus is a relatively strong member of the adductor group of muscles, but it is considered here because it forms a boundary of the femoral triangle (see Figs. 18-7B, 18-8A, and 18-8B).

Proximal Attachment. To the pubis near its crest and tubercle.

Distal Attachment. To the middle third of the medial lip of the linea aspera of the femur.

Nerve Supply. The anterior branch of the obturator nerve (L2, ***L3***, and L4).

Action. The adductor longus adducts and flexes the hip and may act as a fixator of the hip joint in flexion of the knee.

FEMORAL TRIANGLE (see Fig. 18-8A)

The femoral triangle is an important anatomical landmark on the anterior aspect of the thigh. Its base is formed by the ***inguinal ligament***, its medial border by the medial edge of ***adductor longus***, and its lateral border by the medial border of ***sartorius***. Its apex is the point where the medial border of sartorius crosses the medial border of adductor longus. The floor of the triangle (from lateral to medial side) is formed by iliopsoas, pectineus, and adductor longus. The contents of the triangle are (from lateral to medial) the femoral nerve, femoral artery, femoral vein, and certain branches and tributaries of these structures. The roof is the fascia lata. The femoral triangle contains the ***deep inguinal lymph nodes*** which drain the deeper tissues of the lower limb and the glans penis (clitoris in the female).

At the apex of the femoral triangle the femoral vessels enter the ***adductor*** (subsartorial) ***canal*** to pass downward and medially to supply the rest of the thigh, the leg, and the foot. Some branches of the femoral nerve accompany the femoral vessels into the adductor canal, but their area of supply does not extend far below the knee (with the exception of the saphenous nerve).

Femoral Sheath. The iliac fascia and the transversalis fascia extend from the abdomen, behind the inguinal ligament, to form the femoral sheath. This sheath is a firm tube of fibrous tissue that contains the femoral artery, the femoral vein, and the ***femoral canal***, from lateral to medial. The femoral nerve is ***not*** included in the femoral sheath.

Femoral Nerve. The femoral nerve passes from the abdomen into the thigh posterior to the inguinal ligament, just lateral to the femoral sheath. It has a short course in the thigh (about 2.5 cm) before it ends by breaking up into numerous terminal cutaneous and unnamed muscular branches. The cutaneous branches are the intermediate and medial cutaneous nerves of the thigh and the saphenous nerve. The femoral nerve sends muscular branches to the components of the quadriceps femoris muscle, the sartorius and pectineus muscles.

Femoral Artery. The femoral artery is the main artery of the lower limb. It is the continuation of the external iliac artery and runs from the midpoint of the inguinal ligament to the lower end of the adductor canal.

The femoral artery gives off some major branches in the thigh (Fig. 18-9).

1. ***Femoral Circumflex Arteries.*** There are two circumflex arteries: a medial and a lateral. They are usually branches of the profunda femoris artery but may arise from the femoral artery. They wind around the upper end of the femur, deep to most of the muscles in the upper thigh, and anastomose with branches of the gluteal arteries in the buttock.

2. ***Profunda Femoris Artery.*** The profunda femoris artery is a large branch that arises from the posterior aspect of the femoral artery near the base of the femoral triangle and passes posteriorly between the pectineus and adductor longus muscles to descend near the back of the femur. It gives off four ***perforating branches*** that pierce the ***adductor magnus muscle*** to supply the ***hamstring muscles*** on the back of the thigh. The perforat-

Figure 18-9.
Femoral artery.

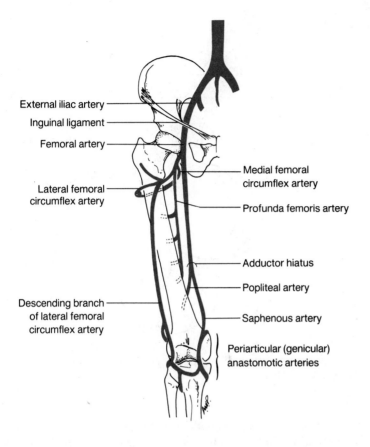

External iliac artery

Inguinal ligament

Femoral artery

Lateral femoral circumflex artery

Medial femoral circumflex artery

Profunda femoris artery

Adductor hiatus

Popliteal artery

Descending branch of lateral femoral circumflex artery

Saphenous artery

Periarticular (genicular) anastomotic arteries

ing arteries anastomose with each other as well as with the cir-cumflex and gluteal arteries.

CLINICAL NOTE

The pulsation of the femoral artery can be felt just below the midpoint of the inguinal ligament. Obstruction of the femoral artery by atheroma, thrombosis, or embolism usually occurs at the point of origin of the pro-funda artery. Thus palpation of a femoral pulse at the midinguinal point is no guarantee of adequate blood flow below the origin of the profunda artery. Young physicians, during their internships or residencies, will fre-quently be called on to take arterial blood samples for analysis from the femoral artery; thus its position is important.

Femoral Vein. The femoral vein lies medial to the artery in the upper part of the femoral triangle, but it soon winds its way to the posterior aspect of the artery. While in the femoral triangle it receives the profunda femoris vein and the great saphenous vein as its major tributaries.

Femoral Canal. The femoral canal is a space medial to the femoral vein. It is formed by the same layers of fascia as the femoral sheath. It is a short, blind potential space that contains some fat and a lymph node (node of Cloquet). The femoral canal is a potential site for a ***femoral hernia***. The upper end of the femoral canal is the ***femoral ring***, which has well-marked boundaries: the anterior boundary of the femoral ring is the ingui-nal ligament; posteriorly lies the pectineal ligament (on the superior ra-

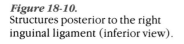

Figure 18-10.
Structures posterior to the right
inguinal ligament (inferior view).

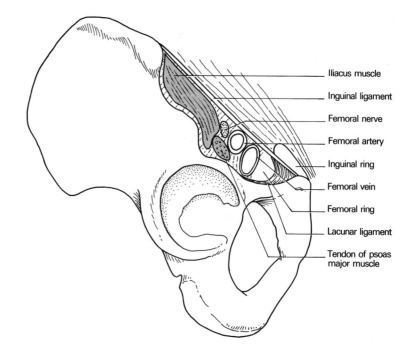

Iliacus muscle

Inguinal ligament

Femoral nerve

Femoral artery

Inguinal ring

Femoral vein

Femoral ring

Lacunar ligament

Tendon of psoas
major muscle

mus of the pubis); medially lies the lacunar ligament; and laterally is the femoral vein (Fig. 18-10).

CLINICAL NOTE

Femoral hernia is not uncommon. It occurs most frequently in females, probably because of the relatively greater width of the superior ramus of the pubis of the female pelvis.

*Unlike the inguinal canal, the femoral canal has no muscular walls to protect it, and, under conditions of increased intra-abdominal pressure, herniation of peritoneum with visceral contents may occur. Femoral hernias are particularly prone to strangulate (**i.e.**, the blood supply of the herniated viscus is compressed and occluded at the **femoral ring**). A large femoral hernia may extend through the deep fascia of the thigh at the **saphenous opening**.*

*The techniques of repair of femoral hernias are beyond the scope of this book, but the reader is reminded that the lacunar ligament may have to be cut to relieve a strangulated femoral hernia and that an **abnormal obturator artery** may be present.*

Inguinal Lymph Nodes. The femoral triangle contains two or three deep inguinal lymph nodes; they drain the deeper tissues of the lower limb. The superficial inguinal lymph nodes are arranged at the base of the femoral triangle, in the superficial fascia, parallel to the inguinal ligament; they drain the skin of the lower limb, the buttock, the perineum, the skin of the scrotum (vulva), the lower half of the anal canal, and the skin of the trunk (front and back) from below the level of the umbilicus.

ADDUCTOR (SUBSARTORIAL) CANAL

The adductor canal accommodates the femoral vessels and the saphenous nerve and extends from the apex of the femoral triangle to the **adductor hiatus**, an opening in the tendon of the **adductor magnus muscle**. The adductor canal is formed by the ***adductor longus*** and ***adductor magnus*** muscles medially and the ***vastus medialis*** laterally. The roof of the canal is formed by the ***sartorius*** muscle and the deep fascia of the thigh. At the adductor hiatus the femoral vessels enter the ***popliteal fossa*** behind the knee joint and change their names to ***popliteal artery and vein*** (see Fig. 18-8B).

THIGH (MEDIAL ASPECT)

The medial aspect of the thigh contains the adductor group of muscles. Adductor longus and pectineus have already been described with the femoral triangle.

The adductor group of muscles has a very large mass, but its exact function is the subject of controversy. These muscles undoubtedly adduct the thigh and act as fixators of the hip joint in flexion of the knee. In addition they appear to be active, although perhaps not as prime movers, during walking. Electromyographic studies show little activity in the adductor muscles when the body is standing easily in the erect posture.

Pectineus

This muscle forms part of the floor of the femoral triangle.

Proximal Attachment. To the pectineal line of the pubis.

Distal Attachment. A line running from the lesser trochanter to the linea aspera, the pectineal line of the femur.

Nerve Supply. The femoral nerve (L2, *L3*).

Action. An adductor of the hip.

Adductor Longus

A triangular muscle, the adductor longus runs from the pubic crest to the linea aspera (see Figs. 18-7B, 18-8A, and 18-8B).

Proximal Attachment. Pubic crest.

Distal Attachment. Linea aspera.

Nerve Supply. The anterior branch of the obturator nerve (L2, *L3*, and L4).

Action. An adductor of the hip.

Adductor Brevis (Fig. 18-11)

Lying on a more posterior plane, the adductor brevis is not seen from the anterior aspect of the thigh unless the adductor longus is displaced.

Proximal Attachment. To the inferior ramus of the pubis.

Distal Attachment. The middle section of the linea aspera.

Nerve Supply. From the obturator nerve (L2, *L3*, and L4).

Action. An adductor of the hip.

Adductor Magnus

A very large muscle, the adductor magnus forms the bed of the hamstring muscles on the back of the thigh and of the other adductors (Fig. 18-11).

Figure 18-11.
Obturator nerve (right side).

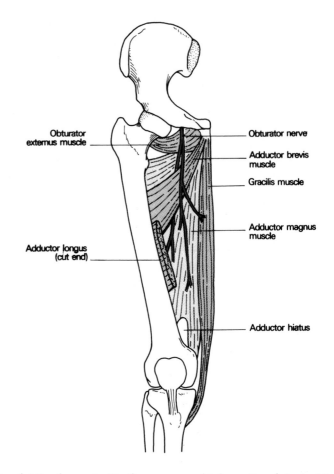

Obturator
externus muscle

Obturator nerve

Adductor brevis
muscle

Gracilis muscle

Adductor longus
(cut end)

Adductor magnus
muscle

Adductor hiatus

Proximal Attachment. To the ramus and tuberosity of the ischium and the inferior ramus of the pubis.

Distal Attachment. From the gluteal tuberosity of the femur to the linea aspera, medial supracondylar line, and the adductor tubercle. There is a gap in the fibers of its lower attachment, the ***adductor hiatus***, through which femoral vessels pass into the popliteal fossa.

Nerve Supply. Adductor magnus is undoubtedly a compound muscle receiving a nerve supply from the posterior branch of the obturator nerve (L2, *L3*, and *L4*) and from the tibial portion of the sciatic nerve (*L4*).

Action. The portion of adductor magnus supplied by the obturator nerve flexes and adducts the hip while the part innervated by the sciatic nerve is an extensor of the hip (like the hamstring muscles).

Gracilis

The gracilis is a comparatively weak member of the adductor group.

Proximal Attachment. From the inferior ramus of the pubis.

Distal Attachment. To the medial aspect of the upper part of the tibia. It is the most medial muscle of the thigh.

Nerve Supply. The obturator nerve (*L2* and L3) (Fig. 18-11).

Action. Adducts and flexes the hip and flexes the knee.

Obturator Externus

The curiously shaped obturator externus is ideally situated to be a lateral rotator of the hip.

Medial Attachment. It winds around the inferior aspect of the hip joint from the external surface of the obturator membrane (and surrounding bone).

Lateral Attachment. To the ***trochanteric fossa*** on the medial surface of the greater trochanter of the femur.

Nerve Supply. The posterior branch of the obturator nerve (L3 and **L4**).

Action. It is a lateral rotator of the hip, but its proximity to the hip joint is reminiscent of the rotator cuff muscles of the shoulder joint (Fig. 18-11).

OBTURATOR NERVE

The ***obturator nerve*** is a branch of the anterior (ventral) division of the lumbar plexus. It runs through the obturator foramen, where it divides into anterior and posterior branches; the adductor brevis intervenes between these two branches. It supplies all of the adductor muscles except pectineus and the "hamstring" portion of adductor magnus. The obturator nerve supplies the hip joint, a strip of skin on the medial aspect of the thigh and knee joint, and gives a sensory branch to the knee joint.

CLINICAL NOTE

John Hilton, a surgeon and anatomist of the 19th century, formulated **Hilton's Law**, *which states, "The same trunks of nerves, whose branches*

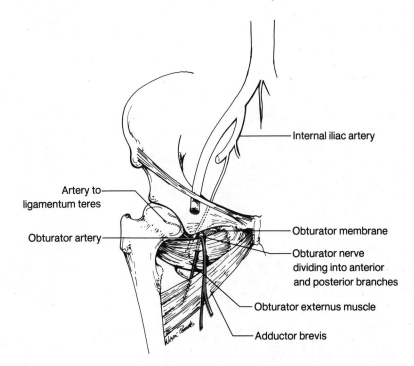

Figure 18-12.
Oburator nerve and obturator artery.

supply the groups of muscles moving a joint, furnish also a distribution of nerves to the skin over the insertion of the same muscles; and–what at this moment more especially merits attention–the interior of the joint receives its nerves from the same source."

The often cited clinical example of this is the patient who complains of pain in the knee when the true pathologic origin of the pain is in the hip joint. This is another example of **referred pain.**

The adductor muscles, the function of which is the subject of much controversy, may go into severe, painful, and deforming spasm in certain paralytic conditions in the central nervous system. They can be divided surgically to relieve this condition.

VESSELS

The small *obturator artery* normally passes through the obturator foramen to supply the muscles surrounding the foramen. The obturator artery usually sends a branch to the head of the femur through the *ligamentum teres* (Fig. 18-12). The *abnormal obturator artery* has already been mentioned.

The Gluteal Region and Posterior Aspect of the Thigh

GLUTEAL REGION

BONES

The main features of the os coxae have already been discussed. However, a few bony landmarks must now be added and others reviewed to allow a fuller understanding of the gluteal region.

> *Ilium*: inferior, anterior, and posterior gluteal lines; the auricular surface (Fig. 19-1).
> *Ischium*: spine and tuberosity. If the subject is in the anatomical position, the ischial spine is on the same horizontal plane as the greater trochanter.
> *Sacrum*: foramina, median crest, lateral mass, and the position of the sacroiliac joint, including the attachment of the interosseous ligament and the auricular surface.

LIGAMENTS

Note the **sacrospinous** and **sacrotuberous** ligaments closing the sciatic notches to produce the greater and lesser sciatic foramina (see Fig. 15-1). In general the greater sciatic foramen allows structures to pass into or out of the **pelvis**; the lesser sciatic foramen allows structures to pass into or out of the **perineum**.

MUSCLES

Gluteus Maximus

This is by far the largest muscle in the lower limb, indeed in the whole body. It has very coarse red fasciculi and is a powerful extensor of the hip *and* a stabilizer of the hip joint in the erect posture (Fig. 19-2).

Proximal Attachments. The outer surface of the ilium posterior to the posterior gluteal line, the posterior aspect of the lateral mass of the sacrum, and the sacrotuberous ligament.

Distal Attachments. Its most important distal attachment is into the **ilio-tibial tract**, which receives three quarters of the muscle fibers of gluteus

19

Figure 19-1.
Skeleton of right gluteal region and back of thigh (posterior view).

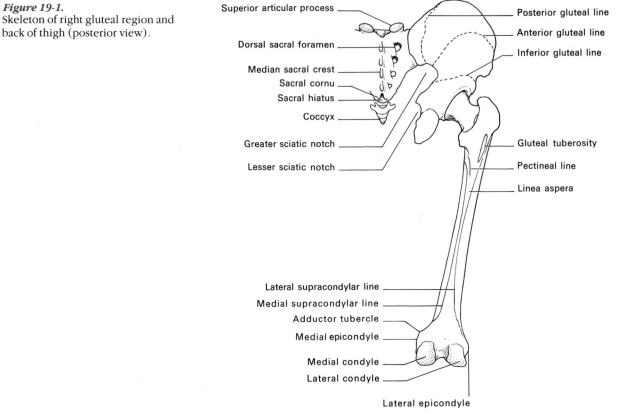

Superior articular process

Dorsal sacral foramen

Median sacral crest

Sacral cornu

Sacral hiatus

Coccyx

Greater sciatic notch

Lesser sciatic notch

Posterior gluteal line

Anterior gluteal line

Inferior gluteal line

Gluteal tuberosity

Pectineal line

Linea aspera

Lateral supracondylar line

Medial supracondylar line

Adductor tubercle

Medial epicondyle

Medial condyle

Lateral condyle

Lateral epicondyle

maximus; the remaining quarter of the muscle is attached to the gluteal tuberosity of the femur.

Nerve Supply. From the inferior gluteal nerve (L5, *S1*, and *S2*).

Action. The gluteus maximus extends the hip, particularly by tilting the pelvis backward on the femur when getting up from a stooped position or in walking upstairs, and it helps to steady the gait as weight is shifted from one leg to the other in walking. The gluteus maximus is inactive in standing erect but will exert a strong pull on the iliotibial tract when the knee is in extension with the quadriceps muscles relaxed.

Note that the iliotibial tract steadies the leg on the thigh and the thigh on the hip. The ***gluteus maximus*** and the ***tensor fasciae latae*** entering the iliotibial tract from opposite directions act as adjustable guy wires (see Fig. 18-5).

Bursae. A bursa is a small sac with a synovial lining that normally contains a very small amount of fluid. Its purpose is to reduce the friction on a tendon passing over a bone or some other area of resistance; bursae will also be found where skin or muscles are subjected to pressure against bony prominences.

1. One large bursa lies between the gluteus maximus and the greater trochanter of the femur.
2. A second bursa lies between the muscle and the ischial tuberosity.
3. A third bursa is found between the gluteus maximus and the vastus lateralis. These bursae permit the muscle to slide over these prominences.

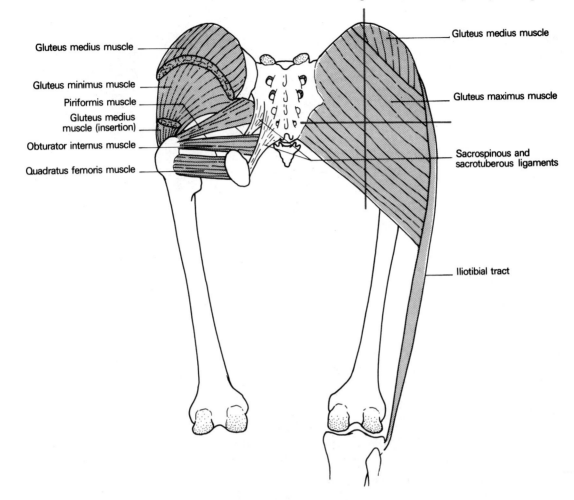

Gluteus medius muscle

Gluteus minimus muscle

Piriformis muscle

Gluteus medius muscle (insertion)

Obturator internus muscle

Quadratus femoris muscle

Gluteus medius muscle

Gluteus maximus muscle

Sacrospinous and sacrotuberous ligaments

Iliotibial tract

Note that the gluteus maximus is quadrilateral in shape with its upper and lower borders almost parallel. The lower border, crossed by a horizontal fold of skin, helps to form the fold of the buttock.

Figure 19-2.
Muscles of the gluteal region. The gluteus maximus has been removed on the left side.

Gluteus Medius

The pelvis is steadied at each step by this important abductor of the hip joint (Fig. 19-2). In this action it is assisted by the gluteus minimus.

Proximal Attachment. The outer surface of the ilium between the posterior and the anterior gluteal lines (Figs. 19-2 and 19-3).

Distal Attachment. The lateral aspect of the greater trochanter of the femur.

Nerve Supply. The superior gluteal nerve (*L5* and S1).

Gluteus Minimus

This is the smallest of the gluteal muscles (Figs. 19-2 and 19-3).

Proximal Attachment. The ilium between the anterior and inferior gluteal lines.

Figure 19-3.
Some structures of the right gluteal region and posterior thigh.

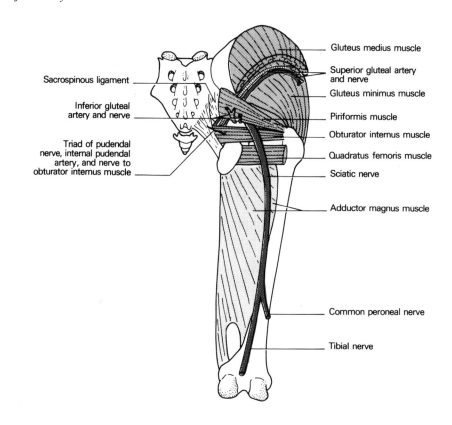

Sacrospinous ligament

Inferior gluteal artery and nerve

Triad of pudendal nerve, internal pudendal artery, and nerve to obturator internus muscle

Gluteus medius muscle
Superior gluteal artery and nerve
Gluteus minimus muscle
Piriformis muscle
Obturator internus muscle
Quadratus femoris muscle
Sciatic nerve
Adductor magnus muscle
Common peroneal nerve
Tibial nerve

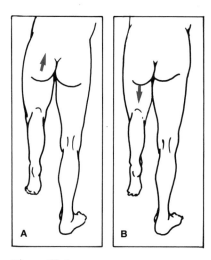

Figure 19-4.
Trendelenberg's sign.
A. Trendelenberg's sign negative. Patient is able to abduct pelvis on right femur. **B.** Trendelenberg's sign positive. Patient is unable to abduct pelvis on right femur; therefore left buttock droops.

Distal Attachment. To the anterior part of the greater trochanter antero-medial to the attachment of gluteus medius.

Nerve Supply. Superior gluteal nerve (*L5* and S1).

Actions. Acting from their pelvic attachments the gluteus medius and minimus abduct the thigh, and their anterior fibers are medial rotators of the hip. Acting from below they tilt the pelvis laterally to maintain the body balance when the ***opposite*** foot is off the ground, for example, in walking or standing on one leg.

CLINICAL NOTE

It has just been stated that the gluteus medius and minimus abduct the pelvis on the femur (i.e., they tilt the pelvis) in walking and in standing on one leg. To be able to perform this function the muscles, their levers, and the fulcrum of the movement must be intact; the levers are the head and neck of the femur, and the fulcrum is the hip joint.

*When any of these structures is not functioning normally the pelvis will not tilt to the affected side when the opposite foot is raised from the ground. This is best observed by watching the naked patient from behind and seeing whether the fold of the buttock tilts upward at its medial end when the weight is borne on that side. Failure of the fold of the buttock to tilt in the appropriate manner is called **Trendelenberg's sign** and is, of course, indicative of failure of one of the three components of the pelvic tilting mechanism on that side (Fig. 19-4).*

The gait resulting from the above-mentioned defects will be one in which the affected side dips with each step.

Piriformis

This small muscle is an important landmark for the vessels and nerves of the gluteal region (see Figs. 19-3 and 19-5).

Proximal Attachment. The anterior aspect of the second to fourth segments of the sacrum, inside the pelvis.

Distal Attachment. It leaves the pelvis through the greater sciatic foramen and attaches to the top of the greater trochanter of the femur (see Figs. 13-4 and 19-3).

Nerve Supply. From the sacral plexus (L5, *S1*, and *S2*).

Action. The piriformis is a lateral rotator of the extended thigh and an abductor of the flexed thigh.

Obturator Internus

The obturator internus was encountered in the pelvis (Figs. 19-3 and 19-5).

Proximal Attachment. The greater part of the internal surface of the ilium, pubis, and ramus of the ischium, surrounding the obturator foramen. Its fibers pass out of the pelvis through the lesser sciatic foramen and run horizontally behind the capsule of the hip joint.

Distal Attachment. The distal tendon makes a sharp bend as it slides over the ischium below the ischial spine, leaving a smooth area on that bone. It then attaches to the medial surface of the greater trochanter. A pair of small muscles are found on the superior and inferior borders of the distal obturator internus: the superior and inferior *gemelli*.

Nerve Supply. Nerve to obturator internus (one of three structures crossing the ischial spine from the greater to the lesser sciatic notch) (L5, *S1*).

Action. Lateral rotation of the hip and stabilizing the hip joint.

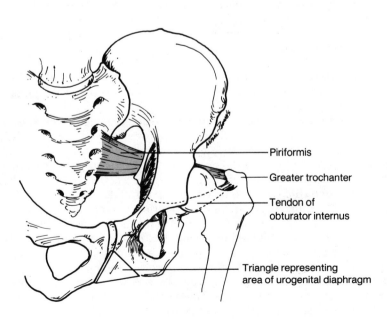

Piriformis

Greater trochanter

Tendon of obturator internus

Triangle representing area of urogenital diaphragm

Figure 19-5.
Left pelvis showing piriformis and obturator internus.

CLINICAL NOTE

Nélaton's line is a line joining the anterior superior iliac spine to the most prominent part of the ischial tuberosity. The tip of the greater trochanter should lie on this line or just below it. If the tip of the greater trochanter is above this line it indicates an upward displacement of the trochanter due to either dislocation of the hip joint or a fracture of the neck of the femur; the latter is common in elderly people and is often incorrectly referred to as a "fractured hip."

NERVES

Superficial Nerves

The superficial (cutaneous) nerves of the gluteal region are relatively unimportant. They come from the posterior primary rami of the lumbar and sacral nerves and from the sacral plexus. The only cutaneous nerve of significant size is the ***posterior cutaneous nerve of thigh***, which comes from the sacral plexus (S2 and S3), passes below the lower border of the gluteus maximus, and supplies the skin of the buttock, the posterior aspect of the thigh, and the upper part of the posterior surface of the leg; at the lower border of the gluteus maximus it gives off a much mentioned, but relatively unimportant, perineal branch that winds over the origin of the hamstring muscle and supplies the scrotum (vulva in the female).

Deep Nerves

The deep nerves all come from the sacral plexus. Their locations are best understood and remembered in their relationships to the piriformis muscle, which acts as a handy guide to their positions.

Surface Marking. The position of the lower border of piriformis can be found as follows: find the midpoint of a line joining the tip of the coccyx to the posterior superior iliac spine. Join this point to the tip of the greater trochanter. This line forms the surface marking of the lower border of piriformis.

   The superior gluteal nerve emerges from the greater sciatic foramen ***above*** the piriformis muscle. It runs between gluteus medius and minimus to supply both of them and then terminates in the tensor fasciae latae, which it also supplies. The superior gluteal nerve is accompanied by branches of the superior gluteal artery (see Fig. 19-3).

   The inferior gluteal nerve passes through the greater sciatic foramen, appears at the ***lower*** border of piriformis, and supplies the gluteus maximus (see Fig. 19-3).

   The sciatic nerve passes through the greater sciatic foramen inferior to piriformis and then descends midway between the ischial tuberosity and the greater trochanter of the femur. It runs down the back of the thigh and usually splits into the ***tibial*** nerve and the ***common peroneal*** nerve somewhere above the popliteal fossa (see Fig. 19-3). The sciatic nerve has two constituents: the common peroneal nerve and the tibial nerve. They differ in that the tibial nerve arises from anterior (ventral) divisions of the sacral plexus whereas the common peroneal nerve originates from posterior (dorsal) divisions of that plexus (Fig. 19-6). These two nerves can be easily separated by blunt dissection up into the gluteal region. Occasionally the two components leave the greater sciatic foramen already separated; in that case the tibial nerve will emerge below the piriformis muscle and the common peroneal nerve will pierce that muscle. In its descent into the thigh the sciatic nerve travels behind the ischium, obturator internus,

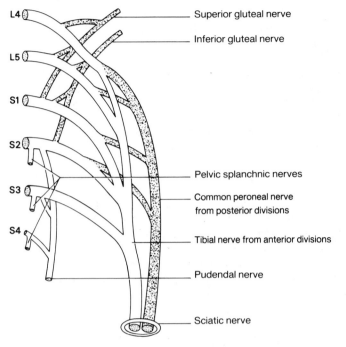

Figure 19-6.
Sacral plexus (anteroinferior view),
showing anterior and posterior
divisions.

quadratus femoris, and adductor magnus muscles under the cover of the
gluteus maximus muscle.

CLINICAL NOTE

*The sciatic nerve may be damaged in traumatic dislocations of the hip
joint; its powers of regeneration are minimal. The term **sciatica**, that is,
pain passing from the buttock to the distal distribution of the sciatic nerve
(e.g., into the calf or ankle) has been ascribed incorrectly to trapping or
irritation of the sciatic nerve. Sciatica is nearly always caused by irritation of nerve roots of the cauda equina in the spinal canal or as they pass
through intervertebral foramina.*

The nerve to quadratus femoris leaves the greater sciatic foramen
below piriformis and passes deep to the sciatic nerve and obturator internus muscle to enter the deep surface of the quadratus femoris muscle.

The pudendal nerve passes inferior to piriformis and posterior to
the ischial spine or sacrospinous ligament (see Figs. 15-6 and 19-3). It
enters the perineum through the lesser sciatic foramen (medial to the internal pudendal artery, which is medial to the nerve to the obturator internus) and supplies the structures of the perineum. It is an important nerve
because it supplies the voluntary anal sphincter, the external sphincter
urethrae, and sensation to the external genitalia.

ARTERIES

It will be seen from the following description of the arteries (and accompanying veins) of the gluteal region that the vessels to be mentioned form
an extensive anastomotic network around the posterior aspect of the hip
joint. Participating in these anastomotic channels are branches of the internal iliac artery and branches of the femoral (and profunda femoris) artery.

Such anastomotic channels are found around all joints (Fig. 18-9);
they maintain the circulation of blood to the limb when the vessels passing

over a joint are partially obstructed by extreme flexion of the joint. It can be imagined that when the thigh is in contact with the anterior abdominal wall (as in squatting) the femoral artery will be acutely angulated at the flexed hip joint. The circulation to the distal limb is then maintained by the anastomotic branches from major vessels that are not angulated in this particular joint configuration. This principle will be encountered repeatedly around joints and can be called a ***periarticular anastomotic system***.

The ***superior gluteal artery*** comes from the internal iliac artery, passes through the greater sciatic foramen (see Figs. 15-6 and 19-3) superior to the piriformis, and supplies the gluteus maximus, medius, and minimus; it gives a large ***nutrient artery*** to the ilium of the os coxae. It anastomoses with the inferior gluteal and medial femoral circumflex arteries.

The ***inferior gluteal artery*** arises from the internal iliac artery, passes through the greater sciatic foramen inferior to piriformis, and supplies the gluteus maximus (see Figs. 15-6 and 19-3). It anastomoses with the superior gluteal, medial, and lateral femoral circumflex arteries and with the first perforating branch of the profunda femoris artery to form the cruciate anastomosis (see Fig. 18-9). The inferior gluteal artery also supplies the ***artery to the sciatic nerve***.

CLINICAL NOTE

The artery to the sciatic nerve (arteria nervi ischiaticae) is a prominent example of the arteries that supply every nerve. If a nerve is cut and the surgeon wishes to suture it to restore continuity so that regeneration can occur, it is vitally important that the two ends to be sutured together are not twisted along their long axes. Matching the two cut ends of the artery aids this correct orientation.

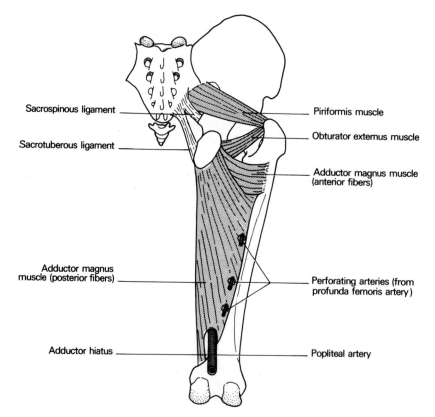

Sacrospinous ligament

Sacrotuberous ligament

Adductor magnus muscle (posterior fibers)

Adductor hiatus

Piriformis muscle

Obturator externus muscle

Adductor magnus muscle (anterior fibers)

Perforating arteries (from profunda femoris artery)

Popliteal artery

Figure 19-7.
Right adductor magnus muscle (posterior view).

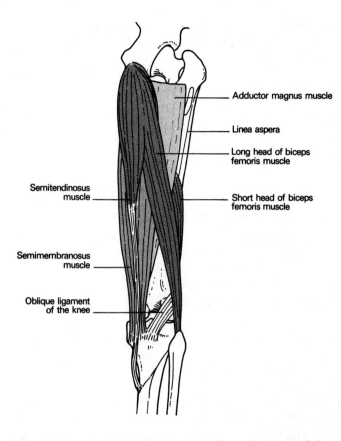

Figure 19-8.
Right hamstring muscles (posterior view).

Adductor magnus muscle

Linea aspera

Long head of biceps femoris muscle

Short head of biceps femoris muscle

Semitendinosus muscle

Semimembranosus muscle

Oblique ligament of the knee

The **internal pudendal artery** arises from the internal iliac artery, passes lateral to the pudendal nerve, posterior to the ischial spine or sacrospinous ligament, and through the lesser sciatic foramen to the perineum (see Fig. 15-6). It supplies the erectile tissues of the external genitalia and other perineal structures.

The **medial femoral circumflex artery** appears between the muscles in the posterior region of the thigh. This artery, a branch of the profunda femoris artery, passes posteriorly between the pectineus and psoas major muscles and, at the back of the thigh participates in the **cruciate anastomosis** behind the hip (see Fig. 18-9). The **perforating arteries** are branches of the profunda femoris artery that pierce the adductor magnus muscle to reach the back of the thigh and form a longitudinal anastomotic network close to the shaft of the femur (Fig. 19-7).

CLINICAL NOTE

The gluteal region is a suitable place for the administration of intramuscular injections. Many of the medications to be injected are irritants to tissue, and care must be taken that they are not injected into the sciatic nerve. The upper and outer quadrant of the gluteal region (see Fig. 19-2) is a safe site.

THIGH (POSTERIOR ASPECT)

The collective name for the muscles of the back of the thigh is **hamstrings** (Fig. 19-8). The proximal attachment of a hamstring muscle is the ischial tuberosity. The distal attachments are the tibia or fibula. One hamstring muscle (the biceps femoris) has some superior attachment to the posterior aspect of the femur. The hamstring muscles are supplied by the *sciatic nerve*.

MUSCLES (HAMSTRINGS)

Semitendinosus

This hamstring muscle has a very long, thin tendon that gives the muscle its name (Fig. 19-8).

Proximal Attachment. The semitendinosus shares its upper attachment with the biceps femoris muscle at the tuberosity of the ischium; at this level the two muscles are inseparable.

Distal Attachment. The lower attachment of the semitendinosus is to the medial surface of the upper part of the shaft of the tibia behind (and blended with) the lower attachments of sartorius and gracilis. The semitendinosus is fleshy above but tendinous below.

Nerve Supply. The tibial division of the sciatic nerve (*L5*, *S1*, and S2).

Action. Extension of the hip and flexion of the knee. When the knee is flexed semitendinosus can medially rotate the tibia on the femur.

Semimembranosus

This muscle has a long, flat aponeurosis-like upper tendon that gives the muscle its name (Figs. 19-8 and 19-9).

Proximal Attachment. From the tuberosity of the ischium by a membrane-like tendon that partially wraps around the fleshy bellies of the long head of biceps and semitendinosus.

Distal Attachment. The muscle passes down the back of the thigh to the medial condyle of the tibia and is attached to a horizontal groove on the posteromedial aspect of the condyle.

Nerve Supply. From the tibial division of the sciatic nerve (*L5* and *S1*).

Action. Extension of the hip and flexion and medial rotation of the tibia at the knee.

Biceps Femoris

This major member of the hamstring group of muscles has two heads or bellies; the composite muscle runs downward to be attached to the fibula (Figs. 19-8 and 19-9).

Proximal Attachment. The ischial tuberosity (long head), the posterior aspect of the shaft of the femur, the lateral supracondylar line of the femur, and the lateral intermuscular septum (short head).

Distal Attachment. The lower tendon is attached to the head of the fibula. The tendon is split at this attachment by the *fibular collateral ligament* of the knee joint.

Nerve Supply. The tibial division of the sciatic nerve (*L5* and *S1*). The portion of the muscle attached to the femur is supplied by the common peroneal division of the sciatic nerve (L5 and S1).

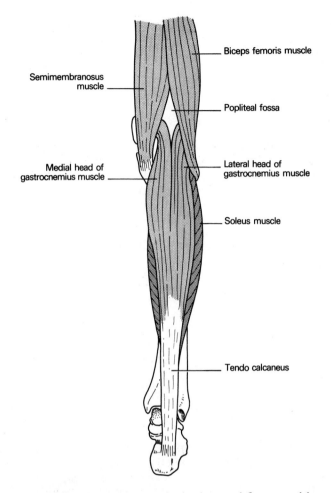

Biceps femoris muscle

Semimembranosus muscle

Popliteal fossa

Medial head of gastrocnemius muscle

Lateral head of gastrocnemius muscle

Soleus muscle

Tendo calcaneus

Figure 19-9.
Right gastrocnemius muscle (posterior view).

Action. The biceps femoris extends the hip and flexes and laterally rotates the tibia at the knee joint.

Actions of the Hamstring Muscles. The actions of the individual members of this group of muscles were described above. Acting together they extend the hip and flex the knee joint.

The hamstring muscles are so located that if the hip is flexed (thereby stretching the hamstrings) the knee cannot be fully extended. This may be easily proved by extending the knee and then trying to raise the lower extremity so that the thigh is in contact with the anterior abdominal wall. You will note that it is difficult to flex the hip beyond 90 degrees with the knee extended. However, if the knee is flexed it is quite possible to flex the hip so that the thigh is in contact with the anterior abdominal wall, that is, the hip can now be flexed through nearly 180 degrees.

NERVE

Sciatic

The sciatic nerve is the thickest nerve in the body (about 2 cm in diameter). It appears in the gluteal region at the lower border of piriformis and then lies successively on obturator internus, quadratus femoris, and adductor magnus (see Fig. 19-3). The short head of the biceps femoris is attached to the linea aspera of the femur lateral to the sciatic nerve. The only structure that must be cut to expose the whole length of sciatic nerve is the long head of the biceps, which passes from the ischial tuberosity to the head of the fibula by crossing posterior to the sciatic nerve.

ARTERIAL SUPPLY

The multiple anastomotic vessels of the gluteal region were described above. Multiple (four) perforating arteries from the profunda femoris artery pierce the adductor magnus muscle (see Fig. 19-7) and form a chain of arteries between the gluteal arteries and the popliteal artery. This chain supplies the muscles of the region, the sciatic nerve, and the femur.

POPLITEAL SPACE (POPLITEAL FOSSA)

The popliteal fossa is a diamond-shaped space behind the knee joint (see Fig. 19-10). It has the following *boundaries*:

Superolateral: the biceps femoris muscle.
Superomedial: the semitendinosus and semimembranosus muscles.
Inferolateral: the lateral head of the gastrocnemius muscle.
Inferomedial: the medial head of gastrocnemius muscle.

CONTENTS OF THE POPLITEAL FOSSA   (Fig. 19-10)

Vessels of the Popliteal Fossa

The popliteal artery enters the popliteal fossa through the hiatus in the adductor magnus muscle. At this point the femoral artery changes its name to *popliteal artery*. The popliteal artery passes inferiorly and somewhat laterally to leave the popliteal fossa under cover of the gastrocnemius muscle. The *popliteal vein* lies posterior (superficial) to the artery, and the *popliteal nerve* lies, in turn, posterior to the vein. The immediate anterior relationship of the popliteal artery is the posterior capsule of the knee joint.

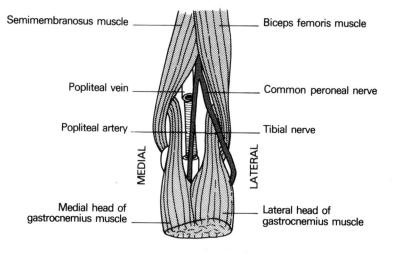

Figure 19-10.
Right popliteal fossa (posterior view).

The branches of the popliteal artery are numerous, and large branches supply the muscles of the region. The artery also gives off five **genicular arteries** that supply the knee joint and pass below the knee joint to anastomose with genicular branches of the anterior and posterior tibial arteries. Just below the popliteal fossa the popliteal artery divides into its terminal branches, the **anterior** and **posterior tibial arteries**.

CLINICAL NOTE

The popliteal artery is deeply placed and although its pulsation may be felt, it is not always easily palpable.

The bifurcation of the popliteal artery may become the site of lodgment of an arterial embolus, resulting in sudden cessation of blood flow to the leg. The artery is vulnerable to damage by laceration or compression in fractures of the femur at the level of the condyles.

Nerves of the Popliteal Fossa

The sciatic nerve usually divides at the upper border of the popliteal fossa into the **tibial nerve** (medial popliteal nerve) and the **common peroneal nerve** (lateral popliteal nerve).

The **tibial** nerve continues inferiorly through the popliteal space to disappear deep to the gastrocnemius. It is the most superficial of the three main structures in the popliteal fossa, and here is an exception to the **VAN** arrangement.

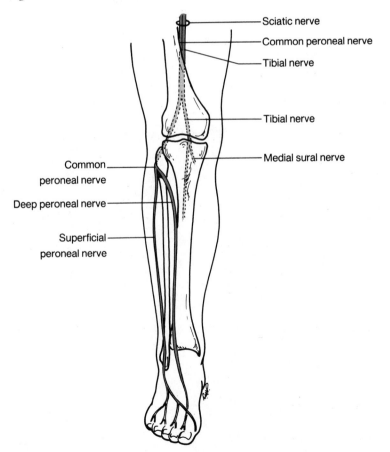

Figure 19-11.
Terminal branches of the sciatic nerve.

Obeying Hilton's law the tibial nerve gives off branches to the knee joint. Other branches arising in the popliteal fossa supply the two heads of the gastrocnemius, plantaris, and popliteus muscles.

The ***common peroneal nerve*** runs inferolaterally along the medial border of the biceps femoris muscle and leaves the popliteal fossa by passing superficial to the lateral head of the gastrocnemius muscle.

The nerve winds around the neck of the fibula, deep to the peroneus longus muscle, and divides into ***superficial*** and ***deep peroneal nerves***. The deep peroneal nerve enters the anterior crural (tibial) compartment of the leg, and the superficial peroneal nerve remains in the peroneal compartment (Fig. 19-11).

CLINICAL NOTE

The common peroneal nerve is vulnerable to injury as it winds around the neck of the fibula. The nerve can be palpated in this area.

Cutaneous Nerves

In the popliteal fossa the tibial nerve gives off the ***sural*** nerve which passes inferolaterally, piercing the deep fascia at about the middle of the leg. The sural nerve is joined by the ***sural communicating branch*** of the common peroneal nerve to supply the lateral side of the ankle and foot. The tibial and peroneal nerves also give off ***medial*** and ***lateral sural cutaneous nerves*** to the skin of the calf.

The Leg and Dorsum of the Foot

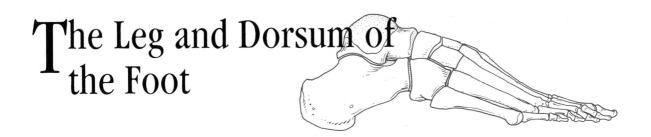

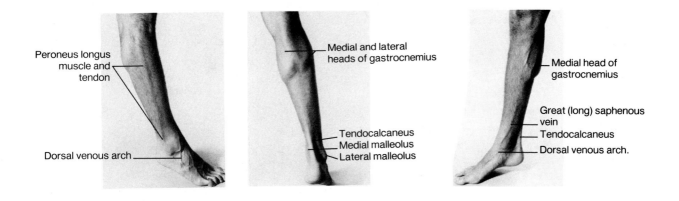

Peroneus longus muscle and tendon

Dorsal venous arch

Medial and lateral heads of gastrocnemius

Tendocalcaneus
Medial malleolus
Lateral malleolus

Medial head of gastrocnemius

Great (long) saphenous vein
Tendocalcaneus
Dorsal venous arch.

Certain bony landmarks must be recognized before an understanding of the leg and foot can be achieved. Many of these landmarks can be palpated. Try this on yourself.

BONES OF THE LEG

Tibia (Fig. 20-1). The tibia is the main bone and weight bearer of the leg. The fibula transmits little weight but has extensive surfaces for muscular and fascial attachments. The superior end of the tibia has two *condyles* that articulate with the femoral condyles and an *intercondylar eminence* that fits between the femoral condyles in the intercondylar notch. Note that the medial tibial condyle is larger than the lateral tibial condyle. On the posterolateral extremity of the inferior surface of the lateral condyle is the *facet* for articulation with the head of the fibula. Anteriorly there is the *tibial tuberosity* to which the ligamentum patellae is attached; this tuberosity is subcutaneous and easily palpable.

The shaft of the tibia has a sharp anterior subcutaneous border: the shin, which many of us have banged and bruised. The medial surface of the tibia is subcutaneous, the lateral surface is covered by muscular attachments, and the posterior surface is deeply placed. The lateral border of the tibia is the *interosseous border* for the attachment of the interosseous membrane that connects the tibia to the fibula.

The expanded lower end of the tibia has an articular surface for the *talus*. Medially the lower end of the tibia projects downward as the *medial malleolus*. The medial malleolus also articulates with the talus.

Photos show lateral view of right leg (*left*), posterior view of right leg (*center*), and medial view of right leg (*right*).

20

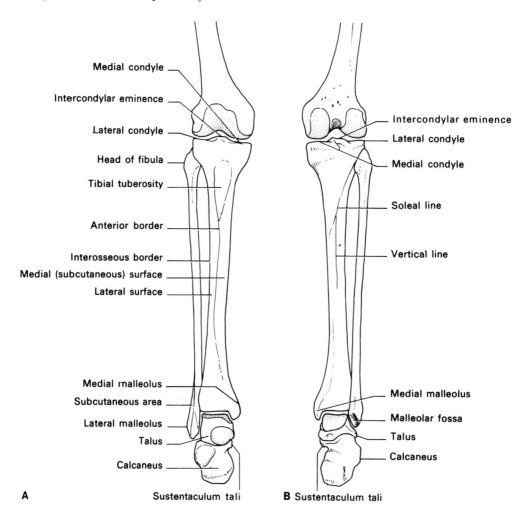

Medial condyle

Intercondylar eminence

Lateral condyle

Head of fibula

Tibial tuberosity

Anterior border

Interosseous border

Medial (subcutaneous) surface

Lateral surface

Medial malleolus

Subcutaneous area

Lateral malleolus

Talus

Calcaneus

A   Sustentaculum tali

Intercondylar eminence

Lateral condyle

Medial condyle

Soleal line

Vertical line

Medial malleolus

Malleolar fossa

Talus

Calcaneus

B Sustentaculum tali

Figure 20-1.
Skeleton of the right leg. **A.** Anterior view (anterior portion of foot has been removed). **B.** Posterior view.

The posterior surface of the shaft of the tibia exhibits a diagonal **soleal line** that passes downward and medially from the articular facet for the fibula to the medial border. A **vertical line** runs inferiorly from the middle of the soleal line. The area above the soleal line gives attachment to the **popliteus muscle**.

Fibula (Fig. 20-1). The fibula bears no weight, and its chief functions are to provide attachments for muscles and fascia and stability for the ankle joint. The proximal end of the fibula is expanded to form the **head**, the superior surface of which bears a **facet** for articulation with the inferior surface of the lateral condyle of the tibia. The **apex** of the head of the fibula gives attachment to the biceps femoris tendon and the lateral (fibular) collateral ligament of the knee joint. The lower end of the fibula forms the **lateral malleolus** of the ankle joint; this projects below the level of the medial malleolus of the tibia. The medial surface of the lateral malleolus bears a facet for articulation with the talus, and behind the facet there is an irregular depression, the **malleolar fossa**.

The shaft of the fibula has a marked *__interosseous border__* for the attachment of the interosseous membrane. This border ends inferiorly in a rough triangular surface.

Tibiofibular Union. Superiorly the fibula is joined to the tibia by a synovial joint, inferiorly by a fibrous union, and, in between the superior and inferior tibiofibular joints, the shafts of the two bones are joined by the *__interosseous membrane__*.

BONES OF THE FOOT (Figs. 20-2A and 20-2B)

The bones of the foot are shaped so that there is relatively little movement between individual bones but, conversely, there is maximum stability. The bones have irregular outlines, and they are normally held in proper alignment by ligaments. The bones and ligaments together are chiefly responsible for the maintenance of the *__arches__* of the foot.

Talus. The talus consists of a *__body__* that is connected to the *__head__* of the talus by the *__neck__*. The body has three facets, which articulate with the facet on the inferior surface of the tibia, the facet on the medial surface of the lateral malleolus, and the facet on the lateral surface of the medial malleolus. The head and neck project anteriorly and medially, and the head has a large convex facet for articulation with the *__navicular bone__*. The inferior aspect of the body has two (sometimes three) facets for articulation with the *__calcaneus__*. The posterior facet is separated from the others by a deep groove, the *__sulcus tali__*. The *__lateral tubercle__* of the posterior process of the talus occasionally does not unite with the body and forms a separate bone, the *__os trigonum__*. On an x-ray film this separation may be mistaken for a fracture.

CLINICAL NOTE

*The talus has no muscular attachments. Inspection of the neck will show multiple nutrient foramina that run in the direction of the body (i.e., proximally). Fractures of the talus may occur when the sole of the foot is heavily compressed against an unyielding surface, as in falling from a height and landing on the heel. In such injuries the talus is fractured through the neck, and disruption of the interosseus ligament in the sinus tarsi occurs. The blood supply of the body may be destroyed, causing **avascular necrosis** of the body of the talus.*

Calcaneus. The calcaneus is the heel bone. It exhibits a posterior *__tubercle__* on its inferior surface. Laterally there is a small tubercle above a groove, the *__peroneal trochlea__*, and medially there is a prominent projection, the *__sustentaculum tali__*, which helps to support the talus. The anterior facet of the calcaneus is for articulation with the *__cuboid__* bone. Superiorly there are three facets for articulation with the talus. One is at the junction of the posterior and middle thirds of the calcaneus, one at the anterior end of the upper surface, and one on the sustentaculum. The posterior facet is separated from the other two by the *__sulcus calcanei__*.

When the talus and calcaneus are fitted together properly, it will be seen that a "tunnel" passes from behind the sustentaculum running anteri-

Figure 20-2.
Skeleton of the right foot. **A.** Superior
view divided into units (concept after
J.C.B. Grant). **B.** Lateral and medial
views indicating arches.

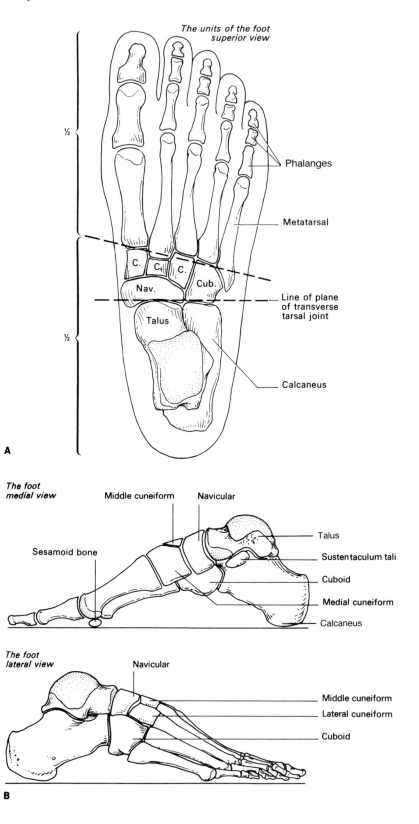

orly, laterally, and inferiorly. This tunnel, called the **sinus tarsi**, contains an interosseous ligament which helps to hold these two bones together. This sinus is formed by the juxtaposition of the sulcus tali and sulcus calcanei.

The inferior portion of the posterior aspect of the calcaneus is roughened for the attachment of the **tendo calcaneus**. The tendo calcaneus is separated from the upper part of this posterior surface by a bursa.

Navicular. The navicular bone has a proximal facet for the head of the talus and three distal facets for the **cuneiform** bones. There may be a small lateral facet for the **cuboid** bone, and medially there is a tuberosity that is easily palpable on the side of the foot.

Cuboid. Proximally, the cuboid has a facet for articulation with the calcaneus; there is also a process of the cuboid that projects posteriorly underneath the anterior end of the calcaneus and helps to maintain the lateral arch of the foot. There are two distal facets for articulation with the lateral two metatarsal bones and a medial facet for the lateral cuneiform and sometimes the navicular.

The inferior surface of the cuboid displays a groove that runs forward and medially for the tendon of peroneus longus. At the proximal end of this groove is a facet for a sesamoid bone in the tendon of peroneus longus (Fig. 20-3).

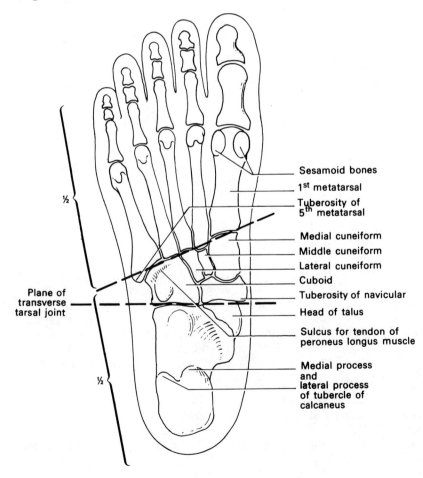

½

Plane of
transverse
tarsal joint

½

Sesamoid bones

1st metatarsal

Tuberosity of
5th metatarsal

Medial cuneiform

Middle cuneiform

Lateral cuneiform

Cuboid

Tuberosity of navicular

Head of talus

Sulcus for tendon of
peroneus longus muscle

Medial process
and
lateral process
of tubercle of
calcaneus

Figure 20-3.
Skeleton of the right foot. Inferior surface.

Cuneiforms. Three wedge-shaped cuneiform bones articulate proximally with the navicular and distally with the medial three metatarsal bones. The cuneiforms articulate with each other, and the lateral cuneiform articulates with the cuboid bone. The bases of the wedges face dorsally, and their sharp ends face the sole of the foot. The three cuneiforms in conjunction with the cuboid form a bony transverse arch for the foot.

Metatarsals. The five metatarsals are "miniature" long bones. They articulate proximally with the bones of the distal row of the tarsus and distally with the phalanges of the toes. The base of the second metatarsal is wedged between the medial and lateral cuneiforms and between the first and third metatarsals. The heads of the metatarsals usually bear the weight of the body in such a way that the lateral two form the anterior pillar of the *lateral arch* of the foot; the medial three form the anterior pillar of the *medial arch*. The head of the first metatarsal articulates with two sesamoid bones.

Phalanges (singular, phalanx). Each of the lateral four toes has three phalanges (proximal, middle, and distal); the great toe has just two phalanges.

ONE FOOT = THREE UNITS

The bones of the foot can be grouped into three units. The posterior unit contains the talus and calcaneus. The middle unit consists of the cuboid, the navicular, and the three cuneiforms. The distal unit is formed by the metatarsals and phalanges (see Fig. 20-2A). Between the proximal and middle units is found the *transverse tarsal joint*, which really is two joints–one between the talus and navicular (and calcaneus) and the other between the cuboid and calcaneus. The middle and distal elements are joined at the joints between the cuneiforms and metatarsals and between the cuboid and metatarsals.

The *transverse arch of the foot* is formed chiefly by the shapes of the cuneiforms and the cuboid. Is it really an *arch*? The reader should articulate the bones of the foot of his or her skeleton and put it flat on the floor; can there be an arch with only one pillar?

MOVEMENTS OF FOOT AND ANKLE

The ankle moves mainly around a horizontal axis so that it can flex and extend. For clarity, instead of speaking of flexion and extension we use the terms *dorsiflexion* and *plantarflexion* (Figs. 20-4 and 20-5). The ankle joint is most stable in the dorsiflexed position when the broader anterior part of the trochlea of the talus makes a tight fit within the mortise of the medial and lateral malleoli. conversely, the joint is unstable in plantar flexion and its ligaments are most easily damaged when the plantar-flexed ankle and foot are twisted ("turned").

Transverse Tarsal Joint. At this joint the movements of *inversion* and *eversion* of the foot occur. Inversion turns the sole of the foot medially and eversion turns it laterally. Normally inversion also includes some adduction and plantarflexion whereas eversion includes abduction and dorsiflexion.

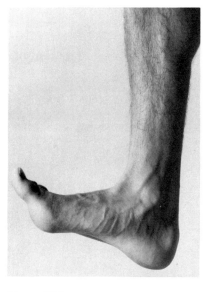

Figure 20-4.
Dorsiflexion of the ankle.

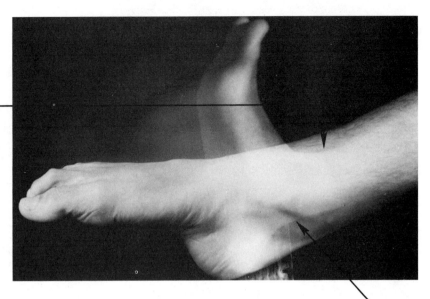

Dorsiflexion

Plantar flexion

Figure 20-5.
Plantar flexion of the ankle moving into dorsiflexion.

CUTANEOUS STRUCTURES OF THE LEG AND FOOT

The *leg* is that portion of the lower limb between the knee and the ankle. The cutaneous structures of importance are two veins and some cutaneous nerves.

CUTANEOUS VEINS

The veins on the dorsum of the foot drain the toes and dorsal interosseous spaces and join to form a ***dorsal venous arch*** (see Fig. 18-4), which drains medially into the ***great (long) saphenous vein*** and laterally into the ***small (short) saphenous vein***.

Great (Long) Saphenous Vein. The great (long) saphenous vein runs upward ***anterior*** to the medial malleolus to pass approximately a hand's breadth posterior to the medial edge of the patella, and from there it ascends into the femoral triangle, where it pierces the deep fascia at the saphenous opening and joins the femoral vein.

Small (Short) Saphenous Vein. The small (short) saphenous vein drains the lateral portion of the dorsal venous arch, passes ***posterior*** to the lateral malleolus, and ascends up the center of the leg posteriorly to pierce the deep fascia of the popliteal fossa and join the popliteal vein.

It must be realized that these two saphenous systems are not isolated and communications exist in the subcutaneous plane.

CLINICAL NOTE

*The great and small saphenous veins often become **varicose**. This condition occurs when the valves in the **deep communicating veins** become incompetent or are destroyed by thrombosis; the deep communicating veins are veins that link the subcutaneous veins to the deep veins. In the condition of varicose veins the pressure of venous blood in the deep veins is transmitted to the superficial veins, and the resulting dilatation and tortuosity of the superficial veins are called **varicose veins**.*

Figure 20-6.
Major structures of anterior
compartment of right leg.

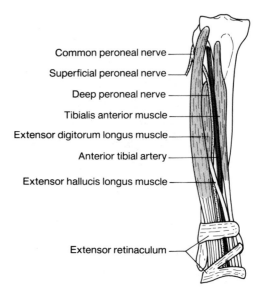

Common peroneal nerve

Superficial peroneal nerve

Deep peroneal nerve

Tibialis anterior muscle

Extensor digitorum longus muscle

Anterior tibial artery

Extensor hallucis longus muscle

Extensor retinaculum

CUTANEOUS NERVES

The saphenous and sural nerves have already been discussed. The remaining important cutaneous nerves of the leg and foot are branches of the superficial and deep peroneal nerves and the cutaneous portions of the tibial nerve.

Superficial Peroneal Nerve. This branch of the common peroneal nerve descends in the peroneal compartment of the leg and pierces the deep fascia in the lower third of the lateral aspect of the leg. It supplies the skin of the dorsum of the foot and toes except for the first interdigital cleft (contiguous surfaces of first and second toes).

Deep Peroneal Nerve. The terminal cutaneous portion of the deep peroneal nerve pierces the anterior deep fascia of the leg just above the ankle joint and supplies the skin of the dorsum of the first interdigital cleft (Fig. 20-6).

Tibial Nerve. The tibial nerve sends cutaneous branches to the heel (calcanean branches) and to the sole of the foot through the ***medial*** and ***lateral plantar nerves***.

ANTERIOR COMPARTMENT OF THE LEG AND DORSUM OF THE FOOT

The leg is surrounded by a dense layer of deep fascia and divided into three compartments by intermuscular septa of the deep fascia which extend from the investing deep fascia to the fibula (Fig. 20-7). The deep fascia joins the tibia on either side of its medial subcutaneous area. Many of the muscles of the leg gain attachment to the deep fascia and to the intermuscular septa.

The musculature of the leg is supplied by three different nerves. The muscles of the ***anterior compartment*** are supplied by the ***deep peroneal nerve***, those of the ***lateral (peroneal) compartment*** by the

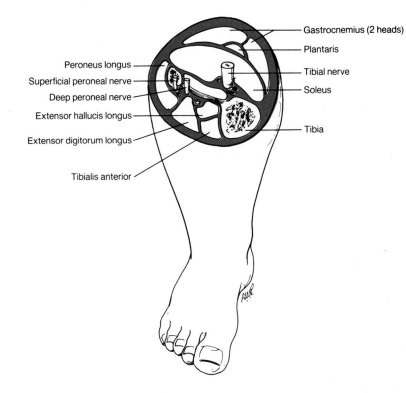

Peroneus longus
Superficial peroneal nerve
Deep peroneal nerve
Extensor hallucis longus
Extensor digitorum longus
Tibialis anterior

Gastrocnemius (2 heads)
Plantaris
Tibial nerve
Soleus
Tibia

Figure 20-7.
Transverse section through leg to show compartments and nerve supply.

superficial peroneal nerve, and those of the *posterior compartment* by the *tibial nerve*.

Extensor Retinaculum. The extensor retinaculum is a condensation of deep fascia above the ankle joint. The superior band passes from the fibula to the tibia above the medial malleolus. The inferior band passes from the calcaneus to the medial malleolus and the medial edge of the plantar aponeurosis. The purpose of the retinaculum is to prevent the tendons passing in front of the ankle joint from *bowstringing* when the ankle is dorsiflexed (Fig. 20-8).

MUSCLES

The anterior compartment of the leg contains *four* muscles, discussed below.

Tibialis Anterior

This is a long, narrow muscle that dorsiflexes the ankle and inverts the foot (Fig. 20-8).

Proximal (Upper) Attachments. The lateral condyle and upper half of the lateral surface of the tibia.

Distal (Lower) Attachments. By means of a long tendon that passes behind the extensor retinaculum to the inferior surface of the medial cuneiform and the base of the first metatarsal.

Nerve Supply. Deep peroneal nerve (*L4* and L5).

Actions. Dorsiflexion of the ankle and inversion of the foot.

Figure 20-8.
Anterior view of muscles of anterior
compartment of right lower extremity.

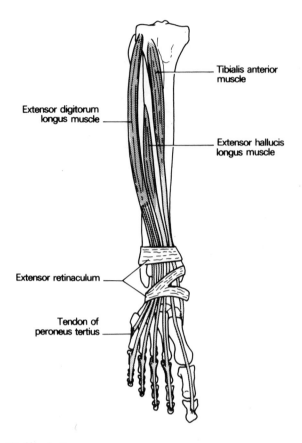

Tibialis anterior
muscle

Extensor digitorum
longus muscle

Extensor hallucis
longus muscle

Extensor retinaculum

Tendon of
peroneus tertius

Extensor Hallucis Longus

This muscle lies between tibialis anterior and extensor digitorum longus
(Fig. 20-8).

Proximal (Upper) Attachment. The central half of the shaft of the fibula and the interosseous membrane.

Distal (Lower) Attachment. By means of a tendon deep to the extensor
retinaculum into the dorsum of the distal phalanx of the great (first) toe.

Nerve Supply. Deep peroneal nerve (L5 and S1).

Actions. Extension of great toe and dorsiflexion of ankle.

Extensor Digitorum Longus

This muscle splits into four tendons that pass behind the extensor retinaculum (Fig. 20-8).

Proximal (Upper) Attachments. The lateral condyle of the tibia and
three quarters of the anterior surface of the fibula and the interosseus
membrane.

Distal (Lower) Attachments. By four tendons to the lateral four toes.
Each tendon expands into a dorsal digital expansion which attaches by two
slips to the sides of the middle phalanx, and into a central slip which attaches to the base of the distal phalanx. The lumbrical and interosseous
muscles of the foot are attached to the dorsal digital expansion.

Nerve Supply. Deep peroneal nerve (L5, S1).

Actions. Extension of toes (for details, see actions of lumbricals and interossei [below]) and dorsiflexion of the ankle.

Peroneus Tertius

This insignificant slip of muscle which belongs to extensor digitorum longus fails to reach the toes and attaches to the fifth metatarsal. It dorsiflexes the ankle and everts the foot (Fig. 20-8).

Extensor Digitorum Brevis

This is a muscle on the lateral side of the dorsum of the foot (Fig. 20-9).

Proximal (Upper) Attachment. The upper surface of the calcaneus.

Distal (Lower) Attachment. By four tendons, one to the base of the proximal phalanx of the great toe, three to the lateral sides of the tendons of extensor digitorum longus to toes two, three, and four.

Nerve Supply. Lateral division of deep peroneal nerve (S1, S2).

Action. Extends toes.

CLINICAL NOTE

This unimportant muscle makes a bulge on the dorsum of the foot, and the skin overlying it is usually a little blue. The patient, and unwary physician, may mistake it for a bruise.

Figure 20-9.
Structures of the dorsum of the right foot.

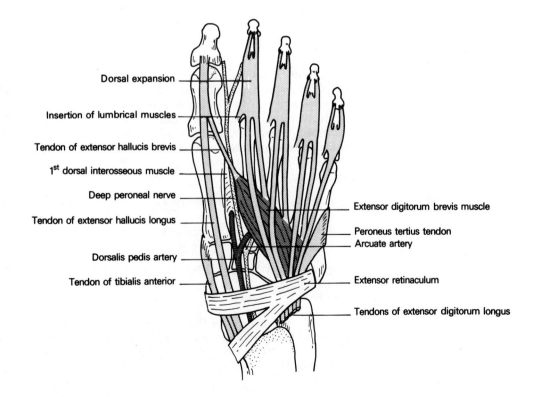

Dorsal expansion

Insertion of lumbrical muscles

Tendon of extensor hallucis brevis

1st dorsal interosseous muscle

Deep peroneal nerve

Tendon of extensor hallucis longus

Dorsalis pedis artery

Tendon of tibialis anterior

Extensor digitorum brevis muscle

Peroneus tertius tendon
Arcuate artery

Extensor retinaculum

Tendons of extensor digitorum longus

NERVES

Common Peroneal Nerve. The common peroneal nerve winds around the neck of the fibula, where it may be palpated and also easily injured. It divides into a ***superficial*** and a ***deep peroneal nerve***. The deep peroneal nerve pierces the anterior intermuscular septum and the attachment of extensor digitorum longus, passes downward on the interosseous membrane, where it runs with the anterior tibial artery, and descends between the tibialis anterior and the extensor hallucis longus. It supplies the muscles of the anterior compartment and gives a cutaneous branch to the first dorsal interdigital cleft.

CLINICAL NOTE

*The common peroneal nerve may be injured by compression or laceration as it winds around the neck of the fibula (see Fig. 20-6). This would cause paralysis of the dorsiflexors of the foot; the result is called **foot drop**. The individual cannot walk on the heel; dragging of the toes can be avoided only by raising the foot higher, and this results in a gait in which the foot is slapped onto the ground in a rather uncontrolled fashion. There will be some anesthesia of the skin over the dorsum of the foot.*

ARTERIES

Anterior Tibial Artery (Fig. 20-6). The anterior tibial artery arises as a terminal branch of the popliteal artery at the lower border of the popliteus muscle. It passes forward between the tibia and fibula above the interosseous membrane and descends with the deep peroneal nerve on the anterior surface of the interosseous membrane between tibialis anterior and extensor hallucis longus. It passes deep to the extensor retinaculum and anterior to the ankle joint; it is crossed by the tendon of extensor hallucis longus. It continues forward onto the dorsum of the foot as the ***dorsalis pedis artery*** (Fig. 20-9).

Dorsalis Pedis Artery. This artery passes forward to terminate in the space between the first two metatarsals. Here it divides into a ***deep branch*** that passes into the sole of the foot and an ***arcuate artery*** that arches laterally over the dorsum of the foot. The arcuate artery gives off ***dorsal metatarsal arteries*** that pass toward the toes. The dorsalis pedis artery supplies ***tarsal branches*** to the dorsum of the foot.

CLINICAL NOTE

The pulsation of the dorsalis pedis artery can be palpated where it passes over the navicular and cuneiform bones. It is to be found lateral to the tendon of extensor hallucis longus (Fig. 20-9).

LATERAL COMPARTMENT OF THE LEG

MUSCLES

The lateral compartment of the leg contains two peroneal muscles supplied by the ***superficial peroneal nerve***. The tendons of the peronei are held in place behind and below the lateral malleolus by the ***peroneal retinaculum*** (Fig. 20-10).

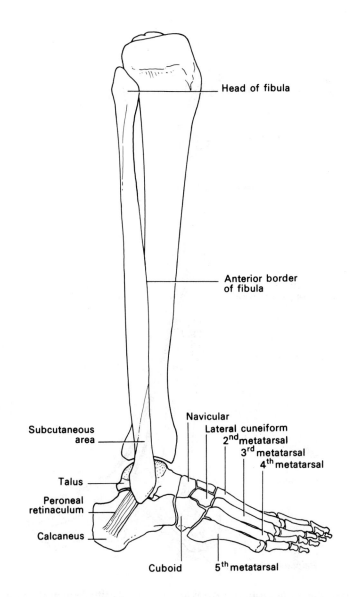

Figure 20-10.
Lateral view of the skeleton of the right leg and foot. Only the superior portion of the peroneal retinaculum is shown.

Peroneus Longus

This is a long narrow muscle running from the fibula to the sole of the foot (Fig. 20-11).

Proximal Attachments. Head and upper two thirds of the lateral aspect of the shaft of the fibula.

Distal Attachments. By a long tendon that passes deep to the peroneal retinaculum and behind the lateral malleolus, inferior to the peroneal tubercle of the calcaneus, inferior to the cuboid bone, and across the sole of the foot in the groove on the inferior surface of the cuboid bone to insert into the base of the first metatarsal and the medial cuneiform bone (compare with attachments of tibialis anterior).

Nerve Supply. Superficial peroneal nerve (*L5*, *S1*, and S2).

Figure 20-11.
The peroneal muscles and nerves of the right lower extremity (lateral view).

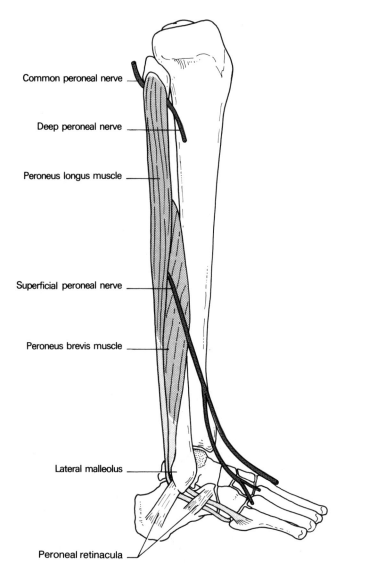

Common peroneal nerve

Deep peroneal nerve

Peroneus longus muscle

Superficial peroneal nerve

Peroneus brevis muscle

Lateral malleolus

Peroneal retinacula

Actions. Plantarflexion of the ankle, eversion of the foot, and support of the lateral arch of the foot by the sling-like action of its tendon as it passes under the lateral aspect of the foot.

Peroneus Brevis

This muscle lies deep to the peroneus longus.

Proximal Attachment. Lower two thirds of the fibula.

Distal Attachment. To the tubercle of the fifth metatarsal (Fig. 20-11).

Nerve Supply. Superficial peroneal nerve (*L5*, *S1*, and S2).

Action. Plantarflexion and eversion of the foot.

NERVES

The ***superficial peroneal nerve***, a branch of the common peroneal nerve, lies between the peroneus longus and brevis, supplies these two muscles, and then pierces the deep fascia in the lower third of the leg anterior to the peroneus longus (Fig. 20-11). It supplies the skin of the dorsum of the foot and toes except for the first interdigital cleft.

POSTERIOR COMPARTMENT OF THE LEG

The muscles of the posterior compartment of the leg are separated into a superficial group and a deep group by the deep transverse fascia of the leg (see Fig. 20-7).

MUSCLES OF THE SUPERFICIAL GROUP

This bulky group of muscles acts on the calcaneus through the tendo calcaneus and is the powerful plantarflexor that enables us to step off in walking, running, and leaping.

Gastrocnemius Muscle

This muscle, together with the **soleus** muscle, forms the **triceps surae** of the calf. These muscles are the principal plantar flexors of the ankle joint.

Attachments. The gastrocnemius muscle has two heads, which are attached to the posterior surface of the femur just above the articular surfaces of the femoral condyles (Fig. 19-10). The fibers of the two heads join at the lower angle of the popliteal fossa to form a single muscle belly which becomes tendinous near its middle. This tendon, when joined by the tendon of the soleus muscle, forms the **tendo calcaneus** (Achilles tendon), which is attached below to the lower part of the posterior surface of the calcaneus. A bursa intervenes between the upper part of the posterior surface of the calcaneus and the tendo calcaneus.

Nerve Supply. The tibial nerve (**S1** and S2).

Action. The gastrocnemius contracts to plantar flex the ankle, for example, as in standing on the tips of the toes or as in walking, running, or leaping.

There is a small **sesamoid bone** (the fabella) in the lateral head of the gastrocnemius close to its femoral attachment. This sesamoid bone is visible on an x-ray film and should not be mistaken for a fracture.

CLINICAL NOTE

Despite its thickness the tendo calcaneus can be ruptured by sudden exertion, such as leaping on the tennis court.

Plantaris

This is a very small muscle with a long tendon.

Proximal Attachment. Lower end of the lateral supracondylar line of the femur.

Distal Attachment. By a long narrow tendon running between the gastrocnemius and the soleus to the lower part of the medial border of the tendo calcaneus.

Nerve Supply. Tibial nerve (*S1* and S2).

Action. Very weak plantar flexor and flexor of knee joint.

CLINICAL NOTE

The plantaris tendon may be ruptured in sudden, forced dorsiflexion of the ankle (e.g., in jumping and landing on the toes). This can cause sudden severe pain in the back of the calf. Many cases of pain ascribed to rupture of the plantaris tendon may actually be due to avulsion of the upper attachment of some fibers of the medial head of the gastrocnemius.

Soleus

This bulky muscle, together with the two heads of the gastrocnemius, forms the ***triceps surae*** muscle that forms the bulk of the calf of the leg (Fig. 20-12).

Proximal Attachments. In a horse-shoe shape from the upper quarter of the posterior surface of the shaft of the fibula, the head of the fibula, a fibrous arch superficial to the tibial vessels and nerve, the soleal line of the tibia, and the middle third of the medial border of the tibia.

Distal Attachment. By the tendo calcaneus into the posterior surface of the calcaneus.

Nerve Supply. Tibial nerve (S1 and *S2*).

Action. Plantar flexion of the ankle, particularly in walking; it also steadies the leg on the foot.

MUSCLES OF THE DEEP POSTERIOR GROUP

Popliteus

This muscle, which is important in "unlocking" the knee, has an intracapsular tendon (Fig. 20-12).

Proximal Attachment. By a cord-like tendon in a depression on the lateral surface of the lateral condyle of the femur. This attachment is ***inside*** the capsule of the knee joint and deep to the lateral (fibular) collateral ligament.

Distal Attachment. Into the triangular area on the posterior surface of the tibia above the soleal line. The inferior border of popliteus, at its tibial attachment, closely parallels the upper border of soleus.

Nerve Supply. The nerve to popliteus (L4, L5, and S1), a branch of the tibial nerve. A trivial piece of information is that the nerve to popliteus also sends a small branch to the inferior tibiofibular joint; is this in accord with Hilton's law?

Actions. Popliteus unlocks and flexes the knee joint by rotating the tibia medially on the fixed femur (see Knee Joint, Chap. 22). It resists forward displacement of the femur when the knee is partially flexed.

The tendons of the remaining three muscles of the deep compartment all pass deep to the ***flexor retinaculum*** (Fig. 20-13) that passes from the medial malleolus to the calcaneus. These three muscles serve to steady the leg on the foot when standing.

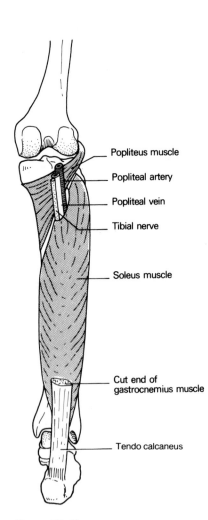

Popliteus muscle

Popliteal artery

Popliteal vein

Tibial nerve

Soleus muscle

Cut end of gastrocnemius muscle

Tendo calcaneus

Figure 20-12.
Popliteus and soleus muscles of the right lower extremity (posterior view).

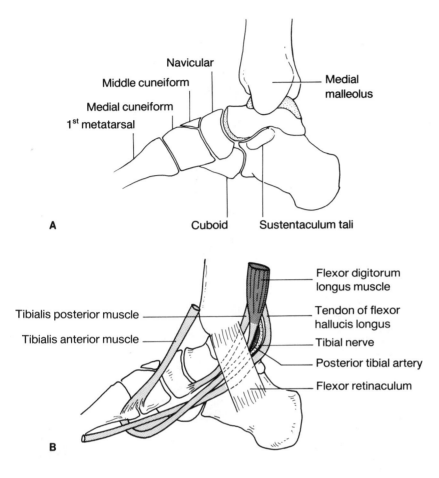

Figure 20-13.
Structures of the medial surface of the right ankle. **A.** Skeletal landmarks. **B.** Soft structures.

Flexor Hallucis Longus

This long muscle is the most lateral of the group (Fig. 20-14).

Proximal Attachments The lower two thirds of the posterior surface of the fibula and the adjacent part of the interosseous membrane. Note that this muscle which is destined to the most medial toe arises from the lateral bone in the leg.

Distal Attachment. The plantar surface of the distal phalanx of the great toe.

Nerve Supply. The tibial nerve (*S2* and S3).

Actions. Flexion of the interphalangeal and metatarsophalangeal joints of the great toe, plantarflexion of the ankle, and support of the long arch of the foot.

Flexor Digitorum Longus

This long, narrow muscle is the most powerful flexor of the lateral four toes (see Fig. 20-14).

Figure 20-14.
Deep structures of the posterior compartment of the right lower extremity.

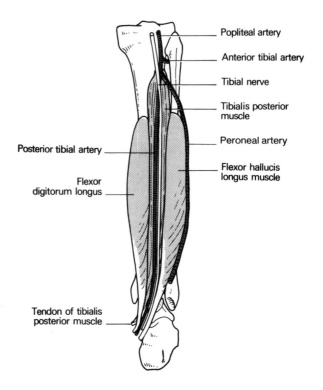

Popliteal artery

Anterior tibial artery

Tibial nerve

Tibialis posterior muscle

Peroneal artery

Posterior tibial artery

Flexor hallucis longus muscle

Flexor digitorum longus

Tendon of tibialis posterior muscle

Proximal Attachment. Posterior surface of the shaft of the tibia medial to the vertical line and inferior to the soleal line.

Distal Attachments. By a tendon that passes behind the medial malleolus, crosses the tendon of flexor hallucis longus in the sole of the foot, and attaches to the bases of the distal phalanges of the lateral four toes.

Nerve Supply. The tibial nerve (**S2** and S3).

Actions. Flexion of the interphalangeal and metatarsophalangeal joints, plantar flexion of the ankle, and support of the longitudinal arch of the foot.

Tibialis Posterior

This is the most deeply placed muscle in the calf and serves as an important invertor of the foot (see Fig. 20-14).

Proximal Attachments. The interosseous membrane and the tibia and fibula between the attachments of flexor hallucis longus and flexor digitorum longus.

Distal Attachments. The tendon passes as the most medial structure behind the medial malleolus. It passes medial to the deltoid ligament to be attached to the tuberosity of the navicular and many other bones of the foot.

Nerve Supply. Tibial nerve (**L4** and **L5**). Note that the two important invertors of the foot are supplied by the same spinal cord segments.

Actions. Plantar flexion of the ankle and inversion of the foot; helps to maintain the longitudinal arches of the foot.

Action of Leg Muscles. The muscles of the leg contract to counteract any tendency of the standing body to deviate from the erect position. They also contract to enable the foot to adapt itself to uneven or sloping ground. Their most important action is in providing the many movements of foot and ankle required in walking. A more detailed analysis of the mechanism of walking will be found on page 309.

ARTERIES

Posterior Tibial Artery (Figs. 20-13 and 20-14). The posterior tibial artery arises as a terminal branch of the popliteal artery at the lower border of the popliteus muscle. Here it passes deep to the arch of soleus and descends on the posterior surface of the tibialis posterior muscle. In the lower quarter of the leg the posterior tibial artery leaves the tibialis posterior and runs between the flexor hallucis longus (laterally) and flexor digitorum longus (medially) tendons; it maintains that relationship to the tendons of these muscles as it runs behind the medial malleolus. As it enters the sole of the foot it lies deep to the abductor hallucis and terminates by dividing into the ***medial*** and ***lateral*** plantar arteries.

CLINICAL NOTE

The posterior tibial artery can be felt (and frequently seen) as it pulsates in its location behind the medial malleolus.

Branches of the Posterior Tibial Artery. The artery gives off multiple muscular and calcanean branches. The only important artery for the student is the ***peroneal artery***. This artery branches off the posterior tibial artery about 2.5 cm below the arch of the soleus, passes toward the fibula, and descends on the interosseus membrane in close relationship to the flexor hallucis longus. At the lower end of the leg it pierces the interosseous membrane and anastomoses with the ***arcuate*** artery on the dorsum of the foot. The peroneal artery usually looks quite small in the embalmed body but is a much larger vessel in living persons; this can be seen in an angiogram. There tends to be a reciprocity in size between the peroneal artery and the anterior tibial artery.

NERVE

Tibial Nerve. The ***tibial nerve*** is one of the two terminal branches of the sciatic nerve. The division of the sciatic nerve can occur at any level below the sacral plexus but is usually found to occur in the popliteal fossa. The nerve descends posterior (superficial) to the popliteal vein and artery and leaves the popliteal fossa by passing deep to the arch of soleus (see Fig. 19-10). It passes inferiorly on the surface of tibialis posterior, crosses the posterior tibial artery, and runs below the medial malleolus between the posterior tibial artery and the tendon of flexor hallucis longus. It supplies muscular branches to the popliteus and the muscles of the back of the calf.

CLINICAL NOTE

Damage to the tibial nerve paralyzes the plantar flexors of the ankle. The patient cannot stand on his toes or curl them. The sole of the foot will be insensitive.

CLINICAL NOTES: LEG

The above account of the anatomy of the leg has focused on some clinical situations as individual structures were described. The reader should now consider the whole leg as a functioning unit and realize that many clinical abnormalities do not occur in isolation.

COMPARTMENT SYNDROMES

The deep fascia of the leg is continuous with that of the thigh and the foot and provides an enveloping sleeve of tough, unyielding fibrous tissue for the limb. Intermuscular septa passing from the investing layer of the fascia to the tibia and fibula, together with the tough interosseous membrane, divide the leg into three fascial compartments (see Fig. 20-7).

Trauma to bone or soft tissues, or both, in these compartments can cause swelling of soft tissues and extravasation of blood, which may raise the intracompartmental pressures to levels that are incompatible with proper vascular perfusion, resulting in ischemia and ultimate death of tissues.

*The clinical findings will be those of extreme pain in the area of the compartment, loss of muscle function and pain on stretching the involved muscles, and loss of neurologic function. It has been estimated that an intracompartmental pressure in excess of 30 mm Hg is sufficient to cause damage, and the reader must realize that this is below the normal arterial pressure; this means that the absence of a peripheral pulse is **not** a prerequisite for a diagnosis of compartmental compression syndromes.*

This information should allow the reader to anticipate the following clinical findings.

ANTERIOR COMPARTMENT SYNDROME

This compartment is confined by the investing layer of deep fascia, the lateral surface of the tibia, the anterior intermuscular septum, the shaft of the fibula, and the interosseous membrane. Its contents have been described. The symptoms and signs may be as follows:

1. ***Severe pain***
2. ***Loss of dorsiflexion of ankle*** *and* ***extension of the toes***
3. ***Sensory loss*** *over the dorsum of the first interdigital cleft*
4. ***Severe pain*** *on* ***passive*** *plantar flexion of the ankle, eversion of the foot, and* ***passive*** *flexion of the toes*

LATERAL COMPARTMENT SYNDROME

This compartment is walled off by the deep fascia surae, the posterior crural intermuscular septum, the interosseous membrane, and the fibula. The symptoms and signs follow:

1. ***Severe pain***
2. ***Inability*** *to plantarflex and evert the foot; there may also be weakness of dorsiflexion and inversion because of involvement of the deep peroneal nerve before it leaves the lateral compartment to enter the anterior compartment*
3. ***Anesthesia*** *over the dorsum of the foot and toes*
4. ***Severe pain*** *on* ***passive*** *dorsiflexion and inversion of the foot*

POSTERIOR COMPARTMENT SYNDROMES

*The posterior compartment is limited by the investing fascia of the leg, the interosseus membrane, tibia, and fibula and the **deep transverse** intermuscular septum that divides the posterior space into a superficial compartment and a deep compartment.*

SUPERFICIAL POSTERIOR COMPARTMENT SYNDROME

1. *Pain*
2. *Inability* to plantarflex the ankle
3. *Anesthesia* over the distribution of the sural nerve
4. *Pain* on *passive* dorsiflexion of the ankle

DEEP POSTERIOR COMPARTMENT SYNDROME

1. *Severe pain*
2. *Weakness* of plantar flexion and flexion of the toes
3. *Anesthesia* involving most of the sole and the plantar surfaces of the toes
4. *Severe pain* on *passive* extension of the toes or *passive* dorsiflexion of the ankle

The Sole of the Foot

The sole of the foot contains muscles and tendons that are arranged in four layers. The arrangement of the bones, muscles, and ligaments enables the foot to support the body on uneven ground. The precise attachments of the muscles are less important than an understanding of their integrated functions.

SKIN AND FASCIA

The skin and fascia are modified to resist the pressure of weight bearing and to provide protection and a stable "grip" of the foot on the ground. The skin is thick and is tied by dense fibers to the deep fascia to keep the skin securely anchored to the underlying tissues. Padding is provided by a connective tissue framework that resembles the structure of a sponge, with fat filling the spaces. This padding is particularly evident under the heel and under the heads of the metatarsals (submetatarsal cushions). The weight of the body is borne by the heel and the metatarsal heads, with the head of the first metatarsal carrying the greatest load (Fig. 21-1).

CUTANEOUS NERVES

The sole is supplied by the ***medial*** and ***lateral*** plantar nerves and by small contributions from the saphenous nerve and the sural nerve along the medial and lateral borders, respectively.

   The medial plantar nerve is a terminal branch of the tibial nerve and supplies the skin over the medial part of the heel, the medial part of the sole, and the medial three and one half toes (Fig. 21-2B).

   The lateral plantar nerve is the other terminal branch of the tibial nerve and supplies the lateral part of the sole and the small toe and the lateral half of the fourth toe.

PLANTAR APONEUROSIS

The plantar aponeurosis is a thick layer of fibrous tissue that stretches from the calcaneus to the front of the foot, where it fans out into slips that are attached to each toe. The plantar aponeurosis is particularly thick in its

21

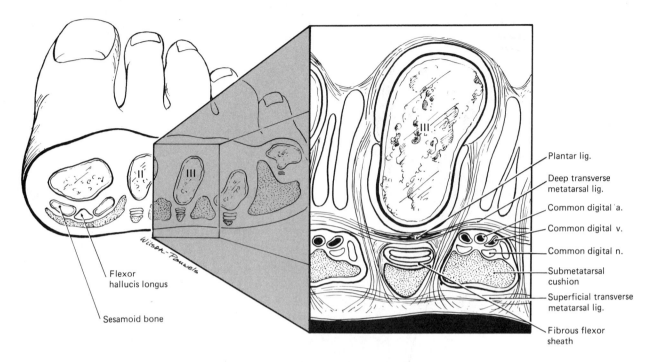

Plantar lig.

Deep transverse
metatarsal lig.

Common digital a.

Common digital v.

Common digital n.

Submetatarsal
cushion

Superficial transverse
metatarsal lig.

Fibrous flexor
sheath

Flexor
hallucis longus

Sesamoid bone

Figure 21-1.
Cross-section of metatarsal heads
illustrating submetatarsal cushions.

central portion and less dense where it underlies the muscles to the first
and fifth toes. Septa pass upward from the central portion to blend with
fascia covering the first and fifth metatarsals, thus dividing the foot into
medial, central, and lateral compartments. Infections of the sole tend to
localize in their respective compartments.

The central portion of the aponeurosis is so tough that it is usually
considered to contribute toward the support of the arch of the foot.

FIRST LAYER OF MUSCLES (Fig. 21-2A)

The muscles of the foot are arranged in four layers. The bones to which
they are attached are seen in Figure 21-3. The most superficial layer consists of three muscles.

Abductor Hallucis

This is a fairly sturdy muscle on the medial side of the foot. Its proximal
attachment covers the division of the posterior tibial artery and the tibial
nerve into their lateral and medial plantar branches.

Proximal Attachments.   Posterior part of the calcaneus and the medial
flexor retinaculum.

Distal Attachment.   The medial side of the proximal phalanx of the
great toe.

Nerve Supply.   Medial plantar nerve (S2 and *S3*). *Note:* all the intrinsic
muscles of the foot are supplied by spinal cord segments S2 and *S3*.

Action.   Flexion and abduction of the first toe. *Note:* in the foot the central
axis of reference for abduction and adduction is the second toe.

(*text continues on page 280*)

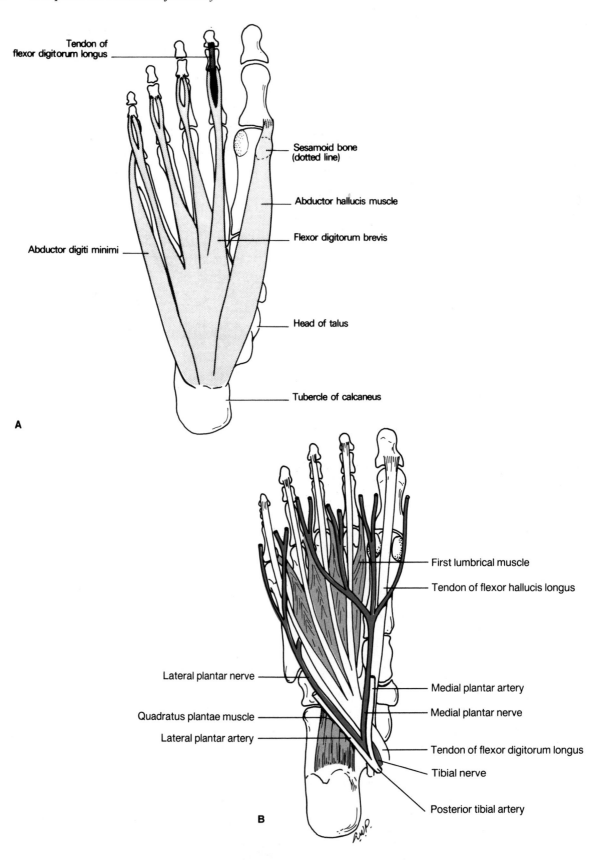

Tendon of
flexor digitorum longus

Sesamoid bone
(dotted line)

Abductor hallucis muscle

Flexor digitorum brevis

Abductor digiti minimi

Head of talus

Tubercle of calcaneus

A

First lumbrical muscle

Tendon of flexor hallucis longus

Lateral plantar nerve

Medial plantar artery

Quadratus plantae muscle

Medial plantar nerve

Lateral plantar artery

Tendon of flexor digitorum longus

Tibial nerve

Posterior tibial artery

B

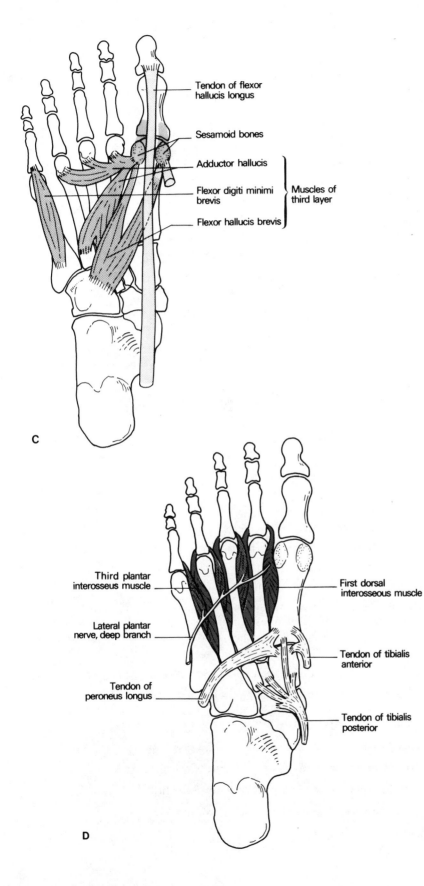

Tendon of flexor hallucis longus

Sesamoid bones

Adductor hallucis

Flexor digiti minimi brevis

Flexor hallucis brevis

Muscles of third layer

C

Third plantar interosseus muscle

Lateral plantar nerve, deep branch

Tendon of peroneus longus

First dorsal interosseous muscle

Tendon of tibialis anterior

Tendon of tibialis posterior

D

Figure 21-3.
Skeletal structures of the inferior
surface of the right foot, divided into
units as suggested by J.C.B. Grant.

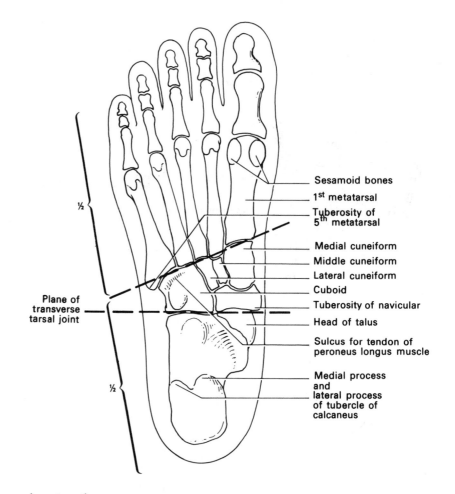

Sesamoid bones
1st metatarsal
Tuberosity of
5th metatarsal

Medial cuneiform
Middle cuneiform
Lateral cuneiform
Cuboid
Tuberosity of navicular
Head of talus

Sulcus for tendon of
peroneus longus muscle

Medial process
and
lateral process
of tubercle of
calcaneus

Plane of
transverse
tarsal joint

(*continued*)

Flexor Digitorum Brevis

This muscle is similar to the flexor digitorum superficialis of the upper limb.

Proximal Attachment. Posterior part of calcaneus.

Distal Attachments. By four tendons that split to insert into the lateral sides of the middle phalanges of the lateral four toes. Each tendon splits into two slips that permit the passage of the tendons of flexor digitorum longus to the distal phalanges of the toes.

Nerve Supply. Medial plantar nerve (S2 and *S3*).

Action. Flexes middle phalanges of lateral four toes.

Abductor Digiti Minimi

This is the most lateral muscle of the sole of the foot.

Proximal Attachment. Posterior part of calcaneus.

Distal Attachment. Proximal phalanx of the little toe.

Nerve Supply. Lateral plantar nerve (S2 and *S3*).

Action. Abducts little toe.

SECOND LAYER OF MUSCLES (Fig. 21-2B)

Flexor Hallucis Longus and Flexor Digitorum Longus

The tendons of these two muscles cross in the sole of the foot on their way to the distal attachments to their respective digits. They are deep to the first layer of muscles. The tendon of flexor digitorum crosses superficial (inferior) to the tendon of flexor hallucis longus.

Quadratus Plantae (Flexor Digitorum Accessorius)

This small, flat muscle joins the tendons of flexor digitorum longus to the calcaneus.

Proximal Attachment. Body of calcaneus.

Distal Attachment. The posterolateral margin of flexor digitorum longus tendon before it splits.

Nerve Supply. Lateral plantar nerve (S2 and *S3*).

Action. It adjusts the pull of the flexor digitorum longus so that it is more directly in line with the long axes of the digits.

Lumbrical Muscles

These four muscles are attached proximally to the tendons of flexor digitorum longus.

Proximal Attachments. The tendons of flexor digitorum longus after the main tendon has split into its four tendons to the individual digits.

Distal Attachments. The dorsal digital expansions of extensor digitorum longus.

Nerve Supply. The most medial lumbrical is supplied by the medial plantar nerve and the remainder by the lateral plantar nerve (*cf* median and ulnar nerves in the palm of the hand) (S2 and *S3*).

Actions. In conjunction with the interosseus muscles the lumbricals extend the interphalangeal joints when the metatarsophalangeal joints are held in flexion; they enable the toes to push off from the ground in walking.

THIRD LAYER OF MUSCLES (Fig. 21-2C)

Flexor Hallucis Brevis

This is a small muscle with two tendons.

Proximal Attachments. Cuboid and lateral cuneiform bones.

Distal Attachments. By two tendons to the proximal phalanx of the great toe. The tendon of flexor hallucis longus passes between these two tendons to be inserted into the distal phalanx. Each of the two tendons of flexor hallucis brevis contains a sesamoid bone that articulates with the head of the first metatarsal. They help to take the weight of the body and protect the long flexor tendon in walking. The medial tendon of flexor hallucis brevis is joined by the tendon of ***abductor hallucis*** at its attachment to the medial side of the base of the first phalanx, and the lateral tendon is joined in a similar manner by the tendon of ***adductor hallucis***.

Nerve Supply.   Medial plantar nerve (S2 and *S3*).

Adductor Hallucis

Comprising an oblique and a transverse head, this is a fairly powerful muscle that keeps the great toe aligned during walking.

Proximal Attachments.   Oblique Head: second, third, and fourth metatarsals; transverse head: from the plantar metatarsophalangeal ligaments of the third, fourth, and fifth toes.

Distal Attachment.   In conjunction with the lateral tendon of flexor hallucis brevis into the lateral side of the base of the proximal phalanx of the great toe.

Nerve Supply.   Lateral plantar nerve (S2 and *S3*).

Action.   Adduction of the first toe.

Flexor Digiti Minimi Brevis

A minor muscle of the little toe.

Proximal Attachment.   Fifth metatarsal.

Distal Attachment.   Proximal phalanx of little toe.

Nerve Supply.   Lateral plantar nerve (S2 and *S3*).

Action.   Flexes the middle phalanx of the little toe.

FOURTH LAYER OF MUSCLES (Fig. 21-2D)

Tendons of Peroneus Longus and Tibialis Posterior

These tendons cross the sole to reach their distal attachments.

Interossei

There are three plantar and four dorsal interossei. The dorsal muscles ***ab***duct the toes and the plantars ***ad***duct the toes. Their attachments can be worked out by remembering that the axis of reference for these movements is the second toe. The interossei are supplied by the lateral plantar nerve (S2 and *S3*) and attach distally into the dorsal digital expansions of the extensor digitorum longus tendons. In conjunction with the lumbricals they extend the interphalangeal joints when the metatarsophalangeal joints are held in flexion.

NERVES OF THE SOLE OF THE FOOT

The tibial nerve divides into lateral and medial plantar nerves deep to the abductor hallucis. The nerves form a V in the sole of the foot (Fig. 21-2B).

Medial Plantar Nerve

The medial plantar nerve passes deep to the abductor hallucis, where it branches and emerges between it and the flexor digitorum brevis as digital

Figure 21-4.
Medial and lateral plantar arteries.

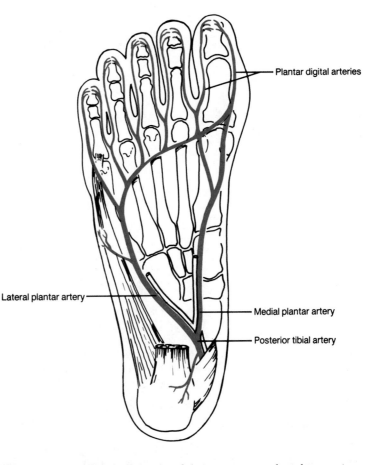

Plantar digital arteries

Lateral plantar artery

Medial plantar artery

Posterior tibial artery

nerves that pass to the medial side of the great toe and to the contiguous sides of the medial four toes. The medial plantar nerve gives off motor fibers to the ***abductor hallucis, flexor hallucis brevis, flexor digitorum brevis***, and the most ***medial lumbrical***.

Lateral Plantar Nerve

This nerve originates deep to the abductor hallucis and passes across the sole toward the tubercle of the fifth metatarsal between the first and second layers of the foot. It supplies the skin of the lateral one and a half digits and all the muscles of the foot that are not supplied by the medial plantar nerve. Its deep, terminal branch passes between layers three and four.

ARTERIES OF THE SOLE OF THE FOOT

The posterior tibial artery divides deep to the abductor hallucis to form the ***medial*** and ***lateral plantar arteries*** that run parallel to the similarly named nerves. They supply the muscles of the sole and anastomose with the dorsal arch near the heads of the metatarsals. At the webs of the toes the arteries anastomose with dorsal metatarsal arteries and give off digital branches to the toes (Fig. 21-4).

Lower Extremity: Joints and Lymphatic System

HIP JOINT

BONY SURFACES

Os Coxae. The bony socket for the head of the femur is the ***acetabulum*** (Fig. 22-1A), but the articulation is confined to the horseshoe-shaped area that covers the interior of the rim of the acetabulum. The central part of the acetabulum is not covered by articular cartilage. The acetabular notch is directed inferiorly and is closed by the ***transverse acetabular ligament*** (Fig. 22-1B).

Femur. The head of the femur is approximately three fifths of a sphere, and most of it, except for the ***fovea*** near its center, is covered by articular cartilage (Fig. 22-2A). The periphery of the fovea provides attachment for the ***ligamentum teres*** (ligament of the head of the femur). The articular surface of the head of the femur is larger than the articular surface of the acetabulum. A fibrocartilaginous ***labrum*** (lip) is attached to the rim of the acetabulum, slightly increasing the depth of the socket (Fig. 22-1B). The joint surfaces are reciprocally curved, and the depth and configuration of the joint give it a lot of inherent stability. The head of the femur is directed upward, medially and forward at an angle of about 15 degrees (the angle of torsion) relative to the transverse axes of the femoral shaft and greater trochanter.

ACETABULAR LABRUM

A fibrocartilaginous ring (the ***labrum***) is attached around the rim of the acetabulum and over the acetabular notch. This completely circular labrum surrounds the femoral head and increases the effective depth of the acetabulum.

CAPSULE

The capsule attaches to the edge of the acetabulum just beyond the labrum and to the transverse ligament (Fig. 22-1B). The obturator artery sends a branch through the acetabular notch, into the ligamentum teres to the head of the femur. The femoral attachments of the capsule follow the bases of the trochanters, the intertrochanteric line, and a line midway between

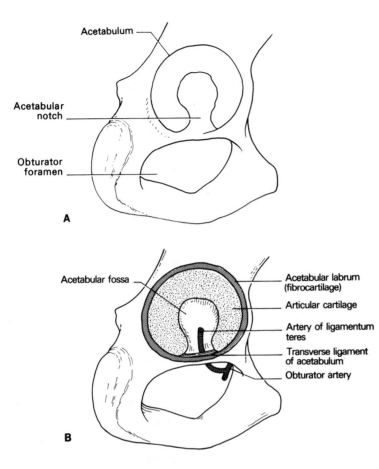

Acetabulum

Acetabular notch

Obturator foramen

A

Acetabular fossa

Acetabular labrum (fibrocartilage)

Articular cartilage

Artery of ligamentum teres

Transverse ligament of acetabulum

Obturator artery

B

Figure 22-1.
The right acetabulum. **A.** Bony structure. **B.** Soft structures.

the femoral head and the intertrochanteric crest posteriorly. Reflections of the capsular attachments (***retinacula***) are continued along the femoral neck, and these ***retinacula*** carry blood vessels from the periarticular anastomosis of vessels to the femoral head.

CLINICAL NOTE

Examination of the attachments of the capsule and retinacula to the femoral neck will show that a fracture through the femoral neck at any point medial to these attachments may leave the head deprived of blood supply. (The artery in the ligamentum teres disappears in many mature people.) Avascular necrosis of the femoral head is a common complication of femoral neck fractures.

SYNOVIAL MEMBRANE

The synovial membrane passes from the edge of the articular cartilage of the femoral head along the neck and reflects to cover the inside of the capsule. From the capsule it passes onto the os coxae and also covers the acetabular labrum. It is attached around the edge of the acetabular fossa, covers the ligamentum teres, and attaches around the margins of the fovea of the femoral head.

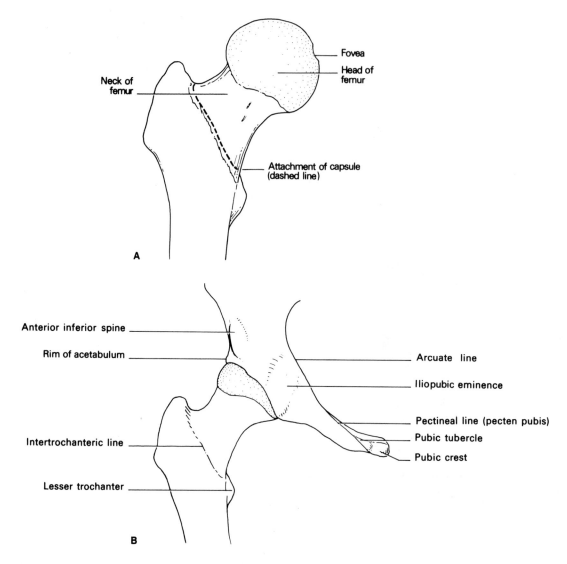

Figure 22-2.
Skeletal elements of the right hip joint.
A. Anterior view of femoral head.
B. Anterior view of hip joint.

LIGAMENTS

The capsule is reinforced by named condensations of connective tissue that constitute the intrinsic ligaments of the hip joint (Fig. 22-3).

Iliofemoral Ligament. This is a very strong Y-shaped intrinsic ligament that is attached to the anterior inferior iliac spine and to the intertrochanteric line. The tendon of iliopsoas muscle passes to the lesser trochanter in such a manner as to cover a deficiency between the iliofemoral and the ***pubofemoral*** ligament, thereby helping to strengthen the hip joint. The iliofemoral ligament is tense in extension.

Pubofemoral Ligament. This rather weak structures runs from the iliopubic eminence to the undersurface of the femoral neck; it is tense in abduction.

Ischiofemoral Ligament. This ligament passes in a spiral fashion from the ischium below the acetabulum to the back of the femoral neck. It is tense in abduction.

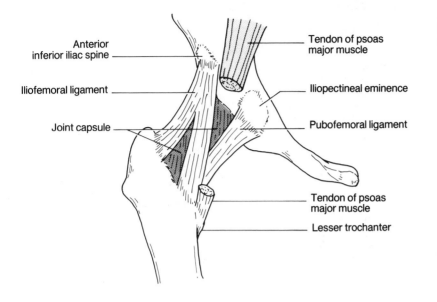

Anterior
inferior iliac spine

Iliofemoral ligament

Joint capsule

Tendon of psoas
major muscle

Iliopectineal eminence

Pubofemoral ligament

Tendon of psoas
major muscle

Lesser trochanter

Figure 22-3.
Ligaments of the right hip joint
(anterior view).

The three above-mentioned ligaments are arranged in a spiral manner that reflects the rotation of the lower limb from the embryonic position. This shape tends to screw the femur into the acetabulum in full extension. They all resist hyperextension. The iliofemoral ligament is particularly strong and can support the full weight of the body in the erect position. In the erect position the body's weight falls just behind the center of the hip joint; in this position little muscle activity occurs, the stability of the joint being sustained by the ligaments and maximum contact, congruence, and compaction of the articular surfaces.

MOVEMENTS (Table 22-1)

Movements possible at the hip joint are flexion, extension, abduction, adduction, lateral rotation, medial rotation, and circumduction. These movements should be thought of in terms of the os coxae moving on the femur with the foot on the ground and movement of the femur on the acetabulum when the foot is off the ground. The muscles producing these various movements have already been described, and a summary, along with the spinal cord segments supplying the groups of muscles, follows.

Flexion. Psoas major and iliacus, assisted by pectineus, rectus femoris, and sartorius (***L1, L2,*** and ***L3***).

Extension. Gluteus maximus and hamstrings (***L5, S1***).

Abduction. Gluteus medius and minimus, tensor fasciae latae (***L5, S1***).

Adduction. Adductors longus, brevis, and magnus, gracilis and pectineus (***L1, L2,*** and ***L3***).

Medial Rotation. Tensor fasciae latae and anterior fibers of gluteus medius and minimus; a weak movement (***L4, L5,*** and ***S1***).

Lateral Rotation. The obturator muscles, the gemelli, quadratus femoris, piriformis, gluteus maximus, and sartorius (***L5, S1***).

Table 22-1.
Hip Joint

Movement	Muscles	Nerves	Spinal Segment
Flexion	Iliopsoas	Lumbar plexus	L1, L2
	Sartorius	Femoral	L2, L3
	Pectineus	Femoral	L2, L3
	Rectus femoris	Femoral	L2, L3
Extension	Gluteus maximus	Inferior gluteal	L5, S1
	Semitendinosus	Sciatic	L5, S1
	Semimembranosus	Sciatic	L5, S1
	Biceps femoris	Sciatic	L5, S1
	Adductor magnus	Obturator and sciatic	L2, L3
Abduction	Gluteus medius	Superior gluteal	L5, S1
	Gluteus minimus	Superior gluteal	L5, S1
	Tensor fasciae latae	Superior gluteal	L4, L5
	(Sartorius)	Femoral	L2, L3
Adduction	Adductor magnus	Obturator and sciatic	L2, L3
	Adductor longus	Obturator	L2, L3
	Adductor brevis	Obturator	L2, L3
	Gracilis	Obturator	L2, L3
Lateral rotation	Obturator externus	Obturator	L3, L4
	Obturator internus	N. to obt. int.	L5, S1
	Piriformis	N. to piriformis	L5, S1
	Quadratus femoris	N. to quad. fem.	L5, S1
	(Gluteus maximus)	Inferior gluteal	L5, S1
	(Sartorius)	Femoral	L2, L3
Medial rotation	Tensor fasciae latae	Superior gluteal	L4, L5
	Gluteus medius	Superior gluteal	L5, S1
	Gluteus minimus	Superior gluteal	L5, S1

CLINICAL NOTES ON THE ANATOMY OF THE HIP: DISLOCATIONS

May be congenital or traumatic.

Congenital Dislocation of the Hip. *Some infants are born with deficient development of the acetabulum with the femoral head riding on the posterior aspect of the ilium. If the condition is not recognized at birth and corrected at an early age, such an individual will walk with a characteristic limp.*

Traumatic Dislocation of the Hip. *The medially rotated, flexed, and adducted hip is most vulnerable to dislocation if a strong thrust is applied along the axis of the femoral shaft. An example is the knee hitting the dashboard in a head-on automobile collision. The dislocation will be posterior and superior; the rim of the acetabulum may be fractured.*

CLINICAL NOTE: FRACTURES OF THE NECK OF THE FEMUR

The "broken hip" of the elderly is usually a fracture of the femoral neck. If the fracture occurs through the intracapsular part of the femoral neck, the blood supply to the femoral head may be destroyed. This problem does not arise if the fracture is extracapsular, for example, intertrochanteric.

KNEE JOINT

The classic anatomical description of the knee as a hinge joint is an over-simplification of a very complex and somewhat unstable joint. The hinging motion that occurs is accompanied by gliding and rotational movements of the tibia relative to the femur.

The knee joint is a composite of three "subjoints": between the patella and the front of the lower end of the femur; between the lateral femoral condyle and the lateral tibial condyle; and between the medial femoral condyle and the medial tibial condyle.

The condyles of the tibia move on the condyles of the femur. The condyles of the femur, although, in general, appearing to be like wheels, are actually so arranged that the hub of the wheel is posterior to the midpoint of the condyle. Thus when the femur is rotated on the wheels the front portion of the femur is stopped against the tibia. This means that any continued rotation puts extreme tension on the collateral ligaments that are attached to the femur at the hubs of the wheels, and hyperextension is prevented provided that the ligaments remain intact.

The movements of flexion and extension of the knee joint actually occur between the femur and the menisci; medial or lateral rotation of the tibia on the femur occurs between the tibia and the menisci.

CAPSULE

The capsule of the knee joint is fairly strong and contains certain named ligaments (Fig. 22-4).

Posteriorly. Superiorly the capsule is attached to the femur just above its condyles and the margin of the intercondylar notch. Inferiorly it is attached to the posterior margins of the tibial condyles and, between these condyles, forward onto the intercondylar eminence of the tibia.

Anteriorly. The capsule is attached around the margins of the patella and superiorly well above the articular surface of the femur. Inferiorly the capsule is attached to the tuberosity of the tibia and blends with the ligamentum patellae.

Laterally. The fibers of the capsule are attached from above the articular surface of the lateral femoral condyle to just below the lateral articular surface of the tibia. The capsule is deficient to allow the exit of the tendon of popliteus from its attachment to the lateral epicondyle of the femur. The upper part of the lateral aspect of the capsule blends with the lateral collateral ligament, but the rest of the lateral collateral ligament stands apart and will be described separately.

Medially. The capsule is attached beyond the articular surfaces of the femur and tibia. Its fibers are thickened to form the medial collateral ligament, which will be described below.

On either side of the patella the capsule consists of the medial and lateral ***patellar retinacula***, which are the aponeurotic continuations of the muscle fibers of vastus medialis and lateralis.

LIGAMENTS

Most of the ligaments are strong and blended with the capsule, that is, they are intrinsic.

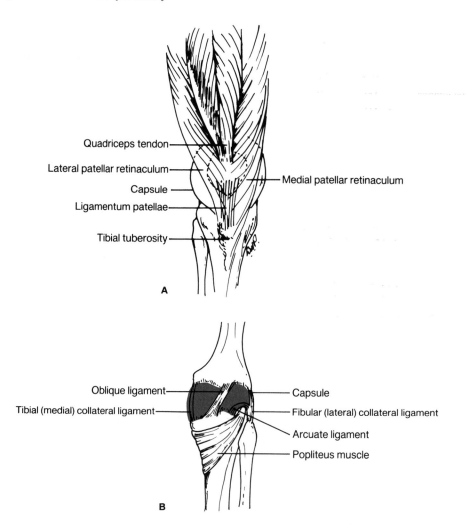

Quadriceps tendon

Lateral patellar retinaculum

Capsule

Ligamentum patellae

Tibial tuberosity

Medial patellar retinaculum

A

Oblique ligament

Tibial (medial) collateral ligament

Capsule

Fibular (lateral) collateral ligament

Arcuate ligament

Popliteus muscle

B

Figure 22-4.
The right knee joint. **A.** Anterior view.
B. Posterior view.

Ligamentum Patellae. This strong ligament (about 8 cm long) connects the apex of the patella with the tibial tuberosity (Fig. 22-4A). It is an intrinsic ligament and is separated from the upper part of the tibial tuberosity by a bursa and a pad of fat.

Fibular (Lateral) Collateral Ligament (Figs. 22-5A and 22-5B). The lateral collateral ligament is cord-like and stands separated from the capsule. (A pad of fat and genicular vessels intervene.) It is attached to the lateral femoral epicondyle and to the head of the fibula. It lies superficial to the tendon of popliteus, which separates it from the lateral meniscus, and splits the tendon of biceps femoris at its attachment to the head of the fibula (Fig. 22-5A). Its lower portion is clearly an extrinsic ligament; its upper part tends to blend with the capsule (see Fig. 22-4B).

Tibial (Medial) Collateral Ligament. The medial ligament of the knee joint is a strong intrinsic ligament that runs from the medial femoral epicondyle to the medial condyle and upper part of the medial surface of the tibia (see Fig. 22-4B). The ligament lies somewhat posterior to the long axis of the joint so that it is tightest when the joint is fully extended. It is attached to the periphery of the medial meniscus.

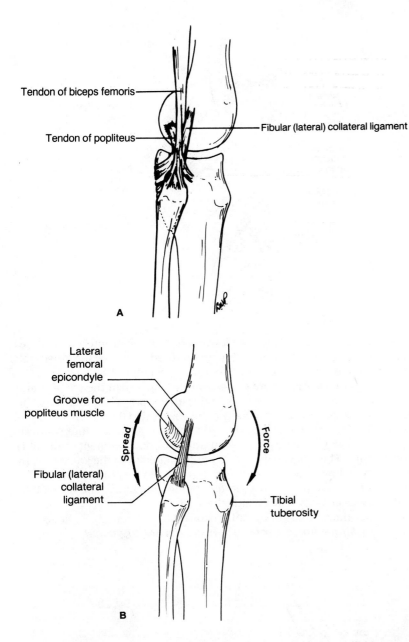

Tendon of biceps femoris

Tendon of popliteus

Fibular (lateral) collateral ligament

A

Lateral femoral epicondyle

Groove for popliteus muscle

Spread

Force

Fibular (lateral) collateral ligament

Tibial tuberosity

B

Figure 22-5.
Function of the fibular collateral ligament of the right knee joint.
A. Lateral view at rest.
B. Hyperextension renders the ligament taut.

Cruciate Ligaments. The cruciate ligaments (Fig. 22-6) are two heavy ligaments found in the intercondylar notch of the femur. They are described as ***anterior*** and ***posterior*** because of the sites of their attachments to the tibia.

The ***anterior cruciate ligament*** is attached to the anterior part of the intercondylar area of the tibia, just posterior to the attachment of the anterior horn of the medial semilunar cartilage. The ligament runs upward, backward, and laterally (for the right knee this is the direction of a pen held in the writing position of the right hand) and attaches to the medial aspect of the lateral femoral condyle. The ligament is tense in hyperextension of the knee and tends to groove the back of the femur.

Figure 22-6.
Sagittal section (schematic) of the right
knee.

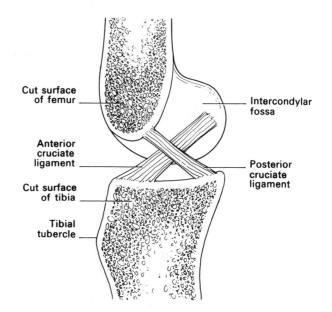

Cut surface
of femur

Intercondylar
fossa

Anterior
cruciate
ligament

Posterior
cruciate
ligament

Cut surface
of tibia

Tibial
tubercle

The anterior cruciate ligament will tend to prevent anterior displacement of the tibia on the femur. It can be ruptured by a force that drives the femur backward on the fixed tibia (football or ski injury).

The ***posterior cruciate ligament*** is stronger than the anterior and is attached to the posterior aspect of the intercondylar eminence of the tibia. It runs anteriorly and medially to be attached to the lateral surface of the medial femoral condyle near the anterior end of the intercondylar notch. This ligament tends to prevent posterior displacement of the tibia on the femur. The cruciate ligaments are separated from the joint cavity by synovial membrane and from a purely anatomical viewpoint are not inside the joint. From the perspective of the injured patient and the orthopedic surgeon, they are very much inside the joint.

Other thickenings in the joint capsule have been given names, for example, ***oblique posterior ligament*** and ***arcuate ligament***.

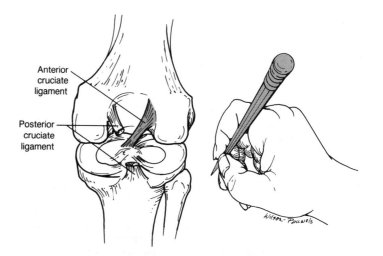

Anterior
cruciate
ligament

Posterior
cruciate
ligament

Figure 22-7.
Cruciate ligaments of the right knee
(posterior cruciate cut to show interior
of knee joint). "Pencil in right hand"
sketch illustrates position of right
anterior cruciate ligament.

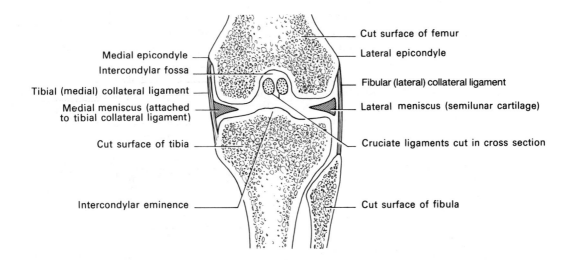

MENISCI (SEMILUNAR CARTILAGES)

The two **menisci** (Figs. 22-7, 22-8, and 22-9) are semilunar-shaped intra-articular discs composed of fibrocartilage. Their tips are firmly attached to the intercondylar eminence of the tibia, and their margins are attached to the periphery of the articular surfaces of the tibia by the coronary ligaments. The coronary ligaments are covered by synovial membrane and contain blood vessels that supply the periphery of the menisci; the more central portions of the menisci are avascular and consequently will not heal after injury. The wedge-shaped structure of the menisci deepens the articular surfaces available for the femoral condyles, whereas the flat inferior aspects of the menisci allow them to slide and rotate on the tibia.

Figure 22-8.
Coronal section (schematic) of the right knee as viewed from the posterior aspect. The synovial membrane and capsule have been omitted.

Figure 22-9.
Superior surface of the right tibia.
A. Landmarks. **B.** With menisci.

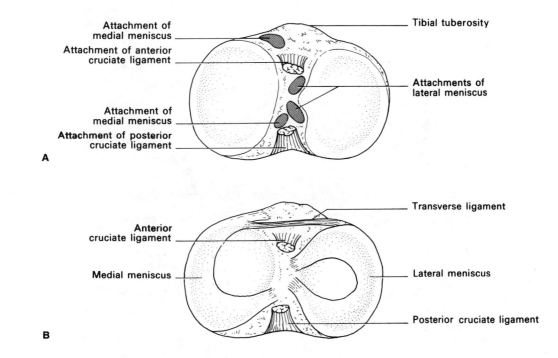

Medial Meniscus. The medial meniscus is a C-shaped cartilage. Its anterior end is the most anterior of several structures attached to the intercondylar eminence of the tibia (Fig. 22-9B). Its posterior attachment is anterior to that of the posterior cruciate ligament and posterior to the attachment of the lateral meniscus. The medial meniscus tends to deepen the articular surface but is more shallow than the lateral meniscus, and a substantial amount of medial femoral condyle articulates with the medial condyle of the tibia. Passing laterally from the anterior tip of the medial meniscus is the transverse ligament of the knee.

Lateral Meniscus. The lateral meniscus is nearly circular in shape, and its two extremities are attached close together to the intercondylar eminence of the tibia.

CLINICAL NOTE

The attachment of the medial meniscus to the medial collateral ligament of the knee may make it more vulnerable to injuries imposed by compressing, shearing, or rotational forces on the medial compartment of the joint. In contrast, the lateral meniscus does not have a firm attachment to the joint capsule and is pulled posteriorly by the attachment of the tendon of popliteus to its posterolateral aspect. Tears of the lateral meniscus are less common; the anatomical differences may be relevant.

SYNOVIAL MEMBRANE

The synovial membrane of the knee joint adapts itself to the configuration of the capsule. If the reader remembers that synovial membrane covers the interior of the capsule and the intracapsular structures (except for the articular cartilages) of any joint, the configuration of the synovial cavity of the knee may be understood (Fig. 22-10).

Attachments of the Synovial Membrane. The synovial membrane is attached around the periphery of the patella. That part of the membrane

Figure 22-10.
Structures of the knee joint.

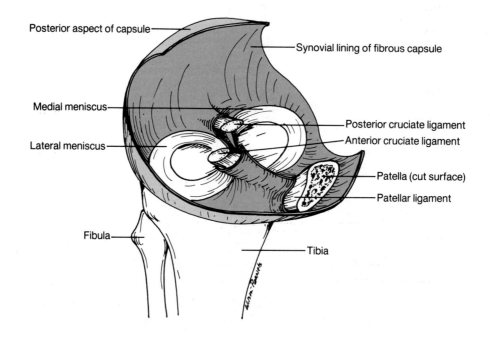

Posterior aspect of capsule

Synovial lining of fibrous capsule

Medial meniscus

Posterior cruciate ligament

Lateral meniscus

Anterior cruciate ligament

Patella (cut surface)

Patellar ligament

Fibula

Tibia

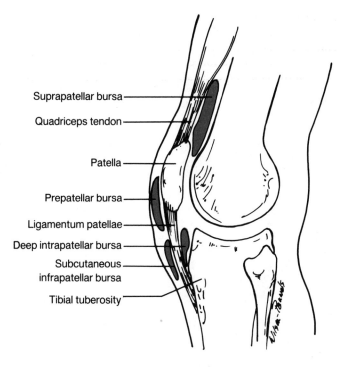

Figure 22-11.
Bursae of knee.

Suprapatellar bursa

Quadriceps tendon

Patella

Prepatellar bursa

Ligamentum patellae

Deep intrapatellar bursa

Subcutaneous
infrapatellar bursa

Tibial tuberosity

that is attached to the lower border of the patella runs posteriorly and up-ward to attach to the femur at the anterior boundary of the intercondylar notch. This attachment forms the ***infrapatellar fold***. Inferiorly the edges of the infrapatellar folds are attached to the intercondylar eminence of the tibia.

Projecting into the anterior aspect of the infrapatellar fold is the infra-patellar pad of fat. The synovial membrane continues backward along the margins of the intercondylar notch of the femur. The membrane from this portion of the notch passes inferiorly to be attached to the tibia on either side of the intercondylar eminence. This means that the medial and lateral cavities of the knee are separated by synovial membrane running from the edges of the intercondylar eminence to the intercondylar notch, but anteriorly the two cavities are joined superior to the infrapatellar fold. There is usually a defect in the infrapatellar fold so that the two subjoints communicate at that level. The synovial membrane then goes on to line the capsule of each "subjoint" posteriorly, laterally (or medially), and an-teriorly. The membrane continues above the patella to form the deep and extensive suprapatellar bursa which lies between the quadriceps tendon and the anterior aspect of the lower end of the femur.

BURSAE

There are many bursae around the knee, and some of them have clinical significance (Fig. 22-11).

Subcutaneous Prepatellar Bursa.   Lies between the patella and the skin. It enables the skin to slide freely over the patella.

Subcutaneous Infrapatellar Bursa.   Lies between the tibial tuberosity and the skin.

Table 22-2.
Knee Joint

Movement	Muscles	Nerves	Spinal Cord Segment
Extension	Quadriceps femoris	Femoral	L2, L3, L4
	Tensor fasciae latae	Superior gluteal	L4, L5
Flexion	Biceps femoris	Sciatic	L5, S1, S2
	Semimembranosus	Sciatic	L5, S1, S2
	Semitendinosus	Sciatic	L5, S1, S2
	Gastrocnemius	Tibial	S1, S2
	Sartorius	Femoral	L2, L3
	Gracilis	Obturator	L2, L3
	Popliteus	Tibial	L4, L5, S1
Medial tibial rotation	Popliteus	Tibial	L4, L5, S1
	Sartorius ⎫ when	Femoral	L2, L3
	Semitendinosus ⎬ knee is	Sciatic	L5, S1, S2
	Semimembranosus ⎭ flexed	Sciatic	L5, S1, S2
Lateral tibial rotation	Biceps femoris (when knee is flexed)	Sciatic	L4, L5
	Tensor fasciae latae (near full extension of knee)	Superior gluteal	L4, L5

Deep Infrapatellar Bursa. Is found between the ligamentum patellae and the tibial tuberosity.

Suprapatellar Bursa. Lying between the femur and the quadriceps muscles and tendon, this large bursa communicates freely with the knee joint. It can conceal a considerable amount of synovial fluid in diseased states of the joint.

CLINICAL NOTE

Kneeling whether from necessity or piety can cause traumatic effusions of fluid into the prepatellar or superficial infrapatellar bursae; involvement of the first is called "housemaid's knee," and the second affliction is called "parson's knee."

MOVEMENTS (Table 22-2)

The most apparent movements of the knee are flexion and extension, but rotation and gliding also occur. Examination of the articular surfaces of the femur and tibia will show that flexion of the femur on the tibia is combined with forward sliding of the femoral condyles, and, conversely, extension involves forward sliding of the tibia on the femur. The medial femoral condyle is larger and more curved than the lateral condyle. In the last 30 degrees of extension of the knee joint the lateral condyle has reached its limit, whereas the medial condyle has not moved into full extension. To accomplish completion of extension (as it were "to satisfy the medial condyle") the femur rotates medially on the tibia around an axis that runs from the femoral head through the lateral femoral condyle. This final rotation is said to "lock" the tibia in full extension. When the joint is flexed from the fully extended position the femur must be "unlocked" by lateral rotation on the tibia; popliteus muscle produces this rotation. Hyperextension of the joint is prevented by the collateral ligaments, the anterior cruciate ligament, and the oblique ligament.

In the erect position the body's weight is in front of the knee joint so that little muscle effort is required (or recorded electromyographically) to maintain this position. The joint will remain in extension supported by its ligaments and the congruence and compression of its joint surfaces.

Extension. The quadriceps muscles with some help from tensor fasciae latae (*L3* and *L4*).

Flexion. Biceps femoris, semitendinosus and semimembranosus, gracilis, sartorius, gastrocnemius, and popliteus (*L4, L5,* and *S1*).

Medial and Lateral Rotation. Medial and lateral rotation of the flexed knee can be produced by the muscles attached to the medial and lateral sides of the tibia, respectively.

TIBIOFIBULAR JOINTS

There are two tibiofibular joints (Fig. 22-12): the superior and inferior. In addition, the shafts of the two bones are connected by an interosseous membrane.

Superior joint (lateral view)

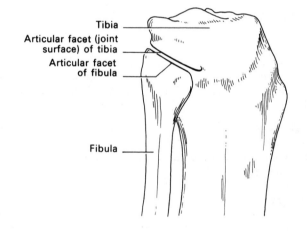

Inferior joint (coronal section)

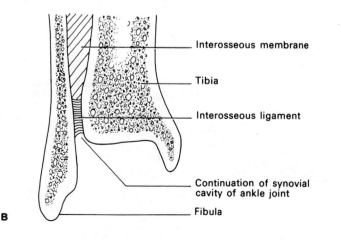

Figure 22-12.
A. Superior tibiofibular joint (lateral view). **B.** Inferior tibiofibular joint (coronal section).

SUPERIOR TIBIOFIBULAR JOINT

This is a synovial joint between the facet on the head of the fibula facing upward, forward, and slightly medially and the facet on the undersurface of the lateral tibial condyle facing in the reciprocal directions (Fig. 22-12A). The capsule is fairly sturdy and contains no named ligaments. Some movement occurs at this joint when forces at the ankle joint cause some lateral movement of the fibular shaft.

INFERIOR TIBIOFIBULAR JOINT

This joint is a *syndesmosis* with a tough *interosseous ligament* and anterior and posterior tibiofibular ligaments that help to deepen the *mortise* of the ankle joint (Fig. 22-12B). There is a small projection of the synovial space of the ankle joint between inferior ends of the tibia and fibula.

TALOCRURAL (ANKLE) JOINT

The ankle joint (Fig. 22-13) is formed by the *tibia* above and medially, the *fibula* laterally, and the *talus* below. The talus fits into a *mortise* formed by the tibia and fibula.

BONY SURFACES

Tibia.   The tibia has two bony surfaces for articulation with the talus. The major one is on the inferior surface of the tibia; this is slightly concave from front to back and wider anteriorly than posteriorly. The medial malleolus has a lateral articular facet for the comma-shaped facet on the medial side of the talus.

Fibula.   The lower end of the fibula, the lateral malleolus, is placed more posteriorly than the medial malleolus and also projects further inferiorly. It has a medial facet for articulation with the lateral surface of the talus.

Figure 22-13.
Skeletal structures of right ankle joint.
A. Lateral view. **B.** Posterior view.

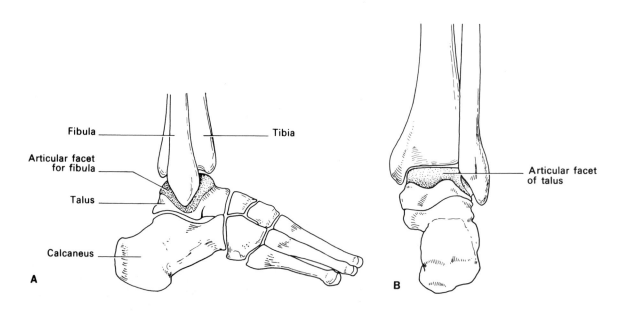

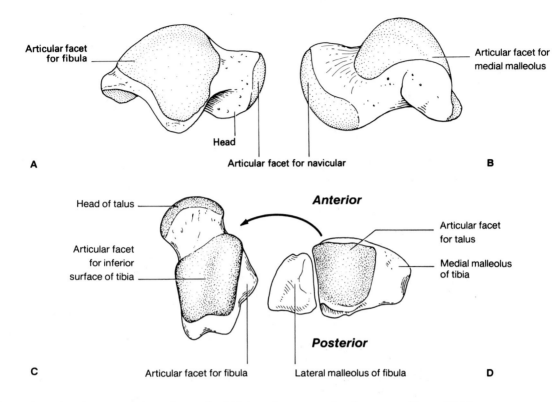

Articular facet
for fibula

Head

Articular facet for navicular

A

Articular facet for
medial malleolus

B

Head of talus

Articular facet
for inferior
surface of tibia

Anterior

Articular facet
for talus

Medial malleolus
of tibia

Posterior

C

Articular facet for fibula

Lateral malleolus of fibula

D

Figure 22-14.
A. Lateral view of talus. **B.** Medial view
of talus. **C.** Superior view of talus.
D. Inferior view of tibia and fibula
forming mortise for superior articular
facet of talus.

Talus. The talus (Figs. 22-14A, 14B, and 14C) has three articular facets
that participate in the ankle joint. The superior (trochlear) surface is convex from anterior to posterior and concave from side to side. It is wider in
front than behind. On the medial side of the body of the talus there is a
comma-shaped articular facet that is widest anteriorly. On the lateral side
there is a triangular (delta-shaped) articular facet.

The mortise formed by the tibia and fibula is wider anteriorly than
posteriorly (Fig. 22-14D). The superior articular surface of the talus is also
wider in front so that in full dorsiflexion of the joint the talus is wedged
firmly in that mortise. This gives the joint great stability in the fully dorsiflexed position; because the body weight falls just in front of the ankle
joint in the erect posture and the joint is not fully dorsiflexed in this position, sustained contraction of the soleus with intermittent bursts of activity
in the gastrocnemii is required to maintain this position. In plantar flexion
the narrow part of the upper surface of the talus is between the more
widely spaced anterior part of the mortise. This allows the talus to tilt side-
ways and causes instability of the joint.

CAPSULE

The capsule surrounds the borders of the articular surfaces except where
it extends forward onto the superior surface of the neck of the talus. The
capsule is thin in front and behind, but at the sides it is thickened into
certain relatively strong, named ligaments.

Anterior and Posterior Tibiofibular Ligaments. These two ligaments
are really part of the inferior tibiofibular joint but are described here because they help to deepen the mortise of the ankle joint. The anterior liga-

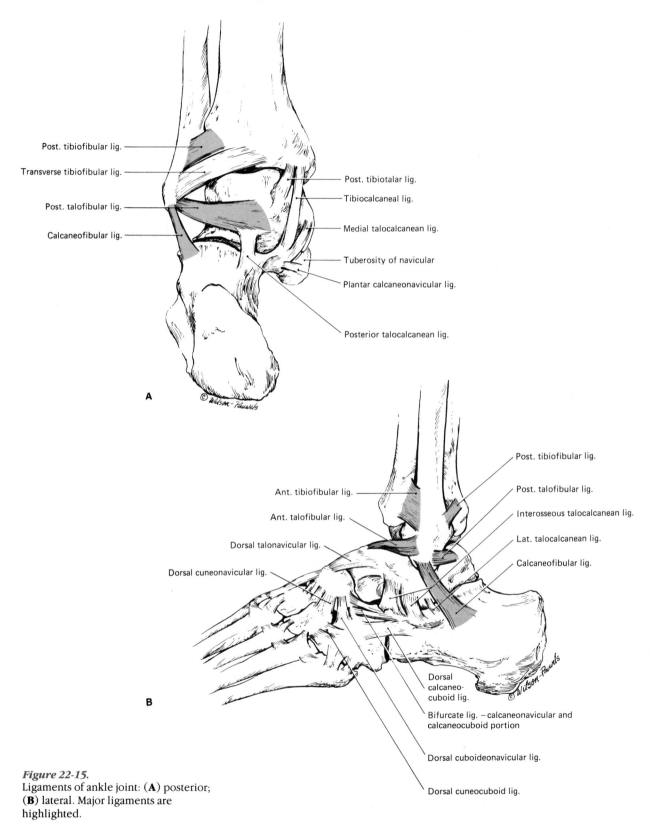

Post. tibiofibular lig.
Transverse tibiofibular lig.
Post. talofibular lig.
Calcaneofibular lig.

Post. tibiotalar lig.
Tibiocalcaneal lig.
Medial talocalcanean lig.
Tuberosity of navicular
Plantar calcaneonavicular lig.

Posterior talocalcanean lig.

A

© Wilson-Pauwels

Ant. tibiofibular lig.
Ant. talofibular lig.
Dorsal talonavicular lig.
Dorsal cuneonavicular lig.

Post. tibiofibular lig.
Post. talofibular lig.
Interosseous talocalcanean lig.
Lat. talocalcanean lig.
Calcaneofibular lig.

Dorsal calcaneo-cuboid lig.

Bifurcate lig. – calcaneonavicular and calcaneocuboid portion

Dorsal cuboideonavicular lig.

Dorsal cuneocuboid lig.

B

© Wilson-Pauwels

Figure 22-15.
Ligaments of ankle joint: (**A**) posterior;
(**B**) lateral. Major ligaments are
highlighted.

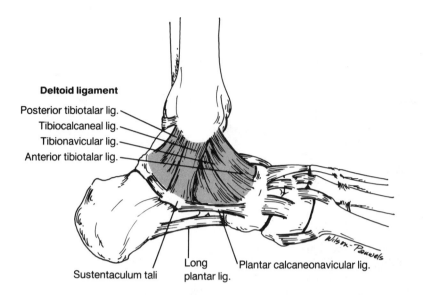

Figure 22-16.
Deltoid ligament of left ankle (medial view).

ment is unremarkable. The inferior part of the posterior ligament runs from the malleolar fossa of the fibula to the posterior inferior edge of the tibia almost as far as the medial malleolus. This ligament extends **distal** to the lower border of the tibia and deepens the ankle joint (Fig. 22-15).

Lateral Ligaments of the Ankle.   There are three named ligaments (Fig. 22-15B) uniting the fibula to the tarsal bones. The **anterior talofibular** ligament runs from the fibular malleolus to the neck of the talus. The **posterior talofibular ligament** runs from the malleolar fossa almost horizontally backward to attach to the lateral tubercle of the talus. The **calcaneofibular** ligament runs from the tip of the malleolus posteriorly and inferiorly to the calcaneus. It is crossed by the tendons of peroneus longus and brevis.

Deltoid Ligament.   This is a tough ligament on the medial side of the ankle joint (Fig. 22-16). It joins the medial malleolus to the **talus**, the **sustentaculum tali**, the medial side of the plantar **calcaneonavicular ligament** (spring ligament), and the **tuberosity** of the navicular bone. The deltoid ligament is crossed by the tendons of tibialis posterior and flexor digitorum longus.

CLINICAL NOTE

The ankle joint has least stability in the plantar flexed position, and an inversion strain in that position is most likely to injure the anterior talofibular ligament. This is the most common injury in a "sprained ankle." If complete rupture of the lateral ligamentous complex should result from an inversion injury the talus can be felt to tilt (and seen to do so on an x-ray film) in passively forced inversion.

The deltoid ligament is so strong that in forced eversion injuries of the ankle, rather than the deltoid ligament rupturing, the medial malleolus of the tibia will be torn off (i.e., fractured).

MOVEMENTS (Table 22-3)

The movements permitted at the ankle joint are dorsiflexion and plantarflexion.

Table 22-3.
Ankle Joint

Movement	Muscles	Nerves	Spinal Cord Segment
Dorsiflexion	Tibialis anterior	Deep peroneal	L4, L5
	Extensor hallucis longus	Deep peroneal	L5, S1
	Extensor digitorum longus	Deep peroneal	L5, S1
	Peroneus tertius	Deep peroneal	L5, S1
Plantarflexion	Gastrocnemius	Tibial	S1, S2
	Soleus	Tibial	S1, S2
	Tibialis posterior	Tibial	L4, L5
	Flexor hallucis longus	Tibial	S2, S3

Dorsiflexion. This movement wedges the wide anterior part of the body of the talus back into the tibiofibular mortise. This is an extremely stable position that is acknowledged in the dorsiflexed cant of the design of ski boots. The main dorsiflexors are tibialis anterior and the extensors of the hallux and toes (L4 and L5).

Plantarflexion. Plantarflexion is limited by tension in the anterior crural muscles and also by tension of the anterior fibers of the deltoid ligament. In plantarflexion the joint is very unstable, and tilting of the talus can take place. This instability may be demonstrated by walking on one's toes. The same phenomenon occurs when wearing high-heeled shoes or dancing on the toes. Inversion of the foot appears more marked in plantarflexion than in dorsiflexion owing to tilting of the talus. The main plantar flexors are all the muscles of the posterior compartment (S1 and S2).

CLINICAL NOTE: POTT'S FRACTURE

Pott's fracture (Fig. 22-17) is a fracture-dislocation of the ankle caused by forced eversion of the foot. The deltoid ligament "pulls off" the medial

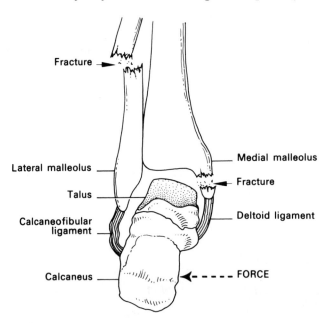

Figure 22-17.
Fracture–dislocation (Pott's fracture) of the ankle caused by forced eversion.

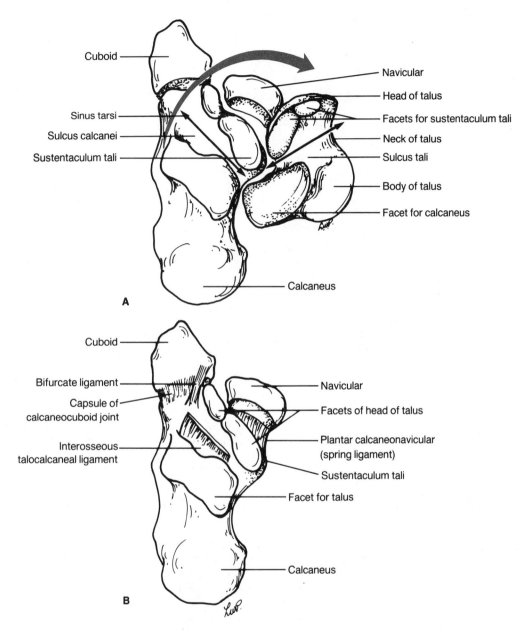

Figure 22-18.
A. Subtalar (talocalcanean joint) with talocalcaneal joint "opened." Sulcus tali + sulcus calcanei = sinus tarsi.
B. Interosseous talocalcaneal ligament in sinus tarsi.

malleolus, there is damage to the inferior tibiofibular syndesmosis, and the fibula fractures at a higher level. Many varieties of this type of fracture can occur.

JOINTS OF THE FOOT

There are many joints in the foot, but only the most important two will be considered in detail.

SUBTALAR (TALOCALCANEAN) JOINT (Fig. 22-18)

This joint is formed between the body of the talus above and part of the calcaneus below. There is a single facet on each bone, and the joint capsule is relatively thin and unremarkable. The strongest bond between talus and calcaneus is formed by the ***interosseous talocalcanean ligament***, which lies in the ***sinus tarsi*** and which contains blood vessels supplying both bones (Fig. 22-18A).

TRANSVERSE TARSAL JOINT

The transverse tarsal joint comprises two joints: the ***talocalcaneonavicular*** joint and the ***calcaneocuboid*** joint. This compound joint permits the basic movements of inversion and eversion; however, some inversion and eversion occur at the talocalcanean joint.

Much of the adjustment the foot makes on uneven ground is the result of movements at the transverse tarsal and subtarsal joints. Inversion is often accompanied by some adduction and plantarflexion of the foot, and eversion is often combined with abduction and dorsiflexion.

Talocalcaneonavicular Joint. This joint is located anterior to the subtalar joint, separated from it by the interosseous talocalcanean ligament. The talus has three facets, one facing primarily anteriorly for articulation with the navicular, and two facing inferiorly for articulation with the calcaneus. One of the latter articulates with the sustentaculum tali and the other is for articulation with the superior surface of the anterior end of the calcaneus. (The last two facets may be fused to form a single facet.) The calcaneus has two facets corresponding to the two facets on the talus, and the navicular has a posterior concave facet for the head of the talus.

The ***plantar calcaneonavicular (spring) ligament*** (Figs. 22-19 and 22-20A) runs from the ***sustentaculum tali*** to the posteroinferior margin of the ***navicular*** bone and forms part of the socket for the head of the talus. In some cases the head of the talus shows a separate facet for this ligament. The capsule of the joint blends with the interosseous talocalcanean ligament posteriorly and with the deltoid ligament of the ankle medially; otherwise it is unremarkable.

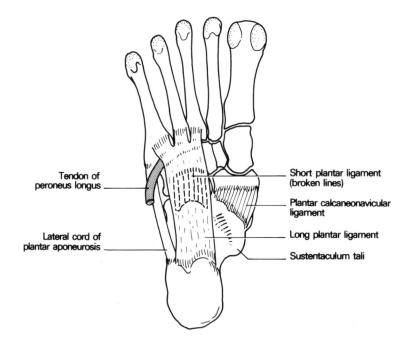

Figure 22-19.
Ligaments of the sole of the right foot. The short plantar ligament is deep to the long plantar ligament.

Tendon of peroneus longus

Lateral cord of plantar aponeurosis

Short plantar ligament (broken lines)

Plantar calcaneonavicular ligament

Long plantar ligament

Sustentaculum tali

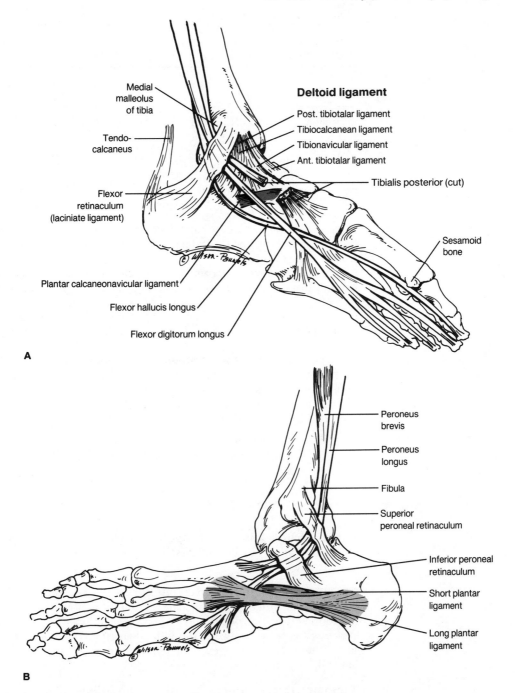

A

B

Dorsolaterally the calcaneus is attached to the navicular by the medial half of the **_bifurcate ligament_**, which passes from the anterior end of the superior surface of the calcaneus to the navicular. (The other half is attached to the cuboid; see Fig. 22-15B.)

Calcaneocuboid Joint. The calcaneocuboid joint is a separate articulation between the calcaneus and the cuboid. There is a projection on the inferior aspect of the cuboid that helps to support the anterior end of the calcaneus.

Figure 22-20.
A. Medial longitudinal arch of left foot showing plantar calcaneonavicular (spring) ligament. **B.** Lateral longitudinal arch of the left foot showing long and short plantar ligaments.

Table 22-4.
Transverse Tarsal Joint

Movement	Muscles	Nerves	Spinal Cord Segment
Inversion	Tibialis anterior	Deep peroneal	L4, L5
	Tibialis posterior	Tibial	L4, L5
Eversion	Peroneus longus	Superficial peroneal	L5, S1, S2
	Peroneus brevis	Superficial peroneal	L5, S1, S2
	Peroneus tertius	Deep peroneal	L5, S1

The dorsal part of the capsule is reinforced by the lateral part of the bifurcate ligament, and the plantar aspect of the joint is covered by the two plantar ligaments.

The ***long plantar ligament*** is a tough band of connective tissue that extends from the tubercle of the calcaneus to the ridge on the plantar surface of the cuboid; from there some of its fibers bridge the groove in the cuboid, forming a tunnel for the tendon of the peroneus longus and attaching to the bases of the (second), third, fourth, and fifth metatarsals (Fig. 22-20B).

The ***plantar calcaneocuboid (short plantar) ligament*** lies deep to the long plantar ligament and runs from the anterior end of the calcaneus to the proximal edge of the cuboid (see Fig. 22-19).

Movements of the transverse tarsal (talocalcaneonavicular) joint (Table 22-4) are essentially inversion and eversion:

Eversion : peroneal muscles (L5, S1)
Inversion: tibialis anterior and posterior (L4 and L5).

OTHER JOINTS OF THE FOOT

Cubonavicular Joint. This is the synovial joint between the cuboid and navicular bones. A strong interosseous ligament is the only notable feature of this joint.

Cuneonavicular Joint, Intercuneiform Joints, and Cuneocuboid Joint. These articulations all share a common synovial cavity. There are appropriately named dorsal and plantar capsular ligaments and also ***interosseous ligaments*** between the individual cuneiforms and between the lateral cuneiform and the cuboid. Relatively minor gliding movements occur at these joints.

Tarsometatarsal Joints. These plane joints allow relatively little movement. The first tarsometatarsal joint is for the first digit and has its own synovial cavity. The base of the second metatarsal is held in a mortise between the three cuneiforms and articulates with all of them. The third metatarsal articulates with the lateral cuneiform and the fourth and fifth metatarsals articulate with the cuboid (see Fig. 21-3).

Intermetatarsal Joints. These joints are between the bases of the metatarsal bones. There are interosseous ligaments as well as dorsal and plantar

Table 22-5.
First Metatarsophalangeal Joint

Movement	Muscles	Nerves	Spinal Cord Segment
Flexion	Flexor hallucis longus	Tibial	S2, S3
	Flexor hallucis brevis	Tibial	S2, S3
Abduction	Abductor hallucis	Medial plantar	S2, S3
Adduction	Adductor hallucis	Lateral plantar	S2, S3
Extension	Extensor hallucis longus	Deep peroneal	L5, S1
	Extensor digitorum brevis	Deep peroneal	S1, S2

metatarsal ligaments. The heads of the metatarsals are all connected by deep transverse ligaments which (unlike in the hand) attach to the first metatarsal; thus little individual movement of the metatarsals is possible.

Metatarsophalangeal Joints (Tables 22-5 and 22-6). The metatarsophalangeal joints are condyloid (knuckle) joints in which the joint surface of each metatarsal head extends onto the dorsal and plantar aspects. This is particularly true of the first metatarsophalangeal joint, since the big toe is extremely dorsiflexed every time a step is taken. The capsules of these joints bear a thickened fibrocartilaginous *plantar plate* on their plantar surfaces, and the metatarsal heads are joined to the bases of the proximal phalanges by cord-like collateral ligaments.

Interphalangeal Joints. The interphalangeal joints are similar in construction to the metatarsophalangeal joints.

ARCHES OF THE FOOT

The foot must be strong in order to bear the weight of the body, but it must also be flexible to absorb the pounding that occurs at every step. It must be able to adapt itself to uneven terrain in standing and walking and must act as a strong lever that can propel the body weight with every step. To meet all of these requirements the foot is constructed of a system of relatively elastic arches.

Table 22-6.
The Toes

Movement	Muscles	Nerves	Spinal Cord Segment
Flexion	Flexor digitorum longus	Tibial	S2, S3
	Flexor digitorum brevis	Medial plantar	S2, S3
Extension	Extensor digitorum longus	Deep peroneal	L5, S1
	Extensor digitorum brevis	Deep peroneal	S1, S2
	Lumbricals and interossei (at interphalangeal joints)	Medial and lateral plantar	S2, S3
Abduction	Dorsal interossei	Lateral plantar	S2, S3
Adduction	Plantar interossei	Lateral plantar	S2, S3

Longitudinal Arches

There are two longitudinal arches in the foot: a medial and a lateral (see Fig. 22-20). Observe your wet footprint after a shower.

Medial Arch. The bony elements of the medial longitudinal arch of the foot are the calcaneus, talus, navicular, three cuneiforms, and the metatarsals of the medial three toes. The posterior pillar is the heel (posterior end of calcaneus), and the anterior pillar is composed of the heads of the medial three metatarsals with the head of the first metatarsal, which rests on two sesamoid bones, bearing the greatest burden.

Lateral Arch. The lateral arch of the foot is lower, the lateral side of the foot usually resting on the ground. The bones of the lateral arch are the calcaneus (heel), cuboid, and the lateral two metatarsals. Note that the heel (the posterior end of the calcaneus) is the posterior pillar of both arches.

Transverse Arch

When the bones of the foot are examined a ***transverse arch*** of the foot appears to be present. It is formed by the cuboid, the appropriately wedge-shaped cuneiforms, and the adjacent bases of the metatarsals. It is a peculiar arch in that it lacks a medial pillar and its functional existence is doubtful. When the two feet are placed side by side it can be seen that the combined arches describe the outline of a dome.

MAINTENANCE OF THE ARCHES OF THE FOOT

The arches of the foot depend for their existence on two main factors: the shapes of the bones and the strength of the ligaments. It has been seen from the description of the joints of the foot that there are some extremely strong ligaments that bind the bones of the foot. These ligaments are particularly strong on the plantar aspect (spring ligament, long and short plantar ligaments, plantar aponeurosis) and appear capable of supporting the weight of the body on the arched foot. It has been calculated that in the resting, standing position each calcaneus carries 10% of the body weight transmitted through the ipsilateral leg, while its metatarsal heads carry the other 40%.

The head of the first metatarsal carries 40% of the metatarsal burden so that the ultimate weight distribution on each foot will be as follows: 10% of body weight on each heel, 40% of body weight on the metatarsal heads of each foot, and a total of 16% of body weight on the head of each ***first*** metatarsal (40% of the 40% distributed on the metatarsals).

The muscles of the leg and foot seem to be of relatively little importance in the support of the normal arch. When the arch tends to "fall" because of overstretching of the ligaments or deformity of the bones, the muscles may, for a time, support the arches. Ultimately, however, the muscles will weaken or tire, and pain in the front of the leg, the calf, or the sole of the foot may be the first symptoms of weakness of the arches.

The importance of the bones and ligaments in the maintenance of the ***static*** arches is displaced by the importance of the leg and foot muscles when it comes to using the foot as a ***mobile*** arched lever in walking.

LIGAMENTS OF IMPORTANCE TO THE ARCHES OF THE FOOT

The various ligaments described with the joints of the foot, especially the interosseous ones, are all of importance in the maintenance of the arches of the foot, but the following four structures are particularly important and should be committed to memory:

> The plantar calcaneonavicular (spring) ligament.
> The long plantar ligament.
> The short plantar (calcaneocuboid) ligament.
> The plantar aponeurosis (particularly the strong band that stretches from the tuberosity of the calcaneus to the tubercle on the base of the fifth metatarsal).

CLINICAL NOTE

Fallen arches or flat feet can produce painful symptoms, but the frequency of occurrence of this condition has probably been exaggerated. Many so-called flat feet present only an apparent loss of the medial arch when the foot is held in some eversion or dorsiflexion.

WALKING

There are many ways in which walking can be described, but it is easiest to understand if it is divided into two distinct phases. Considering the action of one lower limb (using the **right** as an example) during a complete cycle of one step, it is evident that there are two phases:

1. **The swing phase of walking** in which the limb is brought forward.
2. **The stance phase** in which the foot is on the ground.

A complete cycle is more easily understood if it is considered from the moment the heel strikes the ground until the heel of the same limb strikes the ground again. Consider, then, the action of the different joints of the **entire limb** when the step is taken, beginning at **right heel strike**.

HIP OF RIGHT LOWER LIMB

When the heel of the right foot strikes the ground, the right hip is **semiflexed**; then, as the body weight is carried forward over the right foot, the right hip **extends**. When the swing phase starts at takeoff, the hip flexes and then as the lower limb comes forward past the body, the hip starts to extend so that, at heel strike, it is in **semiflexion** (one flexion–extension cycle per step).

At heel strike the hip is in a neutral position with reference to abduction and adduction. In the stance phase, as the body weight passes forward, the opposite (left) limb starts to lift off the ground, which requires tilting of the pelvis and shifting of the center of gravity over the **right hip joint**. This is the action of the right gluteus medius and minimus, which abduct the pelvis at the hip. The glutei remain contracted until the left heel strikes the ground. When this occurs the right hip returns to neutral position, and as the right lower limb is lifted in the swing phase there is slight **adduction** at the **right** hip produced by gravity in counteraction to **abduction** at the **left** hip.

Rotation of the hip also occurs. As the right heel strikes, the right hip is slightly laterally rotated; as the body weight passes over the right foot the hip medially rotates, and it is in medial rotation at the start of the swing phase. During the swing phase the hip again rotates laterally.

This rotation of the pelvis is necessitated by the fact that the pelvis will follow the motion of the swinging limb and therefore must be rotated at the opposite hip.

KNEE OF THE RIGHT LOWER LIMB

At right heel strike the right knee is ***extended*** and locked. As the foot touches the ground the knee unlocks (popliteus rotates femur laterally on tibia), and the knee is then free to ***flex*** as the weight of the body is carried forward over the foot. After the weight of the body has passed over the foot, the knee begins extension and continues to extend until the swing phase starts. At the end of this extension the knee will again be locked by the conjunct medial rotation of the femur on the tibia. Once the swing phase starts the knee will unlock and flex so that the toes can clear the ground, and then as the limb is brought forward the knee extends and locks again at heel strike. There are two flex–extend cycles per step. Depending on individual gait it is probable that complete locking does not occur at each step.

ANKLE OF RIGHT LOWER LIMB

As the heel strikes, the ankle is slightly ***dorsiflexed***; then, as the foot falls to the ground (so that the toes touch the ground), ***plantarflexion*** begins. As the weight of the body comes forward the ankle is again dorsiflexed. Once the weight starts to leave the right limb, the ankle plantarflexes and remains plantarflexed until takeoff has been completed. To prevent the toes from dragging, the ankle dorsiflexes and remains in that position until the foot has gone through the lowest arc of the swing phase, when plantarflexion gradually starts, and continues until the sole of the foot is on the ground. There are two dorsiflex-plantarflex cycles per step.

METATARSOPHALANGEAL JOINTS OF RIGHT LOWER LIMB

At the moment the heel strikes, the metatarsophalangeal joints are ***extended***. As the weight of the body passes forward, over the foot, the metatarsophalangeal joints start to ***dorsiflex*** (extend) so that they are dorsiflexed at the moment of takeoff. Then they start to plantarflex to complete takeoff, and then they dorsiflex again to allow the toes to clear the ground during the swing phase. Toward the end of the swing phase the toes will return to the moderately extended neutral position (two cycles per step).

POINTS TO NOTE IN WALKING

Center of Gravity. The center of gravity (which is supposed to reside just anterior to the sacral promontory) rises each time the center of the stance phase is reached. Since this occurs twice in each cycle the head bobs twice in a cycle.

Height. The limb is seldom fully extended when the body is over it in the stance phase; therefore in walking one is slightly shorter than in standing.

Invertors and Evertors. The muscles that produce inversion and eversion of the foot stabilize and support the arches with each step.

Control by Central Nervous System. The whole cycle of walking is under complex control of the central nervous system. The proprioceptive (stretch receptors) fibers of the muscles and joints are very important in allowing the brain to know the exact position of the individual joints. Little of this knowledge reaches consciousness. In the condition of ***tabes dorsalis*** (a form of third stage neurosyphilis) the proprioceptive pathways in the spinal cord are damaged. The individual is unable to judge the positions of the various joints and muscles, and his gait assumes a curious flail-like, slapping motion.

LYMPHATIC SYSTEM OF THE LOWER LIMB

The lymphatic system of the lower limb drains primarily into the ***inguinal lymph nodes***, although some of the deep structures of the buttock drain into the ***internal iliac lymph nodes***.

The inguinal lymph nodes drain an area that includes all of the lower limb, the perineum, the abdominal wall below the umbilicus and the skin of the back to an equivalent level, the buttock, the natal cleft and the lower half of the anal canal, the lower end of the vagina, and the distal end of the penile urethra. The lymph drainage of the skin of all these areas is into the ***superficial inguinal lymph nodes***.

SUPERFICIAL LYMPH NODES

There is usually a single superficial popliteal lymph node that drains the area drained by the small saphenous vein and is located close to the termination of that vein; its efferents drain to the deep inguinal nodes. The rest of the superficial lymph nodes are located in the inguinal region. The superficial inguinal lymph nodes can be divided into ***proximal*** and ***distal*** groups. The proximal group consists of five or six nodes that lie parallel to the inguinal ligament. The distal group of four or five nodes lies parallel and close to the upper end of the great saphenous vein (Fig. 22-21).

Area Drained by the Superficial Inguinal Lymph Nodes. The superficial inguinal lymph nodes drain the superficial tissues of the entire body below the level of the umbilicus with the exception of the area drained by the superficial popliteal node. The efferent vessels from the superficial inguinal lymph nodes drain into the ***deep inguinal lymph nodes***.

DEEP LYMPH NODES

The deep lymph nodes drain the deep tissues of the entire lower extremity and are found in several locations.

Anterior Tibial Node. This is a single node in the anterior compartment of the leg near the upper margin of the interosseous membrane.

Deep Popliteal Nodes. Five or six lymph nodes are found in the fat of the popliteal fossa around the popliteal vessels. Their efferent vessels pass along the subsartorial canal to the deep inguinal nodes.

Deep Inguinal Nodes. A surprisingly small number of nodes (three or four), the deep inguinal nodes, are clustered around the upper end of the

Figure 22-21.
Anterior and posterior view of
lymphatic drainage of right lower
extremity. Solid lines = superficial
lymphatics; dotted lines = deep
lymphatics.

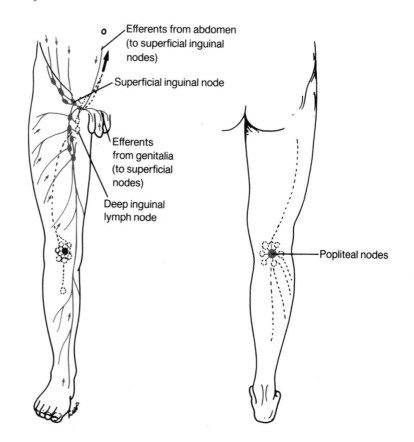

Efferents from abdomen
(to superficial inguinal
nodes)

Superficial inguinal node

Efferents
from genitalia
(to superficial
nodes)

Deep inguinal
lymph node

Popliteal nodes

femoral vein. One of these is often found in the femoral canal. They drain
the deep tissues of the lower limb and receive the efferent vessels from the
anterior tibial node, the deep popliteal nodes, and the superficial inguinal nodes.

The efferent vessels from the deep inguinal nodes drain into the **external iliac** lymph nodes, which are found along the external iliac vessels
in the iliac fossae of the abdomen.

Deep Lymph Vessels of the Gluteal Region

The lymph vessels draining the deep structures of the gluteal region pass
along the gluteal vessels to the **internal iliac lymph nodes** located along
the internal iliac vessels in the pelvis.

CLINICAL NOTE

*Enlargement of the superficial inguinal lymph nodes is extremely com-
mon and in most cases is the result of minor superficial infections. When
the texture and size of enlarged inguinal nodes suggest the possibility of
malignant disease a careful search for the primary lesion has to be undertaken. This means examination of the skin of the whole lower limb, including the clefts between the toes, the skin of the abdomen and back
below the level of the umbilicus, the lower genitals, vulva, perineum, scrotum, and lower anal canal. Similarly when confronted with a lesion in
any of these areas the physician must examine the inguinal nodes to determine whether disease has spread to them.*

The Back

BONES AND JOINTS

The general features of the vertebrae must be understood before attempting to understand the complex anatomy and functions of the back.

THE VERTEBRAE

General Features

Certain features are common to all vertebrae (Fig. 23-1), including the ***body***, ***arch***, ***lamina***, ***pedicle***, and the ***transverse***, ***spinous***, and ***articular processes*** and ***facets***. An ***intervertebral foramen*** for the passage of the spinal nerve is found between two adjacent vertebrae.

The spine is divided into ***cervical***, ***thoracic***, ***lumbar***, ***sacral***, and ***coccygeal*** regions, and the vertebrae of these regions show certain features that make them readily identifiable.

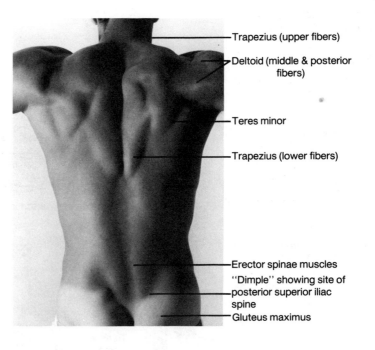

Trapezius (upper fibers)

Deltoid (middle & posterior fibers)

Teres minor

Trapezius (lower fibers)

Erector spinae muscles

"Dimple" showing site of posterior superior iliac spine

Gluteus maximus

Cervical Vertebrae (Fig. 23-2). There are seven cervical vertebrae, and each has a transverse process that contains a ***foramen transversarium***. In the upper six cervical vertebrae this foramen transmits the vertebral artery and vein. The cervical spinous processes are usually bifid. The articular facets of cervical vertebrae are placed so that they face mainly upward and downward, which permits considerable rotation between cervical vertebrae.

Thoracic Vertebrae (Fig. 23-3). There are 12 thoracic vertebrae; these were described in Chapter 3. For comparative purposes some of their salient features will be mentioned again: the sides of their bodies bear costal facets for articulation with ribs; the spinous processes are long and slope downward to project over the arch and spinous process of the vertebra below; the articular processes face mainly forward and backward, allowing some rotation.

23

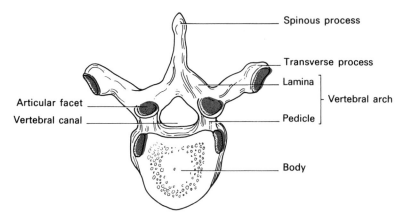

Figure 23-1.
Characteristic features of a typical
vertebra.

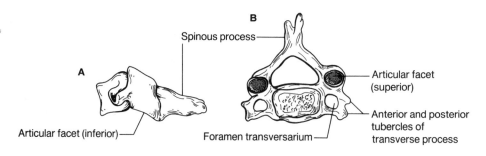

Figure 23-2.
Cervical vertebra. **A.** Left lateral view.
B. Superior view.

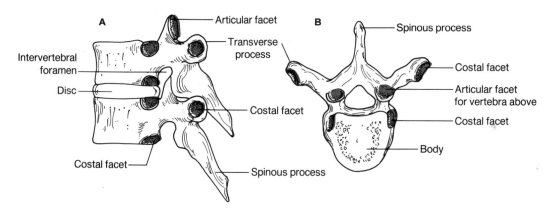

Figure 23-3.
Thoracic vertebrae. **A.** Left lateral view.
B. Superior view.

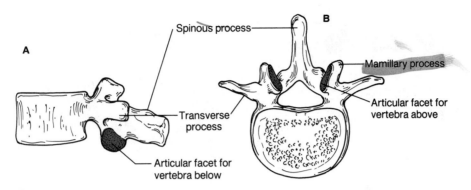

Lumbar Vertebrae (Fig. 23-4). These five bones were described in Chapter 8. Typically they are massive with stubby transverse and broad spinous processes. Their articular processes face laterally and medially, allowing considerable flexion and extension but little rotation.

Specialized Vertebrae. These are vertebrae that show special features quite unlike those of the typical vertebrae. Some are fused units such as the sacrum and coccyx (see Chap. 8), and others are specialized in construction to meet certain functional needs.

The **atlas**, the first cervical vertebra, has donated its body to the second cervical vertebra, the axis. The atlas has an **anterior** and **posterior arch** and two **lateral masses** (Fig. 23-5).

The **axis** has a large **dens** (odontoid process) which is the borrowed body of the atlas. The dens projects posterior to the anterior arch of the atlas and is held in place by the **transverse ligament** of the atlas (Fig. 23-6B).

Figure 23-4.
Lumbar vertebra. **A.** Left lateral view. **B.** Superior view.

Figure 23-5.
Specialized vertebrae (superior view).

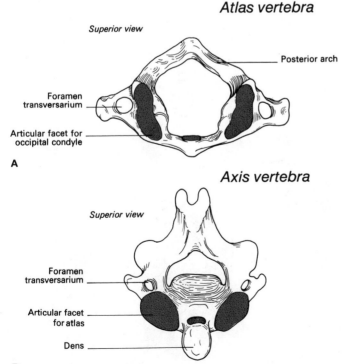

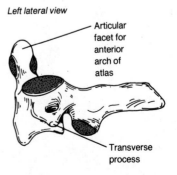

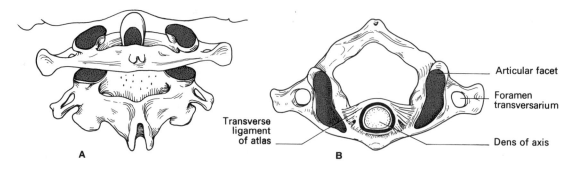

Figure 23-6.
Atlas and axis vertebrae. **A.** Posterior
view. **B.** Superior view.

The **sacrum** and **coccyx** consist of several fused vertebrae and were
described with the walls of the pelvis.

THE SPINE

The embryo lies curled up in the womb in the fetal position, and its spine
forms a C-shaped curve called the **primary curvature**. As the individual
matures, two **secondary curvatures** develop in the cervical and lumbar
regions, with the concavity posteriorly, whereas the primary curvature persists in the thoracic and sacrococcygeal regions (Fig. 23-7). The wedge-
shaped intervertebral discs are largely responsible for these curvatures.

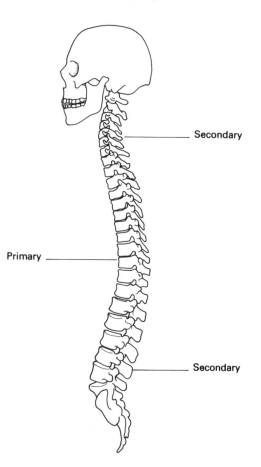

Figure 23-7.
Primary and secondary curvatures of
the vertebral column.

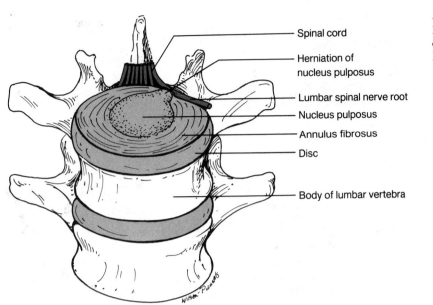

— Spinal cord

— Herniation of
  nucleus pulposus

— Lumbar spinal nerve root

— Nucleus pulposus

— Annulus fibrosus

— Disc

— Body of lumbar vertebra

Figure 23-8.
Intervertebral disc showing herniation
of nucleus pulposus impinging on
lumbar spinal nerve root.

ATTACHMENTS BETWEEN THE VERTEBRAE

The vertebrae are held together by a variety of structures, and basically
two sets of joints exist: the joints between vertebral bodies and the joints
between articular processes.

Intervertebral Discs. A strong fibrocartilaginous disc is found between
the bodies of adjacent vertebrae. These discs are firmly attached to the
vertebrae and form the main bond between them; in the aggregate they
account for about 20% of the height of the vertebral column. The center
of each disc contains a gelatinous ***nucleus pulposus***, a remnant of the no-
tochord (Fig. 23-8). The ***annulus fibrosus*** forms the periphery of the disc
and consists of fibrocartilaginous connective tissue arranged in concentric
circles. The periphery of the disc receives its blood supply from vessels
in the ligaments attached to the vertebral bodies (anterior and posterior
longitudinal ligaments), but the central portion relies on diffusion of nutrients from the cancellous bone of the adjacent vertebrae.

CLINICAL NOTE

*The intervertebral discs are tough structures in the young, and injuries
to the vertebral column will produce fractures of vertebral bodies rather
than disc injuries. After the second decade the discs become more fragile,
and compression injuries may cause rupture of the annulus fibrosus with
posterior or posterolateral protrusion of the nucleus pulposus into the vertebral canal (Fig. 23- 8). This may produce pressure on nerve roots or on
the spinal cord and result in sensory or motor symptoms (pain or muscle
weakness).*

Longitudinal Ligaments. *Anterior* and ***posterior longitudinal ligaments*** are found running the whole length of the vertebral column on the
anterior and posterior aspects of the vertebral bodies and discs.

Ligamenta flava join the laminae of adjacent vertebrae, and *supraspinous* and *interspinous ligaments* join their spines. *Intertransverse ligaments* connect the transverse processes. The *ligamentum nuchae* is a fibroelastic membrane that replaces the supraspinous and interspinous ligaments in the cervical region.

The joints between the articular processes are synovial joints complete with articular cartilages, synovial membrane, and capsule.

MUSCLES

The muscles of the back are responsible for the maintenance of posture and for movements of the vertebral column (Fig. 23-9). They are supplied by the *posterior primary rami* of the regional spinal nerves. There are some muscles located at the back that are superficial muscles belonging to the limbs; these muscles have migrated to the back and are supplied by *anterior primary rami*.

The muscles of the back are enclosed in a fascial sheath that gives attachment to the muscles and is particularly dense in the lower half of the back. This dense fascial sheet is the *thoracolumbar fascia*.

The muscles of the back are identifiable by the direction of their fibers and by the depth at which they are located. The following account is a brief summary of these muscles. The reader should note that there is very little practical value in learning these details.

Figure 23-9.
Muscles of the back (posterior view).
A. Deep layers. **B.** Superficial layers.

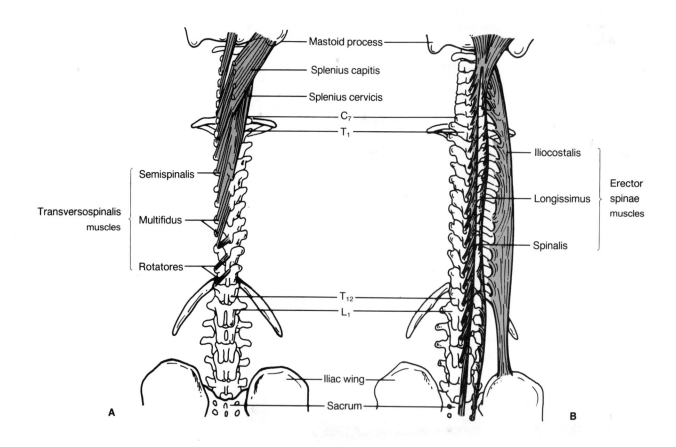

SUPERFICIAL LAYER OF MUSCLES

The muscles, from superficial to deep, are arranged in groups identifiable by the direction in which their fibers run. The most superficial run from *spinous processes* upward and laterally, and the deepest run from *transverse processes* upward and medially to spinous processes.

Splenius Capitis and Cervicis. Splenius capitis is a thin muscle wrapped like a "bandage" around the posterior part of the neck. Fibers run from the ligamentum nuchae and spinous processes of C7 to T6 to the mastoid process and adjacent occipital bone. The fibers of splenius cervicis run from the spinous processes of the upper thoracic vertebrae to the transverse processes of the upper three cervical vertebrae. Acting together, they turn the head from side to side and extend the cervical spine (Fig. 23-9A).

INTERMEDIATE LAYER OF MUSCLES (ERECTOR SPINAE MUSCLES)

The muscles of this layer run parallel to the spinous processes of the vertebrae, and the whole layer forms the *erector spinae muscle* (Fig. 23-9B). Some of the constituents of the erector spinae group are noted below:

= sacrospinalis

Iliocostalis, which is subdivided into lumborum, thoracis, and cervicis portions.

Longissimus, which is subdivided into thoracis, cervicis, and capitis portions.

Spinalis, which is also subdivided into thoracis, cervicis, and capitis portions.

DEEP LAYER OF MUSCLES (TRANSVERSOSPINALIS MUSCLES)

These fibers run upward and medially from transverse process to spinous process and collectively are called *transversospinalis*. The following subdivisions are illustrated in Figure 23-9A):

Semispinalis, divided into thoracis, cervicis, and capitis; these are the most superficial muscles and span five or six vertebrae.

Multifidus lies deep to semispinalis and spans about three vertebrae.

Rotatores, the deepest muscle group, spans single vertebral segments.

SUBOCCIPITAL REGION

The suboccipital region is the area containing the articulation between the skull and the upper end of the cervical spine. The bones that are important in this area are the *occipital condyles* and the adjacent portion of the occipital bone, the *atlas* and the *axis*.

The region contains two joints: the *atlanto-occipital* joint between the atlas and the skull and the *atlantoaxial* joint between the first and second cervical vertebrae.

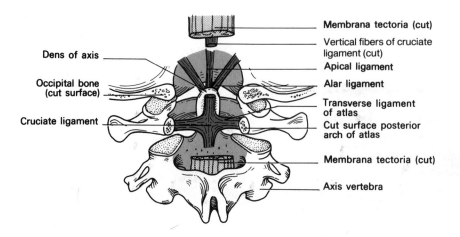

Membrana tectoria (cut)

Vertical fibers of cruciate ligament (cut)

Dens of axis

Apical ligament

Occipital bone (cut surface)

Alar ligament

Transverse ligament of atlas

Cruciate ligament

Cut surface posterior arch of atlas

Membrana tectoria (cut)

Axis vertebra

Figure 23-10.
Atlanto-occipital and atlantoaxial joints (posterior view).

JOINTS OF THE SUBOCCIPITAL REGION

Atlanto-occipital Joint (Fig. 23-10)

This is the joint at which nodding of the head (the *yes* movement) occurs. It possesses two joint cavities, each between an occipital condyle and a lateral mass of the atlas. The ligaments of this joint are the *anterior* and *posterior atlanto-occipital* membranes, which are equivalent to and continuous with the anterior longitudinal ligament and the ligamentum flavum of the rest of the vertebral column.

Certain ligaments run from the axis to the occipital bone and in effect are ligaments of both the atlanto-occipital and the atlantoaxial joints. Two *alar* ligaments run from the dens upward and medially to the internal surface of the occipital bone. The *apical* ligament runs from the apex of the dens to the internal rim of the foramen magnum. The continuation of the posterior longitudinal ligament, the *membrana tectoria*, runs from the body of the axis to the internal surface of the occipital bone covering the alar and apical ligaments.

Atlantoaxial Joint (Fig. 23-10)

This joint, which permits rotation of the head (the *no* movement), has three joint cavities: two between the articular facets of the atlas and the axis, and one between the dens of the axis and the anterior arch of the atlas. The *cruciate ligament* consists of the transverse ligament of the atlas plus fibers that run superiorly to the internal surface of the occipital bone and inferiorly to the body of the axis. The cruciate ligament is covered posteriorly by the membrana tectoria (Fig. 23-10).

CLINICAL NOTE

Rupture of the transverse ligament can drive the dens into the spinal cord; fracture of the dens can cause dislocation of the second cervical vertebra on the first cervical vertebra, which can produce transection of the spinal cord. For this reason patients with suspected neck injuries (e.g., the victim of a whiplash injury in a car accident) must be moved with extreme care and support of the head until clinical and radiologic evaluations have been carried out.

VERTEBRAL ARTERY

The vertebral artery passes superiorly from its origin from the subclavian artery through the foramina transversaria of the upper six cervical vertebrae (Fig. 23-11). As it leaves the foramen transversarium of the atlas it turns posteriorly and medially, making a groove on the lateral mass and posterior arch of the atlas, and pierces the posterior atlanto-occipital membrane. Once inside the vertebral canal (really the foramen magnum at this level) the two vertebral arteries pass anterior to the spinal cord or medulla oblongata to unite to form the ***basilar artery,*** an important source of blood to the brain.

MUSCLES OF THE SUBOCCIPITAL REGION

In the suboccipital region certain small paired muscles are important in nodding and rotation of the skull. The ***rectus capitis posterior minor*** originates on the spinous process of the atlas and inserts into the occipital bone lateral to the midline, whereas the ***rectus capitis posterior major*** originates on the spinous process of the axis and inserts into the occipital bone lateral to the foregoing muscle.

The ***obliquus capitis inferior*** originates on the spine of the axis and inserts into the transverse process of the atlas, whereas the ***obliquus capitis superior*** originates on the transverse process of the atlas and inserts into the occipital bone posterior to the mastoid process of the temporal bone.

The Suboccipital Triangle. The suboccipital triangle lies between the rectus capitis posterior major and the two oblique muscles. The vertebral artery runs through the triangle just before piercing the posterior atlanto-occipital membrane.

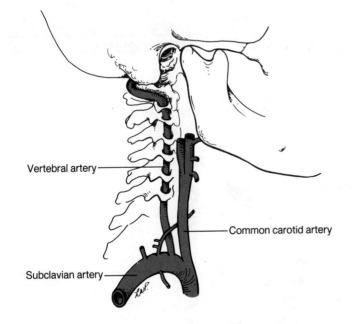

Figure 23-11.
Path of right vertebral artery.

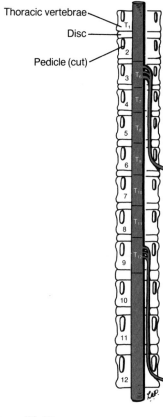

Thoracic vertebrae
Disc
Pedicle (cut)

Figure 23-12.
Schematic posterior view of spinal cord and vertebrae with laminae and spines removed and pedicles cut.

VERTEBRAL CANAL AND SPINAL CORD

The spinal cord lies in the vertebral canal protected by three layers of **meninges** and bathed in **cerebrospinal fluid**. A layer of extradural (epidural) fat and part of the vertebral venous plexus separate the cord and the meninges from the bony walls of the vertebral canal.

Different Growth Rates and Their Consequences

The spinal cord starts at the lower end of the medulla oblongata just below the foramen magnum. In the fetus the spinal cord is almost as long as the vertebral canal, but as the individual grows the vertebral column lengthens faster than the cord. By adulthood the cord terminates at the level of the *first* lumbar vertebra. With differential growth of the vertebral column and the spinal cord the emerging roots of the spinal nerves have to assume an increasingly oblique (downward and lateral) path from the cord to their respective intervertebral foramina. In the adult the cord is so placed that the upper two cervical nerves run almost horizontally, but more inferiorly the obliquity becomes progressively more evident; for example, spinal nerve **C8** leaves the cord at approximately the level of the **sixth** vertebra.

Further down, nerve **T6** leaves the cord at vertebral level **T3** (Fig. 23-12), and nerve **T12** leaves the cord at vertebral level **T9**.

All **lumbar nerves** leave the **cord** at vertebral levels T10–T12, and all **sacral nerves** leave at **vertebral level L1**.

It can be seen that all the **lumbar** and **sacral** nerves leave the cord from its lower end and run a downward course in a mass of descending nerve roots aptly termed the **cauda equina** (horse's tail) (Fig. 23-12).

Each spinal nerve is formed by the junction of a **dorsal** (posterior) and **ventral** (anterior) root. Each root consists of a variable number of rootlets. The dorsal root exhibits the prominent **dorsal root ganglion** (the home of the cell bodies of the peripheral sensory nerves) close to its junction with the ventral root to form the **spinal nerve** (see Figs. 23-13 and 23-14).

The **spinal cord** is a collection of nerve fibers and cells about the diameter of the subject's little finger. It is flattened anteroposteriorly. The cord consists of a central H-shaped column of (mainly) nerve cells called the **gray matter** and a peripheral mass of (mainly) nerve fibers called **white matter** (Fig. 23-14).

THE MENINGES

The spinal cord and the brain are covered by three protective layers (membranes), each with its own characteristics and structure. The layers, from

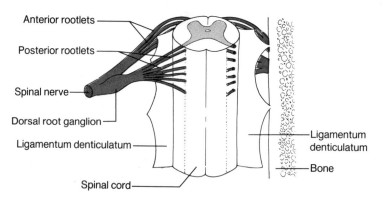

Anterior rootlets
Posterior rootlets
Spinal nerve
Dorsal root ganglion
Ligamentum denticulatum
Ligamentum denticulatum
Bone
Spinal cord

Figure 23-13.
Typical spinal nerve (posterosuperior view).

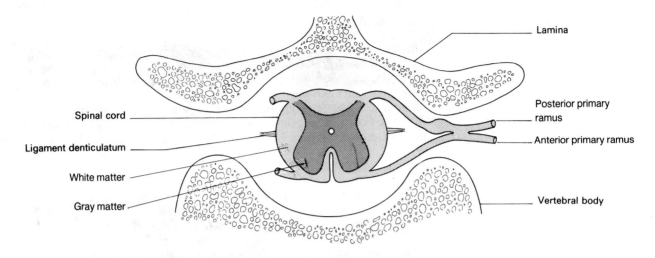

Lamina

Spinal cord

Ligament denticulatum

White matter

Gray matter

Posterior primary ramus

Anterior primary ramus

Vertebral body

the spinal cord outward, are the ***pia mater***, the ***arachnoid mater***, and the ***dura mater***. Collectively these layers form the meninges.

Pia Mater

This is a fine, delicate, vascular membrane (Fig. 23-15) covered on its superficial surface by mesothelium, which is in contact with the ***cerebrospinal fluid (CSF)***. The pia mater covers the surface of the brain, the spinal cord, the roots of the spinal nerves, and the blood vessels of the brain and cord. The ***ligamentum denticulatum*** is a thin membrane of pia mater projecting laterally from the spinal cord into the subarachnoid space. Its lateral edge is scalloped, each projection attaching to the dura mater between intervertebral foramina. The ligamentum denticulatum may provide a stabilizing support for the cord, preventing displacement and torsion. From the inferior end of the cord a fine thread of pia mater, the ***filum terminale***, projects downward to attach to the coccyx.

Figure 23-14.
Cross-section of the vertebral canal. The finely stippled area represents gray matter.

Figure 23-15.
Meninges of the spinal cord (cross-section).

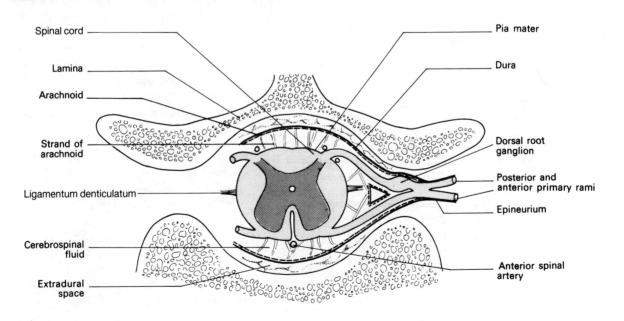

Spinal cord

Lamina

Arachnoid

Strand of arachnoid

Ligamentum denticulatum

Cerebrospinal fluid

Extradural space

Pia mater

Dura

Dorsal root ganglion

Posterior and anterior primary rami

Epineurium

Anterior spinal artery

Arachnoid Mater

The arachnoid is a delicate membrane separated from the pia by the ***subarachnoid space***, which contains the CSF (Fig. 23-15). The arachnoid lines the ***dura mater*** and is closely applied to it, but a potential space exists. The dura and the arachnoid are continued along the spinal nerve roots as far as the intervertebral foramina (sometimes just a little beyond them) so that the subarachnoid space surrounds the nerve roots up to their point of junction (Fig. 23-15).

Dura Mater

The dura mater is a tough, fibrous protective sheath of connective tissue lined by the arachnoid (Fig. 23-16). The dura is separated from the bodies and arches of the vertebrae by the ***extradural space***, which contains a mixture of fat, connective tissue, and blood vessels. The latter includes a venous plexus that forms part of the ***vertebral venous plexus***.

The dura mater is continuous with the intracranial dura and is attached around the margins of the foramen magnum. It descends as a dural sac as far inferiorly as the second sacral vertebra, where it narrows to become a single strand surrounding the pia of the filum terminale and attaching to the periosteum on the posterior surface of the coccyx. The dura is lined by the arachnoid; thus it can be seen that the subarachnoid space descends as far as the sac of the dura, to the vertebral level of *S2* (Fig. 23-16). Below the termination of the spinal cord at the ***conus medullaris***, at vertebral level *L1*, the spinal nerve roots of the cauda equina float in the CSF.

The dura is evaginated along the ventral and dorsal roots of the spinal nerves and continues as a sleeve surrounding them as far as the intervertebral foramen, where it becomes continuous with the ***epineurium*** of the spinal nerve.

CLINICAL NOTE: LUMBAR PUNCTURE, SPINAL TAP

*Access to the CSF may be required for purposes of analysis of the fluid, manometry (measuring of CSF pressure), or the administration of antibiotics or spinal anesthetic. The subarachnoid space can be easily reached by a needle inserted in the midline between **L3** and **L4** or **L4** and **L5**. At*

Figure 23-16.
Median section of the vertebral canal.

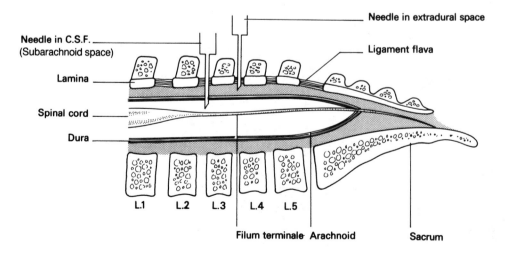

Needle in extradural space

Needle in C.S.F.
(Subarachnoid space)

Ligament flava

Lamina

Spinal cord

Dura

L.1 L.2 L.3 L.4 L.5

Filum terminale Arachnoid

Sacrum

these levels there is no risk of injury to the spinal cord because only the fibers of the cauda equina occupy the subarachnoid space. The patient is bent forward to widen the spaces between the spinous processes, thus allowing the needle to enter easily. The relatively longer cord of the infant may be injured by this procedure, and CSF may be obtained from cisternae at the base of the brain.

Epidural (Extradural) Anesthesia. *It is possible to inject anesthetic around the roots of the sacral and lower lumbar nerves without entering the subarachnoid space. The needle is placed in the extradural space.*

SPINAL NERVES

The dorsal and ventral rootlets of each spinal nerve arise from the cord posterior and anterior to the ligamentum denticulatum, respectively. The rootlets, covered by pia, come together to enter the sleeve of dura, bathed in CSF. These rootlets unite as roots in the intervertebral foramen to form the spinal nerve. On the dorsal root will be found the ***dorsal root (sensory) ganglion***, also covered by pia mater.

VESSELS OF THE SPINAL CORD

Arteries

There is one ***anterior spinal artery*** in the anterior longitudinal sulcus of the spinal cord (see Fig. 23-15). Superiorly it arises from the vertebral arteries. It is an anastomotic artery that is reinforced at most segmental levels by branches from the segmental arteries (*e.g.*, intercostal arteries and lumbar arteries).

There are usually as many as four ***posterior spinal arteries***; they are branches from the vertebral arteries and are also reinforced segmentally. An important reinforcement of the arterial supply of the spinal cord is found in the left (66% of cases) lower thoracic and upper abdominal regions (arteries of Adamkiewicz). If occlusion of the aorta is necessary during surgery, the blood supply to the cord will be compromised and can result in permanent spinal cord damage.

Veins

The veins of the vertebral region form the important ***vertebral venous plexus***, which is an anastomotic pathway for blood returning to the heart. The plexus consists of veins within and outside the vertebral canal, veins inside the bodies of vertebrae, and veins surrounding vertebrae. This plexus has no valves. It is believed that spread of some cancers (*e.g.*, prostate and lung) to vertebrae may be by means of this plexus.

The Head and Neck

Magnetom (nuclear magnetic resonance) shows midsagittal section of head.

The following chapters will describe the somewhat complex anatomy of the head and neck. The complexity of this region of the body is more apparent than real and is caused by the difficulty of understanding the written text without immediate access to the dissected specimen. This handicap can be mitigated by a careful study of the skull and facial bones. The reader who has an adequate three-dimensional image of the skull can synthesize and visualize the arrangement of many of the soft tissues and gain a valuable insight into the anatomy of this region. Unlike the dissected cadaver the skull has the convenience of portability. ***"Don't leave home without it."***

Be aware that in the study of the head and neck the reader will be introduced to the ***cranial nerves***. These nerves occur in pairs and are summarized in Chapter 34.

By convention the 12 pairs of cranial nerves are numbered using Roman numerals; the following is a list of their names and numbers:

Olfactory Nerve, I	*Facial Nerve, VII*
Optic Nerve, II	*Vestibulocochlear Nerve, VIII*
Oculomotor Nerve, III	*Glossopharyngeal Nerve, IX*
Trochlear Nerve, IV	*Vagus Nerve, X*
Trigeminal Nerve, V	*Accessory Nerve, XI*
Abducens Nerve, VI	*Hypoglossal Nerve, XII*

THE FACE AND SCALP

The bones of the skull should be examined together with the following text. Many of the features of the skull to be described can be palpated or seen by the reader on his or her own face and head.

BONES OF THE CRANIUM (Fig. 24-1)

Frontal Bone

The frontal bone forms the forehead and the anterior part of the roof of the cranial vault. In adults the frontal bone is unpaired and shows certain

24

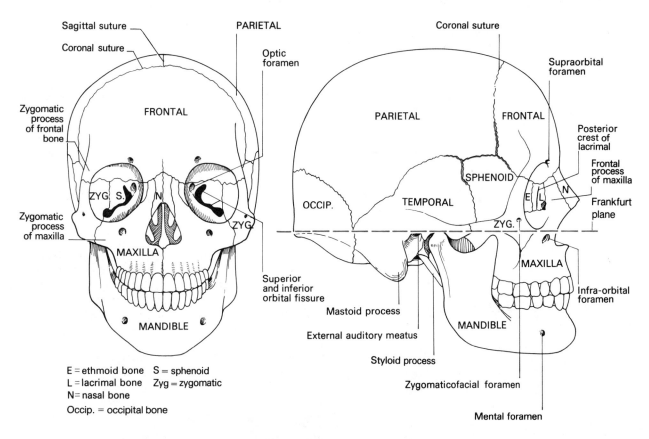

E = ethmoid bone   S = sphenoid
L = lacrimal bone   Zyg = zygomatic
N = nasal bone
Occip. = occipital bone

Figure 24-1.
The skull (anterior and lateral views).

visible and palpable features: the ***frontal eminence***, the ***superciliary arch***, the ***glabella***, and the ***supraorbital notch*** (foramen) are easily visible on the skull and palpable on the living head. The frontal bone forms the roof of the ***orbit*** and the ***supraorbital margin***. It unites, at sutures, with the ***parietal*** bones (***coronal*** suture), the nasal bone, the maxilla (by a maxillary process), the sphenoid, the ethmoid, the zygomatic (by a zygomatic process), and the lacrimal bones (Fig. 24-1).

Parietal Bones

The parietal bones form the upper parts of the side walls of the cranium and about the middle two fourths of the top of the cranium. Each parietal bone has a visible and palpable ***eminence*** and joins its fellow of the opposite side at the ***sagittal*** suture. It is joined to the occipital bone at the ***lambdoid*** suture. It is also joined to the temporal and sphenoid bones. The point where the two (fetal) frontal bones and the parietal bones come together is called the ***bregma***, and the point of junction with the occipital bone is the ***lambda*** (Fig. 24-2).

Temporal Bones

The temporal bone contains the ear and also takes part in the formation of the cranial vault. It articulates by means of sutures with the parietal, occipital, sphenoid, and zygomatic bones. It exhibits the ***external auditory meatus*** (opening to the middle ear), the ***mastoid process***, the ***zygomatic process***, and the ***mandibular fossa*** and ***articular eminence*** for articulation with the head of the mandible (Figs. 24-3 and 24-4).

Figure 24-2.
Skull of newborn showing anterior and posterior fontanelles.

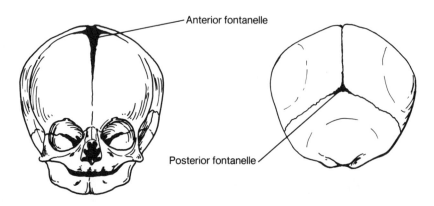

Sphenoid Bone

This is a complex (unpaired) wedge-shaped bone (Fig. 24-4). It articulates with the occipital, temporal, frontal, and parietal bones. Careful examination of the sphenoid and temporal bones will enable the reader to form a three-dimensional image of many of the complex regions to be described.

Inspection of the lateral aspect of the skull will reveal an approximately H-shaped outline of sutures (see Fig. 24-1) where the squamous temporal bone, the greater wing of the sphenoid, and the frontal and parietal bones form the ***pterion***, an important landmark that overlies the course of the middle meningeal artery on the inside of the skull (see Fig. 24-4).

CLINICAL NOTE

In newborns the sutures are open, which allows some gliding movements between the bones of the skull. This permits the head to be molded to the shape of the birth canal during delivery. The bregma (anterior fontanelle) and lambda (posterior fontanelle) of the newborn are easily palpable and can be used to make a rough estimate of intracranial pressure.

Figure 24-3.
Lateral aspect of left temporal bone.

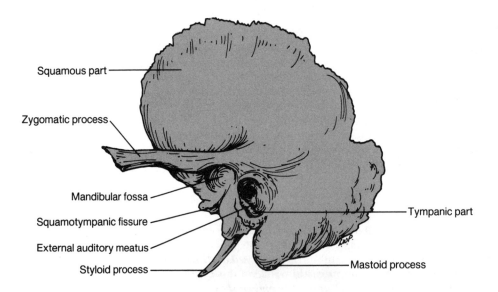

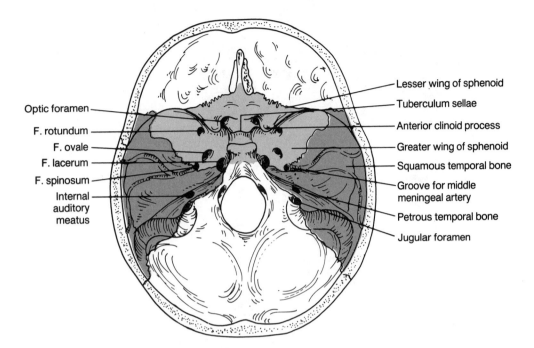

Optic foramen
F. rotundum
F. ovale
F. lacerum
F. spinosum
Internal
auditory
meatus

Lesser wing of sphenoid
Tuberculum sellae
Anterior clinoid process
Greater wing of sphenoid
Squamous temporal bone
Groove for middle
meningeal artery
Petrous temporal bone
Jugular foramen

Figure 24-4.
Interior of skull showing sphenoid and temporal bones.

Other Bones of the Cranium

The paired *nasal* bones are small and form the bridge of the nose. The *lacrimal* bone is a small sliver that sits in the medial wall of the orbit. Its *sulcus* lodges the *nasolacrimal canal.* Posteriorly the lacrimal bone articulates with the *ethmoid* bone, a complex fragile structure that forms a considerable part of the medial wall of the orbit and the lateral wall of the nose.

BONES OF THE FACE (see Fig. 24-1)

Maxilla

The maxilla forms the lower margin and floor of the orbit and the upper jaw. It articulates with the zygoma through the zygomatic process, the frontal bone through the frontal process, and the nasal, lacrimal, sphenoid, and palatine bones. It has an *infraorbital groove, canal,* and *foramen* for the infraorbital nerve; the upper teeth are lodged in the *alveolar process.*

Mandible (Fig. 24-5)

The mandible (lower jaw) is bilaterally divided into a horizontal portion–the *body*–and a vertical portion–the *ramus.* The body and ramus join at the *angle* that measures about 100 degrees. The body has an *alveolar* process for the lower teeth; below the second premolar tooth is the *mental foramen* for the passage of the mental nerve. The posterosuperior aspect of the ramus exhibits the *condylar* process, which is separated from the anteriorly placed *coronoid* process by the *mandibular notch* (Fig. 24-5A). The *mandibular foramen* on the medial side of the ramus transmits

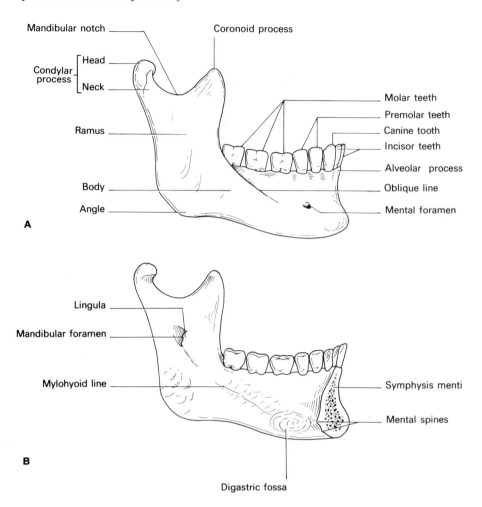

Figure 24-5.
The mandible. **A.** Right mandible,
lateral view. **B.** Bisected left mandible,
medial view.

the inferior alveolar nerve and vessels; a spicule of bone, the ***lingula***, projects from the anterior margin of the mandibular foramen (Fig. 24-5B).

Zygomatic Bone

The zygomatic bone forms the most lateral bony part of the cheek and articulates with the maxilla, sphenoid, and temporal bones. It exhibits the ***zygomaticofacial foramen*** on its anterolateral surface and the ***zygomaticotemporal foramen*** on its posterior surface. The zygomaticofacial and zygomaticotemporal nerves pass through these foramina. They are branches of the maxillary division of the trigeminal nerve that have entered the zygoma through a foramen on its orbital surface (see Fig. 24-1).

Frankfurt Plane. This is a horizontal plane joining the superior border of the external auditory meatus to the inferior margin of the orbit (see Fig. 24-1).

MUSCLES OF THE FACE

The ***muscles of facial expression*** are numerous, and their names are more memorable for their complexity than for their importance. The muscles of facial expression are subcutaneous muscles, that is, they are situ-

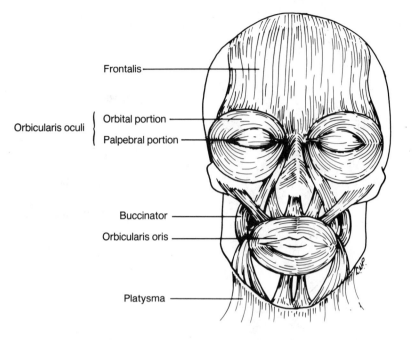

Figure 24-6.
Principal muscles of facial expression.

Frontalis

Orbicularis oculi { Orbital portion

Palpebral portion

Buccinator

Orbicularis oris

Platysma

ated in the subcutaneous plane with attachments to bone and skin. They are all supplied by the ***facial nerve (cranial VII)*** and wrinkle the skin of the face into its various expressive poses. They are continuous over the mandible with the ***platysma*** muscle of the neck. Three major facial muscles will be mentioned (Fig. 24-6):

The ***orbicularis oris*** surrounds the mouth and acts like a sphincter for closing the mouth and pursing the lips. Other muscles attach to the skin around the mouth and move the lips, thereby allowing opening of the mouth, raising and lowering the angles of the mouth (happy face and sad face), pursing of lips, and so forth.

The ***buccinator*** is a rather special muscle of the mouth; it is attached to the alveolar margins of the upper and lower jaws and to the pterygomandibular raphé of the pharynx posteriorly. It controls the cheeks, holds food against the teeth, and blends with the orbicularis oris muscle. The buccinator is also supplied by the facial nerve.

The ***orbicularis oculi*** surround the eye and each acts as a sphincter, permitting tight closure of the eyes.

Muscles That Move the Jaw

These muscles are used for chewing and talking; their collective name is the ***muscles of mastication***. They are all supplied by the mandibular division (V_3) of the ***trigeminal nerve*** (***cranial V***) and will be described later.

NERVES OF THE FACE (Fig. 24-7 and Cranial Nerves V and VII, Chap. 34)

Important Generalization. The seventh cranial nerve supplies the muscles of facial expression. The fifth cranial nerve supplies the muscles of mastication and is otherwise ***sensory*** to the skin of the face, mouth, teeth, nose, and eyes, including the conjunctiva and cornea (for details, see Chap. 34, Cranial Nerves).

Figure 24-7.
General sensory nerves of the face.

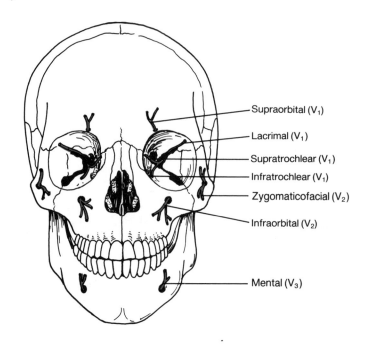

Supraorbital (V₁)
Lacrimal (V₁)
Supratrochlear (V₁)
Infratrochlear (V₁)
Zygomaticofacial (V₂)
Infraorbital (V₂)
Mental (V₃)

Facial Nerve (Cranial VII)

The terminal part of this nerve leaves the skull through the ***stylomastoid*** foramen and enters the deep part of the posterior aspect of the ***parotid*** salivary gland. It runs forward and superficially in the parotid gland and divides within the parotid into five branches that spread out over the face and upper neck. These branches are called temporal, zygomatic, buccal, mandibular, and cervical. An additional branch, the posterior auricular, goes to the occipitalis muscle of the scalp.

CLINICAL NOTE

*Damage to the facial nerve results in paralysis of the muscles of facial expression, known as **Bell's palsy**. Its manifestations are loss of normal wrinkles of the face and forehead; inability to shut the eye, resulting in tears trickling down the face and loss of protection for the cornea; inability to close the mouth, with drooping of the angle of the mouth and spillage of saliva; and inability to chew properly because of buccinator paralysis.*

Trigeminal Nerve (V)

This important nerve divides into three divisions—***ophthalmic, maxillary***, and ***mandibular***—according to the regions they supply. The three divisions leave the skull by separate foramina and supply three areas of the face (see Fig. 24-7).

Ophthalmic Nerve (V₁)

This division of the trigeminal is entirely sensory in function. After supplying structures in the orbit it appears on the face as a series of small branches. The ***supraorbital*** nerve passes through the supraorbital foramen (notch) to supply the skin of the forehead above the orbit and the upper part of the conjunctiva and cornea. The ***supratrochlear*** and ***infra-***

trochlear nerves pass above and below the *trochlea* of the orbit (see below). In addition, *nasal* branches are found.

Maxillary Nerve (V₂)

This division of the trigeminal nerve has extensive distribution to the nose, mouth, and teeth, and its terminal branches reach the skin of the face by way of the *infraorbital*, *zygomaticofacial*, and *zygomaticotemporal* foramina.

Mandibular Nerve (V₃)

This nerve carries motor and sensory fibers. The motor fibers supply the muscles of mastication. The three sensory branches that reach the face are the *auriculotemporal* nerve, which emerges behind the mandible to supply the temporal area, the *buccal* nerve, which supplies the mucosal and cutaneous surfaces of the cheek, and the *inferior alveolar* nerve, which supplies the teeth of the lower jaw and emerges as the *mental* nerve through the mental foramen to supply the skin of the chin and lower lip.

ARTERIES OF THE FACE

The face and scalp enjoy an abundant blood supply from several arteries, which explains the profuse bleeding that may occur from relatively small lacerations and the relatively short healing time for facial wounds.

Facial Artery

This major artery is a branch of the *external carotid* artery (Fig. 24-8). It arises just above the upper border of the hyoid bone and passes close to the posterior border of the submandibular salivary gland deep to the lower border of the mandible. At the anterior border of the masseter muscle the facial artery emerges from behind the mandible and crosses the inferior border of the mandible to pass toward the medial *canthus* (angle) of the eye.

Superficial Temporal Artery

This is one of the two terminal branches of the external carotid artery. The other terminal branch is the *maxillary artery*, which runs deep to the ramus of the mandible. The superficial temporal artery arises deep to or in the substance of the parotid gland. It becomes superficial anterior to the auricle (external ear) and ascends to the temple with the auriculotemporal nerve.

CLINICAL NOTE

Both the facial and superficial temporal arteries can be used for feeling the pulse. Clench your teeth and locate the facial artery where it crosses the lower border of the mandible just anterior to the masseter muscle. Also feel the superficial temporal artery's pulsation where it crosses the zygomatic process just in front of the auricle. Both these pulses are accessible to the anesthetist when other parts of the body may be covered by surgical drapes.

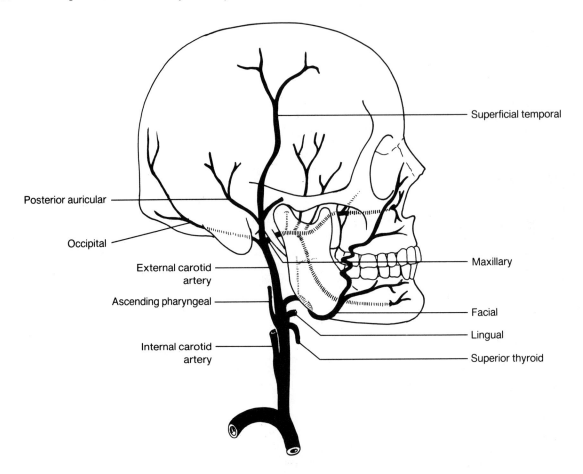

Figure 24-8.
Branches of the external carotid artery.

VEINS OF THE FACE

Facial Vein

The facial vein begins at the medial canthus (inner angle) of the eye and runs obliquely backward and downward across the face, parallel to the facial artery. Below the mandible it is joined by the anterior branch of the ***retromandibular vein*** and then enters the ***internal jugular vein*** deep to the anterior border of the sternocleidomastoid (sternomastoid) muscle.

Retromandibular Vein

This vessel is formed by the junction of the superficial temporal vein and the maxillary vein posterior to the ramus of the mandible. The vein divides into an anterior branch that joins the facial vein and a posterior branch that joins with the ***posterior auricular vein*** to form the ***external jugular vein***.

CLINICAL NOTE

*The veins of the face communicate with the intracranial venous sinuses by means of perforating (emissary) veins. Specifically the anterior facial vein communicates at the medial canthus with the ophthalmic veins, which drain backward into the **cavernous sinus**. It is possible for infec-*

tions of the skin of the face to spread to the cavernous and other intra-cranial sinuses with very serious consequences, such as thrombosis of the sinus or spread of infection through the sinuses to the internal jugular veins and hence to the general circulation.

PAROTID GLAND

The parotid gland (Fig. 24-9) is the largest of the ***salivary*** glands. It lies below and anterior to the auricle and between the mandible and the upper end of the sternomastoid muscle. It passes forward onto the masseter muscle, overlapping it. From its most anterior point the ***duct*** of the parotid gland continues forward over the surface of the masseter and, turning around the anterior border of that muscle, pierces the buccinator muscle to open into the oral cavity opposite the second upper molar tooth. Where the duct enters the mouth there is a small ***papilla*** on the inside of the cheek, referred to by many clinicians as the papilla of Stensen's duct.

The parotid duct can be felt through the skin as it crosses the masseter, especially if the teeth are clenched by contracting the masseter.

SCALP

The scalp is a musculoaponeurotic structure formed by the blending of the skin of the head with the deep fascia of the head. The fascia is aponeurotic and contains the frontalis muscle in front and the occipitalis muscle at the back. Contraction of these two muscles moves the scalp and wrinkles the forehead (Fig. 24-10).

The skin is bound to the aponeurosis (epicranial aponeurosis) by septa of fibrous tissue with globules of fat and blood vessels between the

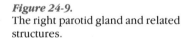

Figure 24-9.
The right parotid gland and related structures.

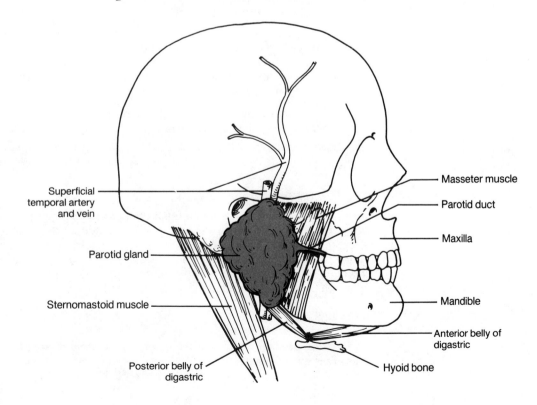

Superficial temporal artery and vein

Parotid gland

Sternomastoid muscle

Posterior belly of digastric

Masseter muscle

Parotid duct

Maxilla

Mandible

Anterior belly of digastric

Hyoid bone

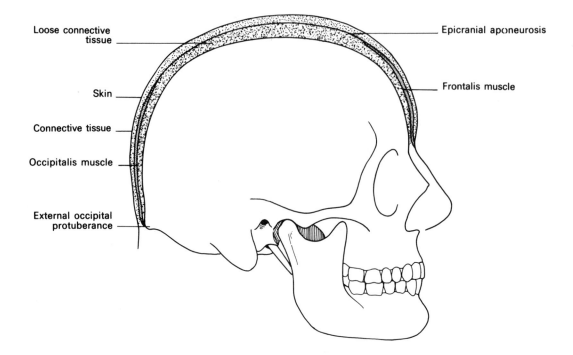

Loose connective tissue

Skin

Connective tissue

Occipitalis muscle

External occipital protuberance

Epicranial aponeurosis

Frontalis muscle

Figure 24-10.
The scalp (sagittal section).

septa. The walls of the vessels of the skin are adherent to these connecting septa so that the vessels cannot contract when severed. This accounts for the profuse bleeding that may occur with scalp lacerations and which is best controlled by the application of pressure to the edges of the scalp around the laceration.

The scalp consists of five layers.

1. **S**kin
2. **C**onnective tissue densely woven between skin and aponeurosis
3. **A**poneurosis (epicranial)
4. **L**oose connective tissue deep to the epicranial aponeurosis
5. **P**eriosteum of the skull

Layer 4 permits layers 1, 2, and 3 (an inseparable unit) to slide over layer 5. Scalping injuries, which are not uncommon in industrial accidents and Western movies, occur through layer 4.

BONY LANDMARKS

The two *mastoid* processes form the lateral extremities of the area under consideration. Midway between them is the *external occipital protuberance*. These three landmarks can be easily palpated on one's own head. The palpable *superior nuchal line* passes laterally from the external occipital protuberance to the mastoid processes. The *inferior nuchal line* lies deep to the muscles at the base of the skull and cannot be palpated, although the reader should identify it on the skull.

MUSCLES

The superior nuchal line gives attachment to the *sternomastoid* and the *trapezius* muscles, which will be encountered in the description of the posterior triangle of the neck (see Chap. 36). It also gives attachment to the epicranial aponeurosis and the ***occipitalis*** muscle.

Passing forward between the occipitalis and the frontalis muscles is the ***epicranial aponeurosis***, which passes anteriorly to about 8 cm above the superciliary ridge. At this point it gives rise to the ***frontalis*** muscle, which is about 5 to 8 cm long and inserts into the skin at about the level of the eyebrows. The two muscles (frontalis and occipitalis) with their aponeurosis form the ***occipitofrontalis***, and between them they move the scalp and wrinkle the forehead. They are muscles of facial expression and are innervated by the ***facial*** nerve (temporal and posterior auricular branches).

The epicranial aponeurosis is blended laterally with fascia over the ***temporalis*** muscle and gains attachment through this fascia to the zygomatic arch.

The Interior of the Skull and the Brain

BONES AND BONY LANDMARKS

The bones that make up the cranium as seen from the exterior should be reviewed before reading the following account. The **calvaria** is the vault of the skull and is that part usually removed to expose the interior of the skull and the brain. The calvaria consists of a double layer of compact bone (the **diploë**). Cancellous bone containing the diploic veins lies between the inner and outer layers of compact bone.

CRANIAL FOSSAE

The interior of the base of the skull is divided into three **cranial fossae** (Fig. 25-1A). The **anterior cranial fossa**, which contains the **frontal lobes** of the brain, is formed largely by the frontal bone, with the **cribriform plate** and **crista galli** of the **ethmoid** occupying a small but important central area anteriorly. The anterior fossa terminates posteriorly at the **lesser wing** of the **sphenoid** bone.

    Next, one steps down into the **middle cranial fossa**, located posterior to the anterior fossa and bounded posteriorly by the **petrous part** of the **temporal** bone. The **greater wing** of the **sphenoid** bone forms the anterior wall and part of the lateral wall, and the **squamous** part of the temporal bone helps to form part of the lateral wall together with a part of the **parietal** bone.

    Another step down is the **posterior cranial fossa**, composed largely of the **occipital** bone with some participation by the **temporal** bone. Note the **internal occipital protuberance** on the interior of the occipital bone; it matches the location of the external occipital protuberance.

SELLA TURCICA

This depression in the cranial surface of the body of the sphenoid bone, located between the two halves of the middle cranial fossa, contains the **hypophysis** (pituitary gland). The **sella turcica** (Turkish saddle) is limited anteriorly by the **tuberculum sellae** and anterolaterally by the **anterior clinoid processes**. Posterolaterally are the two **posterior clinoid processes** with the **dorsum sellae** (back of the saddle) between them.

25

FORAMINA

The skull contains many foramina (holes) for the passage of nerves and blood vessels (Fig. 25-1B). The most important foramina are described below, grouped by their locations.

ANTERIOR CRANIAL FOSSA

The multiple tiny foramina of the ***cribriform plate*** transmit the tiny twigs of the olfactory nerves (cranial I) and some small branches of the ophthalmic artery (anterior and posterior ethmoidal arteries).

MIDDLE CRANIAL FOSSA

A large number of important structures leave and enter the middle cranial fossa. The ***superior orbital fissure*** and the ***optic foramen*** communicate between the middle cranial fossa and the orbit. The optic foramen transmits the ***optic nerve*** (cranial II) and its meninges as well as the ***ophthalmic artery***. The superior orbital fissure transmits the following structures, which will be described in greater detail with the eye and orbit: lacrimal, frontal, trochlear, oculomotor, nasociliary, and abducens nerves and one or more ophthalmic veins.

Posteroinferior to the superior orbital fissure is the ***foramen rotundum***, which transmits the ***maxillary*** division of the trigeminal nerve (V_2). Posterolateral to this is the ***foramen ovale*** for the ***mandibular***) division (V_3) of the trigeminal nerve, and posterolateral to the foramen ovale the ***middle meningeal artery*** enters the middle cranial fossa through the small but important ***foramen spinosum***. Medial to these foramina is the large and ragged looking ***foramen lacerum***, which is really a canal the lower part of which is occupied by a plug of cartilage and which "transmits" nothing, although some important structures (including the internal carotid artery) "skirt" its upper opening. The small but functionally important ***greater petrosal nerve*** leaves the middle cranial fossa through the upper opening of the foramen lacerum; the equally important ***lesser petrosal nerve*** passes through the foramen ovale.

POSTERIOR CRANIAL FOSSA

The large ***foramen magnum*** occupies the central area of the posterior fossa. It contains the transition of the medulla oblongata into the spinal cord, the meninges, meningeal spaces, and the vertebral arteries. Immediately anterolateral to the foramen magnum the ***hypoglossal nerves*** (XII) leave the skull through the ***hypoglossal (anterior condylar) foramina***. Lateral to the anterior third of the foramen magnum, the ***jugular foramen*** transmits the confluence of the sigmoid sinus into the bulb of the ***internal jugular vein*** as well as the ***glossopharyngeal*** (IX), ***vagus*** (cranial X), and ***accessory*** (XI) nerves. In the petrous temporal bone immediately superior to the jugular foramen the ***internal auditory foramen*** transmits the facial (VII) and the ***vestibulocochlear*** (VIII) nerves.

MENINGES OF THE CRANIAL CAVITY

The meninges of the spinal cord (see Figs. 23-15, 23-16, and 25-2) should be reviewed before studying their continuations in the cranial cavity.

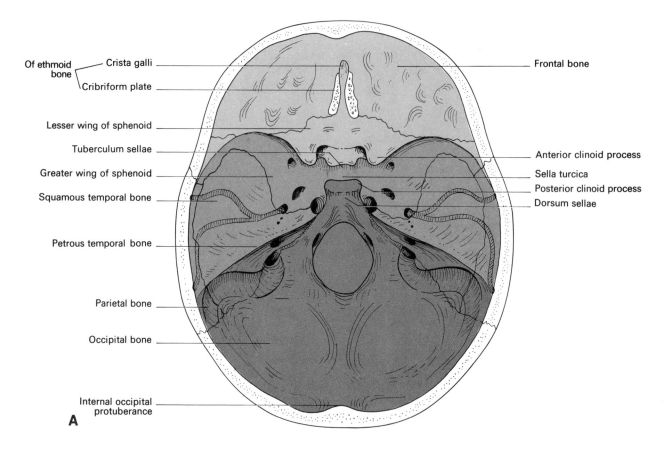

Of ethmoid bone — Crista galli

Cribriform plate

Lesser wing of sphenoid

Tuberculum sellae

Greater wing of sphenoid

Squamous temporal bone

Petrous temporal bone

Parietal bone

Occipital bone

Internal occipital protuberance

Frontal bone

Anterior clinoid process

Sella turcica

Posterior clinoid process

Dorsum sellae

A

Figure 25-1.
Interior of the skull. **A.** Bony features.
B. Foramina.

DURA MATER

This dense layer of connective tissue (dura mater = tough mother) is intimately blended with the periosteum (endosteum) of the cranial bones but constitutes a separate structure. The dura mater of the cranium is continuous with that of the spinal cord and is continued along the cranial nerves as their *epineurium*. In some areas the endosteum and the dura are separated to form the *venous sinuses*. The venous sinuses carry blood from the cerebral and the diploic veins, but the structure of their walls is quite different from that of ordinary veins (Fig. 25-3).

Dural Folds (Fig. 25-3A). The dura is invaginated into the cranial cavity to form three major dural folds that separate parts of the brain. At the bases of these folds are venous sinuses.

The *tentorium cerebelli* is a horizontal fold that separates the cerebrum from the cerebellum in the posterior cranial fossa. Its base contains the two *transverse sinuses*. The free edge of the tentorium surrounds the tentorial notch that contains the brain stem. The free edge of the tentorium is attached to the anterior clinoid process, and the fixed edge is attached to the petrous temporal bone and to the posterior clinoid process.

Falx Cerebri. The falx cerebri is a sickle-shaped fold placed vertically between the two cerebral hemispheres from the crista galli anteriorly to the internal occipital protuberance posteriorly. The base of the falx contains the *superior sagittal sinus*, and the free edge contains the *inferior sagittal sinus*. Posteriorly the falx cerebri blends with the upper surface

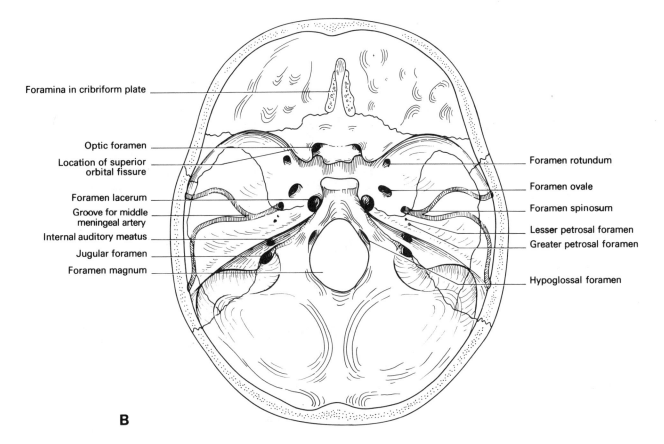

Foramina in cribriform plate

Optic foramen

Location of superior
orbital fissure

Foramen lacerum

Groove for middle
meningeal artery

Internal auditory meatus

Jugular foramen

Foramen magnum

Foramen rotundum

Foramen ovale

Foramen spinosum

Lesser petrosal foramen

Greater petrosal foramen

Hypoglossal foramen

B

of the tentorium cerebelli, and this junction contains the ***straight sinus*** (Fig. 25-3B).

Below the tentorium the ***falx cerebelli*** separates the two hemispheres of the cerebellum. Its base contains the small ***occipital sinus***.

A small separate fold of dura mater covers the hypophysis in the sella turcica. This fold is the ***diaphragma sellae***; it is attached to the periphery of the hypophyseal fossa and leaves a passage in its center for the ***infundibulum*** of the hypophysis (stalk of the pituitary gland).

The dural partitions may be considered to function as "baffle plates" to prevent shifting and distortion of the brain, which has a custard-like consistency in living persons.

Spaces Associated with Cranial Dura Mater

Extradural Space. The extradural space is the potential space between the dura mater and the bones of the cranium. The ***middle meningeal artery***, a branch of the maxillary artery, runs in this space, particularly in the regions of the lateral walls of the anterior and middle fossa. It grooves the inner table of the skull from its point of entry into the skull at the ***foramen spinosum*** (see Fig. 25-1B).

CLINICAL NOTE

The close adherence of the middle meningeal artery to the inner table of the skull makes it vulnerable to laceration from a blow to the skull, particularly (but not necessarily) if the skull is fractured. This rupture

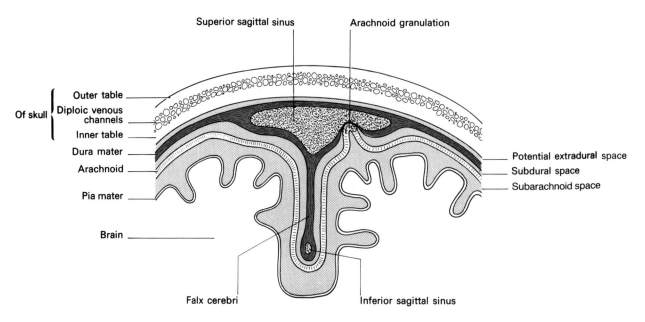

Superior sagittal sinus   Arachnoid granulation

Of skull { Outer table
Diploic venous channels
Inner table
Dura mater
Arachnoid
Pia mater

Brain

Potential extradural space
Subdural space
Subarachnoid space

Falx cerebri   Inferior sagittal sinus

Figure 25-2.
Detail of cranial meninges (coronal section).

*produces an **extradural hematoma**, which results in gradually increasing intracranial pressure leading to coma and death unless the pressure is relieved by evacuation of the accumulated blood clot.*

Subdural Space. This lies between the dura and the arachnoid membranes, and blood from a ruptured cerebral vein can accumulate here to form a ***subdural hematoma***.

ARACHNOID MATER

This delicate membrane is continuous with the spinal arachnoid mater and covers the surface of the brain usually adhering loosely to it, attached to the gyri only by delicate strands of tissue (like a spider's web) and not dipping into the *sulci*. The ***subarachnoid space***, which is filled with cerebrospinal fluid (CSF), lies between the arachnoid and the pia mater and is lined by mesothelium. In some areas the arachnoid is separated from the pia by a considerable gap. These gaps are called ***cisternae***, and they are found in the cerebellomedullary region, between the cerebral peduncles, at the optic chiasma and over the lateral fissure of the cerebral hemispheres.

Arachnoid Granulations (see Fig. 25-2). These structures, which allow CSF to be absorbed into the superior sagittal sinus, are projections of arachnoid into lateral enlargements of the superior sagittal sinus, the ***lacunae laterales***. The CSF is formed in the ventricles of the brain and drains into the subarachnoid space and into the canal of the spinal cord in the region of the medulla oblongata. It then circulates in the subarachnoid space of the cranium and the vertebral canal.

CLINICAL NOTE

*If there is a blockage of flow between its source and the arachnoid granulations, the accumulation of CSF causes swelling of the brain and, in young children, enlargement of the skull; this condition is called **hydrocephalus**.*

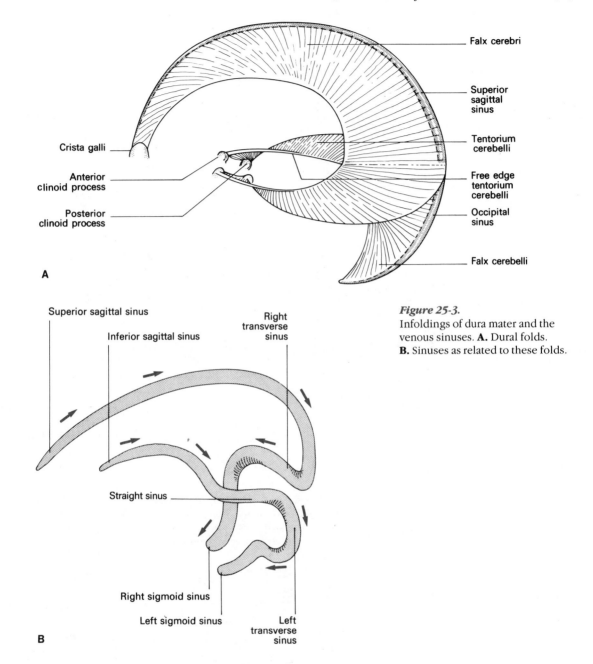

Crista galli

Anterior
clinoid process

Posterior
clinoid process

A

Falx cerebri

Superior
sagittal
sinus

Tentorium
cerebelli

Free edge
tentorium
cerebelli

Occipital
sinus

Falx cerebelli

Superior sagittal sinus

Inferior sagittal sinus

Right
transverse
sinus

Straight sinus

Right sigmoid sinus

Left sigmoid sinus

Left
transverse
sinus

B

Figure 25-3.
Infoldings of dura mater and the
venous sinuses. **A.** Dural folds.
B. Sinuses as related to these folds.

*Subarachnoid Hemorrhage. This is a hemorrhage into the subarachnoid
space usually from rupture of a congenital aneurysm in the **circle of Willis** (Fig. 25-4).*

PIA MATER

The pia mater is a delicate covering that lines all the sulci of the brain and
envelops the blood vessels on the surface of the brain. It follows the blood
vessels into the substance of the brain. Its surface is covered by mesothelium, and it is in contact with the CSF.

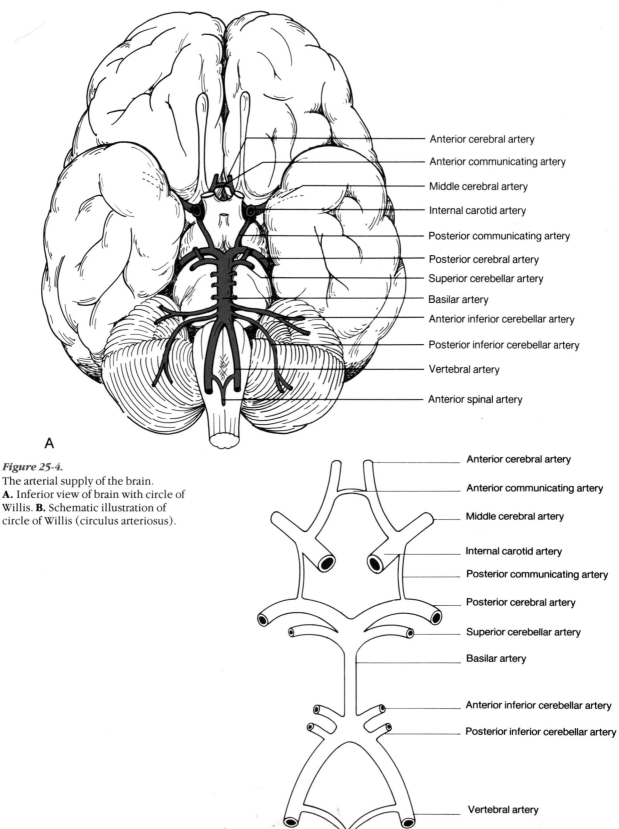

Anterior cerebral artery

Anterior communicating artery

Middle cerebral artery

Internal carotid artery

Posterior communicating artery

Posterior cerebral artery

Superior cerebellar artery

Basilar artery

Anterior inferior cerebellar artery

Posterior inferior cerebellar artery

Vertebral artery

Anterior spinal artery

A

Figure 25-4.
The arterial supply of the brain.
A. Inferior view of brain with circle of
Willis. **B.** Schematic illustration of
circle of Willis (circulus arteriosus).

Anterior cerebral artery

Anterior communicating artery

Middle cerebral artery

Internal carotid artery

Posterior communicating artery

Posterior cerebral artery

Superior cerebellar artery

Basilar artery

Anterior inferior cerebellar artery

Posterior inferior cerebellar artery

Vertebral artery

Anterior spinal artery

B

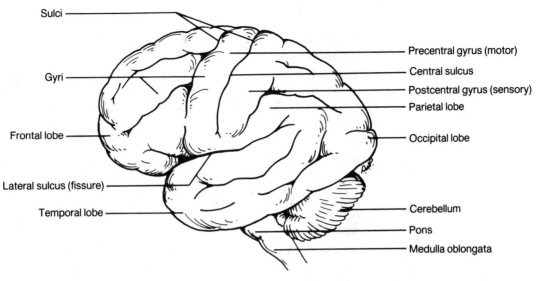

Sulci

Gyri

Frontal lobe

Lateral sulcus (fissure)

Temporal lobe

Precentral gyrus (motor)

Central sulcus

Postcentral gyrus (sensory)

Parietal lobe

Occipital lobe

Cerebellum

Pons

Medulla oblongata

Figure 25-5.
Left lateral surface of brain and brain stem.

BRAIN

Only the gross outlines of the brain will be described in this section. A detailed study belongs to courses and textbooks of neuroanatomy. Thus a brief description follows, with particular stress on the major divisions of the brain and on the intracranial and extracranial paths and components of the cranial nerves (Fig. 25-5).

PARTS OF THE BRAIN

Forebrain

The forebrain is composed of the cerebrum (telencephalon) and diencephalon.

Cerebrum (Telencephalon). This is functionally the most highly developed and sophisticated part of the central nervous system. The cerebral cortex consists mainly of an extensive outer layer of cells (gray matter) and a medullary core of white matter (nerve fibers), in contrast with the spinal cord and brain stem, in which the gray matter (cells) is in the center and the white matter is at the periphery. The cortical cells are sensory and motor. The surface of the brain is very extensively folded, thus increasing the surface area of the cortex. The elevated portions of the folds are the *gyri*, and the grooves are called *sulci*.

The *lateral sulcus* is deep and separates the *temporal lobe* from the *frontal* and *parietal* lobes. The *central* sulcus separates the *precentral* (motor) gyrus from the *postcentral* (sensory) gyrus. The two hemispheres, divided by the median longitudinal fissure, are connected by the *corpus callosum*.

Internal Capsule. This is a large collection of nerve fibers that carries nerve impulses between the cerebral cortex and the rest of the central nervous system.

Diencephalon. This comprises the epithalamus, thalamus, hypothalamus, and subthalamus. It consists of nuclear areas and lies between the telencephalon and the midbrain.

The paired cerebral hemispheres and their lateral ventricles, the thalami, the hypothalamus, and the third ventricle form the forebrain.

Midbrain

The midbrain (mesencephalon) is formed by the cerebral peduncles and the superior and inferior colliculi.

Cerebral Peduncles.   These are two column-like structures in which the fibers of the internal capsule are concentrated as they pass to and from the rest of the brain. The peduncles are attached inferiorly to the *pons*.

Hind Brain

The hind brain is formed by the *pons, cerebellum*, and *medulla oblongata*.

Pons (Metencephalon). The pons is the bridge that carries impulses both vertically and laterally. Inferiorly it is attached to the *medulla oblongata*.

Cerebellum.   This outgrowth of the pons is connected to the medulla and pons; it is the control center for coordination of function, particularly voluntary movements.

Medulla Oblongata (Myelencephalon). This lowest portion of the brain stem is continuous below with the spinal cord and above with the pons. The medulla contains various vital centers.

ARTERIAL SUPPLY OF THE BRAIN

The *circulus arteriosus (circle of Willis)*, which carries blood from the two internal carotid arteries and the two vertebral arteries to all parts of the brain, should be examined in terms of both gross anatomy and neuroanatomy. Figure 25-4 illustrates this arterial circle, the construction of which shows many variations in individual subjects.

CLINICAL NOTE

*Congenital aneurysms may occur on the participating arteries of the circle of Willis. Rupture of such an aneurysm (**berry aneurysm**) may cause severe, and often fatal, subarachnoid hemorrhage in young people.*

Venous Drainage of the Brain

The entire venous drainage of the cranial cavity and its contents is through the *cranial venous sinuses*, which are endothelium-lined spaces usually between the dura and the endosteum of the cranial bones. Most of this drainage goes to the superior and inferior sagittal sinuses before being delivered into the straight, transverse, and sigmoid sinuses. From there drainage continues from the sigmoid sinuses into the internal jugular veins, but some venous blood escapes through small foramina in the skull that contain *emissary veins*, which connect the sinuses with veins in the head and neck regions (Fig. 25-6).

Superior Sagittal Sinus.   This major, unpaired sinus begins at the crista galli, where it may communicate with the veins of the nose by an emissary vein in the foramen cecum. The sinus runs in the attached margin of the falx cerebri to the internal occipital protuberance, where it drains into either *transverse sinus* (usually the right). It receives tributaries from the cerebrum (see Fig. 25-3B).

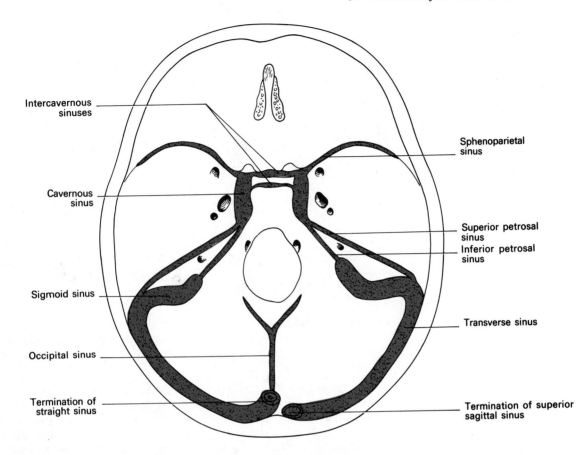

Inferior Sagittal Sinus.   This relatively small longitudinal sinus runs posteriorly in the free edge of the falx cerebri. At the junction of the falx cerebri and the tentorium cerebelli it continues as the ***straight sinus***. The ***great cerebral vein*** (of Galen) joins the anterior end of the straight sinus.

Straight Sinus.   This sinus runs in the junction of the falx and the tentorium to the internal occipital protuberance, where it turns to the left (usually) to become the left transverse sinus. The spot where the superior sagittal, straight, and transverse sinuses meet is called the ***confluence of the sinuses***.

Transverse Sinuses.   These channels, one on each side, run in the attached edge of the tentorium cerebelli from the internal occipital protuberance to the petrous temporal bone. There each turns inferiorly to become the S-shaped ***sigmoid sinus*** (see Figs. 25-3B and 25-6).

Sigmoid Sinus.   This sinus makes a deep groove in the occipital and temporal bones. Each drains into the bulb of the ***internal jugular vein*** on its respective side (see Fig. 25-6).

Internal Jugular Vein.   This vein starts in the jugular foramen as an expanded segment of vein, the ***jugular bulb***. Here the internal jugular vein is closely related to cranial nerves IX, X, and XI. Below the skull the internal jugular vein descends within the carotid sheath through the neck lateral to the internal and common carotid arteries. At the base of the neck it joins with the subclavian vein to become the brachiocephalic vein.

Figure 25-6.
Venous sinuses of the skull (superior view). The tentorium cerebelli has been removed.

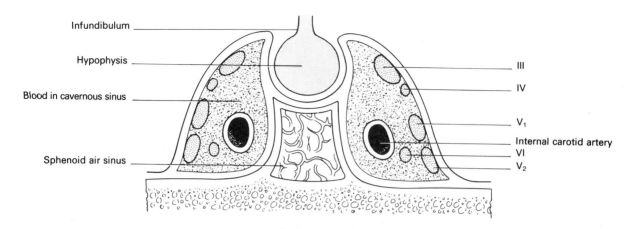

Infundibulum

Hypophysis

Blood in cavernous sinus

Sphenoid air sinus

III

IV

V₁

Internal carotid artery

VI

V₂

Figure 25-7.
Cavernous sinus (coronal section).

Superior Petrosal Sinus. This narrow sinus runs from the cavernous sinus along the superior margin of the petrous temporal bone (in the attached edge of the tentorium cerebelli) to the upper end of the sigmoid sinus (see Fig. 25-6).

Inferior Petrosal Sinus. Another narrow sinus, the inferior petrosal sinus runs from the cavernous sinus inferolaterally over the petrous temporal and occipital bones to join the bulb of the internal jugular vein in the jugular foramen (see Fig. 25-6).

Cavernous Sinus. This sinus differs from the other intracranial sinuses in that it contains connective tissue trabeculae, which give it some resemblance to the erectile tissues of the external genitalia without (we hope) exercising similar functions. The two cavernous sinuses are located on either side of the sella turcica anterior and medial to the cavum trigeminale. They receive blood from the ***superior and inferior ophthalmic veins***, the ***sphenoparietal sinuses***, and the ***pterygoid plexuses*** of veins.

Figure 25-8.
Internal carotid artery in cavernous sinus (sagittal section).

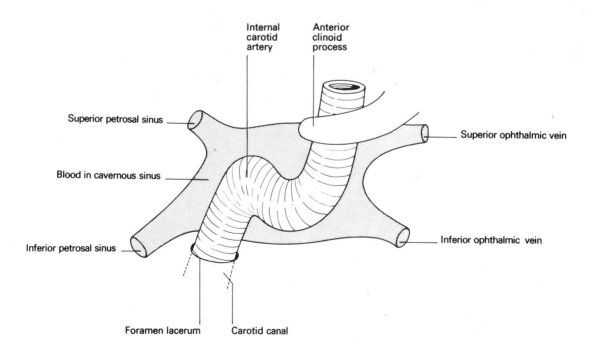

Internal carotid artery

Anterior clinoid process

Superior petrosal sinus

Superior ophthalmic vein

Blood in cavernous sinus

Inferior petrosal sinus

Inferior ophthalmic vein

Foramen lacerum

Carotid canal

They are interconnected by the ***intercavernous sinuses*** and empty into the superior and inferior petrosal sinuses. The cavernous sinuses contain the internal carotid arteries and the following cranial nerves: III, IV, VI, V_1, and V_2 (Figs. 25-7 and 25-8).

The ***sphenoparietal*** and ***occipital*** sinuses are small channels (see Fig. 25-6).

The ***hypophysis*** is located between the two cavernous sinuses. Immediately superior to it is the hypothalamic region of the brain. The ***sphenoidal air sinus*** lies within the sphenoid bone inferior to the hypophysis. The ***internal carotid artery*** takes an S-shaped course through the cavernous sinus (between its posterior and anterior ends).

CLINICAL NOTE

The cavernous sinus and the other sinuses are connected to the veins of the skin of the face, neck, and scalp by emissary veins through which infection can spread into the intracranial sinuses.

Fractures through the bony surroundings of the cavernous sinuses may damage the internal carotid artery so that a communication (fistula) between it and the venous space occurs. This will impede the venous drainage of the eye, resulting in edema and protrusion of the eye (proptosis) and associated tissues. This lesion can also cause damage (permanent or transient) to the cranial nerves in the cavernous sinus with varying sensory loss and squints.

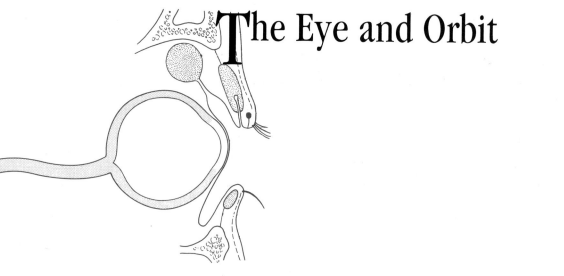

The Eye and Orbit

The orbit is the cavity that contains the eyeball. It is a bony cavity filled with fat, muscles, nerves, blood vessels, lymphatics, and the globe of the eye. Its bony walls are composed of parts of the following bones: frontal, zygomatic, maxilla, sphenoid, lacrimal, ethmoid, and palatine. The contribution of each bone to the parts of the orbit should be noted (Fig. 26-1):

> ***Margin***: maxilla, zygomatic, frontal
> ***Walls***: sphenoid, zygomatic, ethmoid, lacrimal
> ***Roof***: frontal, lesser wing of sphenoid
> ***Floor***: maxilla, zygomatic, palatine.

FEATURES OF THE ORBIT

The easily identifiable features of the bony orbit are as follows: ***superior*** and ***inferior orbital fissures***, ***optic foramen***, ***orbital plates of ethmoid***, ***ethmoidal canals***, ***nasolacrimal canal***, ***infraorbital groove***, ***canal*** and ***foramen***, and ***supraorbital notch*** or ***foramen***.

LACRIMAL BONE

This small bone sits on the anterior aspect of the lower part of the medial wall of the orbit. It has a ***posterior lacrimal crest***, which is a vertical ridge at about the midpoint of the bone. Anterior to the crest is the ***lacrimal groove***, which contains the ***lacrimal sac***, a part of the lacrimal apparatus.

PALPEBRAL LIGAMENTS (Fig. 26-2)

The ***medial palpebral ligament***, a tough band of connective tissue, is attached to the frontal process of the maxilla and to the lacrimal crest. The thinner ***lateral palpebral ligament*** is attached to the lateral wall of the orbit.

ORBICULARIS OCULI (see Fig. 24-6)

This subcutaneous muscle of facial expression arises from the medial palpebral ligament and surrounds the ***palpebral fissure***. Some fibers (the palpebral portion) attach to the lateral palpebral ligament. The remaining fibers surround the fissure and return to their origin. The muscle is used in gentle and forced closing of the eyelids. Its nerve supply is from the ***facial nerve***.

26

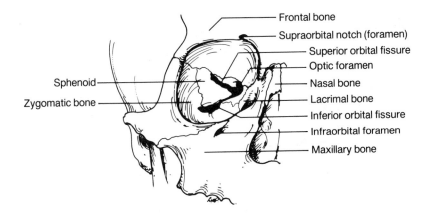

Frontal bone
Supraorbital notch (foramen)
Superior orbital fissure
Optic foramen
Sphenoid
Nasal bone
Zygomatic bone
Lacrimal bone
Inferior orbital fissure
Infraorbital foramen
Maxillary bone

TARSAL PLATES (Fig. 26-2)

The upper and lower tarsal plates consist of tough fibrous connective tissue and form the "skeleton" of the eyelids. They are about 1 cm in width, and the superior tarsal plate receives the attachment of the ***levator palpebrae superioris*** muscle. This muscle has a dual innervation.

1. Superior division of ***oculomotor*** (III) nerve
2. ***Sympathetic*** fibers from the first thoracic segment via the sympathetic trunk and the ***superior cervical ganglion***

CLINICAL NOTE

*Interruption of the sympathetic nerve supply to the orbit at any point between the first thoracic spinal cord segment and the head will produce **Horner's syndrome** (Fig. 26-3) in which the upper eyelid droops, the pupil is constricted (from the unopposed action of the parasympathetic nerve supply to the sphincter pupillae) (myosis), and the eyeball is less prominent (ptosis). If the lesion occurs proximal to the course of sympathetic fibers to the arteries of the face, the face will be flushed (arterial dilatation) and dry from lack of sweat secretion (anhydrosis).*

Figure 26-1.
Anterior view of right bony orbit.

Figure 26-2.
Right orbit (anterior view showing superficial structures).

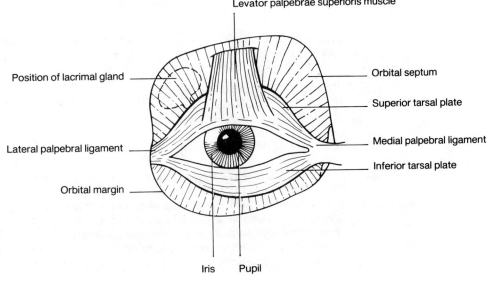

Levator palpebrae superioris muscle
Position of lacrimal gland
Orbital septum
Superior tarsal plate
Lateral palpebral ligament
Medial palpebral ligament
Inferior tarsal plate
Orbital margin
Iris   Pupil

Figure 26-3.
Horner's syndrome.

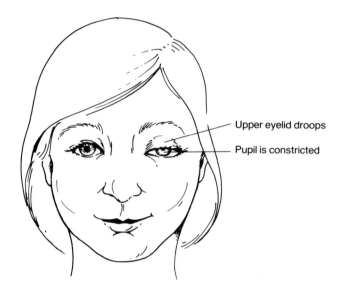

Upper eyelid droops

Pupil is constricted

ORBITAL SEPTUM

The orbital septum is a thin sheet of connective tissue that is continuous with the orbital periosteum at the orbital margins and which finds attachments to the tarsal plates of both eyelids, including a slip of attachment to the posterior lacrimal crest behind the lacrimal sac.

LACRIMAL GLAND (see Fig. 26-2)

The lacrimal gland, which produces the tears that cleanse and lubricate the conjunctival sac, lies in the anterior part of the lateral portion of the superior wall of the orbit; it measures about 1.5 cm in its long axis. The production of tears is regulated by *parasympathetic* fibers from the facial nerve (cranial nerve VII; see Chap. 34) that reach the gland by means of the lacrimal nerve.

CONJUNCTIVA

The conjunctiva is a thin membrane that lines the inner aspects of the eyelids and the front of the eyeball except over the cornea with which it blends. When the eyelids are closed the conjunctiva forms the *conjunctival sac*. The conjunctiva and cornea are exquisitely sensitive and are supplied by the *infratrochlear, maxillary*, and *lacrimal* nerves, which are all branches of the *trigeminal* nerve.

EYELIDS (PALPEBRAE)

The eyelids are made up of the fibrous skeleton of their tarsal plates and are covered by skin and lined internally by conjunctiva. They contain muscle fibers of the orbicularis oculi. The edges of the eyelids are covered by skin which contains the hair follicles of the eyelashes, modified sweat glands (ciliary glands of Moll), and modified sebaceous glands (glands of Zeis). Larger, sebaceous-type *tarsal glands* (Meibomian glands) are found within the substances of the tarsal plates.

CLINICAL NOTE

Infection of a gland of Zeis is a **stye***, and inflammation of a Meibomian gland is a* **chalazion***.*

The student will always be disappointed in the dissecting laboratory by the impossibility of properly examining the eye of the embalmed cadaver, and, indeed, in these days of public awareness of organ donor programs, many cadavers arrive without eyes.

The reader should realize that a valuable resource is the use of a mirror and his or her own eye (or, perhaps, gazing fondly into that of one's beloved). Examination will reveal the following easily visible structures.

Structures Visible on Self-Examination

1. ***Canthus***, lateral and medial, is where the eyelids meet in the "corner of the eye." The medial canthus exhibits a pink swelling, the ***lacus lacrimalis***, in which there is a small bump, the ***lacrimal caruncle***. Just lateral to the caruncle is a fold of tissue, the ***plica semilunaris***. Two small black dots on the medial ends of the edges of the eyelids (puncta) are the openings of the two ***lacrimal canals*** that lead to the invisible ***lacrimal sac*** (Fig. 26-4).

   The lacrimal sac has some fibers of the orbicularis oculi lying behind it so that when the eyelids are closed the lacrimal sac is squeezed and tears run down the ***nasolacrimal duct*** into the inferior meatus of the nasal cavity (see below).

   Tears are the secretion of the ***lacrimal gland*** and are produced when the gland is stimulated by ***parasympathetic impulses*** from the seventh cranial nerve. Tears enter the conjunctival sac at its superolateral corner and are propelled by the action of the eyelids. With every blink they wash across the surface of the eyeball and keep it moist. The fluid is drained through the ***lacrimal puncta*** into the two ***lacrimal canals*** at the medial ends of the eyelids and from them into the ***lacrimal sac***. From the sac the tears drain through the nasolacrimal canal into the inferior meatus of the nasal cavity. Tears protect, cleanse, and lubricate the conjunctiva and cornea and permit easy opening and closing of the eyelids. In facial nerve paralysis the eyelids cannot close; as a result the tears overflow and the conjunctiva and cornea are poorly lubricated. A dry eye with possible ulceration of the conjunctiva and cornea may occur. Obstruction of the lacrimal ducts (as in a cold in the nose) will cause spilling of tears, as will overproduction of tears (as in weeping).

2. ***Epicanthal folds*** are folds of skin in the upper eyelid running to the medial canthus. Their presence accounts for some ethnic differences in facial appearance.

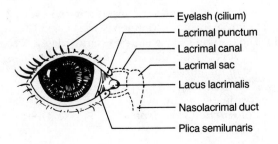

- Eyelash (cilium)
- Lacrimal punctum
- Lacrimal canal
- Lacrimal sac
- Lacus lacrimalis
- Nasolacrimal duct
- Plica semilunaris

Figure 26-4.
Eyelids everted on right eye to demonstrate lacrimal puncta.

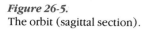

Figure 26-5.
The orbit (sagittal section).

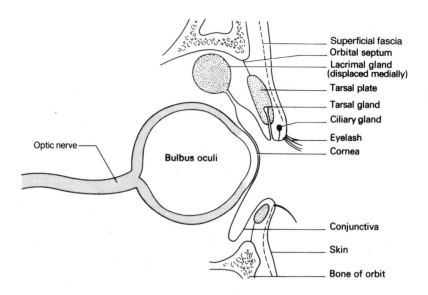

- Superficial fascia
- Orbital septum
- Lacrimal gland (displaced medially)
- Tarsal plate
- Tarsal gland
- Ciliary gland
- Eyelash
- Cornea

Optic nerve

Bulbus oculi

- Conjunctiva
- Skin
- Bone of orbit

CONTENTS OF THE ORBIT

The contents of the orbit include fat, which forms the flotation cushion for the eye, the eyeball (***ocular bulb***), its muscles, arteries, veins, nerves, and lymphatics, the ***lacrimal gland***, and the ***bulbar sheath*** (Tenon's capsule) (Fig. 26-5).

EYEBALL (OCULAR BULB OR BULBUS OCULI)

The eyeball, an outgrowth from the brain, remains connected to the brain by the ***optic nerve*** (II). The optic nerve carries with it prolongations of the intracranial meninges to the orbit. The dura forms the ***sheath*** of the optic nerve and the ***sclera*** of the eyeball.

BULBAR SHEATH (TENON'S CAPSULE)

This fascial layer surrounds the eyeball from the optic nerve to the junction of the sclera with the cornea, where it fuses with the bulb. It forms a capsule within which the eyeball can move. The attachments of the extraocular muscles to the eyeball pierce Tenon's fascia.

MUSCLES OF THE ORBIT (EXTRAOCULAR MUSCLES)

Levator Palpebrae Superioris

This muscle (see Fig. 26-2) does not insert into the eyeball. It runs from the superior margin of the optic foramen to the tarsal plate and skin of the upper eyelid. The portion inserted into the tarsal plate is ***involuntary*** and receives its nerve supply from the sympathetic system. The somatic nerve supply to the rest of the muscle is from the superior division of the oculomotor nerve. This muscle opens the palpebral fissure, and its involuntary part causes the wide-eyed stare of the frightened individual. Loss of this function is part of ***Horner's syndrome*** (see Fig. 26-3).

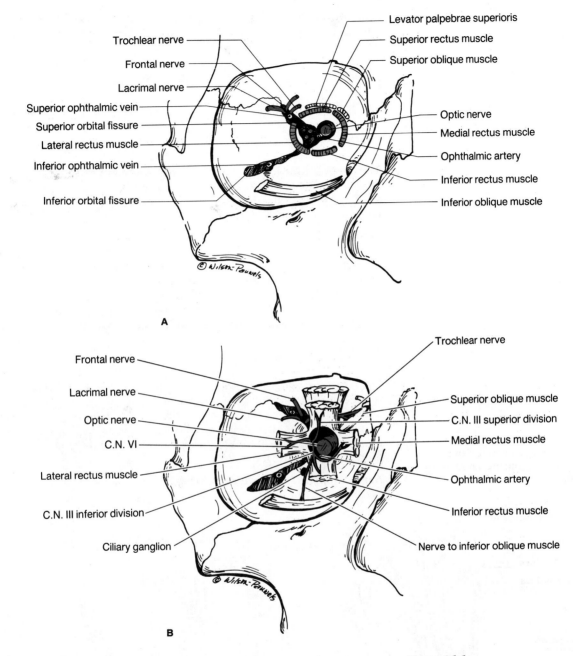

Levator palpebrae superioris
Superior rectus muscle
Superior oblique muscle

Trochlear nerve
Frontal nerve
Lacrimal nerve
Superior ophthalmic vein
Superior orbital fissure
Lateral rectus muscle
Inferior ophthalmic vein
Inferior orbital fissure

Optic nerve
Medial rectus muscle
Ophthalmic artery
Inferior rectus muscle
Inferior oblique muscle

© Wilson-Pauwels

A

Frontal nerve
Lacrimal nerve
Optic nerve
C.N. VI
Lateral rectus muscle
C.N. III inferior division
Ciliary ganglion

Trochlear nerve
Superior oblique muscle
C.N. III superior division
Medial rectus muscle
Ophthalmic artery
Inferior rectus muscle
Nerve to inferior oblique muscle

© Wilson-Pauwels

B

CLINICAL NOTE

The lower lid has no comparable proper muscle and depends on relaxation of the orbicularis oculi for its participation in the opening of the palpebral fissure. In facial nerve palsy the levator palpebrae superioris acts unopposed to keep the eye open, even during sleep.

Figure 26-6.
Right orbit (anterior view of structures in the posterior portion of the orbit).
A. Tendinous ring indicating muscle origins and related structures.
B. Vessels, nerves, and muscles.

COMMON TENDINOUS RING (Fig. 26-6)

This is a ring of fibrous tissue that surrounds the superior, medial, and inferior margins of the optic foramen and the wider medial portion of the superior orbital fissure. It gives proximal attachments to the four rectus muscles, and the oculomotor, abducens, and nasociliary nerves enter the orbit within its circumference.

MUSCLES ATTACHED TO THE TENDINOUS RING

The following four muscles are attached to the tendinous ring:

Rectus Muscles (Medial, Lateral, Superior, and Inferior)

The *medial, lateral, superior, and inferior rectus muscles* all arise from the appropriate segments of the ring and insert into the eyeball just behind the corneoscleral junction. The lateral rectus is supplied by the *abducens* nerve (VI), and the other recti are supplied by the *oculomotor* nerve (III).

Actions. The actions of the medial and lateral recti are simple to understand; they move the pupil medially (adduction) and laterally (abduction), respectively. The actions of the superior and inferior recti are more complex for it must be realized that these muscles lie on the superior and inferior walls of the orbit but that the anteroposterior axis of the orbit deviates laterally from that of the eyeball. Thus the superior and inferior pulls of these muscles will be combined with *abduction* of the eye.

Superior Oblique Muscle

The *superior oblique* passes from the sphenoid bone, superior to the attachment of the superior rectus, to the upper, medial, and superior margin of the orbit, where its tendon passes through the *trochlea* (a fibrous pulley) (Fig. 26-7). At that point the muscle changes direction and its fibers run laterally, posteriorly, and inferiorly to be attached to the eyeball in its postero-supero-lateral quadrant.

Nerve Supply. The *trochlear nerve* (IV; see Chap. 34).

Figure 26-7.
The right orbit (superior view).
A. Superficial structures. **B.** Deep structures.

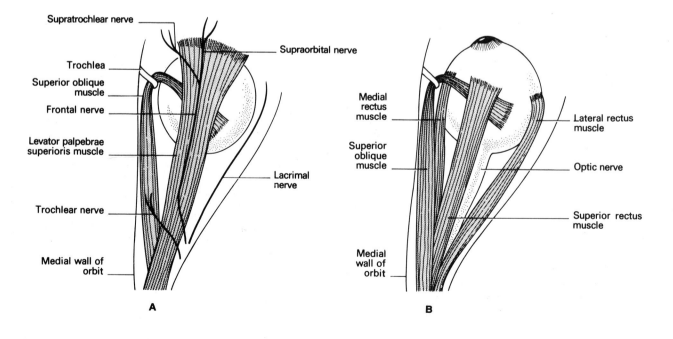

A

B

Action. It turns the pupil downward and medially and rotates the eyeball so that the upper surface turns medially (intorsion).

CLINICAL NOTE

*This description of the isolated action of the superior oblique runs counter to traditional anatomical lore but is confirmed by the clinical finding of double vision on looking down and medially in patients in whom trochlear nerve palsy is the only proven abnormality. These people have a fear of walking downstairs because their double vision (**diplopia**) is accentuated when the gaze is in that direction.*

Inferior Oblique Muscle

The ***inferior oblique*** muscle lies transversely on the floor of the orbit and inserts into the eye in its postero-infero-lateral quadrant.

Nerve Supply. From the inferior division of the ***oculomotor*** nerve.

Action. It rotates the pupil upward and medially.

It is to be hoped that the alert reader will protest the above account of the actions of the extraocular muscles, which is at variance with traditional anatomical teaching and seems to run counter to what can be deduced by studying drawings of the attachments of these muscles. It must be remembered that none of these muscles acts in isolation; rather they produce highly complex and coordinated bilateral movements designed to enable the scanning of a particular ***field of vision***.

CLINICAL NOTE

Despite controversy surrounding the actions of the extraocular muscles, certain clinical observations are pertinent and shed some light on the probable functional actions of these muscles when they work together to enable the individual to scan a normal binocular field of vision. We may not know what combinations of muscle actions produce certain movements, but we can observe the deficits that result from certain nerve and muscle paralyses.

Deficits from Nerve and Muscle Paralysis

1. Lateral rectus paralysis (abducens nerve VI) produces a medial squint.
2. Superior oblique paralysis (trochlear nerve IV) prevents inward and downward looking, as in walking downstairs.
3. Oculomotor nerve (III) paralysis (all the remaining muscles) causes the eye to deviate downward and laterally: "down-and-out." Further details as to loss of visual fields and the locations of false images in diplopia belong to the field of ophthalmology and cannot be pursued in this basic textbook.

STRUCTURES THAT ENTER OR LEAVE THE ORBIT

Various structures enter or leave the orbit by passing through one of the following openings (Table 26-1):

> ***Optic foramen***
> ***Superior orbital fissure*** (within the tendinous ring)
> ***Superior orbital fissure*** (outside the tendinous ring)
> ***Inferior orbital fissure***

Table 26-1.
Structures That Enter or Leave the Orbit

Opening	Structure	Distribution
Optic foramen	Optic nerve	Eyeball, retina
	Ophthalmic artery	Optic nerve, orbital content, eyeball and retina, lacrimal gland, eyelids, forehead, ethmoid sinuses, nose
Superior orbital fissure (through tendinous ring)	Oculomotor nerve	Medial rectus, inferior rectus, inferior oblique, superior rectus, levator palpebrae superioris, ciliary ganglion
	Abducens nerve	Lateral rectus
	Nasociliary nerve	Eyeball, pupil, ciliaris, ethmoid sinuses, nasal mucosa, skin of nose, conjunctiva, cornea, face around lower eyelid
Superior orbital fissure (outside tendinous ring)	Frontal nerve	Orbital content, upper eyelid, cornea, conjunctiva, skin of forehead
	Lacrimal nerve	Lacrimal gland, cornea, conjunctiva, upper eyelid
	Trochlear nerve	Superior oblique
	Superior ophthalmic vein	Drains orbital tissues, eyeball and retina into cavernous sinus
Inferior orbital fissure	Inferior ophthalmic vein	Drains orbital tissues into cavernous sinus
	Maxillary nerve	Nose, palate, upper teeth, cheek and upper lip

STRUCTURES PASSING THROUGH THE OPTIC FORAMEN

The following structures must pass through the tendinous ring because the latter surrounds the foramen (see Fig. 26-6A).

Optic Nerve. The optic nerve with its meningeal coverings enters the orbit through the optic foramen and passes laterally and forward to reach the optic bulb (eyeball).

Ophthalmic Artery. The ophthalmic artery passes through the optic foramen and gives off the following branches:

Lacrimal artery, which runs along the lateral wall of the orbit to the lacrimal gland.
Supraorbital and *supratrochlear* arteries, which accompany the nerve of the same name to the superior orbital margin.
Ethmoidal arteries, which run through the ethmoidal foramina in the medial wall of the orbit into the nose.
Ciliary arteries, which run to the sclera, choroid, and ciliary body.
Palpebral arteries, which supply branches to the eyelids.
Arterial branches to the extraocular muscles.

Central Artery of the Retina. An end artery, the central artery of the retina receives no collateral anastomotic branches after it enters the substance of the optic nerve near the optic foramen. It supplies the optic nerve and the retina, and its branches in the choroidal layer of the retina can be visualized with an *ophthalmoscope*.

CLINICAL NOTE

*Because the central artery of the retina is an **end artery**, blockage means blindness.*

STRUCTURES THAT PASS THROUGH THE SUPERIOR ORBITAL FISSURE INSIDE THE TENDINOUS RING (Fig. 26-6)

The following structures pass *through the tendinous ring*:

Oculomotor Nerve (III). Divides into a superior and an inferior division just before entering the orbit. The superior division supplies the superior rectus and the levator palpebrae superioris. The fibers to the latter muscle contain postganglionic sympathetic fibers from the superior cervical ganglion (which have traveled along the internal carotid artery and its branches).

The inferior division supplies the medial rectus, inferior rectus, and the inferior oblique muscles. It carries parasympathetic fibers that synapse in the *ciliary ganglion*. Postganglionic fibers from the ciliary ganglion constrict the pupil and alter the shape of the lens (cranial nerve III; see Chap. 34).

Abducens Nerve (VI; see Chap. 34). Reaches the lateral aspect of the orbit to supply the lateral rectus muscle.

Nasociliary Nerve. A branch of the ophthalmic division (V_1) of the trigeminal nerve. It passes between the superior rectus muscle and the optic nerve to reach the medial wall of the orbit. It supplies the eyeball, the ethmoidal sinuses, and skin around the lower eyelid and nose (see Chap. 34).

STRUCTURES THAT PASS THROUGH THE SUPERIOR ORBITAL FISSURE BUT OUTSIDE THE TENDINOUS RING (see Fig. 26-6)

Frontal Nerve (Branch of V_1). Passes through the orbit between the orbital periosteum and the levator palpebrae superioris muscle and divides into the supraorbital and supratrochlear nerves at the superomedial angle of the anterior orbital opening. They supply the conjunctiva and the skin of the forehead (see Chap. 34).

Lacrimal Nerve (Branch of V_1). Runs along the lateral wall of the orbit toward the lacrimal gland. It carries postganglionic parasympathetic fibers to the lacrimal gland and is a sensory nerve to the conjunctiva and skin of the upper eyelid (see Fig. 26-7).

Trochlear Nerve (IV). A small nerve, the trochlear nerve passes medially, superior to the levator palpebrae superioris to enter and supply the superior oblique muscle (see Fig. 26-7).

Superior Ophthalmic Vein. Anastomoses anteriorly with the anterior facial vein near the medial canthus. It passes posteriorly from the orbit to drain into the cavernous sinus.

STRUCTURES THAT PASS THROUGH THE INFERIOR ORBITAL FISSURE

Inferior Ophthalmic Vein. Drains the orbit into the ***pterygoid plexus of veins*** and into the cavernous sinus.

Maxillary Nerve (V₂; see Chap. 34). The maxillary nerve, the second division of the trigeminal nerve, passes through the inferior orbital fissure but remains deep to the periosteum of the orbit. It continues in the infraorbital groove as the ***infraorbital nerve*** and exits through the infraorbital foramen to supply the skin of the face. It gives off a zygomatic branch that conveys postganglionic parasympathetic fibers from the pterygopalatine ganglion to the lacrimal nerve. The zygomatic branch also supplies the skin of the face. The remainder of the branches of the maxillary nerve will be described in Chapter 34.

The Neck

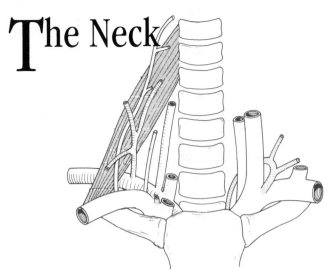

For ease of description and study the neck is divided into various triangles. The posterior triangle will be considered in Chapter 36 in the context of the upper limb.

ANTERIOR TRIANGLE

The *anterior triangle* (Fig. 27-1) is defined by the anterior border of the sternomastoid muscle, the lower border of the mandible, and the midline of the neck.

LANDMARKS

The following reference points should be identified.

Mandible. On the inner surface of the mandible the mylohyoid line receives the attachment of the mylohyoid muscle, which effectively separates the neck from the floor of the mouth. The *digastric fossa* is a depression on the medial surface of the mandible below the anterior end of the mylohyoid line (see Fig. 24-5).

Hyoid Bone. This is a horseshoe-shaped bone (Fig. 27-2) with a body and two horns (cornua). It helps to form the root of the tongue and the floor of the mouth.

Larynx. The voice box bearing the prominent "Adam's apple" consists of three cartilages and occupies the midline of the neck from the level of the second cervical vertebra to the sixth cervical vertebra. Its skeleton comprises the following (Fig. 27-3):

1. The *thyroid cartilage*, which is made up of two flat *laminae*, joined anteriorly in the midline. Each lamina has a superior and inferior *horn* (cornu), and each lamina displays an *oblique line* on its lateral surface for attachment of the sternothyroid and thyrohyoid muscles. The thyroid cartilage makes up the visible and palpable prominence of the larynx (Adam's apple).
2. The *cricoid cartilage* is shaped like a signet ring with the band in front and the shield posteriorly. It articulates with and hinges on the inferior horns (cornua) of the thyroid cartilage.
3. The *arytenoid cartilages* ride on the upper border of the posterior part of the cricoid and will be described with the larynx and vocal cords (see Chap. 32).

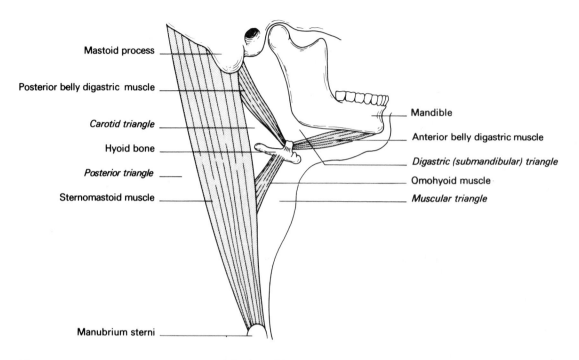

Figure 27-1.
The anterior triangle of the neck (right side).

The hyoid bone is connected to the thyroid cartilage by the ***thyrohyoid membrane***, and the thyroid cartilage is connected to the cricoid by the ***cricothyroid membrane*** (see Fig. 32-1).

Trachea. The lower end of the trachea was studied in Chapter 4. In the neck the trachea is the continuation of the air passage below the cricoid cartilage.

Mastoid Process. This prominent bony landmark has already been studied with the skull. It is easily palpable and is an important reference landmark for structures in the neck.

The reader should take the time to identify all of the above-mentioned landmarks on his or her own body.

Platysma Muscle

The platysma is a thin subcutaneous muscle that covers the anterior triangle and adjacent structures (see Fig. 24-6). It is easily demonstrable by **snarling** in front of the mirror. It tenses the skin of the neck and possibly, in wild animals, forms a protective layer for the deeper structures of the

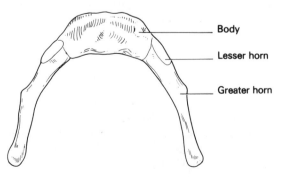

Figure 27-2.
The hyoid bone (superior view).

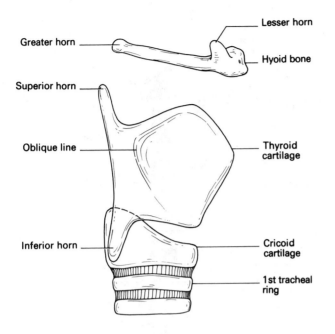

Figure 27-3.
The basic skeleton of the larynx (right lateral view).

neck. In humans it seems to be relegated to appearing as a background prop for razor advertisements. It is supplied by the cervical branch of the *facial* nerve.

DIVISIONS OF THE ANTERIOR TRIANGLE

The anterior triangle can be subdivided into three lesser areas by the digastric and omohyoid muscles. It must be reemphasized that these divisions are merely for ease of description and orientation and have no other significance (see Fig. 27-1). The resultant subdivisions are the *submandibular*, *carotid*, and *muscular triangles*.

Muscles of the Anterior Triangle

Omohyoid Muscle

This thin strap-like muscle runs from the scapular notch on the superior border of the scapula (near its medial angle) across the neck to the junction of the body and greater cornu of the hyoid bone (see Fig. 27-1). It consists of two bellies joined by an intermediate tendon that passes through a sling of the investing layer of the deep cervical fascia. The muscle may act to tense the cervical fascia during deep forced inspiration and thus give some support to the cervical pleura. It is supplied by branches from the *ansa cervicalis* (C1, C2, and C3). Its main importance is as an anatomical landmark.

Digastric Muscle

This is another muscle with an intermediate tendon and two bellies (see Fig. 27-1). It has a *posterior belly* that is attached to a groove on the medial aspect of the base of the mastoid process and an *anterior belly* that fixes the muscle to the digastric fossa on the inner aspect of the anterior part of the body of the mandible. The intermediate tendon is tied down to the

Figure 27-4.
Strap muscles of neck separated to show details of larynx and trachea (right sternohyoid removed).

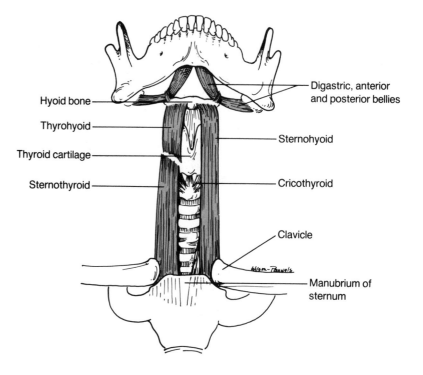

Hyoid bone

Thyrohyoid

Thyroid cartilage

Sternothyroid

Digastric, anterior and posterior bellies

Sternohyoid

Cricothyroid

Clavicle

Manubrium of sternum

hyoid bone by a sling of fascia. The posterior belly is supplied by the facial nerve and the anterior belly by the mylohyoid branch of the mandibular nerve (V_3). The digastric muscle helps to elevate and retract the hyoid bone during swallowing.

Strap Muscles

Strap muscles is a collective term for the strap-like ***sternohyoid, sternothyroid***, and ***thyrohyoid*** muscles (Fig. 27-4). These muscles are shaped like a strap with parallel borders and are about 2 cm wide. Their names reveal their attachments. The strap muscles are innervated by the ansa cervicalis (C1, C2, and C3). All three muscles can depress the hyoid and hence aid the opening of the jaw and movement of the floor of the mouth.

ARTERIES OF THE ANTERIOR TRIANGLE

The major arteries of the anterior triangle are the common, the external, and the internal carotid arteries. The last named has no branches in the neck.

Common Carotid Artery. This major vessel is a branch of the ***brachiocephalic*** artery on the right and of the arch of the ***aorta*** on the left. In company with the ***internal jugular vein*** and the ***vagus nerve*** it is surrounded by a condensation of the deep cervical fascia, the ***carotid sheath*** (Fig. 27-5).

The common carotid artery ascends vertically in the neck under cover of the sternomastoid muscle and in front of the transverse processes of the cervical vertebrae. Medially it is related to the trachea, larynx, esophagus, and thyroid gland and laterally to the internal jugular vein. The vagus nerve runs within the carotid sheath between the common carotid artery and the internal jugular vein but in a somewhat posterior position. Behind the carotid sheath the cervical part of the ***sympathetic trunk*** ascends to the base of the skull.

The common carotid artery divides into the ***internal*** and ***external carotid arteries*** at the upper border of the thyroid cartilage. The point of division, or the first part of the internal carotid artery, is dilated to form the ***carotid sinus***, which is an important ***baroreceptor*** supplied by sensory fibers of the ***glossopharyngeal*** nerve (IX; see Chap. 34).

A small brownish nodule, the ***carotid body***, is lodged in the division of the common carotid artery; it is a ***chemoreceptor***, also supplied by the glossopharyngeal nerve.

CLINICAL NOTE

Pressure on the neck over the carotid sinus can stimulate the baroreceptor to elicit a vagal reflex that will slow the heart beat and lower the blood pressure. This may cause the individual to faint.

Internal Carotid Artery. This large and vital terminal branch of the common carotid artery is a conduit for blood destined for the head and has no branches in the neck. It ascends vertically to the base of the skull to enter through the ***carotid canal*** in the petrous temporal bone. The internal carotid artery is surrounded by a plexus of postganglionic sympathetic nerve fibers from the ***superior cervical ganglion***. After it leaves the carotid canal inside the skull it divides into cerebral and ophthalmic branches.

External Carotid Artery. This major vessel runs from its origin at the upper border of the thyroid cartilage medial to the ramus of the mandible where, often within the substance of the parotid gland, it divides into its two terminal branches, the ***superficial temporal artery*** and the

Figure 27-5.
The right external carotid artery and its branches.

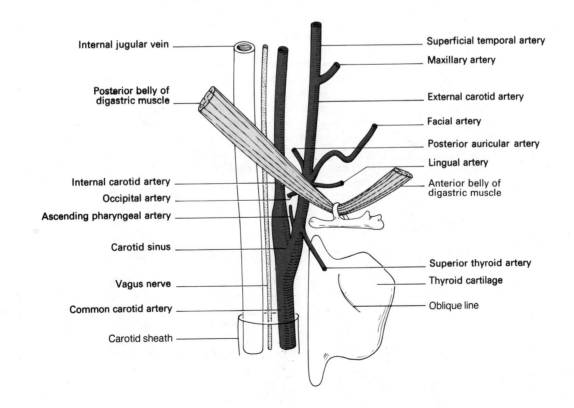

- Internal jugular vein
- Posterior belly of digastric muscle
- Internal carotid artery
- Occipital artery
- Ascending pharyngeal artery
- Carotid sinus
- Vagus nerve
- Common carotid artery
- Carotid sheath
- Superficial temporal artery
- Maxillary artery
- External carotid artery
- Facial artery
- Posterior auricular artery
- Lingual artery
- Anterior belly of digastric muscle
- Superior thyroid artery
- Thyroid cartilage
- Oblique line

maxillary artery. The superficial temporal artery runs upward over the root of the zygomatic process and in front of the temporomandibular joint and the external auditory meatus, where its pulse can be palpated (see Figs. 24-8 and 27-5).

Branches of the External Carotid Artery. The external carotid artery has eight branches of varying degrees of importance (listed below). The more important ones will be mentioned in more detail (see Figs. 24-8 and 27-5).

1. Ascending pharyngeal artery
2. Superior thyroid artery
3. Lingual artery
4. Facial artery
5. Occipital artery
6. Posterior auricular artery
7. Superficial temporal artery
8. Maxillary artery

The ***superior thyroid artery*** originates just below the greater cornu of the hyoid bone. It passes deep to the sternothyroid muscle to the upper pole of the thyroid gland. The external laryngeal branch of the ***superior laryngeal nerve*** accompanies it part of the way.

The ***lingual artery*** originates at the level of the greater cornu of the hyoid bone, makes a little upward loop which is crossed by the hypoglossal nerve, and then passes deep to the hyoglossus muscle to supply the tongue.

The ***facial artery*** arises just above the lingual artery, runs deep to the mandible, grooves the posterior border of the submandibular salivary gland, and then emerges from behind the lower border of the mandible, which it crosses just anterior to the lower fibers of the masseter muscle, and runs obliquely upward toward the medial canthus. It gives large labial branches to the upper and lower lips. Its pulsation can be palpated where it crosses the body of the mandible anterior to the masseter. (Try it!)

The ***superficial temporal artery*** is one of the two terminal branches of the external carotid artery and passes in front of the auricle, where its pulse can be palpated, to supply the skin and scalp of the temple.

The ***maxillary artery*** is the other terminal branch of the external carotid artery. It arises deep to or in the substance of the parotid gland, passes deep to the neck of the mandible, and makes its way through the infratemporal fossa to the pterygopalatine fossa. It will be described later.

The other named branches run in the directions indicated by their names and are relatively unimportant except for the small ascending pharyngeal artery, which gives off a ***tympanic*** branch to the middle ear where it may cause troublesome bleeding during middle ear surgery.

VEINS OF THE ANTERIOR TRIANGLE

Internal Jugular Vein. This vein is formed in the jugular fossa at the base of the skull as the termination of the ***sigmoid sinus*** at the ***jugular bulb*** (the expanded upper part of the jugular vein). In the jugular foramen the ***glossopharyngeal***, ***vagus***, and ***accessory*** nerves lie anterior to the vein, in close contact. Look at the base of the skull and note that the jugular foramen is just behind the opening of the carotid canal and that the jugular foramen is very close to the floor of the tympanic cavity. The internal jugular vein, internal carotid artery, and the vagus nerve become enveloped in the **carotid sheath** of fascia and descend through the neck together.

CLINICAL NOTE

*The anterior and lateral aspects of the internal jugular vein are in close contact with the **jugular lymph nodes** (part of the deep cervical group of nodes). In fact the connecting lymph channels between jugular nodes run within the adventitia of the jugular vein so that the vein has to be removed as part of a surgical procedure that aims to remove the deep cervical lymph nodes in the treatment of malignant disease of the region.*

CERTAIN NERVES OF THE ANTERIOR TRIANGLE

Hypoglossal Nerve (Cranial XII). This important nerve passes through the anterior triangle, appearing from deep to the digastric muscle and then disappearing deep to it.

Ansa Cervicalis. This loop of nerve tissue supplies the strap muscles. The superior root of the loop comes from the first cervical nerve and, for a distance, runs with the hypoglossal nerve. It then leaves the hypoglossal nerve and descends superficial to the carotid sheath to join the inferior root from cervical nerves 2 and 3.

BASE OF SKULL AND SUBMANDIBULAR REGION

There are many landmarks on the base of the skull and in the submandibular region that the reader should identify in order to facilitate the visualization of the three-dimensional disposition of structures in this region. Using Figure 27-6, identify the following features of the skull.

Individual Bones. Examine the shape, structure, and relationships of the **maxilla**, **occipital bone**, **palatine**, **sphenoid**, and **temporal bones** and the **vomer** and **zygomatic** bones.

Holes in the Head (Foramina and Openings). The following holes carry important structures and should be examined **now**: auditory tube, hypoglossal (anterior condylar) canal, choanae, incisive foramen, jugular foramen, foramen lacerum, foramen magnum, foramen ovale, foramen spinosum, greater and lesser palatine foramina, stylomastoid foramen, and squamotympanic fissure.

BONY LANDMARKS

Certain spines, bumps, processes, fossae, and tubercles cannot be ignored and must be added to this apparently formidable list. The reader is encouraged to tackle this review with care; once these features have been identified and learned, the subsequent examination and study of the deeper anatomical structures will become a lot easier.

The following should be identified: mandibular fossa, pharyngeal tubercle, pterygoid processes of the sphenoid bone with medial and lateral pterygoid plates, spine of sphenoid, styloid and mastoid processes of the temporal bone, and the occipital condyles.

Mandible. The mandible should be reviewed and the ramus, body, and angle noted. The ramus bears the condylar process with its head and neck,

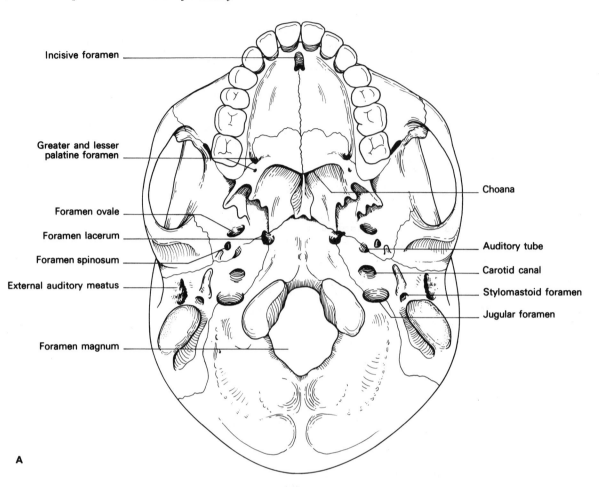

Incisive foramen

Greater and lesser
palatine foramen

Foramen ovale

Foramen lacerum

Foramen spinosum

External auditory meatus

Foramen magnum

Choana

Auditory tube

Carotid canal

Stylomastoid foramen

Jugular foramen

A

Figure 27-6.
Inferior view of the base of the skull.
A. Foramina. **B.** Bony features.

the coronoid process, the mandibular notch, the mandibular (inferior alveolar or dental) foramen, and the lingula. Identify the alveolar process, the mental foramen (mental=pertaining to chin), the mylohyoid line, the digastric fossa, the symphysis menti, the mental spines (genial tubercles), and the oblique line. Take a particularly close look at the mylohyoid line; this line serves for the attachment of the mylohyoid muscle and divides the inner surface of the mandible into the region that belongs to the mouth (above the line) and the part related to the neck (below the line).

MUSCLES AND CONTENTS OF SUBMANDIBULAR (DIGASTRIC) TRIANGLE

The submandibular (digastric) triangle is bordered by the mandible and the two bellies of the digastric muscle.

MUSCLES

Mylohyoid Muscle

This flat muscle is often misrepresented in diagrams as running in a vertical direction. In fact it is almost *horizontal*. Its fibers run from the mylohyoid line of the mandible posteriorly and medially to join those from the oppo-

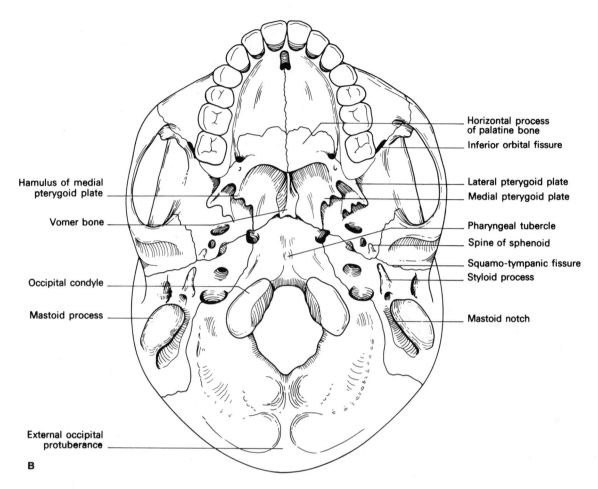

Horizontal process
of palatine bone

Inferior orbital fissure

Lateral pterygoid plate

Medial pterygoid plate

Pharyngeal tubercle

Spine of sphenoid

Squamo-tympanic fissure

Styloid process

Mastoid notch

Hamulus of medial
pterygoid plate

Vomer bone

Occipital condyle

Mastoid process

External occipital
protuberance

B

site side to form a median raphe and to attach to the hyoid bone (Fig. 27-7). The mylohyoid muscle forms the ***floor of the mouth***.

Nerve Supply. The mylohyoid muscle is supplied by the mylohyoid branch of the inferior alveolar nerve (a branch of the mandibular division of the trigeminal nerve).

Action. It forms the floor of the mouth, stabilizes the base of the tongue, and raises the floor of the mouth in swallowing. These actions can be confirmed by palpating the muscle from the exterior during swallowing or by forcing the tongue against the incisor teeth with the mouth closed.

Stylohyoid Muscle

This is a small muscle that helps to elevate the hyoid bone. It has a postero-superior attachment to the styloid process of the temporal bone and an anteroinferior attachment to the greater cornu of the hyoid bone. Its ***nerve supply*** is from the facial nerve.

Base of Tongue. The hyoid bone forms the osseous base of the tongue, and its relationship to the mandible is maintained primarily by the mylohyoid and geniohyoid muscles. The hyoid bone can be depressed by the strap muscles and by gravity acting on the larynx and pharynx.

Figure 27-7.
The floor of the mouth (inferior view).

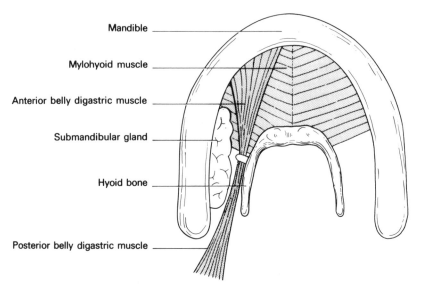

Mandible

Mylohyoid muscle

Anterior belly digastric muscle

Submandibular gland

Hyoid bone

Posterior belly digastric muscle

Hyoglossus Muscle

The hyoglossus passes from the body of the hyoid bone directly into the lateral aspect of the substance of the tongue. It is visible behind the posterior border of the mylohyoid muscle, and its lateral surface is related to the deep part of the submandibular gland, the hypoglossal nerve, the lingual nerve, the submandibular ganglion, and the submandibular duct (Wharton's duct).

Nerve Supply. The hyoglossus is supplied by the ***hypoglossal nerve*** (cranial XII; see Chap. 34).

Action. The hyoglossus depresses the tongue.

SUBMANDIBULAR GLAND

The submandibular gland is a major salivary gland that can be palpated between a finger in the floor of the mouth and a finger under the chin over the posterior part of the mylohyoid muscle opposite the first molar tooth (Figs. 27-7 and 27-8). The gland is shaped like a U in horizontal cross-section, with the larger superficial part lying below the mylohyoid muscle

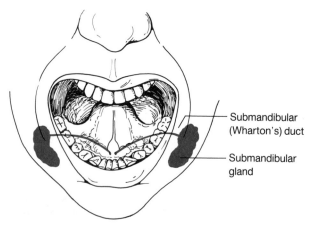

Submandibular (Wharton's) duct

Submandibular gland

Figure 27-8.
Submandibular gland and duct (Wharton's) (tongue elevated).

and the deep portion wedged between mylohyoid and hyoglossus. Posteriorly the gland is separated from the parotid gland by the stylomandibular ligament, a band of fibrous tissue running from the styloid process to the angle of the mandible. The gland may be sufficiently large for its superficial part to overlap both bellies of the digastric muscle.

The ***duct of the submandibular gland (Wharton's duct)*** runs from the deep part of the gland, under the mucosa of the mouth and on the lateral aspect of hyoglossus to a papilla just lateral to the frenulum of the tongue. This papilla can be seen, with saliva issuing from it if the gland is stimulated by sucking a piece of lemon.

CLINICAL NOTE

Just below the neck of the third molar tooth the submandibular duct is crossed by the lingual nerve. This relationship is important if the duct is to be incised through the oral mucosa for the removal of a salivary calculus.

The secretions of the submandibular gland are more viscid than those of the parotid, and thus calculi of the submandibular gland are much more common than stones in the parotid gland.

Nerve Supply. The submandibular salivary gland receives its parasympathetic secretomotor fibers from the ***chorda tympani*** branch of the facial nerve, which joins the lingual nerve in the infratemporal fossa. The preganglionic fibers synapse in the ***submandibular ganglion*** (see below), and the postganglionic fibers rejoin the lingual nerve.

NERVES

Hypoglossal Nerve. The hypoglossal nerve enters the submandibular triangle by passing deep to the posterior belly of the digastric muscle or its tendon. It runs on the surface of the hyoglossus, deep to the superficial part of the submandibular gland, and then passes deep to the posterior edge of the mylohyoid muscle. It remains lateral to the hyoglossus running below the deep part of the submandibular gland. It ends on the tongue and supplies the intrinsic and extrinsic muscles of the tongue. In its course through the submandibular triangle the hypoglossal nerve is flanked by two venae comitantes, which give it a characteristic appearance in living persons (cranial nerves; see Chap. 34).

Lingual Nerve. This sensory branch of the mandibular (V_3) division of the trigeminal nerve can be seen in the submandibular triangle well under cover of the body of the mandible. It runs across the hyoglossus above the hypoglossal nerve and crosses the submandibular duct twice on its way to the tongue. Its proximity to the floor of the mouth can be appreciated by feeling it with the tip of the tongue as a ridge that crosses the mandibular alveolar process about 1 cm below the neck of the third molar tooth. It supplies somatic sensation and sense of taste (from VII) to the anterior two thirds of the tongue and the floor of the mouth, and preganglionic parasympathetic fibers from the chorda tympani to the submandibular ganglion. As the lingual nerve passes across the hyoglossus, the ***submandibular ganglion***, a parasympathetic ganglion for the supply of the salivary glands in the floor of the mouth, is seen suspended from the inferior aspect of the nerve.

ARTERY

Lingual Artery. This branch of the external carotid artery passes ***deep*** to the hyoglossus, just above the hyoid bone, to supply the tongue.

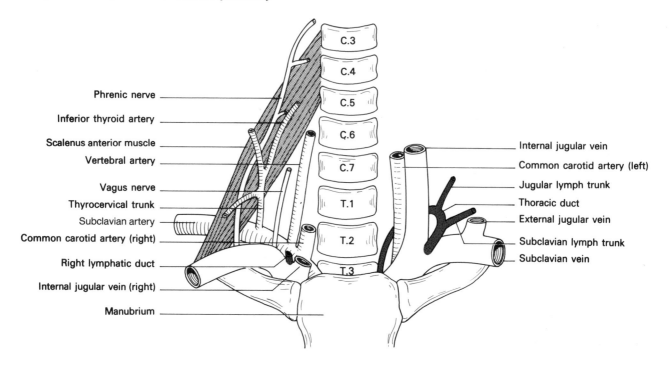

Phrenic nerve

Inferior thyroid artery

Scalenus anterior muscle

Vertebral artery

Vagus nerve

Thyrocervical trunk

Subclavian artery

Common carotid artery (right)

Right lymphatic duct

Internal jugular vein (right)

Manubrium

C.3
C.4
C.5
C.6
C.7
T.1
T.2
T.3

Internal jugular vein

Common carotid artery (left)

Jugular lymph trunk

Thoracic duct

External jugular vein

Subclavian lymph trunk

Subclavian vein

Figure 27-9.
The base of the neck. The formation of brachiocephalic veins is displaced slightly superiorly and laterally.

ROOT OF THE NECK AND THYROID GLAND

The study of the root of the neck requires a review of the clavicle, manubrium sterni, and the first rib. These were described in Chapter 3.

SUBCLAVIAN ARTERY

The right subclavian artery is a terminal branch of the brachiocephalic artery. The left subclavian artery is a direct branch from the arch of the aorta. Both arteries pass upward and laterally over the cervical pleura and first rib to become the *axillary artery* at the lateral border of the first rib. In this course the subclavian artery passes behind the lower attachment of the *scalenus anterior* muscle which separates the artery from the more anteriorly placed subclavian vein.

Branches of the Subclavian Artery (Fig. 27-9)

Vertebral Artery. This artery arises close to the origin of the subclavian artery and passes upward and medially to enter the *foramen transversarium* of the *sixth* cervical vertebra. It runs upward through successive foramina transversaria until it reaches the suboccipital triangle at the base of the skull, where it enters the vertebral canal and runs through the foramen magnum to supply the brain.

Internal Thoracic Artery. The internal thoracic artery was encountered in the thorax.

Thyrocervical Trunk. The thyrocervical trunk arises just medial to the scalenus anterior muscle. It passes upward giving off the *inferior thyroid artery* (see below). The thyrocervical trunk has two other branches: the *suprascapular artery*, which runs across the posterior triangle of the neck to reach the scapula where it makes some important anastomoses, and the ***transverse cervical artery***, which supplies the muscles of the posterior triangle.

Costocervical Trunk. The costocervical trunk passes over the cervical pleura to furnish the intercostal arteries for the first two intercostal spaces; it also supplies muscular branches to the deeper muscles of the neck.

SUBCLAVIAN VEIN

The subclavian vein is formed at the lateral border of the first rib. It runs anterior to the scalenus anterior and over the cervical pleura to join with the internal jugular vein, behind the sternoclavicular joint, forming the brachiocephalic vein. Its only important tributary is the external jugular vein (Fig. 27-9).

THORACIC AND RIGHT LYMPHATIC DUCTS

These two major terminal lymph ducts enter the venous system at the junction of the subclavian and internal jugular veins on the left and right sides, respectively (Fig. 27-9).

The ***thoracic duct*** was encountered in the thorax, where it ran posterior to the esophagus into the left side of the neck. It runs behind the carotid sheath and anterior to the sympathetic trunk and subclavian artery to enter the junction of the subclavian and internal jugular veins.

The ***right lymphatic duct*** ends in a similar manner on the right.

VAGUS NERVE

The cervical part of the vagus nerve lies within the carotid sheath between the carotid (internal and common) arteries and the internal jugular vein, in a somewhat more posterior plane (Fig. 27-10). It enters the thoracic inlet in front of the subclavian artery.

Figure 27-10.
The thyroid gland and related structures. The sympathetic trunk on the right has been displaced laterally.

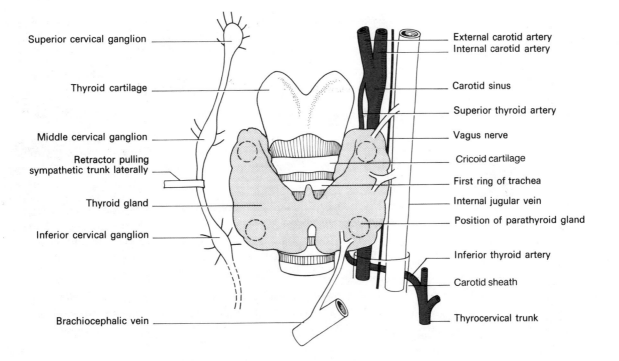

The *recurrent laryngeal nerve*, a branch of the vagus, passes around the subclavian artery on the right side. On the left side it passes around the arch of the aorta in the thorax and runs into the neck along the left side of the trachea. In the neck both recurrent laryngeal nerves run upward in the groove between the trachea and the esophagus to supply all of the *intrinsic* muscles of the larynx and laryngeal sensation below the level of the vocal cords (see Fig. 7-9 and Cranial Nerves, Chap. 34).

CLINICAL NOTE

In their ascent in the neck the recurrent laryngeal nerves lie just posterior to the posterior border of the lateral lobes of the thyroid gland, and in this position they are in danger of being injured during thyroidectomy. Tumors of the thyroid gland may damage the nerves by compression, and the vocal cords should always be examined **before** *and* **after** *thyroid surgery.*

SYMPATHETIC TRUNK

The sympathetic trunk passes upward from the thorax to the base of the skull lying posteromedial to the carotid arteries (common and internal) but outside the carotid sheath. The trunk bears three cervical ganglia, which are the result of coalescence of eight cervical segmental ganglia. All the preganglionic fibers come from white rami communicantes in the thorax (mainly from T1) and synapse at one of the three cervical ganglia (see Figs. 27-10 and 27-11).

Superior Cervical Ganglion. This ganglion sits at about the level of the atlas or axis vertebra. It is 1.5 to 2 cm long, and branches from it pass along the *internal carotid artery* into the cranial cavity. It also send fibers along the *external carotid artery* and its branches. It sends a descending *superior cardiac branch* to the cardiac plexus and sends gray rami communicantes to the upper four cervical nerves.

Middle Cervical Ganglion. This smaller ganglion is found at the level of the cricoid cartilage and gives gray rami to the fifth and sixth cervical nerves and branches to the cardiac plexus and the thyroid gland.

Inferior Cervical Ganglion. This ganglion is often fused with the first thoracic ganglion to form the *stellate ganglion*. Gray rami pass to the seventh and eighth cervical nerves, and other branches pass to the cardiac plexus, and to the vertebral artery with which they ascend into the skull.

Ansa Subclavii. The ansa subclavii is a loop of sympathetic fibers from the stellate ganglion to the middle cervical ganglion. The loop runs in front of the subclavian artery, hooks around its inferior aspect, and ascends behind it.

Important Note: the sympathetic trunk in the neck receives *no* white rami, but from it come *all the sympathetic fibers (postganglionic) to the head and neck*. Thus impulses from the brain must pass down the spinal cord to the thoracic region, along white rami communicantes to the upper thoracic sympathetic trunk, and then upward to reach the head and neck by passing *up* the sympathetic trunk into the cervical region (Fig. 27-11).

From the trunk branches pass to the neck either as gray rami to the cervical spinal nerves or as direct visceral branches to structures such as the thyroid gland.

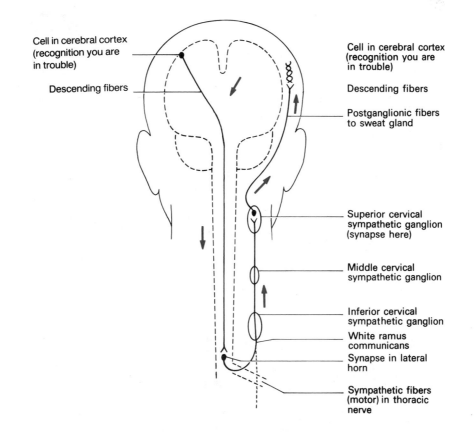

Cell in cerebral cortex (recognition you are in trouble)

Descending fibers

Cell in cerebral cortex (recognition you are in trouble)

Descending fibers

Postganglionic fibers to sweat gland

Superior cervical sympathetic ganglion (synapse here)

Middle cervical sympathetic ganglion

Inferior cervical sympathetic ganglion

White ramus communicans

Synapse in lateral horn

Sympathetic fibers (motor) in thoracic nerve

Branches to structures in the head pass within the adventitia of the appropriate arteries, especially the internal and external carotid arteries and their branches.

Figure 27-11.
Schematic pathway of stimulus from the cerebral cortex affecting sweat glands of the forehead.

CLINICAL NOTE

If the sympathetic pathway on one side of the head is cut, **Horner's syndrome** *will result on that side of the head and neck (see Chap. 26, Fig. 26-3).*

MIDLINE STRUCTURES IN THE NECK

The cartilaginous skeleton of the larynx and trachea forms a set of landmarks to which other structures of the midline of the neck should be related.

Pharynx. The pharynx is the midline tubular structure that serves both respiratory and alimentary functions. It extends from the base of the skull behind the nose (nasopharynx) to the level of the lower border of the cricoid cartilage. The pharynx will be described in detail in Chapter 29.

Esophagus. The esophagus begins as the continuation of the pharynx at the lower border of the cricoid cartilage. It descends through the neck, posterior to the trachea, to enter the thoracic inlet.

THYROID GLAND

The thyroid gland is a major endocrine gland. It secretes *thyroxine*, which controls the body's metabolic rate. It is the "thermostat" of metabolism.

The gland consists of two *lobes* and an *isthmus* (see Fig. 27-10). The isthmus connects the two lobes across the front of the trachea so that the thyroid gland may be said to straddle the trachea. The location of the isthmus is usually at the second and third cartilaginous rings of the trachea. The lobes rest on either side of the trachea and larynx and are covered by the sternothyroid and sternohyoid muscles and the pretracheal fascia. This arrangement prevents the lobes from rising (or expanding) above the thyroid cartilage and is also responsible for the fact that the gland moves up and down during deglutition (swallowing). Occasionally a third *pyramidal lobe* connects the isthmus with the hyoid bone by a thread of connective tissue which represents the remnant of the embryonic *thyroglossal duct*.

CLINICAL NOTE

A cyst may develop along the course of the thyroglossal duct, and the whole of the thyroglossal duct remnant must be removed up to the foramen cecum area of the tongue.

*The capsule of the thyroid gland consists of connective tissue which contains an ample supply of vessels to the gland. The gland is surrounded by the **pretracheal** layer of the deep cervical fascia, and the tissue plane between this fascia and the capsule must be defined during thyroid surgery to avoid injury to the capsule and its abundant blood vessels. Since the four **parathyroid** glands are usually embedded in the lobes of the thyroid gland, subtotal removal of the thyroid is usually performed.*

PARATHYROID GLANDS

The parathyroid glands are (usually) four small (4–5 mm) endocrine glands that secrete a hormone essential for calcium metabolism. They are usually embedded in the posterior aspects of the lobes of the thyroid gland, but one or more may be found lower in the neck or in the mediastinum. They are brownish in color and can be quite difficult to visualize in the living and almost impossible to visualize in the embalmed cadaver (see Fig. 27-10).

Blood Supply of the Thyroid and Parathyroid Glands

The arterial supply comes from two sources (see Fig. 27-10). The *superior thyroid artery* is a branch of the external carotid artery that descends to the superior pole of the lobe of the thyroid gland. The *inferior thyroid artery* is a branch of the thyrocervical trunk of the subclavian artery. The inferior thyroid artery ascends behind the common carotid artery to reach the lateral aspect of the lobe of the thyroid. In so doing it may branch before reaching the gland, and its branches may intermingle with the recurrent laryngeal nerve. This nerve must *always* be clearly identified before anything resembling the inferior thyroid artery is cut or clamped during surgery.

The venous drainage of the thyroid is through three sets of veins. The superior and middle thyroid veins drain into the internal jugular vein, and the inferior thyroid veins drain into the brachiocephalic veins.

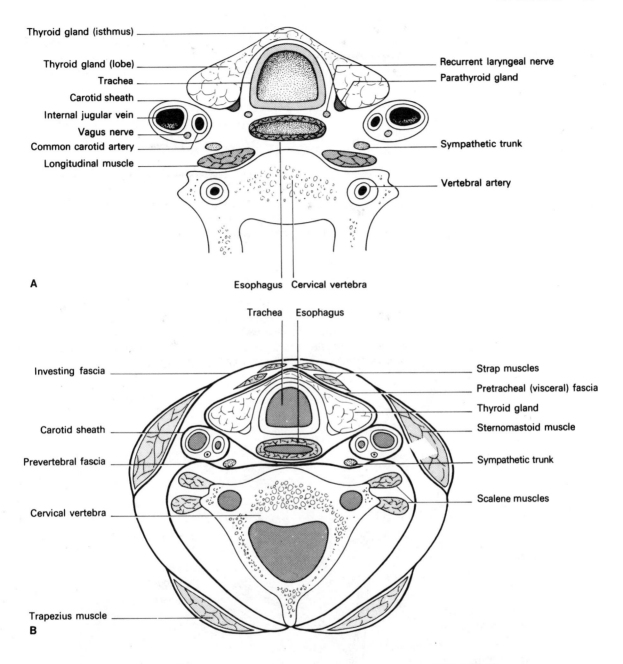

Thyroid gland (isthmus)

Thyroid gland (lobe)
Trachea
Carotid sheath
Internal jugular vein
Vagus nerve
Common carotid artery
Longitudinal muscle

Recurrent laryngeal nerve
Parathyroid gland

Sympathetic trunk

Vertebral artery

A

Esophagus   Cervical vertebra

Trachea   Esophagus

Investing fascia

Carotid sheath

Prevertebral fascia

Cervical vertebra

Trapezius muscle
B

Strap muscles
Pretracheal (visceral) fascia
Thyroid gland
Sternomastoid muscle

Sympathetic trunk

Scalene muscles

FASCIAL LAYERS OF THE NECK

Some of the fascial layers of the neck have already been mentioned and will be summarized here (Fig. 27-12).

Investing Fascia. This layer surrounds the neck, being attached superiorly to the superior nuchal line, mastoid process, zygomatic arch, hyoid bone, and the lower border of the body of the mandible. Inferiorly it is attached to the clavicle, manubrium sterni, and the acromion process of the scapula. The investing layer splits to enclose the sternomastoid and trapezius muscles.

Figure 27-12.
Transverse section of structures of the neck. **A.** Without fascial layers. **B.** With fascial layers.

Prevertebral Fascia. This layer surrounds the vertebral column and the muscles attached to it. Therefore it is superficial to the scalene muscles, levator scapulae, and splenius capitis (see below). It is continued into the upper limb as the ***axillary sheath***, which surrounds the axillary artery and brachial plexus.

Pretracheal Fascia. This is a thin layer of fascia, more evident in living bodies, that runs from the thyroid cartilage downward to blend with the fascia of the anterior mediastinum in the thorax. It encloses the thyroid gland and blends with the carotid sheath laterally.

Retropharyngeal Space. The fascia surrounding the pharynx (bucco-pharyngeal fascia) is separated from the prevertebral fascia by the ***retropharyngeal space***, which contains some loose areolar tissue. This space may be site of a ***retropharyngeal abscess*** arising either from the pharynx or from the vertebral bodies (after penetrating the prevertebral fascia).

LYMPHATIC DRAINAGE OF THE HEAD AND NECK

There are several chains of lymph nodes in the head and neck. All drain directly or indirectly into the ***jugular trunk*** (Fig. 27-13). On the left side the jugular trunk joins the ***thoracic duct*** (see Fig. 27-9), whereas on the right side the jugular trunk joins with the ***subclavian*** and ***bronchomediastinal*** trunks to form the ***right lymphatic duct***.

Figure 27-13.
Lymphatic drainage of the head and neck (modified from Duffy).

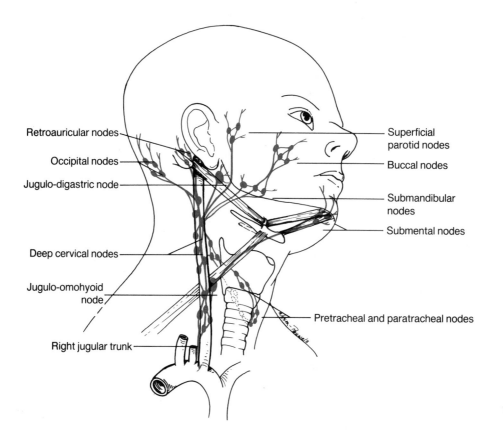

Retroauricular nodes

Occipital nodes

Jugulo-digastric node

Deep cervical nodes

Jugulo-omohyoid node

Right jugular trunk

Superficial parotid nodes

Buccal nodes

Submandibular nodes

Submental nodes

Pretracheal and paratracheal nodes

LYMPH NODES OF THE HEAD AND NECK

The lymph nodes of the head and neck are divisible into superficial and deep groups that form more or less complete rings around the pharynx.

Superficial Ring

The superficial ring consists of four groups of lymph nodes that are all accessible to palpation:

Occipital Nodes. The occipital nodes are found along the upper attachment of the trapezius muscle. Their afferents come from the scalp, and their efferent vessels drain to the deep cervical nodes. They are often enlarged in German measles (rubella).

Retroauricular Nodes. These are clustered around the upper attachment of the sternomastoid muscle. They receive drainage from the ear, temporal region, and scalp, and drain into the deep cervical nodes.

Superficial Parotid Nodes. The superficial parotid nodes are located on the surface of the parotid gland. They receive afferents from the frontal and temporal regions, the root of the nose, the eyelids, the external auditory meatus, and the tympanic and nasal cavities. Their efferents drain into the deep cervical nodes.

Buccal (Mandibular) Nodes. These are located on the lateral surface and along the anterior edge of the ramus of the mandible and on the external surface of the buccinator muscle. They receive afferents from the eyelids and conjunctiva and from the skin and mucous membranes of the nose and cheek. Their efferents drain into the submandibular and deep parotid nodes.

Deep Ring

Five groups of nodes are found in the deep ring.

Retropharyngeal Nodes. These nodes are located posterior to the pharynx in the retropharyngeal space. They drain the posterior part of the nasal cavity, the nasopharynx, and the auditory tube. Their efferents drain into the deep cervical nodes.

Deep Parotid Nodes. The deep parotid nodes are located in the parotid gland tissue. Their afferents come from the eyelids, side of the nose, lips, gums, anterior part of the tongue, and the (superficial) submental and submandibular nodes. Their efferent vessels drain to the deep cervical nodes.

Submandibular Nodes. These nodes are located in the deep part of the submandibular triangle between the mandible and the submandibular salivary gland. They receive afferents from the side of the face, the tip and sides of the tongue, the mandibular alveolar process, the mouth and cheek, and the mandibular and submental nodes. Their efferents drain into the deep cervical nodes.

Pretracheal and Paratracheal Nodes. These nodes, though deep to the deep fascia, have a superficial location, being near the surface and easily palpable if enlarged. They drain the larynx, trachea, thyroid gland, and other adjacent structures. Their efferents drain into the deep cervical nodes.

All of the members of these two rings drain into the ***deep cervical (jugular) nodes***. Efferents from the deep cervical nodes form the jugular lymphatic trunk.

Deep Cervical Lymph Nodes. These nodes lie lateral to the carotid sheath in proximity to the internal jugular vein, with their connecting lymph vessels embedded in the adventitia of the internal jugular vein. According to their location with reference to the digastric and omohyoid muscle they may be divided into ***jugulodigastric*** and ***jugulo-omohyoid*** groups. The jugulodigastric nodes are often called the tonsillar nodes because they are frequently enlarged in young persons who have had several bouts of tonsillitis.

The lowest group of deep cervical nodes lies in the supraclavicular part of the posterior triangle of the neck. They are continuous with the axillary nodes and on the left side are intimately connected to the thoracic duct. Abdominal malignancy may spread along the thoracic duct and invade these nodes. Such enlargement of these "sentinel nodes" of Virchow may be the first evidence of intra-abdominal malignant disease.

Temporal and Infratemporal Regions

Once again the reader is urged to take skull in hand and to examine and visualize the bony territory to be described (Fig. 28-1). Particular attention should be paid to the inferior aspect of the temporal bone, the zygomatic, sphenoid, frontal, and parietal bones, and the mandible. The following features should be identified: mandibular fossa, squamotympanic fissure, articular tubercle, spine of sphenoid, temporal line, and temporal fossa.

Re-examine the styloid process and visualize the stylohyoid ligament running from the styloid process to the lesser horn (cornu) of the hyoid bone.

MUSCLES

Certain previously mentioned muscles should be reviewed. These are the buccinator, the digastric, and the sternomastoid. The masseter will be mentioned with the muscles of mastication later in this chapter.

PAROTID GLAND

This large salivary gland was described in Chapter 24. It must be understood that the parotid gland is a bulky structure of variable size that occupies the available space between adjacent bones and muscles. It may be thought of as being like a piece of putty that has been squeezed into the area bounded superiorly by the zygomatic process of the temporal bone and external auditory canal; inferiorly by the digastric muscle; anteriorly by the ramus of the mandible (and attached muscles), which it overlaps both superficially and deeply; posteriorly by the mastoid process and sternomastoid muscle; and medially by the wall of the pharynx. Indeed the gland must be removed in order to reveal the structures in the infratemporal region. It should be remembered that it is the largest salivary gland and that its function differs from that of the submandibular gland in that its secretions are of a more serous (watery) nature.

The facial nerve traverses the parotid gland and divides it into superficial and deep portions that can be separated surgically by dissecting along the plane of the facial nerve. The bed of the gland includes some important structures that are listed below (from deep to superficial) and shown in Figure 28-2:

> ***Auriculotemporal nerve (branch of V₃)***
> ***The external carotid artery*** and its two terminal branches, the
> ***superficial temporal artery*** and the ***maxillary artery***

28

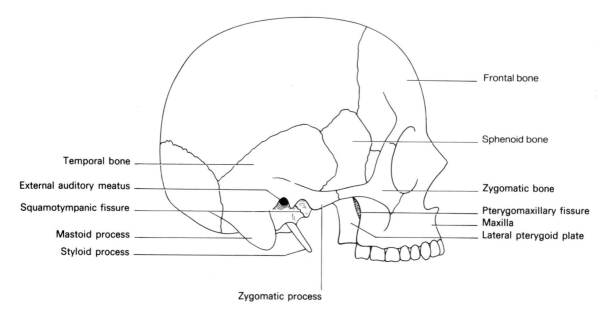

Figure 28-1.
Bony structures of the right temporal region.

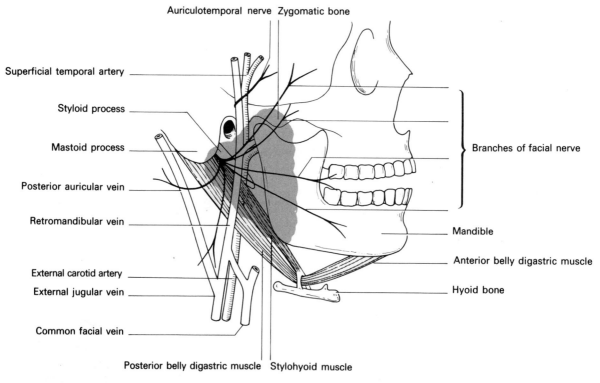

Figure 28-2.
The bed of the right parotid gland.

The retromandibular vein and its tributaries
Branches of the facial nerve
The great auricular nerve

The great auricular nerve is usually superficial to the parotid gland, which surrounds the other above listed structures. The parotid gland has a tough fibrous capsule, and there are lymph nodes within the substance of the gland.

CLINICAL NOTE

Mumps is an infection of the parotid gland, and the pain experienced in this disease is the result of the stretching of the fibrous capsule by the swollen gland.

Nerve Supply.   The parotid gland is a salivary gland that receives its nerve supply from sympathetic and parasympathetic sources.

The **parasympathetic** fibers come from the glossopharyngeal nerve by way of the **lesser petrosal nerve**, which joins the **otic ganglion** on the mandibular nerve (V₃) after it leaves the skull via the **foramen ovale**. Postganglionic parasympathetic fibers arise in the otic ganglion and are distributed to the parotid gland by way of the **auriculotemporal nerve**, which passes through the gland.

The **sympathetic** postganglionic fibers come from the superior cervical ganglion by way of the plexus around the external carotid artery and its terminal branches.

TEMPOROMANDIBULAR (TM) JOINT

The temporomandibular joint is the synovial articulation between the head of the mandible and the articular fossa of the temporal bone. The joint contains an **intra-articular fibrocartilaginous disc** that is of considerable importance. The capsule of the joint is not remarkable (Fig. 28-3).

Three fibrous structures, located at some distance from the joint, play some part in its actions. The **stylomandibular ligament** runs from the styloid process to the angle of the mandible and separates the parotid gland from the submandibular gland. (It is merely a thickening of the investing layer of deep fascia of the neck which splits to enclose the parotid

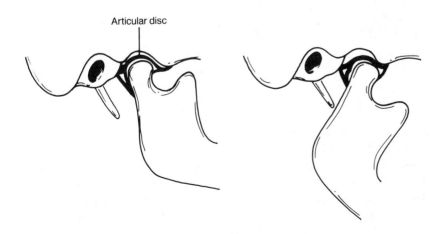

Articular disc

Figure 28-3.
Movements of the right temporomandibular joint.

gland; its influence on the function or stability of the temporomandibular joint has not been established.) The ***pterygomandibular raphe*** runs from the hamulus of the medial pterygoid plate to the upper end of the mylohyoid line and provides some attachment for the superior constrictor muscle of the pharynx and the buccinator muscle of the cheek. The ***sphenomandibular ligament*** joins the spine of the sphenoid to the lingula of the mandible (medial surface) and may play a pivotal role in temporomandibular joint movements.

MOVEMENTS OF THE TEMPOROMANDIBULAR JOINT (Fig. 28-3)

This is not the simple hinge joint that examination of the skull and mandible would suggest. When the mouth is opened the head of the mandible first hinges in the mandibular fossa and then moves forward ***with the articular disc*** onto the articular tubercle. The axis of movement is through the mandibular foramen. One of the muscles that produces this movement of ***protraction*** is the ***lateral pterygoid***, which finds attachment to the articular disc and pulls it forward. In side-to-side chewing movements one mandibular head moves forward while the other head remains in the mandibular fossa or moves backward (retraction) from the protracted position. The axis of this movement runs vertically through the center of the hyoid bone.

The excursions of the mandibular heads can be easily confirmed by palpation of the region on the reader's own body.

CLINICAL NOTE

Very wide opening of the mouth may cause the head of the mandible to slip anterior to the articular tubercle and thereby produce dislocation of the jaw. To reduce the dislocation strong pressure is applied with the thumbs to the crowns of the last lower molar teeth; with backward and downward pressure the mandibular head will slip back into the mandibular fossa.

The articular disc may be torn in violent movements of the temporomandibular joints or it may degenerate as a result of malalignment of the joint surfaces, such as may occur if there is malocclusion of the teeth (the biting surfaces do not fit). This condition may cause annoying clicking sounds in the joint which are only too audible to the patient whose ear is only millimeters away from the source of the sound.

MUSCLES ACTING ON THE TEMPOROMANDIBULAR JOINT

There are four paired muscles of mastication. All are supplied by the mandibular division (V_3) of the trigeminal nerve.

Masseter Muscle

The masseter is a powerful closer of the jaws (Fig. 28-4).

Proximal Attachment. Zygomatic arch.

Distal Attachment. The lateral surface of the ramus and body of the mandible near its angle.

Nerve Supply. A branch of the mandibular division of the trigeminal nerve passes through the mandibular notch to enter its deep surface.

Action. Elevation of the mandible. The contracted masseter can be palpated on the lateral aspect of the ramus and angle of the mandible.

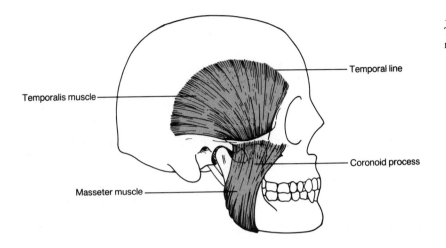

Figure 28-4.
The right temporalis and masseter muscles.

Temporalis Muscle

This is a large and powerful muscle of the temporal region that acts as a powerful closer of the jaw (Fig. 28-4).

Proximal Attachment. Temporal fossa (the region bounded by the zygomatic arch, the zygomatic process of the frontal bone, and the temporal line).

Distal Attachment. The muscle passes deep to the zygomatic arch to insert into the coronoid process and anterior border of the ramus of the mandible right down to the location of the third molar tooth. The posterior fibers can be seen to run in almost a horizontal direction and are responsible for the powerful action of retraction of the protruded mandible.

Nerve Supply. Two deep temporal branches from the mandibular nerve.

Action. Elevation and retraction of the mandible.

Lateral Pterygoid Muscle

The lateral pterygoid muscle lies deep to the ramus of the mandible (Fig. 28-5).

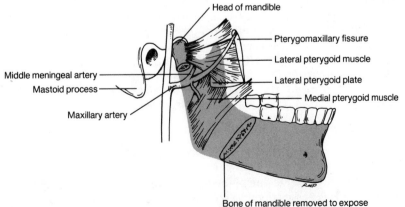

Figure 28-5.
Muscles of the right infratemporal region.

Proximal Attachment. From the lateral surface of the lateral pterygoid plate and from the infratemporal surface and crest of the greater wing of the sphenoid bone.

Distal Attachment. The neck of the mandible and the ***articular disc*** of the temporomandibular joint.

Nerve Supply. A branch from the mandibular nerve enters its deep surface.

Actions. Protrusion (protraction) and depression of the mandible.

Medial Pterygoid Muscle

The medial pterygoid muscle (Fig. 28-5) is located deep to the ramus of the mandible.

Proximal Attachment. From the medial surface of the ***lateral*** pterygoid plate and the adjacent portion of the maxilla just behind the upper third molar tooth.

Distal Attachment. The medial surface of the ramus and angle of the mandible.

Nerve Supply. The nerve to the medial pterygoid from the mandibular nerve; this branch has the parasympathetic ***otic ganglion*** attached to it.

Action. Elevation and protrusion (protraction) of the mandible. The lateral and medial pterygoid muscles acting together produce the side-to-side (mandibular) movement of chewing.

It can be seen from the above description of the muscles of mastication that the muscles act in concert in the important activity of biting and chewing. The power of the human bite requires no elaboration. The mandible deserves some attention in that it is the mobile component of the jaws and the principal lever in the act of biting and chewing. It bears the lower teeth, which number 10 in the young child and 16 in the adult. As the teeth are lost in old age the alveolar processes of both jaws recede and the body of the mandible in the edentulous (toothless) becomes quite narrow in the vertical diameter.

CLINICAL NOTE

The mandible is a ring-like structure which, if it fractures, does so usually in two places on opposite sides. The weak spots are the necks and the canine fossae (the areas of the body that bear the roots of the canine teeth [third from the front]). Thus a fracture may go through both canine fossae, through both necks, or through one canine fossa and the opposite neck. In fractures through the mandibular neck the unopposed action of the lateral pterygoid muscle will tilt the upper fragment, and alignment can be difficult, particularly if the patient has no teeth that can be wired together to achieve and maintain reduction of the fracture.

INFRATEMPORAL REGION

ARTERIES OF THE INFRATEMPORAL REGION

The ***maxillary artery*** is one of the two terminal branches of the external carotid artery. It passes deep to the neck of the mandible and runs deep or superficial to (or between the two heads of) the lateral pterygoid muscle through the *pterygomaxillary fissure* (Fig. 28-5) to enter the *pterygopalatine fossa*.

The maxillary artery is divided into three parts for descriptive purposes. The first part lies deep to the neck of the mandible, the second part is in relation to the lateral pterygoid muscle, and the third part is the portion in the pterygomaxillary fissure and beyond. The only virtue of attempting to remember this separation of parts is that it facilitates the memorizing of the more important branches of the artery.

The branches of the first part all enter *boles in the bead*, the branches of the second part are all *muscular*, and the third part again distributes its branches to *boles in the bead*.

Branches of the First Part of the Maxillary Artery. The *middle meningeal artery*, a small but very important vessel, enters the *extradural space* of the skull through the foramen spinosum. The *inferior alveolar artery*, another small artery, crosses the medial pterygoid muscle and enters the mandibular foramen in company with the inferior alveolar nerve. It supplies the mandible and all of the lower teeth.

CLINICAL NOTE

The inferior alveolar artery has no anastomoses once it has entered the mandibular foramen and is enclosed in a rigid bony canal. If infections around the roots of the teeth are neglected, the artery may become occluded by inflammatory exudate within this unyielding canal. This can cause extensive damage to the mandible and the other teeth.

Branches of the Second Part of the Maxillary Artery. These arteries all supply muscles in the temporal and infratemporal region and are named according to their destination, for example, deep temporal arteries and masseteric artery.

Branches of the Third Part of the Maxillary Artery. These arise in the pterygomaxillary fissure and pterygopalatine fossa and go through bony foramina to supply the maxilla, upper teeth, palate, nose, and nasopharynx.

NERVES OF THE INFRATEMPORAL REGION

All but two of the nerves (lesser petrosal, chorda tympani) of the infratemporal region arise from the mandibular division of the trigeminal nerve shortly after it passes through the foramen ovale. The lesser petrosal nerve (a branch of IX) also runs through the foramen ovale and joins the otic ganglion.

Auriculotemporal Nerve. This small nerve passes posteriorly through the parotid gland, medial to the neck of the mandible, and in front of the external auditory meatus to supply sensory fibers to the external ear and the skin of the temple. At its origin it picks up postganglionic parasympathetic secretomotor fibers from the otic ganglion for distribution to the parotid gland.

Nerve to Medial Pterygoid. This small nerve passes deeply from the parent trunk. Attached to it by small branches is the *otic ganglion* (Fig. 34-11).

Nerve to Lateral Pterygoid. This small twig passes directly from the parent trunk to the muscle.

Nerve to Masseter. This nerve passes through the mandibular notch to enter the deep surface of the muscle (Fig. 34-10).

Nerves to Temporalis. Two fine nerves pass upward to supply the temporalis muscle (Fig. 34-10).

Buccal Nerve. This is a *sensory* nerve that runs between the two heads of the lateral pterygoid to reach the surface of the buccinator muscle. It supplies the skin and mucous membrane of the cheek as well as the lateral (labial) surface of the gum of the lower jaw (Fig. 34-10).

Lingual Nerve. The lingual nerve is a long sensory nerve to the tongue and floor of the mouth (Fig. 34-11). Close to its origin it is joined by the *chorda tympani* branch of the facial nerve. The chorda tympani passes taste fibers and preganglionic parasympathetic efferent fibers to the lingual nerve. The lingual nerve runs deep to the lateral pterygoid muscle and appears at its lower border to pass superficial to the medial pterygoid muscle and then onto the hyoglossus muscle of the tongue. It runs medial to the third inferior molar tooth and can be palpated just deep to the oral mucosa at this point. It is vulnerable to damage in operative or dental procedures in this region. It supplies sensory and taste fibers to the anterior two thirds of the tongue.

Submandibular Ganglion. The submandibular ganglion is a small parasympathetic ganglion that is suspended from the inferior aspect of the lingual nerve as the latter lies on the lateral surface of hyoglossus (Fig. 34-11). The preganglionic fibers for the ganglion come from the chorda tympani and are carried in the lingual nerve. The postganglionic fibers rejoin the lingual nerve and are distributed to the submandibular and sublingual salivary glands and the small individual salivary glands in the mucosa of the floor of the mouth and cheek.

Inferior Alveolar (Dental) Nerve. This branch of the mandibular nerve passes deep to the lateral pterygoid and superficial to the medial pterygoid muscles (Fig. 34-11). It lies above and in front of the lingual nerve and enters the mandibular foramen. It passes along the mandibular canal to supply the lower teeth and ends by emerging through the mental foramen as the mental nerve that supplies sensation to the chin and lower lip.

The *mylohyoid nerve* is a branch of the inferior alveolar nerve and arises just before the latter enters the mandibular foramen. It runs inferiorly and anteriorly below the mylohyoid to supply that muscle and the anterior belly of the digastric muscle.

CLINICAL NOTE

Anesthesia of all of the teeth of one half of the lower jaw can be obtained by injecting local anesthetic agent into the area where the inferior alveolar nerve enters into the mandibular foramen. This procedure will also anesthetize the lower lip because the mental nerve is part of the inferior alveolar nerve. The proximity of the lingual nerve often causes it to be anesthetized by the same procedure.

CERTAIN NERVES AND VESSELS AT THE BASE OF THE SKULL

The external appearance of the base of the skull should be reviewed, noting particularly the relationships of the jugular foramen, carotid canal, and hypoglossal canal (anterior condylar foramen). Cranial nerves IX, X, and XI pass through the jugular foramen anterior to the internal jugular vein and lie posterior to the carotid artery, where it enters the carotid canal (see Chap. 34). Cranial nerve XII passes through the hypoglossal canal and then runs medial to the internal jugular vein to appear between that vein and the internal carotid artery.

Glossopharyngeal Nerve

Cranial nerve IX (see Chap. 34) is sensory to the posterior third of the tongue and pharynx and motor to the stylopharyngeus muscle.

Vagus Nerve

The vagus nerve is the principal parasympathetic nerve to the organs of the thorax and abdomen (see Chap. 34). The vagus nerve gives off certain important branches in the neck:

1. The ***pharyngeal branch*** participates in the formation of the pharyngeal plexus. This nerve passes to the pharynx by running superficial to the internal carotid artery but deep to the external carotid artery.
2. The ***superior laryngeal nerve*** passes deep to both the carotid arteries and divides into an ***internal laryngeal nerve***, which supplies sensory fibers to the larynx down to and including the vocal cords, and an ***external laryngeal nerve***, which supplies motor fibers to the cricothyroid and inferior constrictor muscles.
3. The ***recurrent laryngeal nerves*** were mentioned in Chapter 27 and will be mentioned again in Chapter 34.

Accessory Nerve (XI)

This nerve supplies motor fibers to the sternomastoid muscle and (after traversing the posterior triangle of the neck) to the trapezius muscle (see Chap. 34).

Hypoglossal Nerve (XII)

The hypoglossal nerve leaves the skull through the hypoglossal canal and then runs between the internal jugular vein and the internal carotid artery (see Chap. 34). It is joined by fibers from the first cervical nerve shortly after it leaves the skull. It then runs superficial to most structures of the area to supply the muscles of the tongue. Before it reaches the tongue, fibers from the first cervical nerve pass inferiorly in the submandibular triangle and join with fibers from the second and third cervical nerves to form the ***ansa cervicalis***, which supplies the strap muscles.

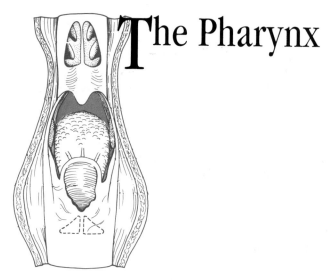

The Pharynx

EXTERIOR OF THE PHARYNX

The pharynx is the combined respiratory and alimentary passage that begins behind the nose at the base of the skull and descends to the cricoid cartilage. Its walls comprise layers of striated muscle covered with a tough sheet of fascia, the ***pharyngobasilar fascia***. The interior is lined by a mucous membrane of squamous epithelium. To visualize the configuration and extent of the pharynx, the reader is advised to review the base of the skull with particular attention to the ***pharyngeal tubercle*** of the occipital bone and the ***hamulus*** of the medial pterygoid plate (Fig. 29-1).

CONNECTIVE TISSUE STRUCTURE OF THE PHARYNX

The ***pharyngeal raphe*** is a band of connective tissue that runs in the posterior midline of the pharynx from the pharyngeal tubercle to the beginning of the esophagus just below the cricoid cartilage. This raphe provides the posterior attachment for the constrictor muscles that form the wall of the tube. The ***pterygomandibular raphe*** provides part of the anterior attachment of the superior constrictor muscle. The ***stylohyoid ligament***, which runs from the styloid process to the lesser cornu of the hyoid bone, provides some of the attachment of the middle constrictor muscle (Fig. 29-1).

The pharynx has four layers, listed below, from inside to outside:

1. The ***mucous membrane***
2. The ***pharyngobasilar fascia***, which forms a lining for the muscles
3. The ***constrictor muscles***
4. The ***buccopharyngeal fascia***, which surrounds the exterior of the pharynx

The ***retropharyngeal space***, a potential space, separates the buccopharyngeal fascia from the prevertebral fascia. Abscesses have been known to form in this space.

MUSCLES OF THE PHARYNX

There are six paired muscles of the pharynx: three of them are constrictors (superior, middle, and inferior), and three are longitudinal muscles. The

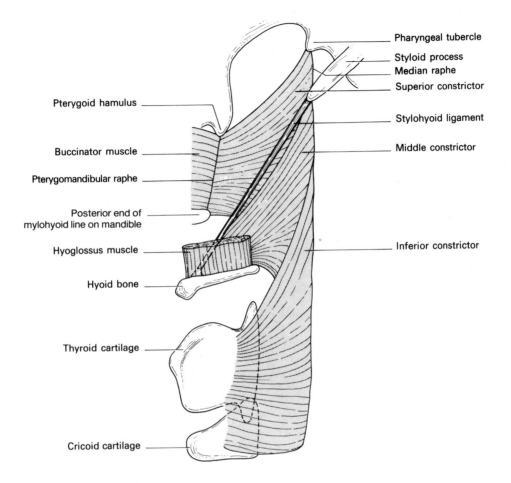

Pharyngeal tubercle
Styloid process
Median raphe
Superior constrictor
Stylohyoid ligament
Middle constrictor
Pterygoid hamulus
Buccinator muscle
Pterygomandibular raphe
Posterior end of mylohyoid line on mandible
Hyoglossus muscle
Hyoid bone
Inferior constrictor
Thyroid cartilage
Cricoid cartilage

Figure 29-1.
Muscles of the pharynx (left side).

latter three are the ***stylopharyngeus***, the ***palatopharyngeus***, and the ***salpingopharyngeus*** (Fig. 29-2). The constrictors surround the pharynx, and the upper border of each overlaps the lower border of the muscle above to a considerable extent. The constrictors are all attached posteriorly to the pharyngeal raphe. Thus the pharynx is constructed like a sleeve with overlapping muscle layers that extends from the base of the skull to the esophagus and is open anteriorly in three places where it connects with the nose, the mouth, and the larynx.

SUPERIOR CONSTRICTOR

The superior constrictor is attached to the pterygoid hamulus, the pterygomandibular raphe, the posterior end of the mylohyoid line of the mandible, and the side of the tongue (see Fig. 29-1). From these extensive anterior attachments the muscle fibers sweep backward to meet each other (one constrictor from each side) in the pharyngeal raphe posteriorly. Between the upper anterior attachment of the superior constrictor to the pterygoid hamulus and its posterior attachment to the pharyngeal tubercle, there is a space into which the ***auditory tube*** passes. The remainder of this space is covered by the ***levator palati*** and ***tensor palati*** muscles and by pharyngobasilar fascia (see Fig. 29-2).

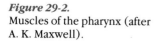

Figure 29-2.
Muscles of the pharynx (after
A. K. Maxwell).

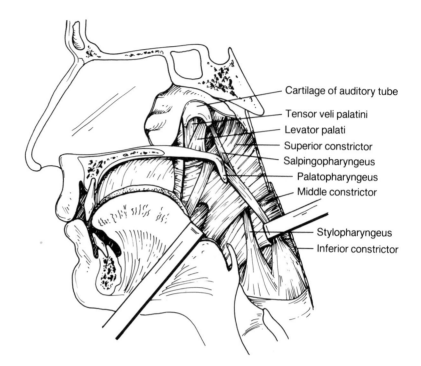

Cartilage of auditory tube
Tensor veli palatini
Levator palati
Superior constrictor
Salpingopharyngeus
Palatopharyngeus
Middle constrictor
Stylopharyngeus
Inferior constrictor

MIDDLE CONSTRICTOR

The middle constrictor is attached to the lower part of the stylohyoid ligament and the two horns (cornua) of the hyoid bone in front and to the
pharyngeal raphe posteriorly. Its anterior fibers lie deep to the hyoglossus
muscle. The upper fibers of the middle constrictor overlap the lower fibers
of the superior constrictor. This overlapping arrangement is repeated by
the inferior constrictor in relation to the middle constrictor. These overlapping muscular walls of the pharyngeal tube facilitate the passage of
food much as layered shingles on a roof facilitate the runoff of rain water
(see Fig. 29-1).

INFERIOR CONSTRICTOR

The inferior constrictor takes its anterior attachment from the thyroid and
cricoid cartilages, and its fibers, overlapping the lower part of the middle
constrictor, run posteriorly to be attached to the pharyngeal raphe (see
Fig. 29-1).

CLINICAL NOTE

*Close examination of the disposition of the muscle fibers of the inferior
constrictor will show that the lower fibers are arranged in a concentrical
circular manner, in continuity with the esophageal musculature,
whereas the upper fibers slant upward (see Figs. 29-1 and 29-3). A small
gap exists between the sloping and the horizontal fibers. The horizontal
fibers are given the separate name of* **cricopharyngeus,** *and if this muscle
goes into repeated spasm the mucosa of the pharynx can bulge outward
between the slanting and the horizontal fibers. This is called a* **pharyngeal (Zenker's) diverticulum.**

Innervation of the Pharynx.   The nerves that supply the pharynx form a *pharyngeal plexus* on the surface of the pharynx. The motor fibers come from the accessory nerve via the pharyngeal and superior laryngeal branches of the vagus nerve. The glossopharyngeal nerve supplies the stylopharyngeus muscle and receives sensory fibers from the pharyngeal lining (cranial nerve IX; see Chap. 34).

SPACES OF THE PHARYNX

Above the superior border of the superior constrictor the pharyngobasilar fascia blends with the buccopharyngeal fascia (external) and with the mucosa to form the thin wall of the *pharyngeal recess* (fossa of Rosenmüller).

> The glossopharyngeal nerve enters into the space between the superior and middle constrictor along with the stylopharyngeus.
> The internal laryngeal nerve passes into a space between the middle and inferior constrictors.
> The recurrent laryngeal nerve passes upward into the larynx deep to the inferior fibers of the inferior constrictor.

INTERIOR OF THE PHARYNX

MEDIAN SECTION OF THE PHARYNX

The study of a median section of the pharynx (Fig. 29-4) will show that the *nasopharynx* lies posterior to the nose and the *oropharynx* lies posterior to the mouth. The *laryngopharynx* is posterior to the larynx. The

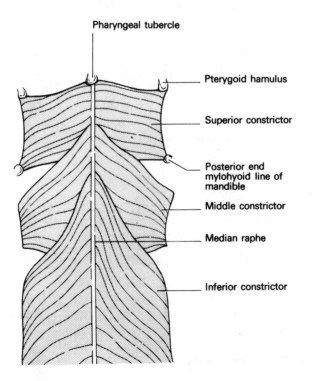

Pharyngeal tubercle

Pterygoid hamulus

Superior constrictor

Posterior end mylohyoid line of mandible

Middle constrictor

Median raphe

Inferior constrictor

Figure 29-3.
Muscles of the pharynx (posterior view).

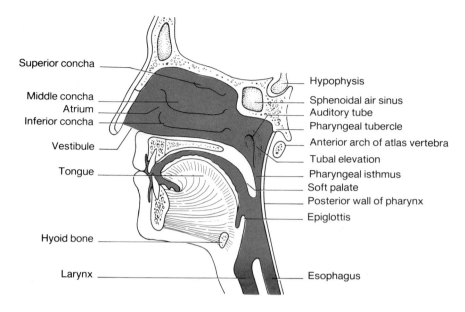

Superior concha
Middle concha
Atrium
Inferior concha
Vestibule
Tongue
Hyoid bone
Larynx

Hypophysis
Sphenoidal air sinus
Auditory tube
Pharyngeal tubercle
Anterior arch of atlas vertebra
Tubal elevation
Pharyngeal isthmus
Soft palate
Posterior wall of pharynx
Epiglottis
Esophagus

Figure 29-4.
Sagittal section of head showing the right lateral wall of the nose.

pharyngeal isthmus joins the oropharynx to the nasopharynx. It is closed during swallowing by the *soft palate* elevating posteriorly to meet a special thickening of the posterior wall of the pharynx. This prevents food and drink from getting up your nose.

Nasopharynx

The nose opens into the nasopharynx through the posterior *choanae*, one on each side of the nasal septum (Fig. 29-4). The opening of the *auditory tube* (pharyngotympanic, Eustachian tube) is on the lateral wall of the nasopharynx at the level of the palate (Fig. 29-4). Posterior and superior to the opening of the auditory tube is the tubal elevation (torus, eminence), which tails downward from its posterior end as the *salpingopharyngeal fold*. This fold contains the salpingopharyngeus muscle deep to its mucosa. The *tubal tonsil* is a collection of lymphoid tissue in the submucosa around the tubal opening.

The *pharyngeal recess*, as seen from the interior of the pharynx, is a fossa behind the tubal elevation. The *pharyngeal tonsil* is a collection of lymphoid tissue in the posterior and superior wall of the nasopharynx (see Fig. 29-8).

CLINICAL NOTE

The pharyngeal tonsil is called the adenoid by clinicians. Adenoidal enlargement from acute or chronic infections can occlude the opening of the auditory tube and pharyngeal isthmus. The former can cause hearing impairment and the latter a "nasal" voice and difficulties in the acquisition of normal speech in young children.

Oropharynx

The oropharynx is separated from the oral cavity (mouth) by the *oropharyngeal isthmus*. The passage itself is called the *fauces* and is bounded superiorly by the soft palate, inferiorly by the base of the tongue, and laterally by the *pillars of the fauces* (Fig. 29-5).

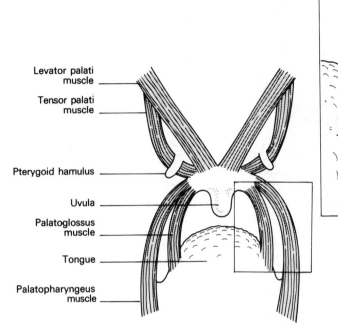

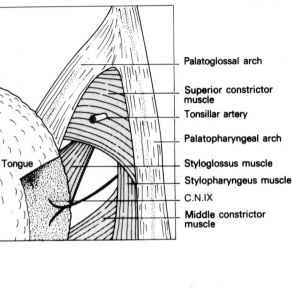

Levator palati muscle

Tensor palati muscle

Pterygoid hamulus

Uvula

Palatoglossus muscle

Tongue

Palatopharyngeus muscle

Tongue

Palatoglossal arch

Superior constrictor muscle

Tonsillar artery

Palatopharyngeal arch

Styloglossus muscle

Stylopharyngeus muscle

C.N.IX

Middle constrictor muscle

The pillars consist of two arches. Anteriorly the ***palatoglossal*** arch (fold) joins the soft palate to the sides of the tongue. The arch is formed by the ***palatoglossus*** muscle, covered with mucosa. The posterior arch is the ***palatopharyngeal*** arch, which runs from the soft palate downward to blend with the wall of the pharynx. The arch is formed by the ***palatopharyngeus*** muscle, covered by mucosa (Fig. 29-5).

Palatine Tonsil (commonly known as ***the*** tonsil). The palatine tonsil (Fig. 29-6) is a collection of lymphoid tissue, covered by epithelium, which is located between the palatoglossal and palatopharyngeal arches.

CLINICAL NOTE

*The palatine tonsil may become inflamed and require surgical removal if infections in it become a recurrent problem. An abscess located in the adjacent connective tissue is called a **peritonsillar abscess** or **quinsy**.*

Tonsillar Bed. The bed in which the tonsil lies is situated between the palatoglossal and palatopharyngeal arches. Its floor is the wall of the pharynx (Fig. 29-6). The upper part of the floor is the superior constrictor muscle, and the lower part of the floor consists of the superior fibers of the middle constrictor muscle. Between these two muscles the ***styloglossus*** passes forward to be inserted into the side of the tongue, and the ***stylopharyngeus*** passes inferiorly to blend with the middle constrictor. The ***glossopharyngeal nerve*** runs with the stylopharyngeus to reach the posterior third of the tongue, to which it supplies sensory fibers for common sensation and taste (cranial nerve IX; see Chap. 34).

Just lateral to the middle constrictor the ascending palatine artery, a branch of the facial artery, gives off the ***tonsillar*** artery, which passes through the superior constrictor to serve as the chief blood supply to the tonsil. The external palatine vein lies in the upper part of the tonsillar bed.

Figure 29-5.
Pillars of the fauces (posterior view). **Inset.** The bed of the right palatine tonsil.

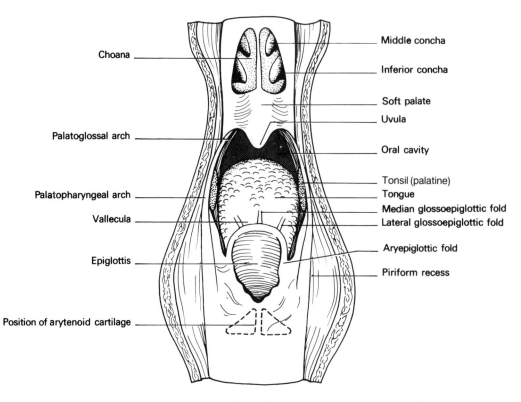

Choana

Palatoglossal arch

Palatopharyngeal arch

Vallecula

Epiglottis

Position of arytenoid cartilage

Middle concha

Inferior concha

Soft palate

Uvula

Oral cavity

Tonsil (palatine)
Tongue
Median glossoepiglottic fold
Lateral glossoepiglottic fold

Aryepiglottic fold

Piriform recess

Figure 29-6.
Posterior view of the openings into the pharynx. (Posterior wall of the pharynx has been removed.)

Damage to either of these vessels during tonsillectomy can cause serious hemorrhage. The ***internal carotid artery*** lies only 2 cm posterolateral to the tonsillar bed. It can be seen that tonsillectomy can be a very dangerous procedure, not to be undertaken for trivial indications or by untrained surgeons (Fig. 29-7).

Lingual Tonsil. Much of the bumpy irregular surface of the posterior third of the tongue is submucosal lymphoid tissue, the lingual tonsil. It

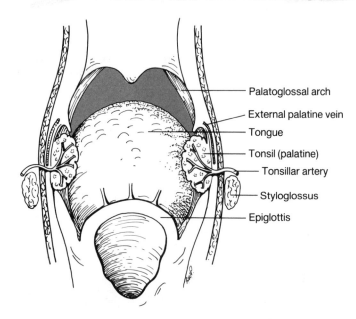

Palatoglossal arch

External palatine vein

Tongue

Tonsil (palatine)

Tonsillar artery

Styloglossus

Epiglottis

Figure 29-7.
Tonsillar bed and artery (posterior view).

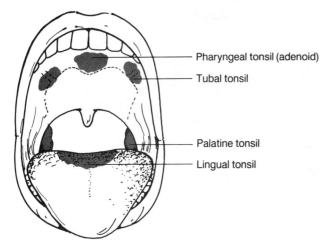

Pharyngeal tonsil (adenoid)

Tubal tonsil

Palatine tonsil

Lingual tonsil

Figure 29-8.
Waldeyer's ring of protection.

can now be seen that the entrance to the pharynx is surrounded by a ring of lymphoid tissue: the **adenoids**, **tubal tonsils**, and **palatine** and **lingual** tonsils. This ring is known as **Waldeyer's ring** (Fig. 29-8). It may be speculated, teleologically, that the purpose of Waldeyer's ring is to guard the entrance to the respiratory passage.

Laryngeal Pharynx

The part of the pharynx posterior to the larynx is the laryngeal pharynx, or laryngopharynx. The opening of the larynx (**aditus**) is in the anterior wall of the laryngeal pharynx. It is bounded anteriorly by the **epiglottis** and laterally by the **aryepiglottic** folds, which join the arytenoid cartilages (posteriorly) to the epiglottis (anteriorly). Lateral to the aryepiglottic fold but medial to the lamina of the thyroid cartilage is a space known as the **piriform recess** (see Fig. 29-6). The piriform recess is of clinical importance because it can be the primary site for a carcinoma that may remain very occult and present symptoms only when there has been gross enlargement of deep cervical lymph nodes by metastases.

The **epiglottis**, an elastic cartilage of the larynx, is joined to the posterior aspect of the tongue by the median and lateral **glossoepiglottic folds**. The depression between the median and lateral glossoepiglottic folds is the **vallecula** (see Fig. 29-6). Foreign bodies (*e.g.*, fish bones) have been known to become lodged in the vallecula, where they can cause much pain and inflammation. The vallecula, like the piriform fossa, may hide a carcinoma.

Soft Palate

In swallowing, the soft palate forms a mobile curtain that is elevated posteriorly to impinge on the thickened muscle of the posterior pharyngeal wall (sphincter of Whillis), thereby blocking the pharyngeal isthmus and preventing regurgitation of swallowed material into the nose. The soft palate is attached anteriorly to the posterior edge of the hard palate (see Fig. 29-4). Its posteroinferior edge hangs free and is prolonged downward at its middle as the **uvula**.

The substance of the soft palate is made of dense fibrous tissue like an aponeurosis and supplemented by the insertion of the muscles described below (see also Fig. 29-2).

Tensor Veli Palatini (Tensor Palati)

This narrow muscle is attached to the outer wall of the auditory tube and the scaphoid fossa of the pterygoid process of the sphenoid bone. It passes downward along the lateral side of the medial pterygoid plate and then hooks medially around the pterygoid hamulus to enter the inferior aspect of the palatine aponeurosis. It tenses the soft palate and is supplied by the mandibular division of the trigeminal nerve.

Levator Veli Palatini (Levator Palati)

This is another narrow muscle that runs from the tip of the petrous temporal bone, just anterior to the inferior opening of the carotid canal to the superior surface of the palatine aponeurosis.

The ***palatoglossus*** and the ***palatopharyngeus*** have already been described. They pull the pillars of the fauces together during swallowing, and the palatopharyngeus helps to elevate the pharynx during swallowing (see Fig. 29-5).

All of these muscles, except for the tensor palati, are supplied by the pharyngeal plexus.

The Nose and Pterygopalatine Fossa

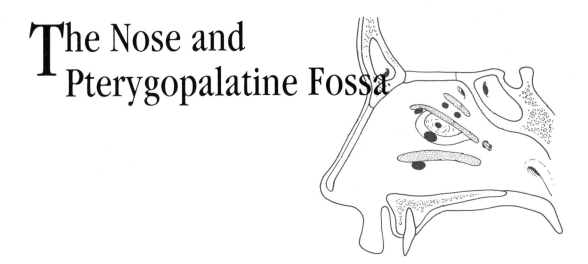

The nose is divided into left and right *nares* by the *nasal septum*. Each naris is divisible into *olfactory* and *respiratory* areas. The olfactory area is located at the roof of the nose, on the lateral wall above the superior nasal concha and on the corresponding area of the nasal septum (see cranial nerve I; Chap. 34). The roof of the nose is formed by the cribriform plate of the ethmoid bone and its covering mucosa. The respiratory area is that part of the nasal cavity between the olfactory area and the floor of the nasal cavity. Its chief function is to warm and moisten inspired air. The floor of the nose is formed by the hard palate, which consists of the palatine processes of the maxillae in front and the horizontal plates of the palatine bones behind (Fig. 30-1). The entrance to the nose is the *vestibule*, which is lined by skin. Posterosuperior to the vestibule is the atrium, or true nasal cavity, which is lined by mucous membrane (respiratory epithelium).

The anterior part of the skeleton of the nose is made of elastic cartilage, which permits mobility of this part of the nose. The rest of the nose is walled by bone.

NASAL SEPTUM

The nasal septum consists of an anterior cartilaginous part and a larger posterior part made up of several bones (Fig. 30-1). The whole septum is covered by very vascular mucosa.

BONES

The bones and parts of bones involved in the composition of the nasal septum are the *vomer* and the *perpendicular plate* of the *ethmoid* and small projections from the maxillary, palatine, frontal, sphenoid, and nasal bones.

The *ethmoid* is a peculiarly shaped bone which is so fragile that parts of it usually break off in the skulls available for study. It may be described as having the appearance of two boxes joined at the top by a perforated horizontal plate, the cribriform plate, and being separated from each other by a perpendicular plate hanging down from the horizontal plate (Figs. 30-1, 30-2, and 30-4). The perpendicular plate forms part of the nasal septum and is continuous above the cribriform plate as the *crista galli*, which was observed in the floor of the anterior cranial fossa.

30

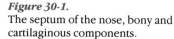

Figure 30-1.
The septum of the nose, bony and
cartilaginous components.

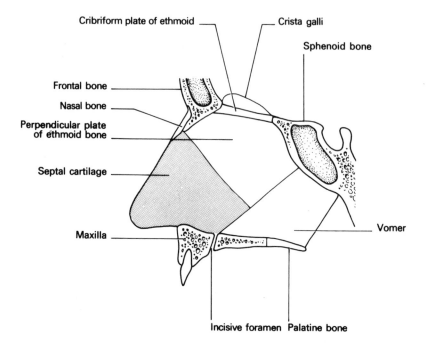

The boxes that form part of the medial walls of the orbit contain the
ethmoidal labyrinth and form part of the lateral walls of the nares. The
labyrinth contains the ***ethmoidal air sinuses***. Projecting medially from
the two boxes are curved bony plates of variable sizes that form the upper
and middle ***conchae***.

The ***vomer*** is a flattened plate of bone located in the posteroinferior
aspect of the nasal septum. It articulates with the perpendicular plate of
the ethmoid and the septal cartilage (see Fig. 30-1). The rest of the vomer
articulates with the body of the sphenoid bone.

Figure 30-2.
Skeletal structures of the right lateral
wall of the nose.

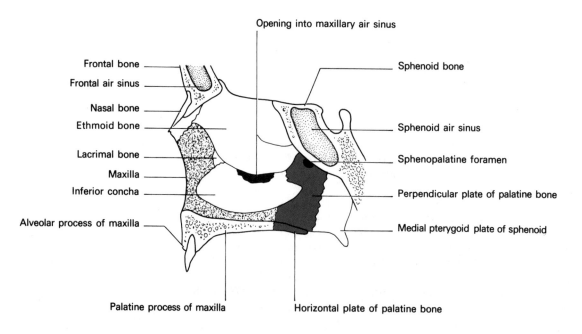

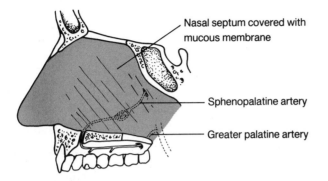

Nasal septum covered with
mucous membrane

Sphenopalatine artery

Greater palatine artery

Figure 30-3.
Blood supply of the mucous membrane
covering the nasal septum.

MUCOUS MEMBRANE

The nasal septum is covered by very vascular and sensitive mucous membrane. The olfactory portion of the septum is supplied by twigs of the *olfactory nerve* (I; see Chap. 34). The anterior part of the septum is supplied by the *anterior ethmoidal nerve*, which is a branch of the *nasociliary* nerve, whereas the rest of the septum is supplied primarily by the *nasopalatine branch* of the maxillary division of the trigeminal nerve (V_2; see Chap. 34). This nerve ends anteriorly by passing through the incisive foramen of the maxilla to supply the anterior parts of the upper gums and the hard palate.

The blood supply of the mucous membrane of the septum comes from the maxillary artery. One branch, the *sphenopalatine* artery, supplies the septum from above and behind, whereas another branch, the *greater palatine artery*, reaches the septum by passing from below the palate through the incisive foramen. These two arteries anastomose in the anteroinferior part of the septum (Little's area) (Fig. 30-3).

CLINICAL NOTE

Little's area is a common site for nose bleeds (epistaxis).

LATERAL WALL OF THE NOSE

Posterior to the vestibule the nose exhibits three lateral scroll-like elevations: the *superior, middle,* and *inferior conchae* (Figs. 30-4 and 30-5). Each concha overhangs a *meatus* named superior, middle, and inferior, respectively. The superior and middle conchae are the mucosa-covered projections from the ethmoid bone, described above; the *inferior concha* is a separate bone that articulates mainly with the maxilla.

Sphenoethmoidal Recess. This is the area superior to the superior concha. The *sphenoidal air cells* open into it (Fig. 30-5).

Superior Meatus. This space on the lateral wall of the nose is overhung by the superior concha (turbinate). The *posterior ethmoidal air cells* open into it (Fig. 30-5).

Middle Meatus. The middle meatus is overhung by the middle concha, and there are some important openings in relation to it (Fig. 30-5):

1. The *sphenopalatine foramen* is in the lateral wall of the nose just posterior to the middle concha. Vessels and nerves from the

Figure 30-4.
The ethmoid bone in coronal section. The dotted line represents a continuation of the medial wall of the orbit.

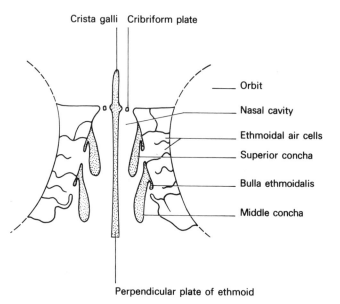

Crista galli   Cribriform plate

— Orbit

Nasal cavity

Ethmoidal air cells

Superior concha

Bulla ethmoidalis

Middle concha

Perpendicular plate of ethmoid

pterygopalatine fossa travel through it to supply the lateral wall and septum of the nose (see Fig. 30-5). This foramen is covered by mucosa so that it does not appear as an opening in the living. It has some clinical importance because local anesthetic applied to the surface of the mucosa over it will succeed in anesthetizing most of the lining of the nose.

2. Bulging into the middle meatus is the rounded ***ethmoidal bulla*** that contains the openings of the ***middle ethmoidal air cells.***
3. Below the ethmoidal bulla is a groove, the ***hiatus semilunaris***. The ***infundibulum***, draining the ***frontal sinus***, opens into the anterior end of the hiatus semilunaris.
4. The large ***maxillary air sinus*** opens into the lowest part of the semilunar hiatus.

Figure 30-5.
Lateral wall of the nose with the conchae removed.

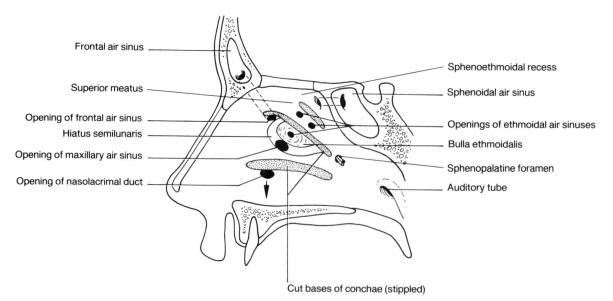

Frontal air sinus

Superior meatus

Opening of frontal air sinus
Hiatus semilunaris

Opening of maxillary air sinus

Opening of nasolacrimal duct

Sphenoethmoidal recess

Sphenoidal air sinus

Openings of ethmoidal air sinuses

Bulla ethmoidalis

Sphenopalatine foramen

Auditory tube

Cut bases of conchae (stippled)

Inferior Meatus. The inferior meatus is the most spacious, and it is overhung by the inferior concha. The ***nasolacrimal duct*** opens into the inferior meatus. The lateral wall of the inferior meatus is formed by the relatively thin medial wall of the maxillary air sinus.

Nerve Supply. The lateral wall of the nose is supplied by branches of the ophthalmic (V_1) and maxillary (V_2) divisions of the trigeminal nerve (see Chap. 34).

The ***infraorbital nerve*** supplies the vestibule; the ***anterior ethmoidal*** branch of the nasociliary nerve supplies the lateral wall and the anterior part of the septum. The ***sphenopalatine*** nerve and ***superior nasal*** branches from the maxillary nerve supply the remainder of the septum and lateral wall of the nose (see Chap. 34).

PARANASAL SINUSES

The paranasal sinuses are lined by ciliated mucous membrane and communicate with the nose (Fig. 30-6). The healthy sinuses contain air, and their location can be shown by transilluminating them by means of a pencil flashlight placed in the mouth or by means of the dark contrast of air-containing spaces on an x-ray.

Maxillary Sinus. The maxillary air sinus is located in the maxilla, which forms all of its walls except for a significant portion of its medial wall where the large opening in the maxilla is encroached upon by the inferior con-

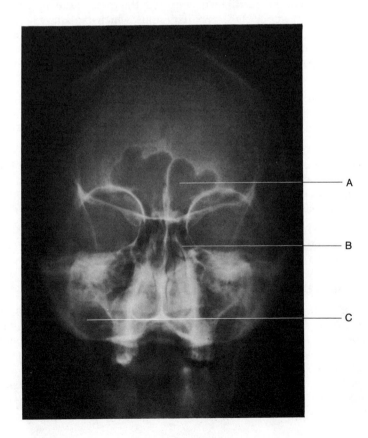

Figure 30-6.
Radiograph showing paranasal sinuses: Left frontal sinus (**A**); left ethmoidal sinus (**B**); right maxillary sinus (**C**).

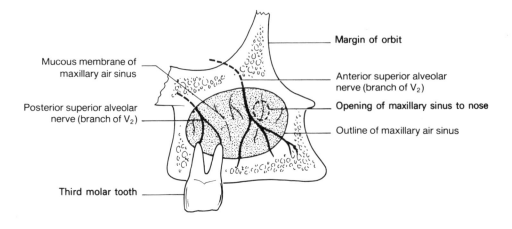

Mucous membrane of
maxillary air sinus

Posterior superior alveolar
nerve (branch of V₂)

Third molar tooth

Margin of orbit

Anterior superior alveolar
nerve (branch of V₂)

Opening of maxillary sinus to nose

Outline of maxillary air sinus

Figure 30-7.
Lateral view of the right maxillary air
sinus with the bony component of the
wall removed. The nerves are in the
lateral wall of the sinus.

cha. Thus the actual opening into the hiatus semilunaris (in the middle
meatus) is quite small and inappropriately placed near the roof of the sinus
(see Fig. 30-5). This placement of the sinus opening prevents its drainage
by gravity; this is one of the factors that makes inflammation of the sinus
so common.

The infraorbital nerve, the continuation of the maxillary nerve, travels in the superior wall of the sinus, between the bone and the mucous
membrane and gives off the posterior, middle, and anterior ***superior alveolar*** nerves that travel between bone and mucosa to the gums and teeth.

Inspection of the maxilla will show that the roots of the upper teeth
are very close to the floor of the maxillary sinus, which explains why sinusitis may cause toothache and why infections of the tooth sockets or dental
trauma may involve the maxillary sinus. Extraction of a tooth may result in
a fistula between the mouth and the sinus, and one's lunch could wind up
in the sinus (Fig. 30-7).

Frontal Sinus.   The frontal air sinus lies within the frontal bone and
drains into the anterior aspect of the hiatus semilunaris in the middle meatus of the nose (see Fig. 30-5).

Ethmoidal Sinuses.   The ethmoidal air sinuses form the labyrinth of the
ethmoid bone and consist of groups of air cells separated from the nose
and from the orbit by very thin plates of bone. They drain into the nose by
three different openings and are therefore divided into anterior, middle,
and posterior sinuses. This division is very variable (see Fig. 30-5).

Sphenoidal Sinus.   The sphenoidal air sinus occupies the body of the
sphenoid bone inferior to the sella turcica. It drains into the sphenoethmoidal recess (see Fig. 30-5).

CLINICAL NOTE

*Infections from the nasal cavity may spread into the paranasal sinuses.
The openings of the sinuses into the nose are quite small, and any swelling
of the mucous membrane of the nose will obstruct their drainage. If bacteria grow in the obstructed sinuses, pus is produced and the painful condition of sinusitis results. The maxillary sinus is poorly equipped to drain
itself because the position of its opening at the highest point in the sinus
defies gravity.*

MAXILLARY NERVE AND PTERYGOPALATINE FOSSA

The reader should review this area on the skull. The palatine bone (one on each side) is shaped like the letter L, in cross-section. It has a perpendicular plate that forms the medial wall of the pterygopalatine fossa and a horizontal plate that forms the posterior 1 cm of the hard palate and floor of the nose (see Fig. 30-2).

The ***pterygopalatine fossa*** is located between the palatine bone medially, the pterygoid process of the sphenoid posteriorly, and the maxilla anteriorly. It has several openings, the largest of which is the pterygomaxillary fissure (Fig. 30-8A) through which it communicates with the infratemporal region. The pterygopalatine fossa contains part of the maxillary nerve, the pterygopalatine ganglion, and the third (terminal) division of the maxillary artery (Fig. 30-8B).

The maxillary nerve enters the pterygopalatine fossa through the foramen rotundum and leaves by passing through the inferior orbital fissure to become the ***infraorbital*** nerve. In the bony orbit it lies deep to the orbital periosteum so that it does not become a true content of the orbit. It enters the infraorbital groove and canal, changes its name to ***infraorbital nerve***, and emerges at the infraorbital foramen to supply the skin of the cheek, upper lip, side of the nose, and lower eyelid. In the infraorbital

Figure 30-8.
A. Skeletal features of the right pterygopalatine fossa, as seen through the pterygomaxillary fissure.
B. Maxillary division (V₂) of the right trigeminal nerve and the right pterygopalatine ganglion.

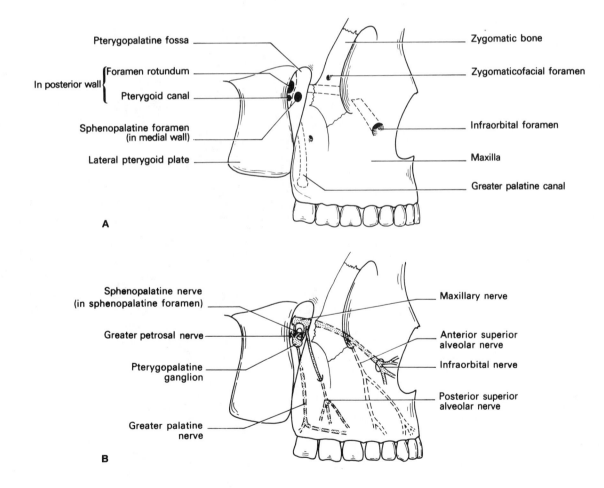

groove it gives off ***superior dental nerves*** to the teeth of the upper jaw (see cranial nerve V_2; Chap. 34) (Fig. 30-8B).

In the pterygopalatine fossa the (parasympathetic) ***pterygopalatine ganlion*** is attached to the nerve, and it and the maxillary nerve give off branches to the palate, pharynx, nose, paranasal sinuses, and zygoma (see Chap. 34).

MAXILLARY ARTERY

The terminal part of the ***maxillary artery*** enters the pterygopalatine fossa by passing through the pterygomaxillary fissure from the infratemporal fossa (see Fig. 24-8). It continues to become the ***infraorbital artery***. The maxillary artery and the infraorbital artery have branches that bear names similar to those of the branches of the maxillary nerve and are distributed in a similar manner.

The Mouth, Teeth, and Tongue

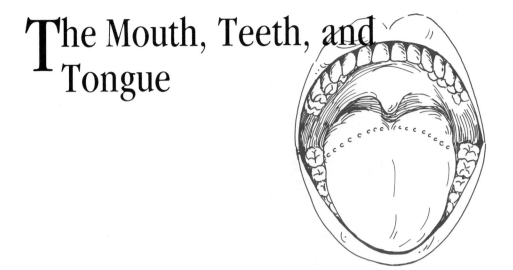

The mouth needs no introduction and little definition. It is the space (also called oral cavity) between the lips in front and the oropharyngeal isthmus behind. Its roof is the palate (hard and soft) and its floor is the floor of the mouth, which is essentially the mylohyoid muscle covered by mucous membrane with some salivary gland tissue in the submucosal plane.

The lips are separated from the gums by the alveololabial sulcus, and the gums bear the teeth which, in adults, consist of a set of eight teeth to each quadrant. The two incisors, one canine, two premolars, and three molars are preceded by the **primary dentition** of the child. The fully erupted primary dentition contains two incisors, one canine, and two molars per quadrant. The teeth of the primary dentition are called deciduous teeth. Each tooth has a crown, a neck, and a root. The crown is covered by a hard and shiny ectoderm derivative, the enamel, and the rest of the tooth is a type of bone. Teeth are living structures with a blood and nerve supply that enter through the apices of the roots.

CLINICAL NOTE

The fact that teeth are extremely sensitive requires no elaboration. Often toothache will be referred to the tonsillar region or to the ear. This is referred pain, which becomes self explanatory if the sensory nerve supply of these structures is remembered.

THE MOUTH

The **lips** are covered by pink skin on the exterior and by mucous membrane on the interior. The line of junction of the pink skin and the mucous membrane is called the **vermilion**. The depression in the center of the upper lip is the **philtrum**. Each lip contains muscle fibers of the orbicularis oris and buccinator muscles, and the submucosa contains labial salivary glands. The center of each lip is bound to the gum by a frenulum.

The lateral walls of the oral cavity are formed by the **buccinator** muscle which is attached to the alveolar processes of the maxilla and mandible opposite the last three teeth. Posteriorly the buccinator meets the superior constrictor of the pharynx at the pterygomandibular raphe. The buccinator is supplied by the facial nerve and is the muscle of blowing and sucking.

The inside of the cheeks receives its sensory nerve supply from the **buccal** branch of the **mandibular nerve** (V_3) and from the **infraorbital nerve** (V_2). The cheeks contain small submucosal salivary glands.

31

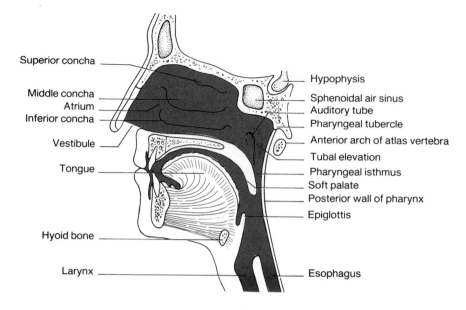

Superior concha

Middle concha
Atrium
Inferior concha

Vestibule

Tongue

Hyoid bone

Larynx

Hypophysis

Sphenoidal air sinus
Auditory tube
Pharyngeal tubercle
Anterior arch of atlas vertebra
Tubal elevation
Pharyngeal isthmus
Soft palate
Posterior wall of pharynx
Epiglottis

Esophagus

Figure 31-1.
Sagittal section of the head showing
the right lateral wall of the nose.

THE PALATE

The soft palate was described in Chapter 29. The hard palate forms the roof of the mouth and the floor of the nasal cavity. It is made up on each side by the palatine or horizontal processes of the maxilla and posteriorly by the horizontal plate of the palatine bone. The hard palate extends posteriorly to just beyond the last molar tooth, and the soft palate hangs from its posterior border. Laterally and anteriorly the hard palate blends with the alveolar process of the maxilla. Examination of the bony palate will show many little pits or depressions which are occupied by submucosal salivary glands in living persons. Note that the mucous membrane of the hard palate is firmly bound to the periosteum so that it cannot move on the bony surface. This allows the palate to participate in mastication with the tongue pressing and grinding food against it (Fig. 31-1).

The sensory nerve supply of the hard palate comes from the palatine branches of the maxillary division (V₂) of the trigeminal nerve (see Chap. 34).

THE TONGUE

The tongue is a mobile muscular organ that can vary greatly in shape. Its chief functions are to push food into the pharynx and to form words and modulate sounds during phonation.

Morphologically the tongue consists of a bag of skin derived from two different sources and the muscles that fill the bag. The skin of the anterior two thirds of the tongue comes from the first branchial arch and the skin of the posterior one third from the third branchial arch. Their nerve supply, to be detailed below, reflects those derivations. The muscles of the tongue have a different origin and migrate to their lingual site from the occipital region.

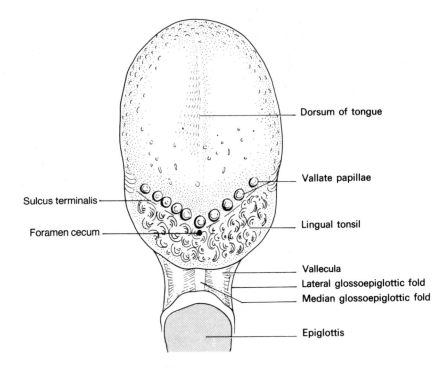

Figure 31-2.
Tongue and epiglottis (superior view).

Dorsum of tongue

Vallate papillae

Sulcus terminalis

Lingual tonsil

Foramen cecum

Vallecula
Lateral glossoepiglottic fold
Median glossoepiglottic fold

Epiglottis

DORSUM OF THE TONGUE

The dorsum of the tongue is divided by a sulcus terminalis into **palatine** and **pharyngeal** parts. The former occupies the anterior two thirds of the dorsum (Figs. 31-2 and 31-3).

Sulcus Terminalis.   The sulcus terminalis is a V-shaped groove, with its apex directed posteriorly. The **vallate papillae**, which contain a large number of taste buds, are arranged along the sulcus terminalis. At the apex of the V the **foramen cecum** marks the embryologic remnant of the thyroglossal duct.

Palatine Part of the Tongue.   The anterior two thirds of the tongue normally comes into contact with the hard palate and receives its somatic sen-

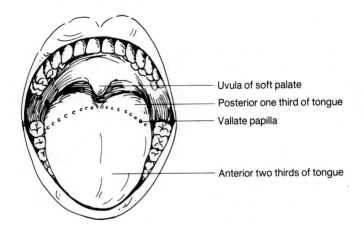

Uvula of soft palate
Posterior one third of tongue
Vallate papilla

Anterior two thirds of tongue

Figure 31-3.
Anterior view of the oral cavity and tongue.

sory supply from the ***lingual*** nerve and its special sensory (sense of taste) innervation from the ***chorda tympani*** branch of the facial nerve (cranial nerve VII; see Chap. 34). This surface of the tongue contains fungiform and filiform papillae that bear taste buds.

Pharyngeal Part of the Tongue. The pharyngeal part of the dorsum of the tongue (posterior third) consists of a large collection of lymphoid tissue (lingual tonsil) covered by mucous membrane. Its sensory innervation for somatic sensation and special sense of taste is from the ***glossopharyngeal*** nerve (see Chap. 34).

Note: Two things must be emphasized in connection with the sensory innervation of the tongue. The tip of the tongue is used to palpate many intraoral structures, and it must have occurred to the reader that it grossly exaggerates the size of things. Second, the reader must remember that taste refers to the appreciation of the tastes of bitter, sweet, salt, and sour only. All other "taste," as appreciated by the gourmet, is really olfaction (smell, aroma).

Inferior Surface of the Tongue

The striking feature on the inferior surface of the tongue is the ***frenulum***, a midline fold of mucous membrane that connects the tongue to the floor of the mouth. On either side of the frenulum is a ***deep lingual vein***, which is easily visible through the mucous membrane in living persons. The duct of the submandibular gland, which runs deep to the mucosa of the floor of the mouth, empties on a papilla located lateral to the frenulum.

Most of the above-mentioned features of the mouth and tongue are visible, and the reader is urged to examine them in front of a mirror.

MUSCLES OF THE TONGUE

The muscles of the tongue originate in the embryonic occipital region and migrate to their lingual location dragging the hypoglossal nerve with them. The pathway of migration can be reconstructed by studying the course and relations of the hypoglossal nerve. The muscles are arranged around a median fibrous septum that forms a partition from the middle of the body of the hyoid bone to the dorsum of the tongue. All the muscles of the tongue are supplied by the ***hypoglossal*** nerve except the palatoglossus, which is supplied by the pharyngeal plexus of nerves.

Extrinsic Muscles

The extrinsic muscles of the tongue originate outside the tongue but insert into it (Fig. 31-4). The ***hyoglossus*** is attached inferiorly and posteriorly to the body and greater cornu (horn) of the hyoid bone and passes upward and forward to the side and dorsum of the tongue. The ***genioglossus*** runs from the mental spine of the mandible to the tongue, extending from its posterior part to its tip. The ***palatoglossus*** runs from the soft palate to blend with the muscles on the side of the tongue, and the ***styloglossus*** runs from the styloid process along the side of the tongue toward its tip.

In addition, the ***geniohyoid*** muscle runs from the mental spine of the mandible to the hyoid bone and assists in pulling the hyoid bone upward and forward. This muscle is supplied by fibers from the first ***cervical*** nerve that run with the hypoglossal nerve.

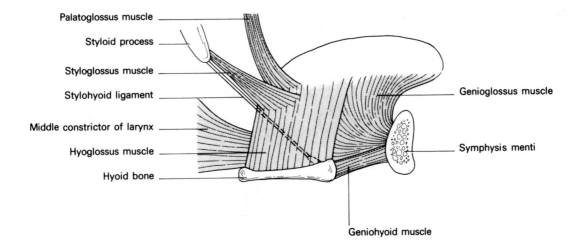

Palatoglossus muscle

Styloid process

Styloglossus muscle

Stylohyoid ligament

Middle constrictor of larynx

Hyoglossus muscle

Hyoid bone

Genioglossus muscle

Symphysis menti

Geniohyoid muscle

Figure 31-4.
Muscles of the tongue (lateral view).

Movements of the tongue such as protrusion or its superoposterior movement in swallowing are accompanied by similar movements of the hyoid bone. This statement can be easily verified by placing a finger on the hyoid bone and moving the tongue.

Intrinsic Muscles of the Tongue

Within the bag of skin that forms the shell of the tongue, an ample collection of striated muscle fibers runs in three directions: vertically, transversely, and longitudinally. These muscles can alter the shape and length of the tongue. The shape of the tongue can be changed quite extensively with much individual variation. The ability to roll the tongue into the shape of a tube is a dominant genetic trait.

NERVES OF THE TONGUE

Three main nerves supply the tongue: one motor and two somatic sensory, with some fibers of special sense (taste) being added (Fig. 31-5).

Hypoglossal Nerve. The hypoglossal nerve was seen in the neck, where it first appeared between the internal jugular vein and the internal carotid artery to pass anteriorly superficial to the external carotid artery and some of its branches. Then it disappears deep to the mylohyoid muscle to run anteriorly on the lateral surface of the hyoglossus muscle between the hyoid bone and the submandibular duct. It ends by entering the genioglossus muscle and being distributed to the intrinsic muscles of the tongue, the styloglossus and genioglossus (Fig. 31-5A).

CLINICAL NOTE

*Damage to the hypoglossal nerve will paralyze one half of the tongue's musculature. Attempted protrusion of the tongue will deviate the protruded tongue to the **paralyzed side**.*

Glossopharyngeal Nerve. The glossopharyngeal nerve swings around the stylopharyngeus muscle to enter the tongue by passing deep to the

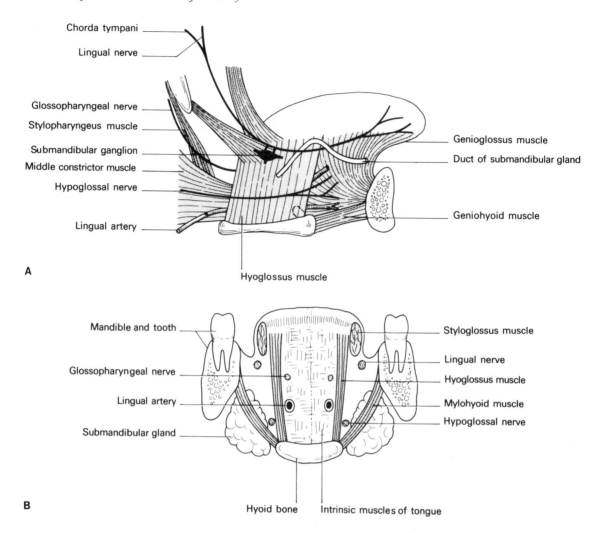

Chorda tympani

Lingual nerve

Glossopharyngeal nerve

Stylopharyngeus muscle

Submandibular ganglion

Middle constrictor muscle

Hypoglossal nerve

Lingual artery

Genioglossus muscle

Duct of submandibular gland

Geniohyoid muscle

A

Hyoglossus muscle

Mandible and tooth

Glossopharyngeal nerve

Lingual artery

Submandibular gland

Styloglossus muscle

Lingual nerve

Hyoglossus muscle

Mylohyoid muscle

Hypoglossal nerve

B

Hyoid bone   Intrinsic muscles of tongue

Figure 31-5.
Tongue: **A.** Lateral view of muscles, nerves, and vessels (right side). **B.** Coronal section.

hyoglossus muscle. It supplies fibers of taste and somatic sensation to the posterior third of the tongue, upper pharynx, and epiglottis (Fig. 31-5A).

Lingual Nerve. The lingual nerve, a branch of the mandibular division of the trigeminal nerve, runs between the medial pterygoid muscle and the mandible and then passes just medial to the third molar tooth along the side of the tongue, on the hyoglossus muscle, to the surface of the tongue. On the hyoglossus muscle it "double-crosses" the submandibular duct (at first superficial and then deep to it) (Fig. 31-5A). The lingual nerve supplies the anterior two thirds of the tongue with somatic sensory fibers. The special sense of taste is carried from the anterior two thirds of the tongue in the chorda tympani nerve, a component of cranial nerve VII that meets and travels with the lingual nerve.

In addition, the lingual nerve carries preganglionic parasympathetic fibers from the chorda tympani which, after synapsing in the ***submandibular ganglion***, supply secretomotor fibers to the submandibular and sublingual salivary glands. The nerves of the tongue are summarized in tabular form on the following page.

Table 31-1.
Nerves of the Tongue

Nerve	Modality	Region of Tongue
XII	Motor	All intrinsic and extrinsic muscles except palatoglossus
IX	Special sensory (taste) Somatic sensory	Posterior one third of tongue
V	Somatic sensory	Anterior two thirds of tongue
VII	Special sensory (taste)	Anterior two thirds of tongue

ARTERIAL SUPPLY OF THE TONGUE

The tongue is a very vascular structure that receives its arterial supply from the ***lingual*** artery, which arises from the proximal part of the external carotid artery and runs deep to the hyoglossus muscle to reach the tongue (Fig. 31-5B).

Venous Drainage of the Tongue. The dorsal veins of the tongue run with the lingual artery and end in the internal jugular vein. The deep veins of the tongue, which are visible on the inferior surface of the tongue, accompany the hypoglossal nerve as its ***venae comitantes*** and end in the facial vein or the internal jugular vein.

LYMPHATIC DRAINAGE OF THE TONGUE

The tongue is a not infrequent site for the occurrence of cancer, and its lymphatic drainage is important to the clinician. Being a midline structure there is much overlap of its lymphatic drainage. The anterior two thirds of the tongue drain into the submental and submandibular groups of nodes which, in turn, drain into the deep cervical nodes ranged along the internal jugular vein. The posterior third of the tongue drains into the upper deep cervical nodes, particularly the ***jugulodigastric*** group of nodes.

THE FLOOR OF THE MOUTH

Submandibular Duct. The mucosa of the floor of the mouth covers the deep part of the submandibular salivary gland and its duct (see Fig. 27-8).

Sublingual Gland. The sublingual gland is another salivary gland that lies in virtually a horseshoe shape around the connective tissue core of the frenulum of the tongue, reaching posteriorly almost as far as the submandibular gland on either side. Its secretions empty onto the floor of the mouth by numerous small independent ducts. The sublingual salivary gland receives its secretomotor nerve supply from the same parasympathetics of the facial nerve as the submandibular gland.

The Larynx

CARTILAGE

The skeleton of the larynx (Figs. 32-1 and 32-2) is cartilaginous, although parts of it may calcify with age.

Cricoid Cartilage. This cartilage is shaped like a signet ring with the shield of the ring posteriorly. It is the lowest of the laryngeal cartilages and is connected to the first ring of the trachea by the ***cricotracheal*** ligament. The inferior horns of the thyroid cartilage articulate with the cricoid cartilage toward its posterior part, and the inferior border of the thyroid cartilage is connected to the cricoid cartilage by the ***cricothyroid*** ligament.

Thyroid Cartilage. The thyroid cartilage is responsible for the visible and palpable prominence of the "Adam's apple" and is the largest of the laryngeal cartilages. It is formed by two flat ***laminae*** joined anteriorly to make a V. The ***thyroid notch***, which is palpable, is found at the superior junction of the laminae. Each lamina has a superior and an inferior horn, the inferior articulating with the cricoid cartilage in a synovial joint that allows the thyroid cartilage to tip backward and forward on the cricoid cartilage. The superior horn is attached to the greater cornu of the hyoid bone by the ***thyrohyoid ligament***. The rest of the superior border of the thyroid cartilage is attached to the hyoid bone by the ***thyrohyoid membrane***.

The thyroid cartilage is more prominent in the adult male. The relatively longer male vocal ligaments account for the deeper voice of males.

Epiglottis. This elastic cartilaginous structure is attached to the inside of the V of the thyroid cartilage, just below the notch (Figs. 32-1 and 32-2). It projects upward behind the hyoid bone, being attached to it by the ***hyoepiglottic ligament***, and then projects farther upward behind the tongue. The epiglottis is covered by mucous membrane. The posterior surface of the epiglottis bears a ***tubercle*** that makes contact with the arytenoid cartilages to complete the closure of the entrance to the larynx during swallowing.

Arytenoid Cartilages. There are two arytenoid cartilages which, although small, are very important. They articulate, by synovial joints, with the upper border of the wide posterior part of the cricoid cartilage in such a manner that they can rotate and also slide apart or slide together

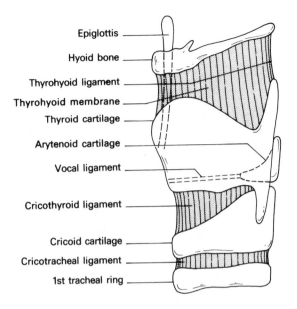

Figure 32-1.
Left side of the skeleton and fibrous structures of the larynx.

(Fig. 32-3). Each arytenoid is shaped like a triangle from every aspect. Each has an *apex* superiorly, a *vocal process* anteriorly, and a *muscular process* laterally. The apex is attached to the *aryepiglottic fold*, the vocal process to the *vocal ligament* (cord), and the muscular processes to a variety of cricoarytenoid muscles that evoke pivotal and gliding movements at the cricoarytenoid joint and influence the positions and tension of the vocal ligament (see Figs. 32-1, 32-2, 32-4, and 32-6).

MEMBRANES AND LIGAMENTS OF THE LARYNX

The cricotracheal, cricothyroid, and thyrohyoid ligaments have already been discussed.

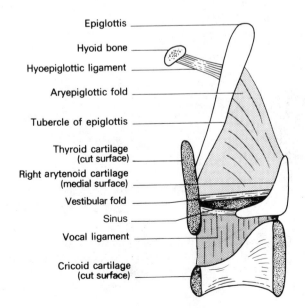

Figure 32-2.
Larynx (median section).

Figure 32-3.
Glottis (superior view).

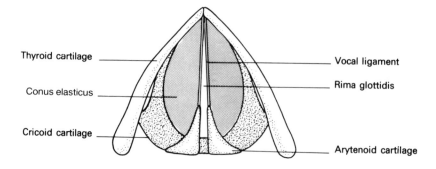

Vocal Ligaments and Vocal Folds. Running from each vocal process of the arytenoid cartilage to the junction of the laminae of the thyroid cartilage is the vocal ligament. It is actually the upper free border of the ***cricothyroid ligament***. Connecting the vocal ligament to the thyroid lamina is the ***conus elasticus*** (see Fig. 32-3), which is really a membrane continuous with the cricothyroid ligament. All the air passing in and out of the larynx has to pass between the two vocal ligaments. The conus elasticus and the vocal ligaments are covered by mucous membrane, but the mucous membrane covering the vocal ligaments contains no submucosa. It is made of stratified squamous epithelium and contains no glands or follicles. This close apposition of the mucosa to the vocal ligament permits vibration without any damping effect and gives the vocal fold (cord) a pearly white appearance.

The vocal apparatus is the ***glottis***, and the space through which the air must pass is the ***rima glottidis***. About two thirds of the margin of the rima is formed by the vocal ligament and the posterior one third by the vocal process of the arytenoid cartilage (Figs. 32-3 and 32-4).

Vestibular Fold. The vestibular fold runs above the vocal fold from the arytenoid cartilage to the thyroid cartilage. It is a fold of mucous membrane (Fig. 32-5). It appears pink in the living body and is lined by mucosa that has an ample connective tissue stroma with mucous glands. It is often called the ***false vocal cord***. The cavity of the larynx above the vestibular folds is called the ***vestibule***, and the space between the vestibular and the vocal folds is called the ***sinus (ventricle)*** of the larynx. An upward running recess from the sinus is called the ***saccule*** of the larynx, and because

Figure 32-4.
Larynx (superior view indicating muscle action).

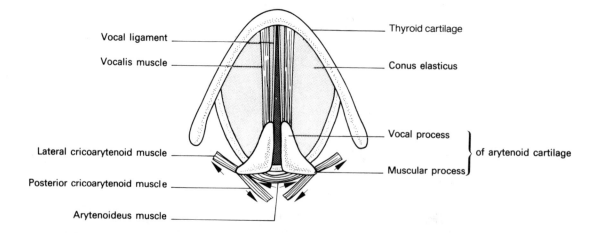

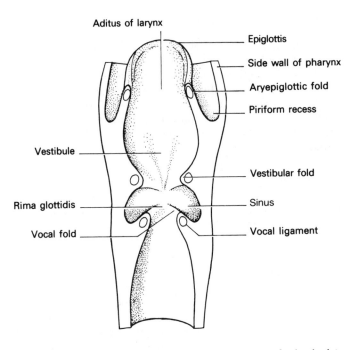

Aditus of larynx

Epiglottis

Side wall of pharynx

Aryepiglottic fold

Piriform recess

Vestibule

Vestibular fold

Rima glottidis

Sinus

Vocal fold

Vocal ligament

Figure 32-5.
Larynx (coronal section, anterior view).

this contains mucus secreting glands, the secretions of which drip down on the vocal folds, it is known as the "oil can" of the vocal cords.

Aryepiglottic Fold. This fold contains some muscle fibers and runs from the apex of the arytenoid cartilage to the side of the epiglottis. The folds come together with the epiglottis to protect the larynx from aspiration of pharyngeal contents during swallowing.

MUSCLES OF THE LARYNX

The muscles of the larynx control the size and shape of the rima glottidis and the tension of the vocal ligaments (see Fig. 32-4). There is one ***extrinsic*** muscle of the larynx, the cricothyroid; the remaining laryngeal muscles are considered to be ***intrinsic*** muscles. The cricothyroid muscle is innervated by the ***external laryngeal*** branch of the ***superior laryngeal*** branch of the ***vagus nerve***, although the true origin of the fibers is from the ***accessory nerve*** (cranial XI).

Cricothyroid Muscle

This muscle runs inferiorly and backward from the lower part of the thyroid cartilage to the cricoid cartilage. It lies superficial to the cricothyroid ligament, and its action is to tilt the thyroid cartilage forward on the cricoid so that the distance between the points of attachment of the vocal ligament is increased and thus the ligament is stretched to increase the pitch of the voice.

Intrinsic Laryngeal Muscles

With one exception these muscles all run between the cricoid cartilage and the arytenoid cartilage. They are all supplied by the ***recurrent laryngeal nerve***, the fibers of which really come from the ***accessory nerve***.

Posterior Cricoarytenoid Muscle

This muscle runs in an upward and lateral direction from the posterior surface of the cricoid cartilage to the muscular process of the arytenoid cartilage. Contraction of this muscle pivots the arytenoid cartilage through a vertical axis so that the vocal process turns laterally and the vocal ligament is abducted, that is, the rima glottidis is widened. In addition the more lateral fibers of this muscle pull the arytenoid cartilage laterally on the cricoid cartilage so that the shape of the rima is converted from a rhomboid to a triangle (Fig. 32-4).

Arytenoideus Muscle

This muscle runs between the posterior aspects of the arytenoid cartilages and brings them closer together, thus narrowing the rima (Figs. 32-4 and 32-6).

Lateral Cricoarytenoid Muscle

The fibers of this muscle run from the lateral aspect of the cricoid cartilage to the muscular process of the arytenoid cartilage. Contraction of this muscle rotates the arytenoids so that the vocal processes are brought together, that is, the vocal ligament is adducted and the rima is narrowed (Fig. 32-4).

Thyroarytenoid Muscle

This is a band of muscle lateral to the vocal ligament running from the thyroid cartilage to the arytenoid. One upper band of this muscle is attached to the vocal process and to the vocal ligament. This part is called the *vocalis*, and its action has been likened to that of the finger on the violin string, modifying the tension of the ligament as a fine tuning adjustment to the pitch of the voice (Fig. 32-4).

NERVES OF THE LARYNX

Vagus Nerve. This nerve supplies motor and sensory fibers to the larynx. The motor fibers, as has been stated before, come from the *accessory nerve* and merely hitch a ride with the vagus.

Superior Laryngeal Nerve. This nerve leaves the vagus high up in the carotid sheath and runs close to the superior thyroid artery to the larynx. It divides into the *external laryngeal nerve*, which supplies the cricothyroid muscle, and the *internal laryngeal nerve*, which is sensory to the upper larynx to and including the vocal folds (cranial nerve X; see Chap. 34).

Recurrent Laryngeal Nerve. This nerve has already been described. It supplies all the intrinsic muscles of the larynx, although its motor fibers really come from the *accessory nerve*. It also supplies sensation to the larynx and trachea below the vocal folds.

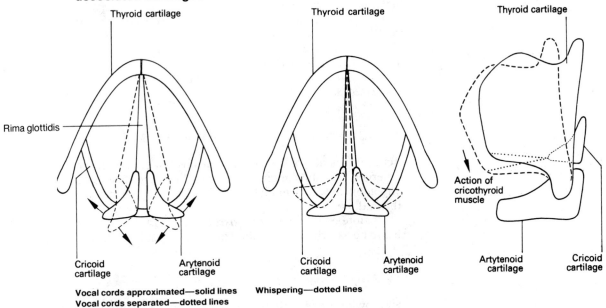

Movements of the vocal ligaments and associated cartilages

Thyroid cartilage

Rima glottidis

Cricoid cartilage

Arytenoid cartilage

Vocal cords approximated—solid lines
Vocal cords separated—dotted lines

Thyroid cartilage

Cricoid cartilage

Arytenoid cartilage

Whispering—dotted lines

Tensing of the vocal ligaments

Thyroid cartilage

Action of cricothyroid muscle

Arytenoid cartilage

Cricoid cartilage

Figure 32-6.
Movements of the arytenoid and thyroid cartilages in phonation.

ARTERIAL SUPPLY OF THE LARYNX

The larynx receives its arterial supply from the ***superior*** and ***inferior*** laryngeal arteries, which are branches of the superior and inferior thyroid arteries.

CLINICAL NOTES

Many of the apparently hidden areas of the larynx and pharynx are accessible to inspection by the clinician by the simple use of a mirror on a long slim handle. This procedure is known as indirect laryngoscopy and should be within the competence of any practicing physician. It enables visualization of the nasopharynx, back of the tongue and epiglottis, valleculae, piriform fossae, vestibular folds, and vocal folds as well as the upper trachea (see Fig. 32-5).

*Certain viral and bacterial infections of sudden onset may cause severe swelling of the epiglottis or vestibular fold, particularly in infants and young children. If all else fails, **tracheostomy** or **laryngotomy** may have to be performed as an emergency, life-saving procedure. The techniques will be described below.*

Disease or surgery may compromise the superior or recurrent laryngeal nerves. Injury to the superior laryngeal nerve(s) will impair or abolish the cough reflex that protects the airway from inhalation of saliva or other pharyngeal content. The vocal fold on the effected side cannot be tensed, and the voice will assume a deep and hoarse timbre.

Damage to the recurrent laryngeal nerve, whether in the neck or, in the case of the left recurrent laryngeal nerve, in the thorax, will paralyze the intrinsic muscles of the larynx on that side (see Chap. 34 for more details).

*The retropharyngeal space continues downward into the thorax to be continuous with the connective tissue of the mediastinum. Air from a wound in the pharynx or cervical esophagus or infections in this space can spread downward and produce a **pneumomediastinum** or **mediastinitis**. Similarly air in the mediastinum, for example, from a bronchial or pulmonary injury (penetrating the mediastinal layer of pleura), can track up into the neck and produce alarming-looking swellings as the air tracks into the soft tissues of the neck and face (subcutaneous emphysema).*

TRACHEOSTOMY AND LARYNGOTOMY

Acute obstruction of the airway at the levels mentioned above may require emergency establishment of an artificial passage. This can be done by laryngotomy in which an incision is made into the cricothyroid ligament, in the midline.

More commonly ***tracheostomy*** (see Fig. 7-1) is used. The incision has to be exactly in the midline over the palpable trachea.

SWALLOWING

Swallowing, or deglutition, is easy to do but much more difficult to analyze and describe. The steps are as follows:

1. Chewing breaks down the size of the morsel and mixes it with saliva so that it becomes a homogeneous slippery ***bolus.***
2. The tongue, hyoid bone, and larynx are elevated. Breathing and chewing cease.
3. The tongue squeezes the bolus against the hard palate and forces it backward into the pharyngeal isthmus.
4. The tensor palati and levator palati elevate the soft palate against the thickened muscle ridge in the posterior pharyngeal wall to close the pharyngeal isthmus and prevent the entry of the bolus into the nasopharynx.
5. The palatoglossus and palatopharyngeus muscles contract to constrict the fauces, helping to squeeze the bolus backward.
6. When the food is in the oropharynx, in contact with the back of the tongue, it has reached the point of no return. The walls of the pharynx are raised by the palatopharyngeus, stylopharyngeus, and salpingopharyngeus, and the larynx also moves upward.
7. The inlet to the larynx is closed by the folding back of the epiglottis and the contraction of the muscle fibers in the aryepiglottic folds. The vocal folds come together in anticipation of a malfunction.
8. The bolus enters the lower pharynx and is propelled by peristalsis.
9. During stages 5 to 7 the pharyngeal constrictor muscles contract in sequence from above, downward.

The Ear

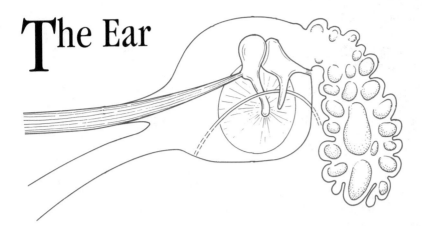

The three divisions of the ear—external, middle, and internal—are described in this chapter.

THE EXTERNAL EAR

The external ear (Figs. 33-1A and 33-1B) consists of the ***auricle***, which is a cartilaginous skeleton covered by skin, and the ***external auditory meatus (canal)***, a canal about 25 mm long. The outer third of the canal is cartilaginous and joins the bony portion of the canal. The meatus is at its narrowest at the osseocartilaginous junction. This is the area where foreign objects introduced into the external meatus become trapped. The meatus curves downward and anteriorly so that the auricle has to be pulled upward and backward to visualize the tympanic membrane (eardrum) through an aural speculum (otoscope). The ***meatus*** is lined with skin that contains ceruminous (wax producing) glands. The meatus ends at the tympanic membrane.

Nerve Supply. The external auditory meatus receives branches of three nerves. The ***auriculotemporal nerve*** (branch of V_3) has already been described. The ***vagus*** nerve supplies a strip along the posteroinferior part of the meatus, and the ***facial*** nerve supplies a similar strip along the anteroinferior aspect of the canal. The corresponding nerves also supply similar areas of the outer surface of the tympanic membrane. The vagal innervation of the meatus accounts for the coughing, nausea, vomiting, or fainting that may occur when the canal is probed or syringed.

The auricle (pinna) has many named parts that are not worth remembering. However the tragus, lobule, and helix should be identified in Figure 33-1.

The skin of the auricle has no subcutaneous tissue stroma except in the lobe, so that infections such as boils will be very tense and hence very painful. Infection of the external meatus is called ***otitis externa***.

TYMPANIC MEMBRANE

The tympanic membrane (eardrum) separates the external ear from the middle ear. It is a connective tissue membrane covered by skin on the outside and by the mucous membrane of the middle ear on the inside.

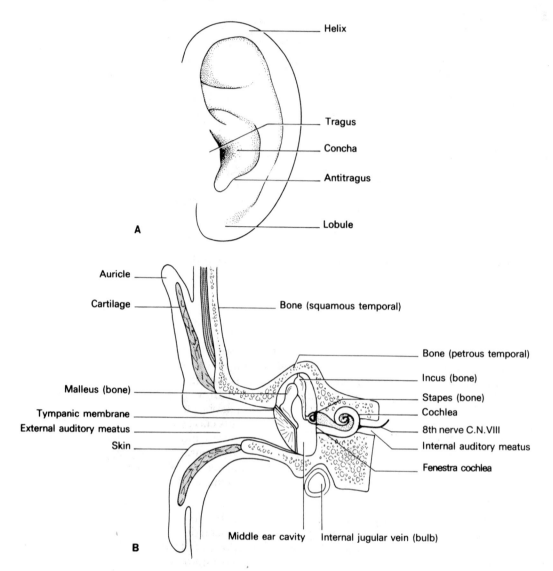

A

B

Figure 33-1.
Ear. **A.** External ear. **B.** Schematic representation of the components of the ear (coronal section).

Its outer surface is slightly concave and directed downward, laterally, and anteriorly.

The distinctive features of the tympanic membrane that can be seen with an otoscope are as follows (Fig. 33-2):

The membrane itself appears pearly white and shiny.

A cone of light, the reflection of the light of the otoscope, is visible on the anteroinferior quadrant of the drum. This light reflex is due to the concavity and the shiny surface of the drum. It will be lost if the drum bulges outward because of accumulated fluid in the middle ear cavity or if the drum loses its gloss because of inflammation.

The handle of the malleus (an ear ossicle) appears as a shadow in the middle of the membrane. The point of greatest concavity of the membrane, the *umbo*, is at the point of attachment of the lower end of the handle of the malleus. The pars flaccida, the upper part of the membrane, appears less tense than the rest of the drum.

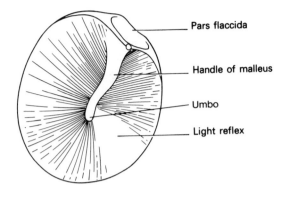

- Pars flaccida
- Handle of malleus
- Umbo
- Light reflex

Figure 33-2.
Right tympanic membrane as seen through the otoscope.

THE MIDDLE EAR

The middle ear is a cavity (Fig. 33-3) in the temporal bone lined by mucous membrane. It is the resonance chamber of the ear. Anteriorly it opens into the ***auditory*** tube. Posterosuperiorly the cavity connects with the mastoid air cells at the ***mastoid antrum***. The cavity itself is shaped like an erythrocyte: 15 mm long, 15 mm high, and 2 mm wide at the narrowest point and 5 mm wide at its widest point. The part of the cavity above the tympanic membrane is the ***epitympanic recess***. The ***ossicles*** are the three small bones of the middle ear; all the ossicles are covered with mucous membrane. They are the malleus, incus, and stapes.

Lateral Wall. The lateral wall of the tympanic cavity consists mainly of the tympanic membrane, some bony wall anterior to the membrane, and the lateral wall of the epitympanic recess (Fig. 33-4).

The tympanic membrane is made of connective tissue, and its medial surface is covered by mucous membrane. The handle of the malleus is fixed to the membrane, with the ***head of the malleus*** extending into the epitympanic recess. The malleus is covered by mucous membrane.

The ***chorda tympani*** runs across the upper part of the tympanic

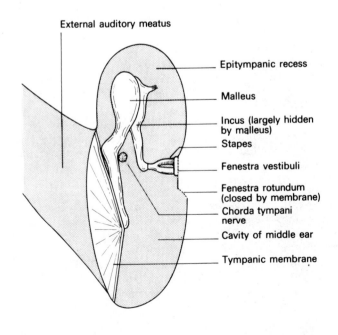

External auditory meatus

- Epitympanic recess
- Malleus
- Incus (largely hidden by malleus)
- Stapes
- Fenestra vestibuli
- Fenestra rotundum (closed by membrane)
- Chorda tympani nerve
- Cavity of middle ear
- Tympanic membrane

Figure 33-3.
Cavity of the middle ear (coronal section).

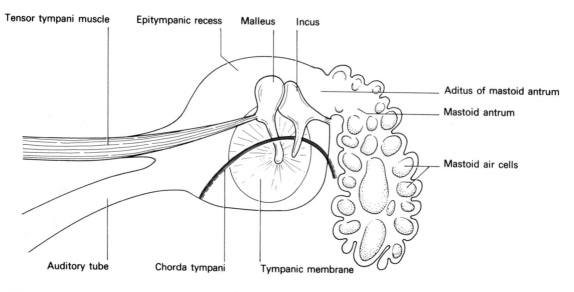

Tensor tympani muscle   Epitympanic recess   Malleus   Incus

Aditus of mastoid antrum

Mastoid antrum

Mastoid air cells

Auditory tube   Chorda tympani   Tympanic membrane

Figure 33-4.
Cavity of the right middle ear (view of lateral wall).

membrane, medial to the handle of the malleus. It leaves the tympanic cavity anteriorly through the petrotympanic fissure.

Floor. The floor of the middle ear cavity is a thin piece of bone that separates the tympanic cavity from the *jugular bulb* posteriorly and the *carotid canal* anteriorly.

Roof. The roof of the middle ear is thin and separates the cavity from the middle cranial fossa. This area can be visualized on the skull by its translucency; it is called the *tegmen tympani*.

Anterior Wall. There are two openings in the anterior wall of the tympanic cavity. At its junction with the floor is the *auditory (Eustachian) tube*. This tube runs downward and medially to open into the nasopharynx behind the inferior meatus of the nose. The proximal one third of the tube is made of bone, and the distal two thirds are made of cartilage and connective tissue. The tube is funnel-shaped with the wide part at the nasopharyngeal end. The tube is lined by mucous membrane. The lateral wall of the cartilaginous part of the tube is membranous, and the tensor palati muscle is attached to the membrane. This muscle will open the tube during swallowing and yawning. This effect on the tube can be felt when the "plugged ear" sensation that one may feel on takeoff or landing of an airplane is relieved by swallowing or yawning. The tube is narrowest at the osseocartilaginous junction.

Running above the tube and parallel to it is a bony canal that contains the *tensor tympani* muscle. The tendon of this small muscle attaches to the handle of the malleus and modulates the tension of the tympanic membrane. The tensor tympani is supplied by a branch of the mandibular division of the trigeminal nerve (see Fig. 33-4).

Posterior Wall. The *aditus* is an opening that leads from the epitympanic recess into the *mastoid antrum* and thence into the *mastoid air cells* (see Figs. 33-4 and 33-5). Also in the posterior wall is a small projection of bone, the *pyramid*, which is hollow and contains the *stapedius* muscle. The tendon of this muscle leaves the canal of the pyramid and is attached to the stapes. This tiny muscle dampens the amplitude of the vibrations of the stapes and is supplied by the facial nerve.

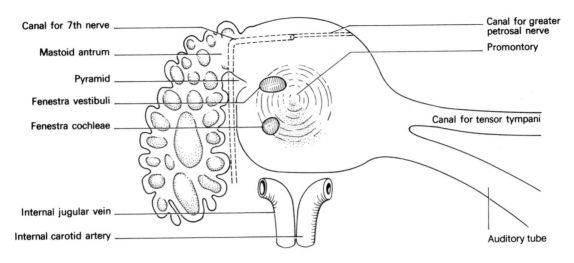

Canal for 7th nerve

Mastoid antrum

Pyramid

Fenestra vestibuli

Fenestra cochleae

Internal jugular vein

Internal carotid artery

Canal for greater petrosal nerve

Promontory

Canal for tensor tympani

Auditory tube

Figure 33-5.
Cavity of the right middle ear (view of medial wall).

Medial Wall. The medial wall of the tympanic cavity exhibits several important features (see Fig. 33-5). The ***promontory*** is a lateral projection from the first turn of the cochlea (the spiral-shaped organ of the inner ear). On the surface of the promontory, small nerve fibers from the glossopharyngeal and facial nerves form the ***tympanic plexus***.

Above the posterior part of the promontory is the ***fenestra vestibuli*** (fenestra ovale, oval window), which accommodates the footplate of the ***stapes***. Through this window vibrations of the ear ossicles are transmitted into the perilymph of the scala vestibuli of the inner ear. Below the fenestra vestibuli on the inferior aspect of the promontory is the ***fenestra cochleae*** (round window), which is closed by a membrane. Through it the waves in the perilymph that are carried along the scala tympani are transmitted to the air in the middle ear. This device simply expends the pressure forced into the perilymph by the vibrations of the stapes (see Fig. 33-3).

Along the medial wall, at its junction with the roof and in its own bony canal, runs the ***facial nerve*** (see Fig. 33-5).

THE OSSICLES

Three small bones (ossicles) form a chain across the tympanic cavity (see Figs. 33-1B and 33-3). All are covered by mucous membrane.

The ***malleus*** has its handle embedded in the tympanic membrane and its head projects into the epitympanic recess. The tendon of the tensor tympani inserts into its handle anteriorly. The head of the malleus articulates (by a synovial joint) with the incus (see Figs. 33-3 and 33-4).

The ***incus*** is shaped like a molar tooth with one long and one short root. The long root articulates with the stapes, and the short root is connected by a ligament to the posterior wall of the tympanic cavity. The malleus articulates with the "tooth" at its crown.

The ***stapes*** is shaped like a stirrup, but the hole in the stirrup is obscured by mucous membrane. The footplate fills the fenestra vestibuli and is held in place by a ligament.

The three bones act as a system of levers that increases the force but not the amplitude of the vibrations transmitted by the tympanic membrane. The tympanic membrane, of course, vibrates in response to sound waves that reach it by way of the external auditory meatus.

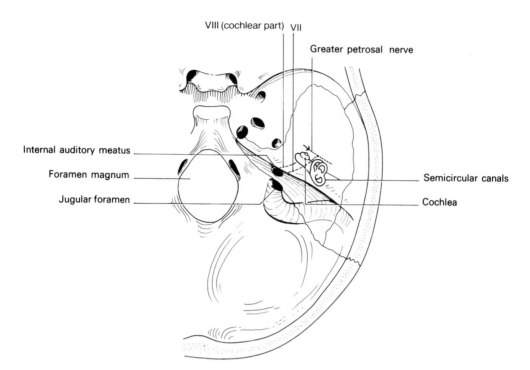

Figure 33-6.
Location of the right inner ear in the petrous temporal bone.

CLINICAL NOTE

Since the pharynx is connected to the middle ear by the auditory tube, inflammation (otitis media) is commonly transmitted to the tympanic cavity from infections of the throat and pharynx. If the infection spreads through the aditus and antrum to the mastoid air cells the result is mastoiditis.

The mastoid air cells vary enormously in their extent and complexity. The newborn child has a very rudimentary mastoid process because the process and air cells do not develop until the age of 2 years. The mas-

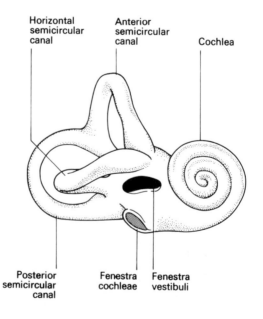

Figure 33-7.
Right inner ear.

toid air cells may extend into the petrous temporal bone, even as far as its anterior end, and they have also been known to extend into the zygomatic process of the temporal bone. The thin layer of bone that separates the middle ear cavity from the middle cranial fossa offers little resistance to repeated infections of the tympanic cavity. In the days before antibiotics, chronic otitis media with perforations of the eardrum was quite common. In some of these unfortunate patients, many of them children, extension of the infection into the middle cranial fossa with resultant meningitis or temporal lobe brain abscesses was not rare.

The stapes may become fixed in the fenestrum vestibuli by a bone disorder called **otosclerosis**. *In this condition the stapes cannot vibrate properly and the patient experiences deafness in that ear. With modern microsurgical techniques the stapes can be replaced by a prosthesis and hearing restored.*

THE INNER EAR

The inner ear will not be discussed in any detail in this textbook because its complexity of structure, related to its complex function, is more appropriately found in a book on neuroanatomy. Its position in the skull should be understood (Fig. 33-6). The **arcuate eminence** of the petrous temporal bone indicates the position of the **anterior semicircular canal**; at right angles to this is the **posterior semicircular canal**. Between them lies the **horizontal semicircular canal**, and anteromedial to these is the **cochlea** (Figs. 33-6 and 33-7).

The Cranial Nerves

*The 12 pairs of cranial nerves should be examined in every routine physical examination and in all cases where neurologic involvement is suspected. Hence their importance to every medical student is unquestionable. Time spent now in learning and understanding these nerves will be rewarded in the clinical years. The function of each nerve should be clearly established. It may be wholly **somatic motor**, **parasympathetic (visceral) motor**, **somatic general sensory**, **visceral sensory**, **special sensory**, or a combination of several of these modalities.*

Housed in the cranium, the brain and brain stem give off modified pairs of spinal nerves that, because of their origin, are called **cranial nerves**. Except for the vagus nerve, the area supplied by these nerves is restricted to the region of the head and neck. They carry **somatic motor** fibers to striated muscles, **parasympathetic motor** fibers to smooth muscle, and **parasympathetic secretomotor** fibers to glands. **General somatic** and **visceral sensation** from the body and the environment is carried into the central nervous system so that the brain can initiate an appropriate response to these incoming data. By convention, the cranial nerves are numbered with Roman numerals, starting with I as the most cranial and XII as the most caudal (Fig. 34-1). The nerves enter and leave the skull through named foramina, which the student is well advised to commit to memory (Fig. 25-1B).

I—OLFACTORY NERVE

This pair of cranial nerves is entirely sensory and carries the **special sense of smell**. Located in the mucosa of the nose where they pick up aromas that are dissolved in its moist atmosphere, the sensory nerve endings are hair-like threads that carry the sensation of smell through the perforations of the cribriform plate of the ethmoid bone to the olfactory bulb (Fig. 34-3). In the olfactory bulb these nerve fibers synapse on neurons whose processes will travel in the olfactory tract to reach the primary and association olfactory areas of cortex in the brain.

Fractures of or blows to the cranium (such as occur when the individual falls and strikes the head on concrete) can result in tearing of the

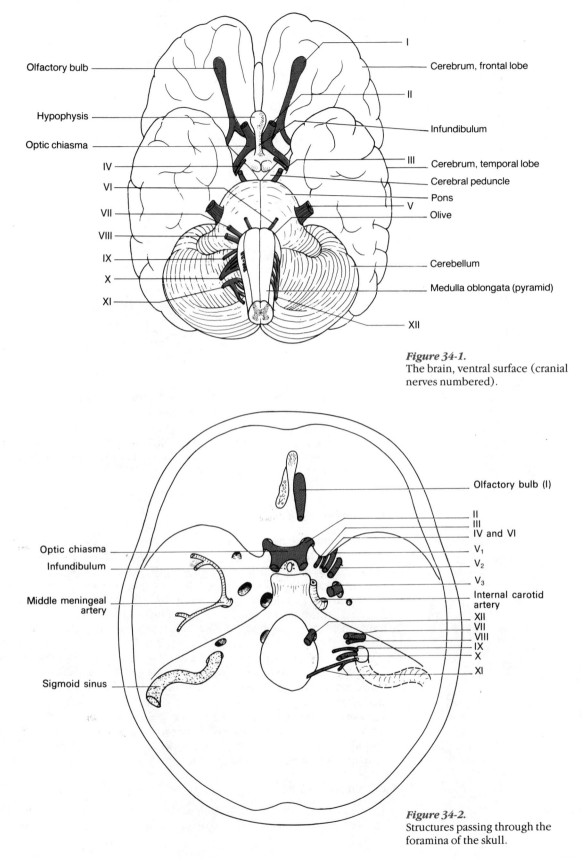

Olfactory bulb

Hypophysis

Optic chiasma

IV

VI

VII

VIII

IX

X

XI

I

Cerebrum, frontal lobe

II

Infundibulum

III

Cerebrum, temporal lobe

Cerebral peduncle

Pons

V

Olive

Cerebellum

Medulla oblongata (pyramid)

XII

Figure 34-1.
The brain, ventral surface (cranial
nerves numbered).

Optic chiasma

Infundibulum

Middle meningeal
artery

Sigmoid sinus

Olfactory bulb (I)

II
III
IV and VI
V₁
V₂
V₃
Internal carotid
artery
XII
VII
VIII
IX
X
XI

Figure 34-2.
Structures passing through the
foramina of the skull.

Figure 34-3.
I: Olfactory nerve (special sense of smell).

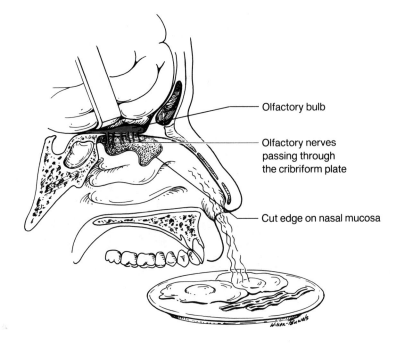

- Olfactory bulb
- Olfactory nerves passing through the cribriform plate
- Cut edge on nasal mucosa

olfactory nerve fibers where they pass through the cribriform plate. There may also be tearing of the meninges and a resultant leakage of cerebrospinal fluid that could become apparent as a runny nose. To test, each nerve is checked independently. While one nostril is blocked, the patient is asked to identify the aroma of substances held to the free nostril. This is repeated on the opposite side.

II—OPTIC NERVE

Like the olfactory nerves, the optic nerves carry only sensory fibers, conducting the ***special sense of vision.*** The ***retina***, which receives the input of visual information from the visual fields, is an outgrowth of the brain. Thus the optic nerve is, in truth, a tract of the brain; however, it is commonly referred to as the ***optic nerve.*** Images from the visual fields are focused on the retina, and, from the retina, nerve fibers form the optic nerve that leaves the orbit through the optic foramen of the skull to enter the cranial cavity on the ventral aspect of the brain (Fig. 34-4). The two optic nerves meet at the ***optic chiasma*** just anterior to the pituitary fossa. At this point each optic nerve divides; fibers from the nasal (medial) retinal fields cross to the opposite side, and fibers from the lateral retinal fields stay on their respective sides. Because of this crossover, each optic tract will carry information from the retinas of both eyes. From the chiasma, most axons travel to the ***lateral geniculate body*** of the thalamus, where they synapse on neurons whose processes form the ***optic radiation*** to the primary visual cortex around the calcarine fissure of the occipital lobe.

CLINICAL NOTE

Visual acuity can be checked using a standard eye chart, testing both eyes together and then testing each eye independently. The range of the visual field can also be checked, and loss of vision in a part of the field can indicate neurologic disorders along the visual pathways.

III—OCULOMOTOR NERVE

This nerve (paired) carries somatic motor fibers to four of the six muscles that move the eyeball. It emerges from the medial side of the cerebral peduncle near its junction with the pons (see Fig. 34-5), emerging between the posterior cerebral and superior cerebellar arteries. It then passes through the dura lateral to the posterior clinoid process and runs in the lateral wall of the cavernous sinus. Just before leaving the skull through the superior orbital fissure, it divides into ***superior*** and ***inferior*** divisions. The superior division supplies the ***superior rectus*** muscle and the ***levator palpebrae superioris*** muscle that elevates the upper eyelid. The fibers to the latter muscle also contain postganglionic sympathetic fibers from the superior cervical ganglion (which have traveled along the internal carotid artery and its branches).

The inferior division gives off a branch that passes inferior to the optic nerve to supply the ***medial rectus*** muscle that ***adducts*** the eye, a branch directly into the ***inferior rectus*** muscle that acts in ***downward gaze,*** and a branch to the ***inferior oblique*** muscle. Just before entering

Figure 34-4.
II: Optic nerve showing relationship to visual fields and visual cortex.

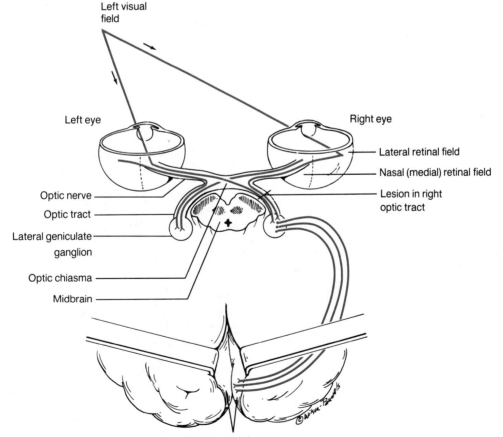

Left visual field

Left eye

Right eye

Lateral retinal field

Nasal (medial) retinal field

Optic nerve

Optic tract

Lateral geniculate ganglion

Optic chiasma

Midbrain

Lesion in right optic tract

Primary visual cortex

the inferior oblique muscle, the nerve gives off a branch to the ***ciliary ganglion. The superior rectus*** and ***inferior oblique*** muscles act together in ***upward gaze.*** The oculomotor nerve also carries proprioceptive fibers from these same muscles (Fig. 34-5). In addition, the oculomotor nerve conveys ***parasympathetic*** fibers to the smooth muscles that control the shape of the lens and the size of the pupil. The preganglionic parasympathetic fibers reach the ***ciliary ganglion***, a small ganglion that lies lateral to the optic nerve very near the apex of the orbit (Fig. 34-5). They synapse in the ganglion, and postganglionic fibers travel in the short ciliary nerves to the eyeball and supply the ***ciliaris*** muscle and the ***sphincter pupillae*** muscle. Contraction of the ciliaris muscle rounds up the lens to accommodate for near vision. Constriction of the pupil is also part of the accommodation reflex. Contraction of the constrictor pupillae muscle constricts the pupil. Sympathetic and proprioceptive fibers run in the short ciliary nerves, having passed through the ciliary ganglion without synapsing in it.

CLINICAL NOTE

The oculomotor nerve emerges from the brain stem between the posterior cerebral artery and the superior cerebellar artery (see Fig. 34-14). Here it is vulnerable to compression if aneurysms develop in either of these vessels. Cranial nerves III, IV, and VI are usually tested as a unit because all supply muscles for eye movement. The range of eye movement is checked by asking the patient to follow the movement of the examiner's fingers as they are moved in all directions of gaze. With oculomotor nerve involvement, the patient will not be able to look up, down, or medially with the affected eye. The patient commonly complains of double vision (diplopia). In addition, the patient will be unable to elevate the upper lid (ptosis), the pupil may be dilated (unable to constrict in response to bright light), and there will be absence of the accommodation reflex owing to the loss of parasympathetic control of the constrictor pupillae muscle.

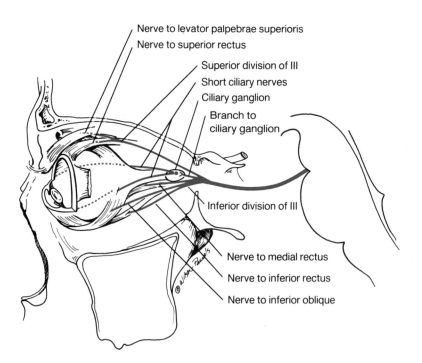

Nerve to levator palpebrae superioris
Nerve to superior rectus
Superior division of III
Short ciliary nerves
Ciliary ganglion
Branch to ciliary ganglion
Inferior division of III
Nerve to medial rectus
Nerve to inferior rectus
Nerve to inferior oblique

Figure 34-5.
III: Oculomotor nerve. Left lateral wall of orbit has been removed.

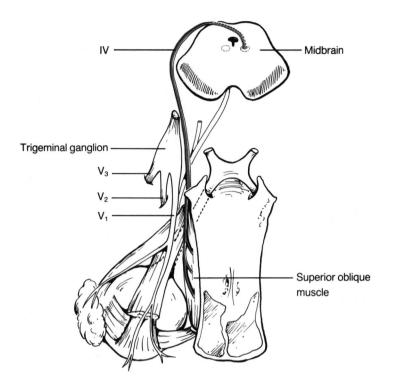

IV ——— Midbrain

Trigeminal ganglion ———

V₃ ———

V₂ ———

V₁ ———

Superior oblique muscle

Figure 34-6.
IV: Trochlear nerve.

IV—TROCHLEAR NERVE

This tiny thread-like nerve is the only one to emerge from the dorsum of the brain stem (Fig. 34-6). It emerges from the superior medullary velum and passes around the cerebral peduncle to pass through the dura between the free and attached edges of the ***tentorium cerebelli***. It then passes along the lateral wall of the cavernous sinus and leaves the skull through the superior orbital fissure. It supplies ***somatic motor*** and possibly ***proprioceptive*** (general sensory) fibers to the superior oblique muscle of the eye. This muscle, acting in isolation, is responsible for turning the eye downward and laterally (see Chap. 26).

CLINICAL NOTE

When the trochlear nerve is damaged, the patient cannot look downward and commonly reports difficulty in walking downstairs.

V—TRIGEMINAL NERVE

The fifth cranial nerve emerges from the anterolateral portion of the pons as a large sensory root and a small motor root. It enters the ***trigeminal cave*** in the dura of the floor of the middle cranial fossa, where it displays a large swelling–the trigeminal (sensory) ganglion. The nerve splits into three divisions: ***ophthalmic (V₁), maxillary (V₂),*** and ***mandibular (V₃)***. The three divisions leave the skull through separate foramina and supply the appropriate areas of the face. The nerve is sensory to the skin of the face and scalp, conjunctiva, nose and nasal sinuses, tongue, mouth, and teeth. It is motor to the muscles of mastication and to the tensor tympani

Figure 34-7.
V: Trigeminal nerve showing divisions
into ophthalmic (V₁), maxillary (V₂),
and mandibular (V₃) nerves.

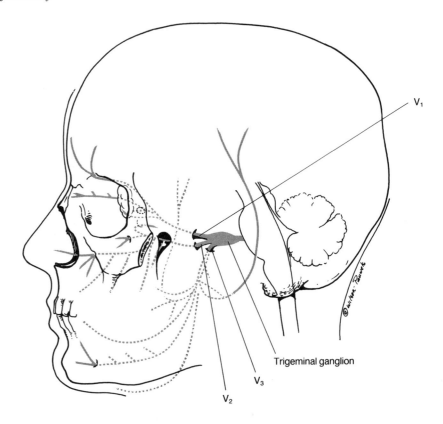

V₁

Trigeminal ganglion

V₃

V₂

and tensor palati muscles, as well as to the anterior belly of digastric and mylohyoid muscles (Fig. 34-7).

The ***ophthalmic*** division (V₁) passes in the lateral wall of the cavernous sinus and emerges through the superior orbital fissure into the bony orbit (Fig. 34-8). It is entirely ***sensory*** in function, and the major branches are the ***lacrimal*** nerve that carries sensation from the lacrimal gland and conjunctiva; the ***frontal*** nerve that divides into ***supratrochlear*** and ***supraorbital*** branches that supply the upper eyelid, conjunctiva, and lower part of the forehead; and the ***nasociliary*** nerve that communicates with the ciliary ganglion, giving off the ***long ciliary*** nerves, the ***infratrochlear*** nerve, and the ***posterior ethmoidal*** branch before ending as the ***anterior ethmoidal*** nerve that supplies the mucosa and skin of the nose. The fibers of the nasociliary nerve that communicate with the ciliary ganglion do not synapse there but travel with the parasympathetic fibers from III that have synapsed in the ciliary ganglion and emerge as the short ciliary nerves. Sympathetic fibers to the dilator pupillae are also carried in the short ciliary nerves. The long ciliary nerves carry general sensation and sympathetic fibers to the ciliary body, iris, and cornea, whereas the infratrochlear nerve supplies the eyelids, lacrimal apparatus, and nose and the posterior ethmoidal nerve supplies the ethmoidal and sphenoidal sinuses (Fig. 34-8).

The ***maxillary*** division (V₂) carries only ***somatic sensory impulses*** (Fig. 34-9). From the trigeminal ganglion it passes in the lateral wall of the cavernous sinus and leaves the skull by passing through the foramen rotundum into the pterygopalatine fossa. Here it gives off the ***zygomatic*** nerve, which divides into the ***zygomaticotemporal*** and ***zygomaticofacial*** branches that, after passing through the foramina of the same names,

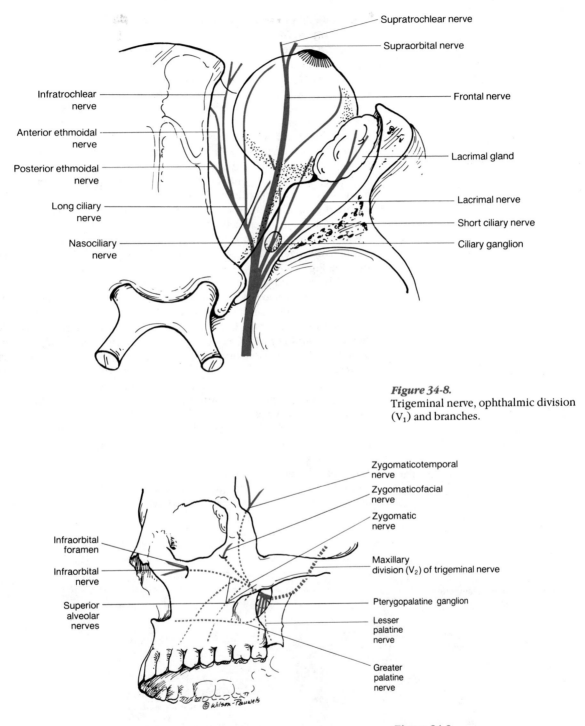

Supratrochlear nerve

Supraorbital nerve

Infratrochlear nerve

Anterior ethmoidal nerve

Posterior ethmoidal nerve

Long ciliary nerve

Nasociliary nerve

Frontal nerve

Lacrimal gland

Lacrimal nerve

Short ciliary nerve

Ciliary ganglion

Figure 34-8.
Trigeminal nerve, ophthalmic division (V₁) and branches.

Zygomaticotemporal nerve

Zygomaticofacial nerve

Zygomatic nerve

Infraorbital foramen

Infraorbital nerve

Superior alveolar nerves

Maxillary division (V₂) of trigeminal nerve

Pterygopalatine ganglion

Lesser palatine nerve

Greater palatine nerve

©Wilson-Pauwels

Figure 34-9.
Trigeminal nerve, maxillary division (V₂) and branches.

supply the lateral side of the cheek and temple. The zygomatic nerve carries **postganglionic parasympathetic** fibers from the **pterygopalatine ganglion** to the lacrimal gland by way of a branch that joins the lacrimal nerve in the lateral wall of the orbit. The **pterygopalatine ganglion** is attached to the maxillary nerve in the pterygopalatine fossa. It is a parasympathetic ganglion that receives its preganglionic fibers from the facial nerve by way of the **greater petrosal** nerve. **Postganglionic** fibers from the ganglion are distributed through the branches of the maxillary nerve. Also in the pterygopalatine fossa, the maxillary nerve gives off the **greater** and **lesser palatine** nerves that supply the palate; the **nasopalatine** nerve to the nasal septum and anterior part of the palate; the superior nasal nerves; and a **pharyngeal** branch to the nasopharynx. From the pterygopalatine fossa, the maxillary nerve continues as the **infraorbital** nerve within the infraorbital groove in the floor of the orbit. It first gives off posterior, middle, and anterior superior alveolar nerves to the teeth of the upper jaw before emerging through the infraorbital foramen to become sensory to the lower eyelid, nose, and the skin and mucosa of the cheek and upper lip (see Fig. 34-7).

The **mandibular** division (V₃) carries both **somatic motor** and **general somatic sensory** fibers. It passes inferior to the cavernous sinus and leaves the skull through the foramen ovale (see Figs. 34-10 and 34-11). It supplies somatic motor fibers to the **muscles of mastication**, the **tensor tympani, tensor palati, mylohyoid**, and the **anterior belly of digastric** muscles. The three sensory branches that reach the face are the **auriculotemporal** nerve, which emerges behind the mandible to supply the external ear and the skin of the temporal area; the **buccal** nerve, which supplies the mucosal and cutaneous surfaces of the cheek; and the **inferior alveolar** (dental) nerve. At the lingula of the mandible the inferior alveolar nerve passes through the mandibular foramen into the mandibular canal. It supplies the teeth of the lower jaw and emerges from the mental foramen as the **mental** nerve which supplies the skin of the lower lip and chin. Note that the auriculotemporal nerve carries postganglionic parasympathetic fibers from the **otic ganglion** for distribution to the parotid gland through which it passes on its way to the temporal area (Fig. 34-11). The

Figure 34-10.
Superficial branches of the mandibular division (V₃) of the right trigeminal nerve.

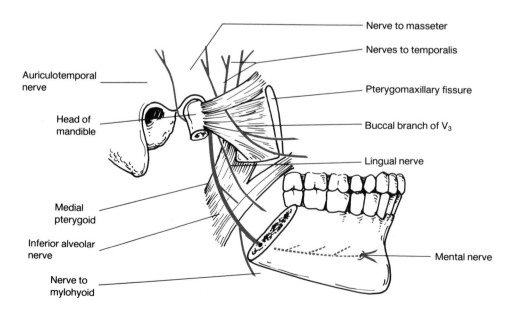

Auriculotemporal nerve

Head of mandible

Medial pterygoid

Inferior alveolar nerve

Nerve to mylohyoid

Nerve to masseter

Nerves to temporalis

Pterygomaxillary fissure

Buccal branch of V₃

Lingual nerve

Mental nerve

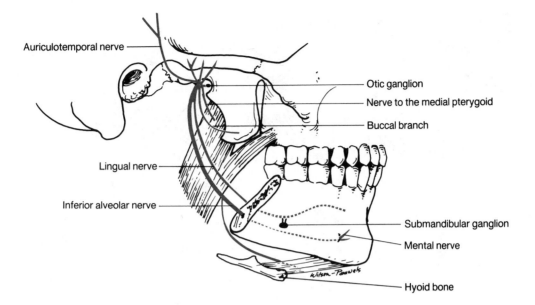

Figure 34-11.
Trigeminal nerve, mandibular division, (V_3) and branches.

otic ganglion is a parasympathetic ganglion attached to the mandibular nerve close to the foramen ovale in the infratemporal region. It receives preganglionic fibers from the glossopharyngeal nerve by way of its **lesser petrosal** branch. Postganglionic fibers travel from the otic ganglion to the parotid gland in the auriculotemporal branch of the mandibular nerve.

The fourth major sensory branch is the **lingual** nerve that supplies general sensation to the mucosa of the mouth and the anterior two thirds of the tongue. Near where it branches from the mandibular nerve, the lingual nerve is joined by the chorda tympani branch of VII that carries special sensory fibers for taste to the anterior two thirds of the tongue and preganglionic secretomotor fibers to the submandibular ganglion that is suspended from the lingual nerve. The preganglionic parasympathetic secretomotor fibers synapse in the ganglion, and postganglionic fibers travel to the submandibular and sublingual salivary glands.

CLINICAL NOTE

Trigeminal neuralgia is an irritation (neuralgia) of the trigeminal nerve or part of it. The pain is very severe and the patient searches desperately for relief.

VI—ABDUCENS NERVE

The abducens nerve carries **somatic motor** and **proprioceptive** impulses (Fig. 34-12). The nerve leaves the brain stem at the junction of the pons and medulla and pierces the dura on the dorsum sellae. It then runs through the cavernous sinus and leaves the skull by passing through the superior orbital fissure within the common tendinous ring from which the rectus muscles of the eye arise. Here it enters the deep surface of the lateral rectus muscle that it supplies.

CLINICAL NOTE

If the nerve is damaged, the patient cannot look laterally with the affected eye.

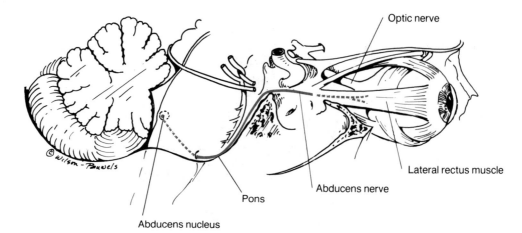
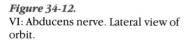

Figure 34-12.
VI: Abducens nerve. Lateral view of
orbit.

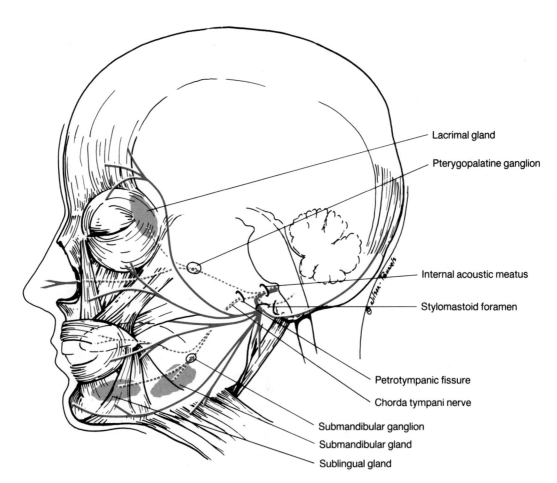

Figure 34-13.
VII: Facial nerve.

VII—FACIAL NERVE

The facial nerve contains **somatic motor, general sensory, special sensory**, and **parasympathetic** fibers (Fig. 34-13). The motor fibers supply the **muscles of facial expression**, the **stylohyoid, posterior belly of digastric**, and **platysma** muscles. The remaining fibers, called the nervus intermedius, provide a few general sensory fibers to the ear and, more importantly, special sensory fibers for **taste** to the anterior two thirds of the tongue and **parasympathetic** fibers to the lacrimal, submandibular, sublingual, and, possibly, the parotid glands. The facial nerve leaves the brain stem at the inferior border of the pons just medial to cranial nerve VIII. It courses laterally to the internal auditory meatus, where it pierces the dura mater to enter the petrous part of the temporal bone with cranial nerve VIII. Here the facial nerve lies in the facial canal between the organs of hearing and balance. In the petrous temporal bone, the nerve displays a small **geniculate ganglion** that contains the nerve cell bodies of the general and special sensory fibers of the nerve. At the genu (knee), the nerve gives off the **greater petrosal nerve** that leaves the skull through the greater petrosal hiatus (foramen) in the petrous temporal bone, passes over the upper part of the foramen lacerum, and enters the pterygoid canal in the base of the pterygoid plates of the sphenoid bone from which it emerges into the pterygopalatine fossa (Fig. 34-14). In the fossa it synapses in the parasympathetic pterygopalatine ganglion that is attached to the maxillary nerve. Postganglionic fibers from the ganglion are distributed through branches of the maxillary nerve to the mucous membranes of the nose, upper pharynx, and paranasal sinuses, to the small salivary glands in the palate, and, through the zygomatic nerve communicating with the lacrimal nerve, to the lacrimal gland.

After giving off the greater petrosal nerve, the facial nerve bends posteriorly to pass through the middle ear cavity still in the facial canal and exits the skull through the stylomastoid foramen. Before reaching the stylomastoid foramen, the facial nerve gives off the **chorda tympani** nerve, which crosses the medial surface of the tympanic membrane and exits the skull through the petrotympanic fissure to travel with the lingual branch

Figure 34-14.
Section through petrous temporal bone at the level of the jugular foramen and internal acoustic foramen showing the path of the greater and lesser petrosal nerves.

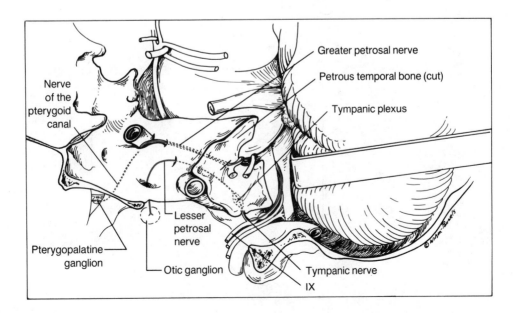

of the mandibular nerve. It carries the ***special sensory fibers for taste*** to the anterior two thirds of the tongue as well as ***preganglionic parasympathetic*** fibers to the submandibular ganglion. The ***submandibular ganglion*** is a parasympathetic ganglion located on the hyoglossus muscle and attached to the lingual nerve. From it postganglionic secretomotor fibers travel in the lingual nerve to reach the submandibular and sublingual glands (see Figs. 34-11 and 34-13).

CLINICAL NOTE

To test the facial nerve, ask the patient to imitate you as you wrinkle your forehead, frown, smile, and raise your eyebrows. Any asymmetry can be observed. Special sensation of taste can be tested by putting salt and then sugar on each side of the anterior part of the protruded tongue. The initial substance must be carefully rinsed away before the subsequent substance is tested.

VIII—VESTIBULOCOCHLEAR (AUDITORY) NERVE

The special senses of ***hearing*** and ***balance*** are carried in this nerve (Fig. 34-15). The vestibulocochlear nerve leaves the brain stem just lateral to the facial nerve at the inferior border of the pons, crosses to the internal

Figure 34-15.
VIII: Vestibulocochlear nerve.

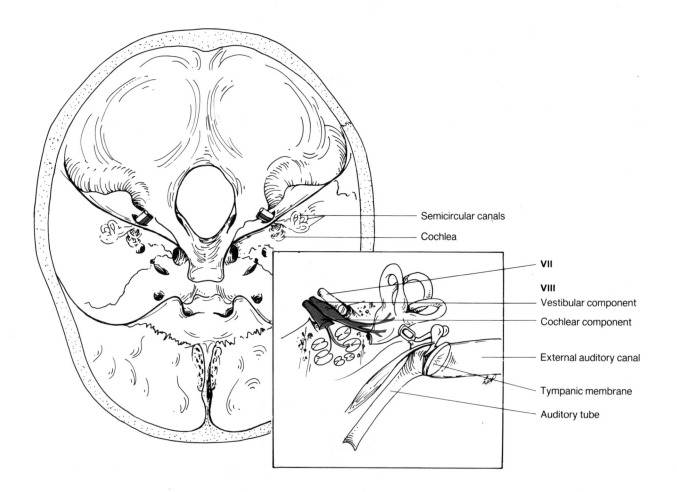

Semicircular canals

Cochlea

VII

VIII

Vestibular component

Cochlear component

External auditory canal

Tympanic membrane

Auditory tube

auditory meatus, pierces the dura, and passes to the internal ear, where it supplies special sensory fibers of hearing and balance.

Receptors in the vestibular (balance) apparatus carry sensory information to the vestibular ganglion in the inner ear. Part of the vestibular apparatus consists of an arrangement of three semicircular canals that contain fluid. When the fluid in the canals shifts in response to changes in body position, the movement is sensed by the vestibular apparatus and carried to the brain, where the interpretation of the movement influences the individual's perception of spatial orientation and balance.

Receptors in the cochlear (hearing) apparatus carry sensory information to the spiral (cochlear) ganglion in the inner ear. From the inner ear, central nerve processes reach the auditory cortex of the brain where sound can be appreciated.

CLINICAL NOTE

Both components of the nerve can be damaged by fractures to the skull involving the petrous temporal bone and by acoustic neuromas (tumors) at the opening of the internal auditory meatus. These can involve the facial nerve as well.

IX—GLOSSOPHARYNGEAL NERVE

Cranial nerve IX is mostly sensory. It transmits impulses that are **somatic motor and parasympathetic motor**, **general somatic**, and **visceral sensory** and fibers for the **special sense of taste** (Fig. 34-16). As the name implies, the glossopharyngeal nerve supplies **general somatic sensation** to the posterior third of the tongue, the tonsil, and pharynx. It also carries **visceral sensory** information from the carotid body (from chemoreceptors measuring oxygen and carbon dioxide tension in the blood) and the carotid sinus (from stretch receptors that monitor arterial blood pressure). It provides fibers for the **special sense of taste** to the posterior third of the tongue and epiglottis, and **parasympathetic motor** fibers to the parotid gland. The glossopharyngeal nerve is **somatic motor** to one muscle (the stylopharyngeus).

This nerve leaves the brain stem as the most cranial three or four rootlets of a series emerging between the olive and the inferior cerebellar peduncle. Before passing through the jugular foramen to exit the skull, the glossopharyngeal nerve gives off the **tympanic** branch. This branch carries preganglionic parasympathetic motor fibers to the **tympanic plexus** on the promontory of the middle ear cavity. The plexus will in turn give rise to the **lesser petrosal** nerve that emerges through the lesser petrosal hiatus (foramen) and continues through the foramen ovale to reach the otic ganglion, from which postganglionic fibers will emerge to travel in the auriculotemporal branch of V$_3$ to supply secretomotor fibers to the parotid gland (see Figs. 34-14 and 34-16). After giving off the tympanic branch, the glossopharyngeal nerve continues through the jugular foramen, where the superior and inferior glossopharyngeal ganglia are located and the nerve is very closely related to cranial nerves X and XI (see Fig. 34-21). From the jugular foramen it passes inferiorly in the neck and gives branches to the **pharyngeal plexus** on the middle constrictor of the pharynx. From the pharyngeal plexus, branches supply somatic sensation to the tonsil and pharynx. The nerve then swings forward along the stylopharyngeus muscle, supplies it and then passes between the superior and middle constrictors of the pharynx to reach the back of the tongue.

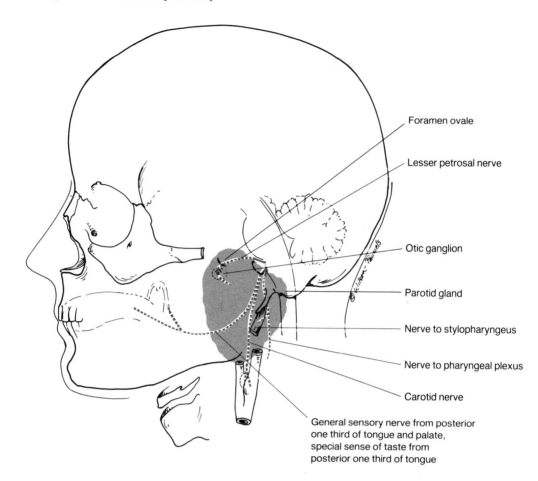

Foramen ovale

Lesser petrosal nerve

Otic ganglion

Parotid gland

Nerve to stylopharyngeus

Nerve to pharyngeal plexus

Carotid nerve

General sensory nerve from posterior
one third of tongue and palate,
special sense of taste from
posterior one third of tongue

Figure 34-16.
IX: Glossopharyngeal nerve.

CLINICAL NOTE

*Because of its close association with cranial nerves X and XI, there is sel-
dom an isolated lesion to cranial nerve IX (see Fig. 34-21). A simple test
for the integrity of the nerve is the* **gag reflex test**. *This is done by stroking
the wall of the pharynx at the back of the throat, which elicits a gag response. If the nerve were damaged there would be no gag response.*

X—VAGUS NERVE

The impulses transmitted by the vagus nerve are **somatic motor, parasympathetic motor**, and **visceral sensory.** The vagus nerve leaves the
brain stem as a series of rootlets in the groove between the olive and the
inferior cerebellar peduncle (see Figs. 34-17). It passes through the jugular foramen, piercing the dura as it leaves the skull. The superior (jugular)
and inferior (nodosum) vagal ganglia (sensory) are located there. The vagus nerve continues through the neck in the carotid sheath between the
internal jugular vein and the internal carotid artery. From the thoracic inlet
the nerve takes a different course on each side of the body to reach cardiac,
pulmonary, and esophageal plexuses (see Figs. 7-9 and 34-18). From the
esophagus, the vagus continues as right and left gastric nerves that supply
the gastrointestinal tract as far as the left colic flexure.

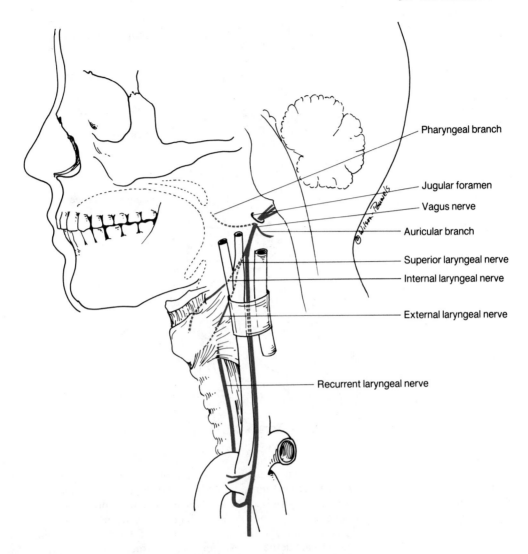

Pharyngeal branch

Jugular foramen

Vagus nerve

Auricular branch

Superior laryngeal nerve

Internal laryngeal nerve

External laryngeal nerve

Recurrent laryngeal nerve

Figure 34-17.
X: Left vagus nerve and left recurrent laryngeal nerve.

As the vagus nerve leaves the skull it gives off the small *auricular* branch, which supplies the posteroinferior part of the external surface of the tympanic membrane and the adjacent strip of skin in the external auditory canal. It then descends in the neck and gives off the *pharyngeal* branch, which enters the pharynx at the upper border of the middle constrictor to supply the majority of muscles of the pharynx. It does not supply stylopharyngeus or tensor palati. The *superior laryngeal* branch is given off next, leaving the vagus high up in the carotid sheath and, after running downward and medially, splits into the *external laryngeal* nerve, which supplies the cricothyroid muscle and the inferior constrictor muscle of the pharynx, and the *internal laryngeal* nerve, which pierces the thyrohyoid membrane and supplies sensory fibers to the mucous membrane of the larynx down to and including the vocal folds. This nerve supply evokes the cough reflex when something foreign tries to enter the larynx. The cough reflex is lost if the superior laryngeal nerve is damaged, whereas it is not affected by damage to the recurrent laryngeal nerve that supplies sensation to the larynx *below* the vocal folds. The *recurrent laryngeal*

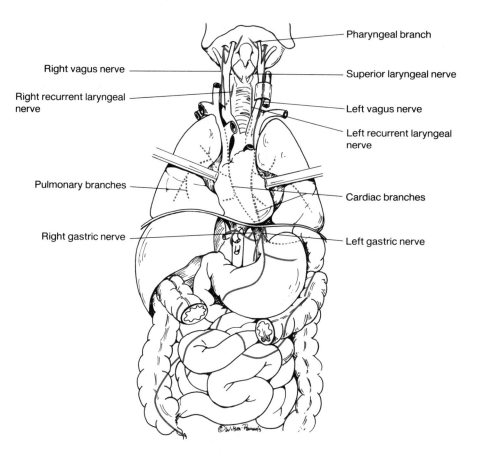

Right vagus nerve

Right recurrent laryngeal nerve

Pulmonary branches

Right gastric nerve

Pharyngeal branch

Superior laryngeal nerve

Left vagus nerve

Left recurrent laryngeal nerve

Cardiac branches

Left gastric nerve

Figure 34-18.
X: Vagus nerve and its distribution in the neck, thorax, and abdomen.

nerve is the next major nerve to branch from the vagus. The ***right recurrent laryngeal*** nerve arises from the vagus in front of the right subclavian artery and loops back behind the artery to ascend in the neck in the groove between the trachea and esophagus. The ***left recurrent laryngeal*** nerve descends into the thorax, loops back under the arch of the aorta, and ascends to reach the groove between the trachea and esophagus. Both nerves enter the larynx by passing deep to the inferior border of the inferior constrictor muscle. They supply all the intrinsic muscles of the larynx except the cricothyroid. ***Cervical cardiac*** branches descend from the neck to the cardiac plexus.

Note: The motor fibers of the vagus nerve to the larynx and pharynx are derived from the cranial portion of the accessory nerve (cranial XI). They hitch a ride on the vagus nerve for distribution to these areas.

CLINICAL NOTE

*Damage to the recurrent laryngeal nerve will result in hoarseness or loss of voice. If the nerve is merely compressed or is in the early stages of invasion by tumor, the **abductors** of the vocal fold will be affected first, and the vocal fold will lie in the adducted position. In complete transection of the recurrent laryngeal nerve the vocal fold will lie immobile in the **cadaveric position** and the opposite fold will move across the midline to enable phonation to occur. The voice may sound relatively normal but will tire easily and will then become weak and husky.*

Hyperactivity of the vagus nerve will cause hypersecretion of gastric acids that may erode the gastroduodenal mucosa, resulting in peptic ulcers.

XI—ACCESSORY NERVE

The accessory nerve is ***somatic motor*** and has cranial and spinal roots (Fig. 34-19). The cranial portion emerges as the inferior rootlets in the groove between the olive and the inferior cerebellar peduncle. They then pass to the jugular foramen, pierce the dura, and travel with cranial nerve X to supply somatic motor fibers to the intrinsic muscles of the pharynx and larynx. The spinal portion arises as rootlets from the side of the spinal cord in cervical segments 1 through 5. It ascends through the foramen magnum to join the cranial portion and leaves the skull through the jugular foramen. These fibers are somatic motor to the sternomastoid and trapezius muscles. The impulses in the accessory nerve are somatic motor and proprioceptive (sensory).

CLINICAL NOTE

The recurrent laryngeal branch of the vagus nerve carries the cranial fibers of the accessory nerve to the muscles of the larynx. It is particularly

Figure 34-19.
XI: Accessory nerve, spinal component.

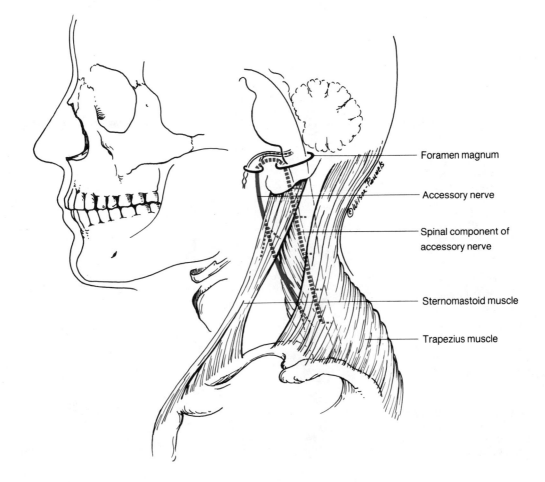

- Foramen magnum
- Accessory nerve
- Spinal component of accessory nerve
- Sternomastoid muscle
- Trapezius muscle

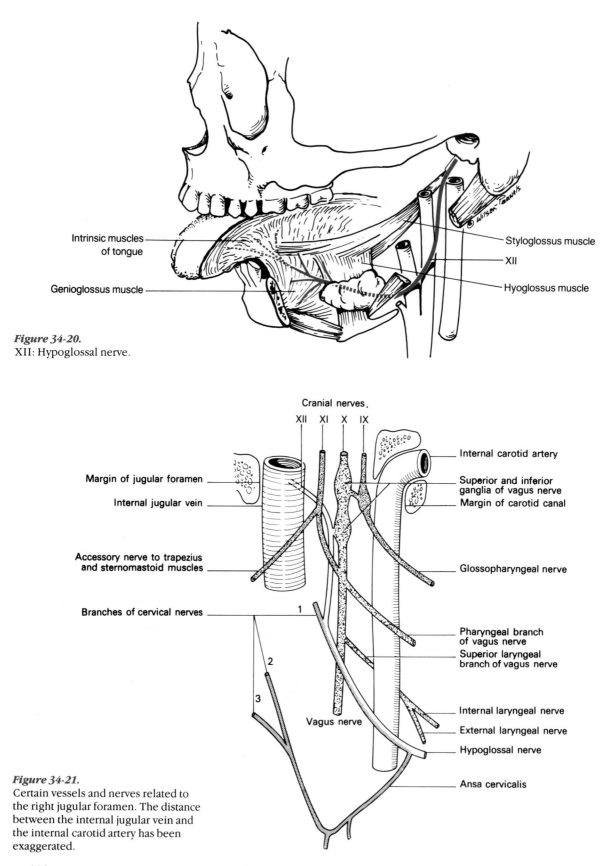

Intrinsic muscles of tongue

Genioglossus muscle

Styloglossus muscle

XII

Hyoglossus muscle

Figure 34-20.
XII: Hypoglossal nerve.

Cranial nerves

XII XI X IX

Margin of jugular foramen

Internal jugular vein

Accessory nerve to trapezius and sternomastoid muscles

Branches of cervical nerves

1

2

3

Vagus nerve

Internal carotid artery

Superior and inferior ganglia of vagus nerve

Margin of carotid canal

Glossopharyngeal nerve

Pharyngeal branch of vagus nerve

Superior laryngeal branch of vagus nerve

Internal laryngeal nerve

External laryngeal nerve

Hypoglossal nerve

Ansa cervicalis

Figure 34-21.
Certain vessels and nerves related to the right jugular foramen. The distance between the internal jugular vein and the internal carotid artery has been exaggerated.

vulnerable to damage during surgery to the neck, for example, carotid endarterectomy or thyroidectomy. A unilateral lesion to the nerve would result in hoarseness. The accessory nerve fibers to the sternomastoid and trapezius muscles may by damaged during radical neck dissections for malignant disease. The patient would have a shoulder drop and weakness in positioning the head.

XII—HYPOGLOSSAL NERVE

This nerve carries **somatic motor** and **proprioceptive** (sensory) fibers to all of the intrinsic and extrinsic muscles of the tongue except palatoglossus (Fig. 34-20). It emerges from the medulla as a series of rootlets in the groove between the olive and the pyramid and passes to the hypoglossal canal in the posterior cranial fossa, where it pierces the dura and leaves the skull. The nerve is medial to cranial nerves IX, X, and XI and passes down in the neck to loop forward anterior to the bifurcation of the common carotid artery to reach the posterior border of the mylohyoid muscle (Fig. 34-21). It passes deep to the mylohyoid to enter and supply the muscles of the tongue.

CLINICAL NOTE

To test for the integrity of this nerve, ask the patient to protrude the tongue. A unilateral lesion to the nerve would result in the tongue protruding to the affected side because the muscles of the damaged side would not be able to oppose the muscles of the intact side.

The Upper Limb

It will be recalled that we introduced the upper limb region in Chapter 2. In that chapter the general structure of skin and subcutaneous tissue was discussed, the breast was described, and some reference was made to the pectoral muscles. In this part of the book we shall describe the anatomy of the upper limb and the pectoral muscles in more detail.

CUTANEOUS STRUCTURES OF THE UPPER LIMB

VEINS

The cutaneous veins of the upper limb are large, prominent, and easily accessible for various clinical procedures, such as the taking of blood samples or the infusion of intravenous fluids.

The Plan of Venous Drainage of the Upper Limb (Fig. 35-1A). Blood from the palm of the hand and fingers drains to the dorsum of the hand either through veins in the interdigital clefts or around the borders of the hand. The veins on the dorsum of the hand form a dorsal arch. The arch drains on each side of the hand into a major cutaneous vein. The lateral vein, the ***cephalic vein***, passes superiorly, anterior to the elbow, to the pectoral region to enter the ***axillary*** vein in the deltopectoral groove. Above the elbow the cephalic vein lies superficial to the groove between the brachioradialis and biceps brachii muscles.

The ***basilic vein***, from the medial border of the hand, passes superiorly in front of the elbow, superficial to the groove between pronator teres and biceps brachii muscles, and becomes the ***axillary*** vein at the lower border of the teres major muscle, where it joins with the brachial vein and the venae comitantes of the brachial artery. In front of the elbow a large superficial vein, the ***median cubital vein***, connects the cephalic to the basilic vein.

NERVES

The cutaneous nerves of the upper limb form a general pattern that is easy to understand if one recalls that the lateral surface of the upper extremity

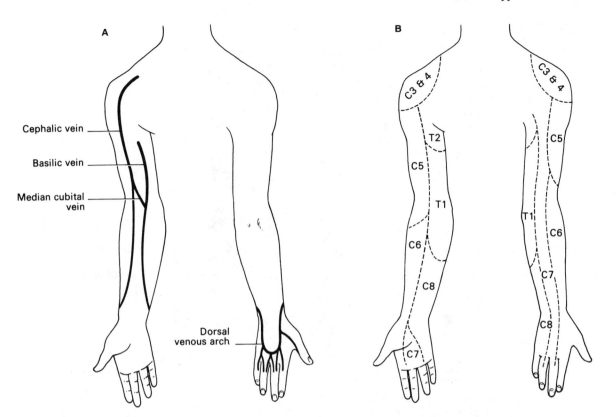

Cephalic vein

Basilic vein

Median cubital vein

Dorsal venous arch

Figure 35-1.
Cutaneous structures of the right upper extremity. **A.** Superficial veins.
B. Segmental cutaneous nerve supply.

is the embryonic *cranial* or preaxial aspect of the limb. With this in mind the reader should note the segmental innervation of the skin of the upper limb (Fig. 35-1B):

C₃ and C₄ supply the region of the shoulder.
C5 supplies the lateral arm.
C6, the lateral forearm, thumb, and index finger.
C7 supplies the middle and ring fingers and also the middle of the posterior surface of the upper limb.
C8 supplies the little finger, medial side of the hand, and medial part of the forearm.
T1 supplies the medial side of the proximal forearm and arm.
T2 supplies the skin of the floor of the axilla and a small portion of the medial aspect of the arm.

Individual Cutaneous Nerves of the Arm and Forearm
Supraclavicular nerves from C3 and C4 pass over the clavicle to supply the skin over the anterior wall of the axilla.
Upper lateral cutaneous nerve of arm comes from the axillary nerve and supplies the skin over the deltoid muscle.
Lower lateral cutaneous nerve of arm is a branch of the radial nerve.
Lateral cutaneous nerve of forearm is the continuation of the musculocutaneous nerve.
Medial cutaneous nerve of arm is a branch of the medial cord of the brachial plexus.
Medial cutaneous nerve of forearm is another branch of the medial cord of the brachial plexus.

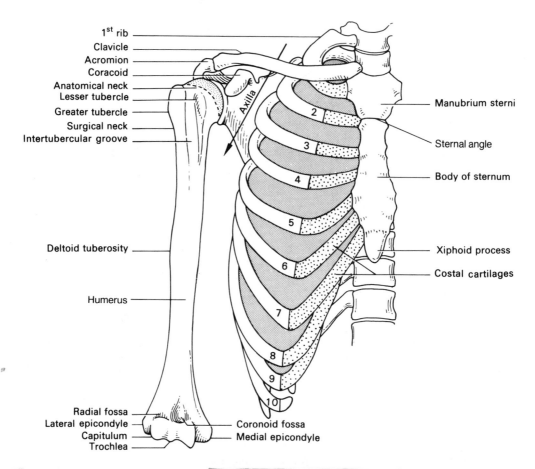

1st rib
Clavicle
Acromion
Coracoid
Anatomical neck
Lesser tubercle
Greater tubercle
Surgical neck
Intertubercular groove
Axilla
Manubrium sterni
Sternal angle
Body of sternum
Deltoid tuberosity
Xiphoid process
Costal cartilages
Humerus
Radial fossa
Lateral epicondyle
Capitulum
Trochlea
Coronoid fossa
Medial epicondyle

Figure 35-2.
Anterior view of the right pectoral region. Note the arrow marking the axilla. (In this specimen, the tenth costal cartilage does not join the ninth.)

Posterior cutaneous nerve of arm is a branch of the radial nerve.
Posterior cutaneous nerve of forearm is a branch of the radial nerve.
Intercostobrachial nerve supplies the axillary floor and the medial side of the upper arm.

THE PECTORAL REGION AND AXILLA

The study of this region requires an understanding of the bony structure of the chest wall (see Chap. 3), the pectoral girdle, and the humerus (Fig. 35-2).

THE PECTORAL GIRDLE

The ***pectoral girdle*** consists of the ***clavicle*** and the ***scapula***. They are considered to be parts of the upper extremity just as the os coxae was considered to be part of the lower limb.

   The clavicle obtains its name from a fancied resemblance to a Latin key. It is an S-shaped long bone (Fig. 35-3) that articulates medially with the manubrium sterni at the ***sternoclavicular joint***. This is a small joint providing the sole articulation between the upper extremity and the axial skeleton. Laterally it articulates with the ***acromion process*** of the scapula.

   The scapula is a flat, triangular bone (Fig. 35-4). Posteriorly it bears a prominent ***spine***, which is easily palpable and which ends laterally in

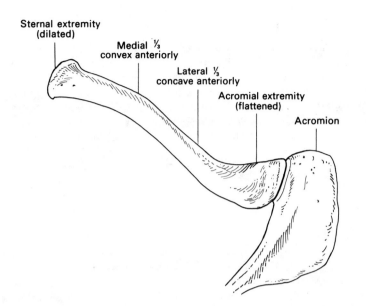

Figure 35-3.
Right clavicle and acromion (superior view).

the flattened **acromion**. The lateral angle of the scapula is truncated and forms the shallow **glenoid cavity** of the shoulder joint. Medial to the glenoid cavity and inferior to the acromion the **coracoid** process juts out in an anterolateral direction. The **scapular notch** in the superior border of the scapula should be noted; it should also be noted that the lateral border of the scapula is thickened to enable it to act in the capacity of a lever in certain movements of the scapula.

HUMERUS

The **humerus** is a long bone with a **head** that articulates with the **glenoid fossa** of the scapula, and the **greater** and **lesser tubercles** continuous inferiorly with the lateral and medial tubercular crests that form the lips of the **intertubercular (bicipital) groove**.

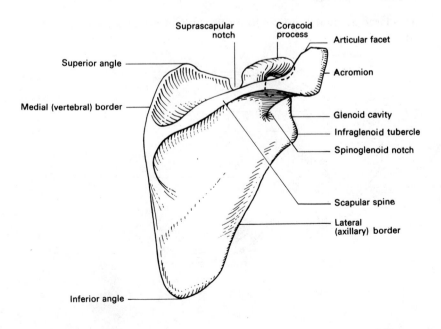

Figure 35-4.
Right scapula (posterior surface).

THE AXILLA

The *axilla* or armpit has the shape of a truncated pyramid with *three* walls, an apex, and a floor. The axilla is the space or passage through which the nerves and vessels from or to the trunk enter or leave the upper limb. The space can be visualized on the articulated skeleton: the thoracic wall forms the *medial* wall of the axilla, the clavicle helps to form the *anterior* wall, and the scapula forms the *posterior* wall. If we dress the bones with muscles and fascia, we shall have reconstructed the axilla. The muscles form the major parts of these walls. The reader should examine the relationships of the clavicle, first rib, and scapula that together form the rather narrow *apex* of the axilla.

BOUNDARIES (Fig. 35-5)

The medial wall is the wall of the thorax covered by the *serratus anterior muscle*. The *anterior wall* is formed by the clavicle and three muscles: *subclavius, pectoralis major*, and *pectoralis minor*.

The posterior wall is formed by the scapula and the muscle that covers nearly all of the anterior surface of the scapula, the *subscapularis*. Inferiorly a small part of the posterior wall is formed by the *teres major* muscle and the tendon of *latissimus dorsi*, which wraps around the teres major on its way to its humeral attachment.

A potential *lateral wall* of the axilla is formed by a narrow part of the humerus, the intertubercular (bicipital) groove, and the tendon of the long head of the *biceps brachii* muscle that lies in that groove.

Anterior Wall

The major components of the anterior wall are the clavicle and the pectoralis major muscle. The subclavius is an insignificant muscle that is almost wholly concealed by the clavicle. Posterior to pectoralis major the anterior wall is reinforced by pectoralis minor muscle and the clavipectoral fascia.

Pectoralis Major (Fig. 35-6)

This large triangular muscle forms the bulk of the anterior wall of the axilla.

Figure 35-5.
Right axilla (transverse section) showing walls and principal contents.

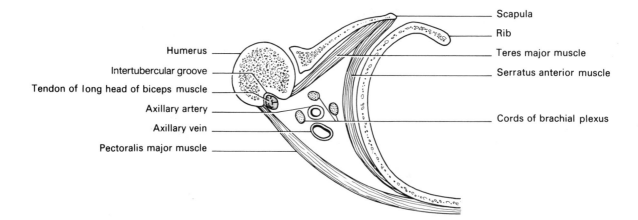

Scapula
Rib
Teres major muscle
Serratus anterior muscle
Cords of brachial plexus

Humerus
Intertubercular groove
Tendon of long head of biceps muscle
Axillary artery
Axillary vein
Pectoralis major muscle

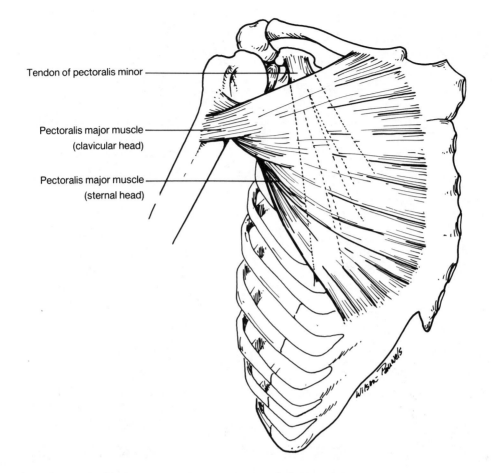

Tendon of pectoralis minor ——————

Pectoralis major muscle ——————
(clavicular head)

Pectoralis major muscle ——————
(sternal head)

Proximal Attachment. The pectoralis major has two sets of fibers which have different attachments and different segmental origins. The clavicular head comes from the anterior aspect of the clavicle and meets the sternocostal head, which comes from the sternum and the upper six costal cartilages, at the pectoralis tendon.

Distal Attachment. The lateral lip of the intertubercular (bicipital) groove of the humerus. At the distal attachment the tendon of pectoralis major undergoes a spiral twist so that the lower fibers are attached to the humerus at a higher level than the upper fibers (Fig. 35-6).

Nerve Supply. The lateral and medial pectoral nerves from the lateral and medial cords of the brachial plexus (C5, *C6*, *C7*, *C8*, and T1).

Action. Both heads can adduct and medially rotate the humerus. The clavicular head can flex the extended shoulder, and the sternocostal head can extend the flexed shoulder joint. Note that here is a muscle that produces two antagonistic movements. This is possible because it is a composite muscle and does not have to contract as a unit. The best way to demonstrate pectoralis major is to put the hands on the hips and push. With the upper limb fixed the pectoralis major can pull the chest wall toward the upper limb, as in pulling yourself up on a beam.

Figure 35-6.
Right pectoralis major muscle.

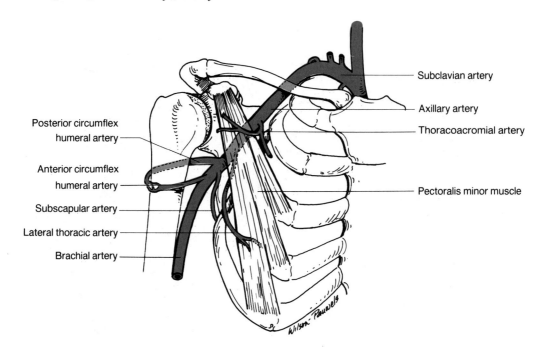

Posterior circumflex humeral artery

Anterior circumflex humeral artery

Subscapular artery

Lateral thoracic artery

Brachial artery

Subclavian artery

Axillary artery

Thoracoacromial artery

Pectoralis minor muscle

Figure 35-7.
Right axillary artery (anterior view).

Pectoralis Minor

This small triangular muscle lies deep to pectoralis major (Figs. 35-6 and 35-7).

Proximal Attachment. Ribs 3, 4, and 5.

Distal Attachment. The *coracoid process* of the scapula.

Nerve Supply. From the medial pectoral nerve (C6, *C7*, and C8).

Action. It draws the scapula downward and forward and steadies it on the chest wall.

Clavipectoral Fascia. This is a thin sheet of deep fascia running from the clavicle to the fascial floor of the axilla, and it therefore supports the floor of the axilla by tying it to the clavicle. When the clavicle is raised, it raises the floor of the axilla. The pectoralis minor muscle is surrounded by the clavipectoral fascia. When the arm is abducted the clavipectoral fascia hitches up the axillary floor.

Cephalic Vein. The cephalic vein (Fig. 35-8) runs in the subcutaneous tissue from the lateral border of the hand to the deltopectoral groove, where it pierces the deep fascia to drain into the axillary vein.

Cutaneous Nerves. The skin over the pectoral region is supplied by the upper intercostal nerves and by the *supraclavicular nerves* (C3 and C4) from the cervical plexus.

CONTENTS OF THE AXILLA

The axilla contains elements of the *brachial plexus* and the *axillary artery* and *vein*, as well as a very important group of *lymph nodes* (see Chaps. 2 and 43). These structures are embedded in fat and fascia.

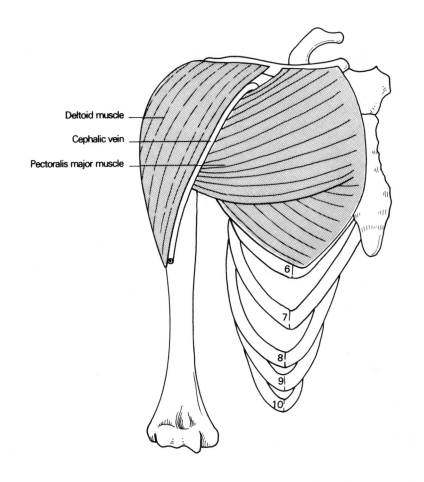

Figure 35-8.
Right deltoid and pectoralis major
muscles (anterior view).

Deltoid muscle

Cephalic vein

Pectoralis major muscle

6

7

8

9

10

Axillary Lymph Nodes

The lymph nodes of the axilla were described in Chapter 2 in connection with the description of the breast (Fig. 35-9) and will be described again in Chapter 43.

Veins

Many veins of varying sizes are found in the axilla, but only one warrants consideration here.

Axillary Vein.   The chief venous structure draining the upper limb is the axillary vein. It is formed at the inferior border of the teres major as the continuation of the **basilic vein,** which drains the medial surface of the forearm and arm. It is joined by many tributaries in its course through the axilla, but its major tributaries are the **brachial** and **cephalic** veins. At the lateral margin of the first rib it continues as the **subclavian vein**.

Arteries

The **axillary artery** (see Fig. 35-7) is the continuation of the subclavian artery and enters the axilla together with the brachial plexus of nerves enclosed in a fascial sheath, the **axillary sheath**, which is a continuation

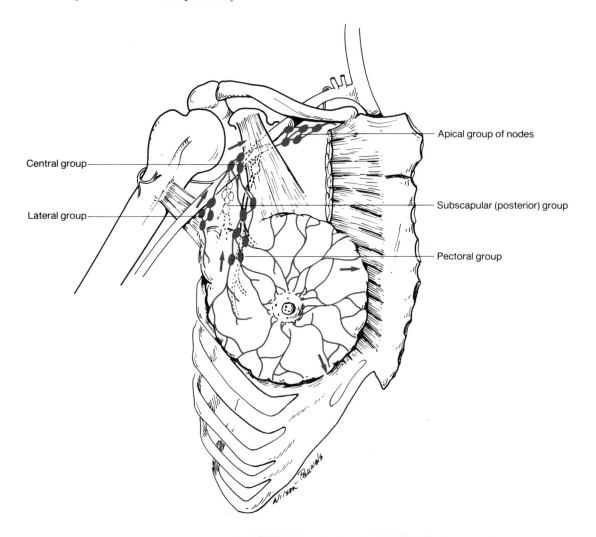

Central group ⎯⎯⎯⎯⎯⎯

Lateral group ⎯⎯⎯⎯⎯⎯

⎯⎯⎯⎯⎯⎯ Apical group of nodes

⎯⎯⎯⎯⎯⎯ Subscapular (posterior) group

⎯⎯⎯⎯⎯⎯ Pectoral group

Figure 35-9.
Axillary lymph nodes.

of the *prevertebral fascia* of the neck. This sheath is important because a local anesthetic may be injected into it at the base of the posterior triangle of the neck for the purpose of anesthetizing most of the upper limb. The axillary sheath does ***not*** enclose the axillary vein.

*The **axillary artery*** begins at the outer border of the first rib and changes its name to ***brachial artery*** at the lower border of the teres major muscle. In its course the axillary artery passes deep to the pectoralis minor muscle; for descriptive purposes it is divided by that muscle into three parts. The first part lies proximal to pectoralis minor, the second part lies posterior to the muscle, and the third part lies distal to it. The axillary artery gives off several branches, the more important ones of which are outlined below (see Fig. 35-7).

1. The ***superior thoracic artery*** runs along the superior border of pectoralis minor and supplies the pectoral muscles and the thoracic wall.

2. The ***thoracoacromial*** (acromiothoracic) artery splits into four branches that are distributed to the pectoral, acromial, coracoid, and deltoid regions.

3. The ***lateral thoracic artery*** follows the lower border of pectoralis minor and supplies the chest wall and ***the mammary gland***. It becomes a very prominent vessel during pregnancy and lactation.

4. The ***subscapular artery*** is a relatively large branch of the axillary artery. It passes downward along the lateral border of the scapula and gives off a large ***circumflex scapular*** branch that winds around the lateral border of the scapula and anastomoses in the supraspinous and infraspinous fossae with the suprascapular artery.

5. The ***anterior and posterior humeral circumflex arteries*** pass around the surgical neck of the humerus and anastomose with each other. The posterior humeral circumflex artery passes with the ***axillary nerve*** through the ***quadrilateral space***, which is bordered by the humerus, teres minor and subscapularis, and teres major and the long head of triceps brachii.

CLINICAL NOTE

The branches of the axillary artery, particularly the lateral thoracic and subscapular arteries, provide a route for collateral circulation between the proximal subclavian artery and the distal axillary artery. Particularly noteworthy are the anastomosis between the subscapular artery and the suprascapular artery, and the potentially large anastomotic channels between the lateral thoracic artery and the intercostal arteries. These form a major secondary circulation which, in a young person, can enlarge to take over the blood supply of the upper extremity if the axillary artery has to be ligated.

Nerves

The large and important ***brachial plexus*** provides innervation for almost the entire upper limb (Fig. 35-10).

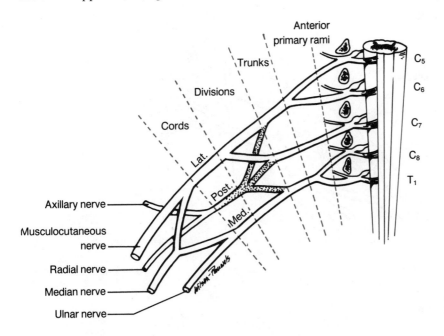

Figure 35-10.
Basic plan of the right brachial plexus.

BRACHIAL PLEXUS

The brachial plexus starts in the base of the neck in an area known as the *posterior triangle*. The distal portion lies in the axilla where the plexus divides into major branches for the upper limb. The whole length of the plexus is about 15 cm.

The brachial plexus is formed by the *anterior primary rami* of cervical nerves *5, 6, 7,* and *8* and the *first thoracic nerve*. Parts of C4 and T2 may also participate. It must be remembered that the first cervical nerve emerges *above* the first cervical vertebra, the second nerve above the second vertebra, and so on until the *eighth* cervical nerve. It emerges *below* the seventh cervical vertebra and the *first thoracic* nerve emerges *below* the first thoracic vertebra.

THE PLAN OF THE BRACHIAL PLEXUS (Fig. 35-10)

The *roots* of the brachial plexus are the anterior primary rami of the nerves involved. The *trunks* are formed by the junction of roots, and the *cords* are formed by the splitting of the trunks into *divisions* that recombine to form the cords. More specifically,

the *anterior primary rami* of C5 and C6 join to form the *upper trunk;*

the *anterior primary ramus* of C7 continues by itself to form the *middle trunk;*

the *anterior primary rami* of C8 and T1 combine to form the *lower trunk.*

Figure 35-11.
Important branches of the right brachial plexus.

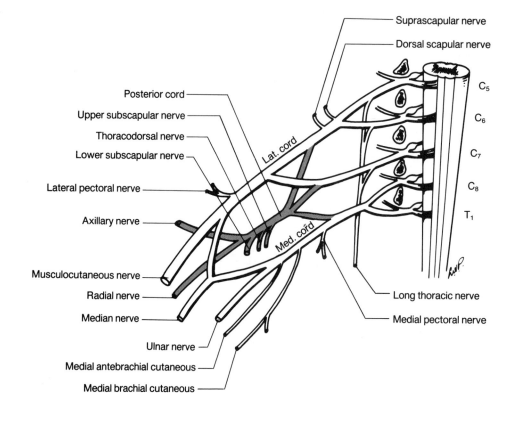

Each trunk divides into an ***anterior division*** and a ***posterior division;***

the ***anterior divisions*** of the ***upper*** and ***middle*** trunks unite to form the ***lateral cord;***

the ***anterior division*** of the ***lower trunk*** continues by itself to form the ***medial cord;***

the ***posterior divisions*** of ***all three trunks*** unite to form the ***posterior cord;***

the cords surround the axillary artery and are named lateral, medial, and posterior according to their relationships to the ***second part*** of the artery.

Each cord of the ***brachial plexus*** divides into two terminal branches (Fig. 35-11);

the lateral cord divides into the ***musculocutaneous nerve*** and the ***lateral root*** of the ***median nerve;***

the ***medial cord*** divides into the ***ulnar nerve*** and the ***medial root*** of the ***median nerve;***

the ***posterior cord*** divides into the ***radial nerve*** and the ***axillary nerve.***

Diagrammatically the formation of the median nerve and the origins of the musculocutaneous and ulnar nerves can be likened to the formation of the letter M, which may be considered the key to remembering the arrangement of these structures.

VARIATIONS OF THE BRACHIAL PLEXUS

Variations of contribution to and distribution of the brachial plexus are so common that they might be considered to be the rule rather than the exception. Thus the upper root of the plexus may be C4, in which case C8 will form the lower root. This is called a ***prefixed plexus***. Alternately the lowest root may be T2, in which case the upper root will be C6. This is called a ***postfixed plexus***.

Communications between the cords and branches of the cords are common; in particular the ulnar nerve and median nerve can often be seen to communicate in the axilla or upper arm so that the nerve supply of the small muscles of the hand may be subject to some variations.

CLINICAL NOTE

*The existence of a **postfixed** brachial plexus or the presence of a **cervical rib** may cause a **thoracic inlet syndrome** (perhaps better named **axillary inlet syndrome**) in which the lowest root to the brachial plexus may be chafed or kinked over the first rib (in the case of a postfixed plexus) or over a cervical rib (in the case of a normally fixed plexus). This may produce sensory changes in the forearm and hand or muscular changes in the small muscles of the hand. Obviously a **prefixed plexus** would not (at least theoretically) encounter any problems even in the presence of a cervical rib. Arterial problems, resulting from obstruction at the junction of the subclavian and axillary arteries, may also occur in this syndrome. The problem can be solved by excision of the offending rib or sometimes simply by division of the scalenus anterior muscle at its attachment to the first rib.*

BRANCHES OF THE BRACHIAL PLEXUS (Fig. 35-11)

The roots and trunks of the brachial plexus give rise to certain branches, and because that part of the plexus is located above the clavicle, in the neck, these branches are often called ***supraclavicular*** branches.

Branches of the Roots and Trunks

1. The ***dorsal scapular nerve*** (***C5***) arises from the anterior primary ramus of C5 and supplies the ***rhomboid*** muscles and sometimes the ***levator scapulae*** muscle.
2. The ***long thoracic nerve*** (of Bell) or ***nerve to serratus anterior*** arises directly from the anterior primary rami of ***C5, C6, and C7*** and passes posterior to the brachial plexus in the neck to its muscle on the lateral thoracic wall.
3. The ***suprascapular nerve*** (***C5 and C6***) arises from the upper trunk and crosses the posterior triangle of the neck before passing through the ***suprascapular notch*** of the scapula to supply the ***supraspinatus*** and ***infraspinatus*** muscles.
4. The unimportant ***nerve*** to the ***subclavius muscle*** (C5 and C6) passes directly to that muscle from the upper trunk of the plexus.

CLINICAL NOTE

*The long thoracic nerve is vulnerable to injury in radical mastectomy and in certain traction injuries of the upper limb. Damage to it will paralyze the serratus anterior, leading to the disability of **winged scapula***.

Branches of the Cords

The lateral cord has three branches, the medial cord has five branches, and the posterior cord has five branches. In each cord two of its branches are terminal branches.

Lateral Cord.   The lateral cord has the following branches:

1. The ***lateral pectoral nerve*** (C5, C6, and C7) pierces the clavipectoral fascia and supplies the pectoralis major muscle. It will often communicate with the medial pectoral nerve (from the medial cord).
2. The important ***musculocutaneous nerve*** ([C5], C6, and C7) is one of the two terminal branches of the lateral cord. It enters the ***coracobrachialis*** muscle and, after emerging from its lateral aspect, continues on to supply the ***biceps brachii*** and ***brachialis*** muscles. The musculocutaneous nerve terminates as the ***lateral cutaneous nerve of the forearm.***
3. The other terminal branch of the lateral cord is the large and important ***lateral root*** (C5, C6, and C7) of the ***median nerve***. The lateral root of the median nerve and its medial root (from the medial cord) unite on the lateral aspect of the axillary artery.

Medial Cord.   The medial cord of the brachial plexus has five branches.

1. The ***medial pectoral nerve*** (C8 and T1) passes through the pectoralis minor muscle into the pectoralis major, supplying both muscles.
2. The ***medial cutaneous nerve of arm*** (C8 and T1) is a small branch that supplies the medial aspects of the upper arm and a portion

of the forearm. In the axilla the medial cutaneous nerve of the arm is joined by the ***intercostobrachial nerve***, which comes from T2 and pierces the second intercostal space to supply the skin over the upper part of the medial aspect of the arm and the floor of the axilla.

3. The ***medial cutaneous nerve of forearm*** (C8 and T1) (medial antebrachial cutaneous nerve) is a relatively large nerve that may be mistaken for the ulnar nerve. It supplies the medial aspect of the forearm.

4. The ***medial root*** (C8 and T1) of the ***median nerve*** is one of the two terminal branches of the medial cord. The root crosses in front of the axillary artery to join the lateral root, and the median nerve descends through the arm, crossing to the medial side of the brachial artery, to be distributed in the forearm and hand.

5. The very important ***ulnar nerve*** ([C7], C8, and T1) is the other terminal branch of the medial cord. It passes down the arm on the medial side of the brachial artery. Halfway down the arm it pierces the medial intermuscular septum to reach the posterior compartment of the arm and continues to descend anterior to the triceps muscle. At the elbow it passes behind the medial epicondyle of the humerus and enters the anterior compartment of the forearm to be distributed in the forearm and hand.

Posterior Cord.   The posterior cord has five branches which, in general, supply muscles that extend joints of the upper limb (as befits the posterior divisions of anterior primary rami of spinal nerves). They also supply cutaneous branches to the extensor surfaces of the limb.

1. The ***upper subscapular nerve*** (C5 and C6) supplies the ***subscapularis muscle.***

2. The ***thoracodorsal nerve*** (nerve to latissimus dorsi) (C6, C7, and C8) runs downward and laterally through the axillary fat and close to the subscapular group of lymph nodes to supply the large ***latissimus dorsi*** muscle.

3. The ***lower subscapular nerve*** (C5 and C6) supplies the ***teres major*** muscle.

4. The important ***axillary nerve*** (circumflex nerve) (C5 and C6) is one of the two terminal branches of the posterior cord. It accompanies the posterior circumflex artery through the ***quadrilateral space***, winds around the "surgical neck" of the humerus, supplies the ***deltoid*** and ***teres minor*** muscles, and terminates by supplying the skin over the deltoid muscle.

5. The large and very important ***radial nerve*** (C5, C6, C7, and [T1]) is the other terminal branch of the posterior cord. It is the major nerve supply to the extensor muscles of the upper limb and also supplies the cutaneous innervation for the extensor aspect of the limb. It leaves the axilla by passing inferior to teres minor and then winds its way around the posterior aspect of the shaft of the humerus, in the ***spiral groove*** of that bone, by passing deep to the long and lateral heads of the triceps muscle.

CLINICAL NOTE

Fractures of the bones of the upper limb pose threats to the integrity of many of the major nerves.

1. *The axillary nerve can be injured in dislocations of the shoulder joint and in fractures through the surgical neck of the humerus. Paralysis of the deltoid and teres minor muscles will result.*

2. *The radial nerve may be injured by fractures through the shaft of the humerus, with resultant loss of extensor muscle function in the arm, forearm, and hand as well as sensory loss over the extensor surface of much of the limb.*

BRACHIAL PLEXUS INJURIES

*Such injuries are fortunately rare but devastating when they occur. Traction on and lateral flexion of the fetal head at delivery may cause damage to C5 and C6 nerve roots. This causes **Erb's palsy** in which abduction and external rotation of the shoulder as well as supination of the forearm and extension of the wrist are lost.*

*Similar traction in the opposite direction on the fetal shoulder and arm during a "breech" delivery can cause damage to T1 root, resulting in a **Klumpke** palsy. In this condition there is loss of function of the small muscles of the hand, and a simian-like **claw deformity** (main en griffe) develops (see Chap. 44, Fig. 44-3).*

Direct trauma from, for example, gunshots as well as traction injuries can cause complete paralysis of most of the upper limb. Because of the complex branching and joining of the elements of the plexus, surgical repair is very difficult and meaningful restoration of function is limited.

Nerve injuries and their consequences will be considered in more detail in Chapter 44.

The Posterior Triangle of the Neck

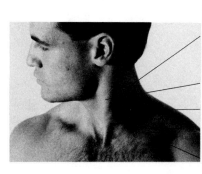

The posterior triangle is a long, relatively narrow area that wraps around the lateral surface and base of the neck like a spiral.

BOUNDARIES OF THE POSTERIOR TRIANGLE

The borders of the posterior triangle are as follows (Fig. 36-1).

> *Inferior Border*. The clavicle.
> *Anterior Border*. The posterior border of the
>   ***sternocleidomastoid muscle***.
> *Posterior Border*. The anterior border of the
>   ***trapezius muscle***.

Sternocleidomastoid Muscle (Sternomastoid Muscle)

This long narrow muscle runs from near the midline to the posterosuperior aspect of the neck.

Lower Attachment. The medial one third of the clavicle and by a separate head from the manubrium sterni.

Upper Attachment. The mastoid process and the lateral half of the superior nuchal line of the skull.

Nerve Supply. The ***spinal*** portion of the ***accessory nerve*** (cranial nerve XI).

Action. Acting individually, each muscle turns the head to the opposite side; acting together, the sternomastoids are flexors of the neck and thrust the head forward or raise it (as from a pillow).

THE ROOF OF THE POSTERIOR TRIANGLE

The roof of the posterior triangle is the ***investing layer of the deep cervical*** fascia which attaches to the clavicle and manubrium sterni below, and to the mastoid process, superior nuchal line, and the lower border of the

Trapezius (upper fibers) forming posterior boundary of triangle

Sternomastoid forming anterior border of triangle

Floor of triangle

Supraclavicular fossa

Clavicle forming inferior border of triangle

Photo shows the posterior triangle of the neck.

36

Figure 36-1.
Boundaries of the right posterior triangle of the neck.

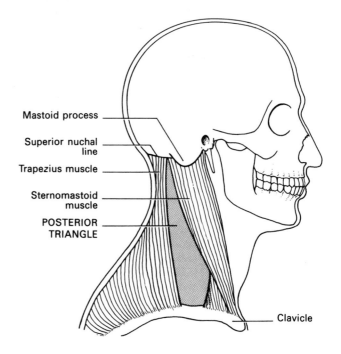

mandible above. The fascia splits to enclose the sternomastoid and trapezius muscles. Part of the superficial fascia of the neck, including the posterior triangle, contains the platysma muscle, which is a muscle of facial expression extending from the deep fascia below the clavicle to the mandible. It is supplied by the facial nerve and will tighten the skin of the neck as in snarling and grimacing. It can be contracted to facilitate shaving.

THE FLOOR OF THE POSTERIOR TRIANGLE

The floor of the posterior triangle consists of four muscles covered by the prevertebral layer of the deep cervical fascia. The reader is advised to review the cervical vertebrae to understand the attachments of these muscles (Fig. 36-2).

SPLENIUS CAPITIS

This muscle's lower attachments are to the ligamentum nuchae and the spines of the upper thoracic vertebrae. Its fibers run upward and laterally to attach to the mastoid process and the lateral one third of the superior nuchal line.

LEVATOR SCAPULAE

This muscle is attached to the transverse processes of the upper four cervical vertebrae, and its fibers pass downward to become attached to the vertebral border of the scapula above its spine.

SCALENUS MEDIUS

This muscle is attached to the posterior tubercles of the transverse processes of all the cervical vertebrae. Below it is attached to the lateral part

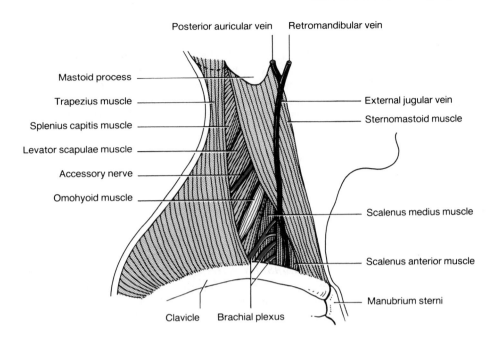

Figure 36-2.
Right posterior triangle (floor and contents).

of the first rib behind the ***groove for the subclavian artery***. Although the precise attachments of muscles to bone are not stressed in this book, the reader is advised to examine the first rib and to identify the area of attachment of the scalene muscles and the grooves made by the subclavian artery and vein. This will facilitate the understanding of the relationships and course from the neck to the axilla of the brachial plexus and the axillary vessels. Note that the scalenus medius is attached to the transverse processes of the cervical vertebrae posterior to the points of issue of the roots of the brachial plexus (Fig. 36-3).

Figure 36-3.
Right brachial plexus.

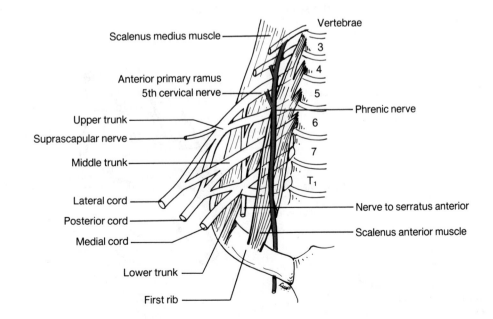

SCALENUS ANTERIOR

This muscle is attached above to the anterior tubercles of the second to sixth cervical vertebrae (*i.e.*, anterior to the roots of the brachial plexus) and attaches below to the scalene tubercle of the first rib. (Note from the markings on the rib that the impression for the subclavian artery is behind the scalene tubercle and that for the subclavian vein is in front of it.)

CONTENTS OF THE POSTERIOR TRIANGLE

VEINS

EXTERNAL JUGULAR VEIN

This subcutaneous vein is formed just below the lobule of the ear by the junction of the ***posterior auricular vein*** and the posterior division of the ***retromandibular vein*** (see Fig. 36-2). It runs downward on the sternomastoid muscle, crossing the muscle from its anterior to its posterior border. It enters the posterior triangle at the posterior border of the sternomastoid a short distance above the clavicle and drains into the ***subclavian vein***.

CLINICAL NOTE

In most patients the external jugular vein is visible through the skin. It can be used as a manometer to judge the pressure of blood in the right atrium. In a patient sitting upright, distension of the vein by blood should not be visible if the right atrial pressure is normal or below normal.

SUBCLAVIAN VEIN

The subclavian vein drains the upper limb and receives several cervical tributaries. It lies posterior to the clavicle, above the first rib and in front of the scalenus anterior muscle (Fig. 36-4). This vein is accessible for the

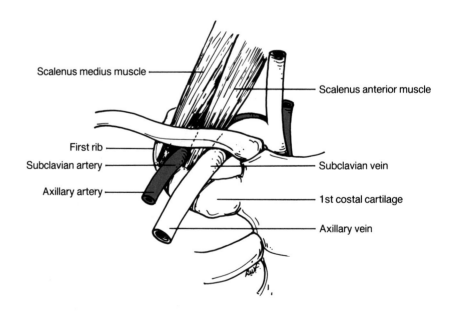

Figure 36-4.
Right first rib and some important related structures.

introduction of a central venous catheter, and the best approach is to insert the catheter by a subclavicular approach, taking care not to puncture the posteriorly placed cervical pleura.

ARTERIES

SUBCLAVIAN ARTERY

The subclavian artery lies low in the posterior triangle and is usually hidden by the clavicle. The artery is in direct contact with the first rib posterior to the scalenus anterior muscle (Fig. 36-4) and inferior to the inferior trunk of the brachial plexus. In an emergency the artery can be compressed against the first rib to control bleeding in the upper limb. The subclavian artery has several branches (Fig. 36-4):

1. ***Vertebral Artery***. This major branch of the subclavian artery lies too far medial to be considered part of the posterior triangle. It was described in Chapter 23.
2. ***Internal Thoracic Artery***. The internal thoracic artery arises from the inferior aspect of the subclavian artery and descends through the thorax lateral to the sternum (see Chap. 3).
3. ***Thyrocervical Trunk***. This artery usually comes off medial to the scalenus anterior muscle. It gives off the ***inferior thyroid artery*** and the ***suprascapular*** and ***transversus colli*** (cervical) arteries.
4. ***Costocervical Trunk***. This branch gives origin to the ***superior intercostal artery***, which passes over the apex of the lung (and cervical pleura) to be distributed to the first two intercostal spaces.
5. ***Dorsal Scapular Artery***. This branch runs between the levator scapulae and rhomboid muscles to the medial border of the scapula.

MUSCLE

One muscle, the ***omohyoid***, traverses the posterior triangle. Inferiorly it is attached to the medial lip of the suprascapular notch of the superior border of the scapula and has a central tendon that joins it to its hyoid portion, which attaches to the hyoid bone.

NERVES

The cervical nerves emerge posterior to the vertebral artery, which ascends in the foramina transversaria. The brachial plexus has already been described (see Chap. 35), but it should be noted that its roots and trunks are located in the posterior triangle of the neck. Certain other nerves are encountered in the posterior triangle.

ACCESSORY NERVE

The portion of the accessory nerve that is seen in the neck is the ***spinal accessory nerve*** (cranial nerve XI). The nerve pierces the anterior border of the sternomastoid muscle near its upper attachment and leaves the muscle at about the midpoint of its posterior border. The nerve then passes obliquely downward and laterally across the posterior triangle (see Fig. 36-2), but within a sleeve of the investing layer of deep cervical fascia, to enter the trapezius muscle at the inferolateral corner of the posterior triangle. It contains fibers from the ***third and fourth cervical nerves***, but these are sensory (proprioceptive).

CLINICAL NOTE

The accessory nerve runs in a very superficial position within the deep cervical fascia and may be injured by operations on superficial structures in that part of the neck. It may become stuck to deep cervical lymph nodes, particularly to those of the jugulo-omohyoid group, and is at risk during surgery on these nodes. Damage to the nerve will paralyze the trapezius muscle and cause severe limitations of upward rotation of the scapula so that the arm cannot be abducted beyond 90 degrees.

CERVICAL PLEXUS

The cervical plexus consists of the ***anterior primary rami*** of cervical nerves 2, 3, and 4. It has the following branches:

Cutaneous Branches
The ***lesser occipital nerve*** comes from C2. It hooks around the accessory nerve and runs upward over sternomastoid to supply the skin posterior to the auricle.
The ***great auricular nerve***, from C2 and C3, passes slightly in front of the lesser occipital nerve and supplies the lower half of the auricle and the skin around it.
The ***transverse colli (cervical) nerve***, from C2 and C3, passes horizontally forward to supply the skin over the anterior aspect of the neck.
The ***supraclavicular nerves*** from C3 and C4 pass downward over the clavicle to supply the skin over the upper shoulder and upper pectoral region.

PHRENIC NERVE

The ***phrenic nerve*** comes from cervical roots 3, 4, and 5. The three roots join in front of the scalenus anterior muscle, and the phrenic nerve leaves the medial edge of the muscle to pass into the thoracic inlet (see Fig. 36-3). Its subsequent path through the thorax to the diaphragm was described in Chapters 3 and 7.

CLINICAL NOTE

*Injury to the phrenic nerve paralyzes the diaphragm. Spinal cord transection or compression above C3 (C4) will cause complete respiratory paralysis, yet a lesion below C5 is compatible with life. Remember, "**C3, 4, and 5 keep the diaphragm alive.**"*

Structures of the Back and Shoulder Region

BONY LANDMARKS

The following bony landmarks should be examined before proceeding (Fig. 37-1). Most of them can be palpated on the living body.

Occipital Bone (of Skull). The ***external occipital protuberance*** and the ***superior nuchal line*** should be re-examined.

Vertebrae. The spinous process of C7 or ***vertebra prominens***, all of the ***thoracic*** spinous processes, the five ***lumbar*** spinous processes, and the ***median crest*** of the sacrum are all palpable. The spinous process of L5 lies at about the same level as the tubercle of the ***iliac crest***.

Scapula. Review the borders and angles of the scapula and note that the ***acromion***, the ***spine***, and often the ***medial border*** and the ***coracoid process*** are palpable with some gentle deep pressure (Fig. 37-2).

Humerus. The ***head***, ***surgical*** and ***anatomical necks***, ***greater*** and ***lesser tubercles***, the ***deltoid tuberosity***, the ***humeral shaft***, and ***spiral groove*** should be identified (Fig. 37-2).

LIGAMENTOUS STRUCTURES

The ***ligamentum nuchae*** is a fibroelastic band (really more like a septum) that joins the spines of the cervical vertebrae, and more particularly the spine of C7, to the external occipital protuberance (Fig. 37-3).

MUSCLES THAT CONNECT THE HUMERUS AND SCAPULA TO THE AXIAL SKELETON

This group contains three large muscles—***trapezius, serratus anterior,*** and ***latissimus dorsi***—and three smaller muscles—***levator scapulae*** and two ***rhomboids*** (major and minor).

Trapezius Muscle

This very large diamond-shaped muscle attaches the upper limb girdle to the skull and spine (Fig. 37-4).

37

Figure 37-1.
Skeleton of the back and shoulder region.

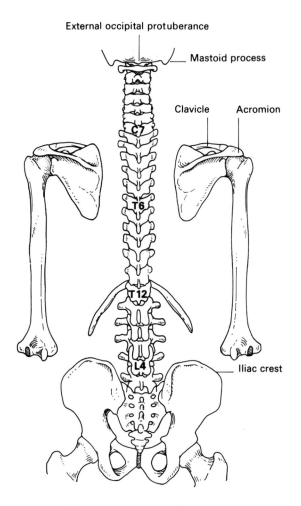

External occipital protuberance

Mastoid process

Clavicle   Acromion

C 7

T 6

T 12

L 4

Iliac crest

Proximal Attachments. The superior nuchal line of the skull, the external occipital protuberance, the ligamentum nuchae, and the vertebral spines of C7–T12.

Distal Attachments. The upper fibers are attached to the lateral third of the clavicle, the middle fibers to the acromion and scapular spine, and the lower fibers to a tubercle at the medial end of the spine of the scapula.

Nerve Supply. The *spinal* part of the *accessory nerve (XI)*.

Actions. The trapezius elevates the scapula (as in shrugging of the shoulder) and keeps the scapula against the chest wall (with the help of serratus anterior). The manner of the muscle's attachment to the spine of the scapula rotates that spine as one might rotate a wing nut, so that the glenoid fossa is rotated superiorly to permit the last 90 degrees of abduction of the arm.

Latissimus Dorsi

This large, flat muscle attaches the humerus to the trunk (Fig. 37-4).

Proximal Attachments. The spinous processes of the lower thoracic vertebrae, the thoracolumbar fascia and iliac crest, and a few slips from the lower ribs (and occasionally a slip from the inferior angle of the scapula).

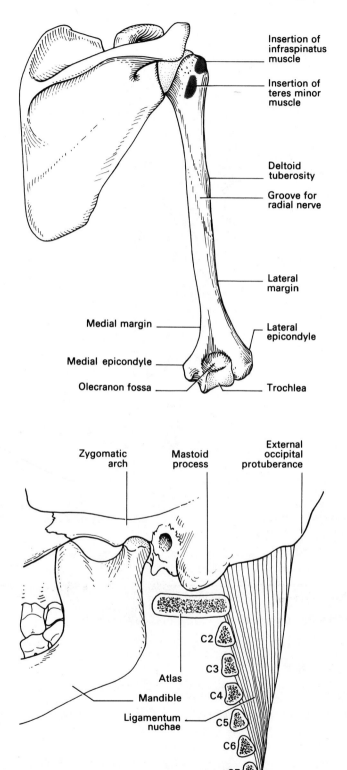

Insertion of
infraspinatus
muscle

Insertion of
teres minor
muscle

Deltoid
tuberosity

Groove for
radial nerve

Lateral
margin

Medial margin

Lateral
epicondyle

Medial epicondyle

Olecranon fossa

Trochlea

Figure 37-2.
Posterior view of the right scapula and
humerus.

Zygomatic
arch

Mastoid
process

External
occipital
protuberance

C2

C3

C4

C5

C6

C7

Atlas

Mandible

Ligamentum
nuchae

Figure 37-3.
Ligamentum nuchae (lateral view).
Numbers indicate cervical spinous
process.

Figure 37-4.
Left latissimus dorsi and right trapezius muscles (posterior view).

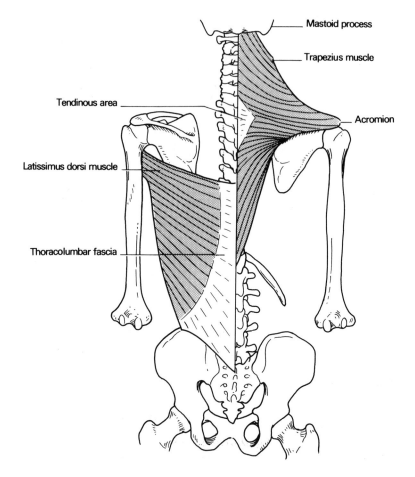

Mastoid process

Trapezius muscle

Tendinous area

Acromion

Latissimus dorsi muscle

Thoracolumbar fascia

Distal Attachment. The tendon of latissimus dorsi wraps around the inferior border of teres major and attaches to the floor of the intertubercular groove of the humerus.

Nerve Supply. The ***thoracodorsal nerve* (*C6, C7,*** and C8), a branch of the posterior cord of the brachial plexus.

Action. It adducts and rotates the humerus medially and extends the shoulder joint; in conjunction with pectoralis major it raises the trunk to the arm, as in performing chin-ups (hoisting yourself up) on an overhead bar.

Levator Scapulae Muscle

This muscle lies deep to trapezius (Fig. 37-5).

Proximal Attachments. The transverse processes of the upper four cervical vertebrae. The muscle wraps around the posterolateral aspect of the neck on its way to its scapular attachment.

Distal Attachment. The upper part of the medial border of the scapula.

Nerve Supply. By direct branches from C3 and C4, and from C5 via the dorsal scapular nerve.

Actions. It rotates the scapula so that the glenoid cavity is depressed and elevates the scapula. It also helps to hold the scapula against the trunk.

Rhomboid Muscles

Two rhomboids are described (major and minor), but they look and behave as one muscle (Fig. 37-5).

Proximal Attachments. The spinous processes of vertebrae C5 to T7, supraspinous ligaments, and ligamentum nuchae.

Distal Attachment. To the medial border of the scapula from the base of its spine to its inferior angle.

Nerve Supply. The ***dorsal scapular nerve*** (C4, ***C5***).

Actions. Retraction of the scapula and downward rotation of the glenoid fossa. The rhomboids assist serratus anterior in holding the scapula against the thoracic wall.

Serratus Anterior Muscle

This large flat muscle is of major importance in ***protracting*** the scapula and in holding it against the chest wall (Boxer's muscle) (Fig. 37-6).

Proximal Attachments. By individual digitations from the upper eight ribs.

Distal Attachment. The whole of the medial border of the scapula.

Nerve Supply. The ***long thoracic nerve*** (nerve of Bell) (C5, ***C6***, and ***C7***).

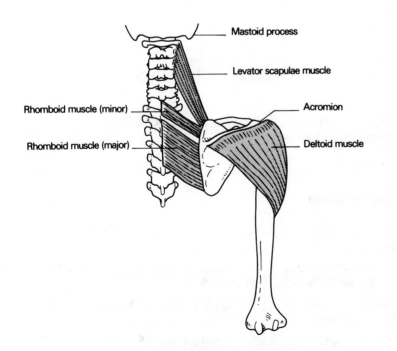

Mastoid process

Levator scapulae muscle

Rhomboid muscle (minor)

Rhomboid muscle (major)

Acromion

Deltoid muscle

Figure 37-5.
Some muscles of the shoulder region (posterior view).

Figure 37-6.
Right serratus anterior and
subscapularis muscles (anterior view).

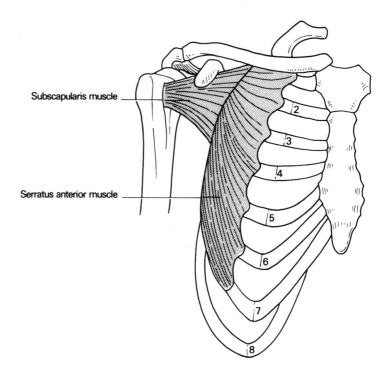

Actions. Serratus anterior wraps around the chest wall and holds the scapula in place. It protracts the scapula, that is, it moves it forward around the chest wall, as in throwing a punch or pushing against an object.

CLINICAL NOTE

*The long thoracic nerve is a long slender nerve that runs along the surface of its muscle. It lies on the medial wall of the axilla and can be damaged in operations that aim to clear out the lymph nodes of the axilla, for example, radical mastectomy. It can also be injured by traction injuries on the upper limb or in severe, forced lateral flexion of the neck to the opposite side. The result is a **winged scapula** (Fig. 37-7).*

MUSCLES THAT ATTACH THE HUMERUS TO THE SCAPULA

All of the following muscles act on and stabilize the shoulder joint.

Deltoid Muscle

This muscle forms the bulk of the round appearance of the shoulder (see Fig. 37-5).

Proximal Attachment. The spine and acromion process of the scapula and the lateral one third of the clavicle.

Distal Attachment. The deltoid tuberosity on the lateral aspect of the humeral shaft.

Nerve Supply. The axillary nerve (*C5* and C6).

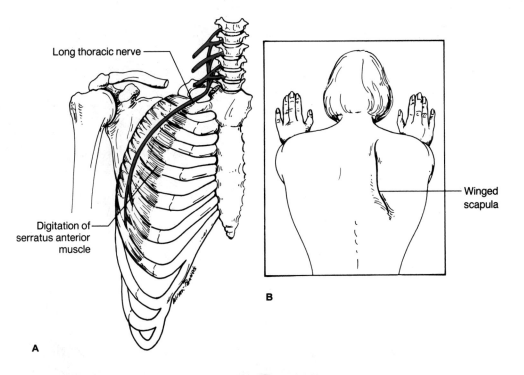

Long thoracic nerve

Digitation of
serratus anterior
muscle

Winged
scapula

A

B

Actions. The posterior fibers extend and laterally rotate the humerus at the shoulder joint; the anterior fibers flex and medially rotate it; and the middle fibers *abduct* the humerus.

Teres Major Muscle

This muscle forms the lower border of the posterior wall of the axilla (with latissimus dorsi wrapped around it) (Fig. 37-8).

Proximal Attachment. The posterior surface of the inferior angle of the scapula.

Distal Attachment. The medial lip of the intertubercular groove of the humerus.

Nerve Supply. The lower subscapular nerve (*C6* and C7).

Actions. Theoretically it medially rotates, extends, and adducts the arm, but electromyographic studies indicate that it stabilizes the shoulder in maintaining normal posture and that it is active in arm swinging during walking. Note that although similar in name, teres major and teres minor have different nerve supplies and belong to different groups of muscles. (Teres simply means round in shape.)

Rotator Cuff Muscles

This is a group of four muscles that joins the scapula to the humerus and forms the *rotator cuff*. Their prime function is to hold the head of the humerus in the glenoid fossa during various ranges of movement but particularly during abduction. Their tendons blend with the capsule of the shoulder joint and afford the joint the benefit of *mobility* with *stability* (Fig. 37-9).

Figure 37-7.
Winging of scapula from loss of serratus anterior muscle. **A.** Anterior view. **B.** Posterior view.

Figure 37-8.
Right shoulder region, deep structures
(posterior view).

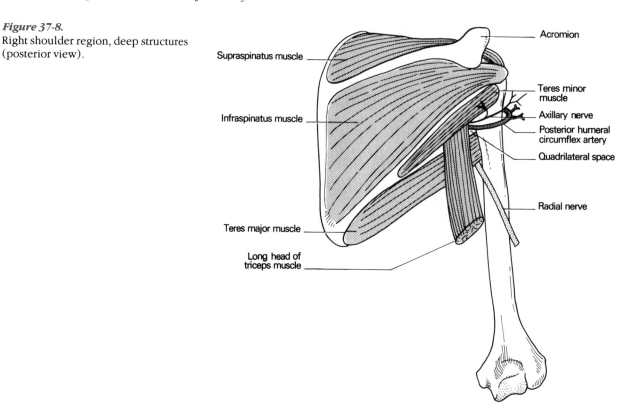

Supraspinatus muscle

Infraspinatus muscle

Teres major muscle

Long head of
triceps muscle

Acromion

Teres minor
muscle

Axillary nerve

Posterior humeral
circumflex artery

Quadrilateral space

Radial nerve

Supraspinatus

Proximal Attachment. The supraspinous fossa of the scapula.

Distal Attachment. By a tendon that passes deep to the acromion process to reach the highest point of the greater tubercle of the humerus.

Nerve Supply. The suprascapular nerve (C4, *C5*, and C6).

Actions. Besides being a part of the rotator cuff, this muscle acts as an abductor of the shoulder, usually in concert with the deltoid muscle. However, if the deltoid is paralyzed, supraspinatus can abduct the upper extremity by itself. It acts strongly when a heavy weight is carried in the hand with the shoulder abducted. It holds the head of the humerus in the glenoid fossa when a heavy weight is carried in the dependent hand. It is a rotator cuff muscle that stabilizes the shoulder joint.

CLINICAL NOTE

*The **subacromial bursa** separates the tendon of supraspinatus from the acromion and the deltoid muscle. Inflammation of the bursa makes abduction too painful to carry out. Rupture of the supraspinatus tendon is not uncommon in patients past middle age; the individual has difficulty initiating abduction.*

Subscapularis Muscle

This muscle covers the anterior aspect of the scapula (Fig. 37-9).

Proximal Attachment. The costal (anterior) surface of the scapula.

Distal Attachment. The lesser tubercle of the humerus.

Nerve Supply. The upper and lower subscapular nerves (C5, *C6*, and C7).

Action. It is a shoulder stabilizer (rotator cuff muscle) and can adduct and medially rotate the humerus.

Teres Minor Muscle

This muscle is almost indistinguishable from the infraspinatus at its proximal end (see Fig. 37-8).

Proximal Attachment. The posterior aspect of the lateral border of the scapula, superior to the attachment of teres major.

Distal Attachment. The lowest part of the posterior aspect of the greater tubercle of the humerus.

Nerve Supply. The axillary nerve (*C5* and C6).

Action. A shoulder stabilizer (rotator cuff) muscle that can also laterally rotate the humerus.

Infraspinatus Muscle

This muscle covers the infraspinous fossa of the scapula (see Fig. 37-8).

Proximal Attachment. The infraspinous fossa.

Figure 37-9.
Rotator cuff muscles.

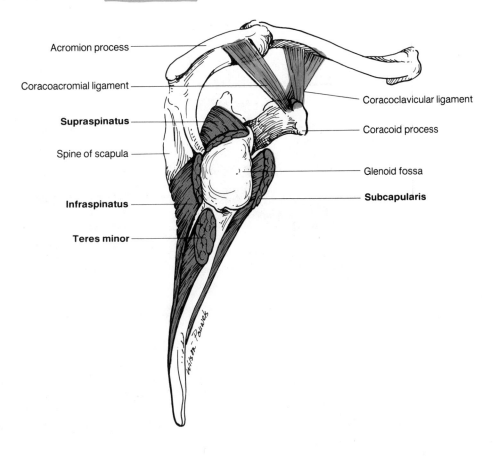

Acromion process

Coracoacromial ligament

Supraspinatus

Spine of scapula

Infraspinatus

Teres minor

Coracoclavicular ligament

Coracoid process

Glenoid fossa

Subcapularis

Figure 37-10.
Right shoulder region, deep vessels
and nerves (posterior view). Shaded
area is the quadrilateral space.

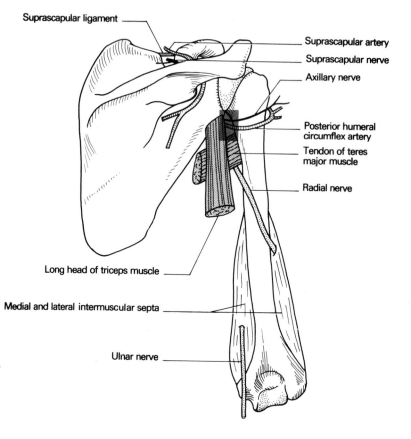

Suprascapular ligament

Suprascapular artery

Suprascapular nerve

Axillary nerve

Posterior humeral circumflex artery

Tendon of teres major muscle

Radial nerve

Long head of triceps muscle

Medial and lateral intermuscular septa

Ulnar nerve

Distal Attachment. The middle part of the posterior aspect of the greater tubercle of the humerus.

Nerve Supply. The suprascapular nerve (*C5* and C6).

Action. A rotator cuff muscle that can also rotate the humerus laterally.

QUADRILATERAL SPACE

This space is an important landmark because the axillary nerve and the posterior circumflex vessels pass through it. The space is formed by the *surgical neck* of the humerus laterally, the *teres minor* and *subscapularis* superiorly, the *teres major* inferiorly, and the *long head* of *triceps muscle* medially (Fig. 37-10). The capsule of the shoulder joint is located superiorly between teres minor and subscapularis.

NERVES AROUND THE SCAPULA

A group of nerves from the brachial plexus supplies the structures around the scapula.

Suprascapular Nerve (C5 and C6). This nerve is a branch from the upper trunk of the brachial plexus; it crosses the posterior triangle of the neck and passes through the suprascapular notch to supply the supraspinatus and, after passing through the spinoglenoid notch, the infraspinatus (Fig. 37-10).

Subscapular Nerves (C5 and C6). The upper and lower subscapular nerves are branches of the posterior cord, and they supply the ***subscapularis*** and ***teres major*** muscles (see Fig. 35-11).

Axillary Nerve (C5 and C6). The axillary nerve is one of the two terminal branches of the posterior cord. It passes through the quadrilateral space, in proximity to the capsule of the shoulder joint and the surgical neck of the humerus, to supply the ***deltoid*** and ***teres minor*** muscles, the shoulder joint, and the ***skin*** over the deltoid muscle (see Fig. 37-10).

ARTERIES AROUND THE SCAPULA

Note that the arteries around the scapula come from the subclavian as well as from the axillary artery (review subscapular, suprascapular, and lateral thoracic arteries). They can form an important anastomosis for maintaining the blood supply to the upper limb if a blockage occurs in the distal subclavian or proximal axillary arteries.

The Arm

The anatomical term **arm** refers to that part of the upper limb between the shoulder and the elbow. The **forearm** is located between the elbow and the wrist. The following bony landmarks around the elbow should be noted.

BONY LANDMARKS

HUMERUS (Fig. 38-1)

1. **The medial epicondyle** can be readily felt on the medial side of the cubital (elbow) region.
2. **The lateral epicondyle** can be felt subcutaneously on the lateral side of the elbow; it is less prominent than the medial epicondyle. A **supracondylar ridge** extends upward along the shaft of the humerus from each epicondyle. These **medial** and **lateral supracondylar ridges** give attachments to intermuscular septa and to muscles.
3. The **trochlea** is a pulley-shaped articular surface at the lower end of the humerus. It articulates with the **trochlear notch** of the **ulna.**
4. The **coronoid fossa** is a depression on the anterior aspect of the lower end of the humerus immediately superior to the trochlea; it accommodates the **coronoid process** of the ulna in extreme flexion of the elbow.
5. The **olecranon fossa** is found on the posterior surface of the lower end of the humerus. It accommodates the **olecranon process** of the ulna in extension of the elbow joint.
6. The **capitulum** is the convex lateral part of the articular surface of the lower end of the humerus. It articulates with the **head** of the **radius**. In the flexed elbow the head of the radius rests in the **radial fossa** of the humerus superior to the capitulum.

RADIUS (Fig. 38-1)

1. The **head** of the radius is a circular expansion of the proximal end of the radius that articulates with the **capitulum** of the humerus. The head of the radius is accommodated by the **radial notch** of the ulna. The **neck** of the radius joins the head to the

38

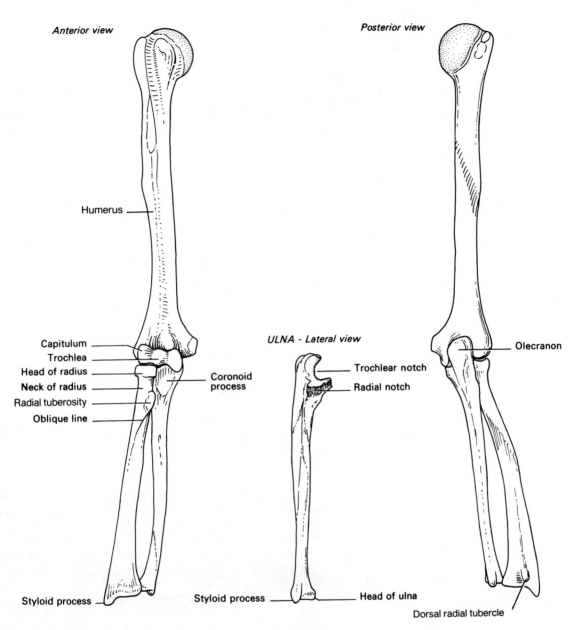

Anterior view

Humerus

Capitulum
Trochlea
Head of radius
Neck of radius
Radial tuberosity
Oblique line

Coronoid process

Styloid process

ULNA - Lateral view

Trochlear notch
Radial notch

Styloid process

Head of ulna

Posterior view

Olecranon

Dorsal radial tubercle

Figure 38-1.
Bones of right arm and forearm.

radial shaft. The head of the radius is easily palpable on the posterior aspect of the elbow.

2. The ***radial tuberosity*** is found on the medial surface of the radius about 6 cm below the head. The posterior part of the tuberosity is roughened for the attachment of the tendon of ***biceps brachii***. The smooth anterior part of the radial tuberosity is separated from the biceps tendon by a bursa.

ULNA (Fig. 38-1)

1. The ***trochlear notch*** of the ulna is a forward-facing opening that articulates with the ***trochlea*** of the lower end of the humerus.

Figure 38-2.
Major structures of the anterior surface
of the right arm.

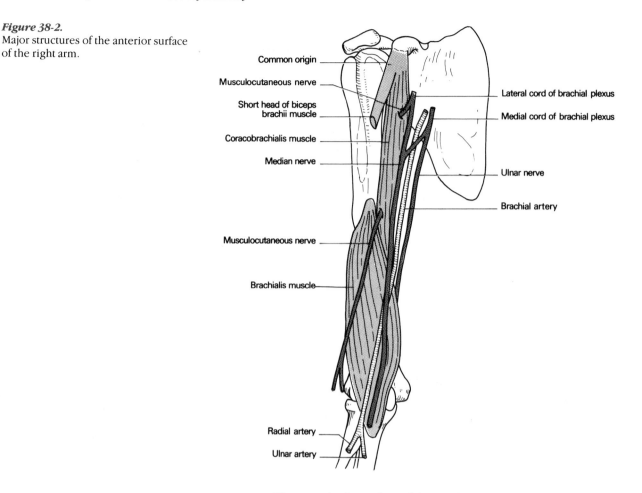

Common origin

Musculocutaneous nerve

Short head of biceps
brachii muscle

Coracobrachialis muscle

Median nerve

Musculocutaneous nerve

Brachialis muscle

Radial artery

Ulnar artery

Lateral cord of brachial plexus

Medial cord of brachial plexus

Ulnar nerve

Brachial artery

The superior boundary of the trochlear notch is the **olecranon process**, and its inferior boundary is the **coronoid process**. Note that the junction of these two processes is at the narrowest diameter of the trochlear notch. This is a weak spot that is not infrequently the site of a fracture. The **coronoid process** receives the attachment of the **brachialis** muscle (Fig. 38-2).
2. The **olecranon process** is the superior boundary of the trochlear notch; the **triceps brachii** muscle is attached to its roughened posterior portion. Anterior to this the olecranon is smooth and separated from the tendon of triceps by a **bursa**. The upper end of the olecranon is the point of the elbow, and a bursa intervenes between bone and skin. The skin over the olecranon is very mobile, elastic, and somewhat redundant in the extended position of the joint.

MUSCLES

The muscles of the arm and forearm receive extensive attachments to the intermuscular septa of the region (Fig. 38-3). Condensations of connective tissue extend from the lateral and medial borders and supracondylar ridges of the humerus to the investing sleeve of deep fascia of the arm. They give attachments to the **triceps** and **brachialis** muscles. The muscles posterior to the septa are supplied by the ***radial*** nerve and those in front of the septa by the ***musculocutaneous*** nerve, with the exception of the ***brachioradialis*** muscles.

The four important muscles of the arm are the ***coracobrachialis***, ***biceps brachii***, ***brachialis***, and ***triceps brachii***.

Brachialis Muscle

This is one of the two major flexors of the elbow joint (Fig. 38-2).

Proximal Attachments. The anterior surface of the lower half of the humerus and the medial intermuscular septum.

Distal Attachment. The coronoid process of the ulna.

Nerve Supply. The Musculocutaneous nerve (C5, *C6*, and C7).

Action. Flexion of the elbow.

Coracobrachialis

This muscle acts on the shoulder joint but is so intimately related to biceps brachii as to warrant description with the arm muscles (Fig. 38-2).

Proximal Attachment. To the coracoid process in common with the short head of biceps brachii.

Distal Attachment. To the medial aspect of the shaft of the humerus near its midpoint.

Nerve Supply. The musculocutaneous nerve (which pierces it) (C5, *C6*, and C7).

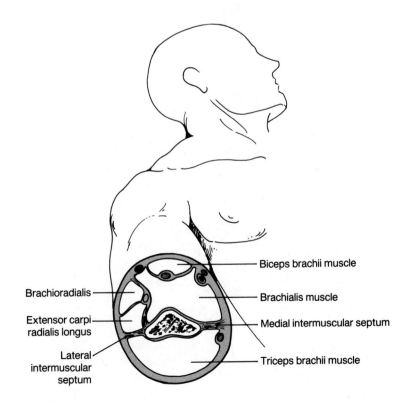

Figure 38-3.
Intermuscular septa dividing the arm into anterior (flexor) and posterior (extensor) compartments.

Action. Draws the arm forward and medially from a position of extension at the shoulder. It acts in synergy with the anterior fibers of deltoid to hold the arm steady in abduction.

Biceps Brachii

This muscle is the other major flexor of the elbow joint, but, because of its distal attachment to the posterior aspect of the radial tuberosity, it imposes a spin action on the radius and acts as a powerful *supinator* of the forearm (Fig. 38-4).

Proximal Attachments. The biceps is attached by two heads. The *short head* is fused to the tendon of coracobrachialis and attaches to the tip of the coracoid process. The *long head* is attached to the supraglenoid tubercle of the scapula and travels through the shoulder joint, above the head of the humerus, and lies in the intertubercular groove before joining the muscle mass.

Distal Attachments. The posterior aspect of the radial tuberosity and by the *bicipital aponeurosis*, which passes from the biceps tendon to blend

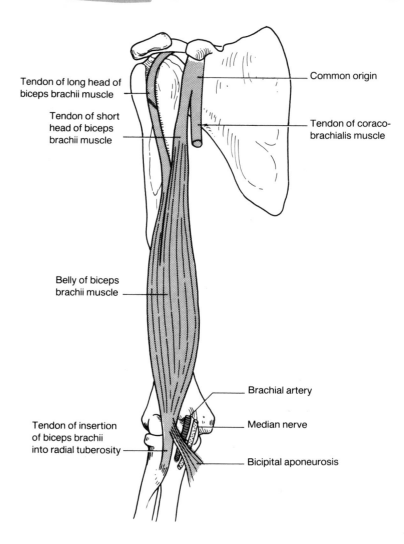

Tendon of long head of biceps brachii muscle

Tendon of short head of biceps brachii muscle

Common origin

Tendon of coraco-brachialis muscle

Belly of biceps brachii muscle

Brachial artery

Median nerve

Tendon of insertion of biceps brachii into radial tuberosity

Bicipital aponeurosis

Figure 38-4.
Biceps brachii muscle of the right arm.

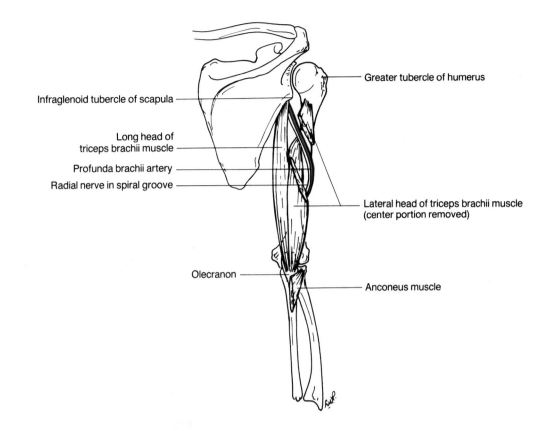

Greater tubercle of humerus

Infraglenoid tubercle of scapula

Long head of
triceps brachii muscle

Profunda brachii artery

Radial nerve in spiral groove

Lateral head of triceps brachii muscle
(center portion removed)

Olecranon

Anconeus muscle

Figure 38-5.
Triceps brachii and anconeus muscles
of right arm.

with the deep fascia on the medial side of the forearm. The bicipital aponeurosis provides some protection for the structures in the cubital fossa.
The bicipital aponeurosis is palpable in the cubital fossa if the elbow is
flexed against resistance.

Nerve Supply. The musculocutaneous nerve (C5 and *C6*).

Actions. Flexion of the elbow joint and supination of the forearm.

    Note: The action of biceps as a supinator adds to the strength of the
supinators of the forearm, which are much stronger than the pronators.
That is why it is easier to screw on the top of the ketchup bottle than to
undo it. (If you are right handed, the more powerful supinators screw it
on and the weaker pronators cannot undo it.)

Triceps Brachii

This powerful muscle makes up the bulk of the back of the arm
(Fig. 38-5).

Proximal Attachments. Triceps brachii originates by three heads. The
long head comes from the ***infraglenoid tubercle*** of the scapula, the ***lateral head*** from the shaft of the humerus above the spiral groove, and the
medial head (better called the deep head) from the shaft of the humerus
below and medial to the spiral groove.

Distal Attachments. The olecranon process of the ulna with a bursa intervening between its tendon and the anterior part of the superior surface
of the olecranon process.

Figure 38-6.
A. The course of the radial nerve. The forearm is pronated. **B.** The course of the ulnar nerve.

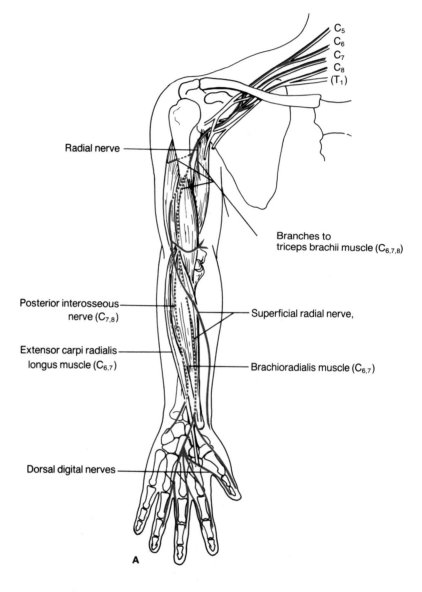

A

Nerve Supply. The radial nerve (C6, *C7*, and *C8*).

Action. Extension of the elbow joint.

Anconeus

This unimportant muscle stretches from the lateral humeral epicondyle to the lateral part of the olecranon and upper shaft of the ulna. It is supplied by the radial nerve and may be considered to be a part of triceps.

ARTERIES

The *anterior* and *posterior humeral circumflex arteries* arise from the axillary artery, but the main blood supply to the arm is the *brachial artery* (see Fig. 38-2), which starts at the lower border of teres major as

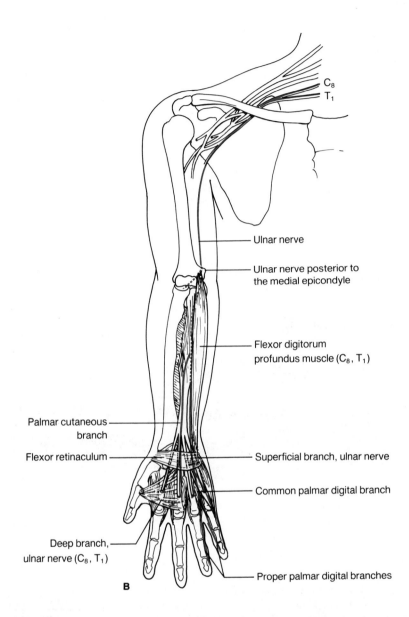

- Ulnar nerve
- Ulnar nerve posterior to the medial epicondyle
- Flexor digitorum profundus muscle (C_8, T_1)
- Palmar cutaneous branch
- Flexor retinaculum
- Superficial branch, ulnar nerve
- Common palmar digital branch
- Deep branch, ulnar nerve (C_8, T_1)
- Proper palmar digital branches

B

the continuation of the axillary artery. It descends in a straight line to the middle of the cubital fossa where, at the level of the neck of the radius, it divides into the ***radial artery*** laterally and the ***ulnar artery*** medially. In the arm it gives off muscular branches and three named arteries that participate in the arterial anastomosis around the elbow joint: the ***profunda brachii artery*** runs with the radial nerve in the spiral groove around the humerus, and two ***ulnar collateral arteries*** (superior and inferior) accompany the ulnar nerve.

The brachial artery lies anterior to the brachialis and triceps muscles. It is overlapped on its lateral side by the coracobrachialis and biceps muscles. The ***median nerve*** starts lateral to the artery and then crosses in front of it to gain its medial side. In the cubital fossa this nerve and the brachial artery are covered by the ***bicipital aponeurosis***, which affords them some protection. Variations in the bifurcation of the brachial artery are quite common.

NERVES

The arm affords passage to four principal nerves, two of which, the *median* and *ulnar nerves*, do not supply any muscles above the elbow. All four nerves are terminal branches of the brachial plexus.

RADIAL NERVE

The radial nerve supplies triceps brachii in the arm. It originates from the posterior cord of the brachial plexus and passes posteriorly and inferiorly on the teres major to reach the *spiral groove* of the humerus, in which it winds around the posterior aspect of the humerus and pierces the lateral intermuscular septum to reappear on the lateral side of the arm between the brachialis and brachioradialis muscles, anterior to the elbow joint (see Figs. 38-5 and 38-6A). In the above-mentioned course it supplies *triceps*, anconeus, *brachioradialis*, and *extensor carpi radialis longus*. In front of the lateral humeral epicondyle it splits into a *superficial branch*, which continues down the forearm to the wrist and hand and is *entirely sensory*, and a *deep branch*, which supplies *extensor carpi radialis longus* and then passes around the upper part of the shaft of the radius *within the substance of the supinator muscle*, which it supplies, to emerge as the *posterior interosseous nerve*, which then supplies muscles of the *extensor origin* of the forearm and the *posterior antebrachial* (forearm) *muscles*.

MUSCULOCUTANEOUS NERVE (see Fig. 38-2)

This is one of the terminal branches of the lateral cord of the brachial plexus. It deviates laterally from its origin, pierces coracobrachialis, which it supplies, and then lies on brachialis, supplying brachialis and biceps brachii before continuing down the arm and forearm as the *lateral cutaneous nerve of forearm*.

ULNAR NERVE (see Fig. 38-2)

This major nerve does not supply any muscles in the arm. It descends from its origin as a terminal branch of the medial cord of the brachial plexus on the medial side of the brachial artery. At the middle of the arm it pierces the *medial intermuscular septum* and descends on the anterior surface of the medial head of triceps to wind around the *posterior aspect of the medial epicondyle* (where it can be easily palpated) to reach the forearm (Fig. 38-6B).

CLINICAL NOTE

The ulnar nerve behind the medial epicondyle is not only easily palpated but also easily jarred (funny bone sensation) or compressed: by resting your forearms and elbows wearily on your desk for hours as you read this book, or lying on your elbow in your sleep. You should be able to determine its sensory distribution in the hand by noting which of your fingers go numb with these experiences.

MEDIAN NERVE (see Fig. 38-2)

This nerve is formed by the junction of the medial and lateral roots from the ***medial*** and ***lateral cords*** of the brachial plexus lateral to the brachial artery. (It often communicates with the ulnar nerve in this situation.) It descends in the arm, crosses anterior to the brachial artery to gain its medial side, and continues on the surface of brachialis to the cubital fossa, where it passes deep to pronator teres muscle to reach the forearm. The median nerve supplies no muscles in the arm but supplies all except one and a half muscles of the front of the forearm (*i.e.*, it supplies pronator teres, flexor carpi radialis, palmaris longus, flexor digitorum superficialis, pronator quadratus, flexor pollicis longus, and the lateral half of flexor digitorum profundus; it ***does not*** supply flexor carpi ulnaris or the medial half of flexor digitorum profundus).

CLINICAL NOTE

*The brachial artery and the median nerve are vulnerable to injury in **supracondylar fractures** of the humerus (see Fig. 38-2). The sharp end of the proximal fragments may compress or sever the artery or nerve, or they may irritate the artery so that it, and its branches, go into spasm and deprive the forearm of blood supply.*

The Cubital Fossa and Flexor Surface of the Forearm

THE CUBITAL FOSSA

The cubital fossa is the anterior depression or hollow of the elbow. Its superior border is a line drawn between the humeral epicondyles, its lateral border is the **brachioradialis muscle**, and its medial border is the **pronator teres muscle** (Figs. 39-1 and 39-2). The floor of the fossa is formed by the **brachialis** (described in Chap. 38) and **supinator** muscles. The roof of the fossa is the deep fascia on which lies the **median cubital vein**. Some important structures lie in the cubital fossa: the **brachial artery** terminates in it by dividing into the **radial** and **ulnar** arteries, and the fossa also contains the **tendon of biceps brachii** and the **median** and **radial** nerves.

Supinator Muscle

This muscle forms the lateral part of the floor of the cubital fossa; it is wrapped around the anterior, lateral, and posterior aspects of the upper third of the shaft of the radius (Fig. 39-1).

Proximal Attachments. The supinator is made up of two layers of muscle: one (superficial) layer is attached to the lateral epicondyle of the humerus, and the other (deep) layer comes from the supinator ridge of the ulna, which is found immediately inferior to the posterior part of the radial notch.

Distal Attachment. The muscle wraps around the radius to be attached between its neck and the distal attachment of the pronator teres.

Nerve Supply. The **posterior interosseus nerve** (C5 and **C6**).

Action. The supinator supinates the forearm, that is, it rotates the radius around the ulna so that the palm of the hand faces forward (Fig. 39-3).

Brachioradialis Muscle

This muscle forms the lateral boundary of the cubital fossa (Fig. 39-1).

Proximal Attachments. The lateral intermuscular septum of the arm and the lateral supracondylar ridge of the humerus.

Distal Attachment. By a long tendon to the base of the styloid process of the radius.

39

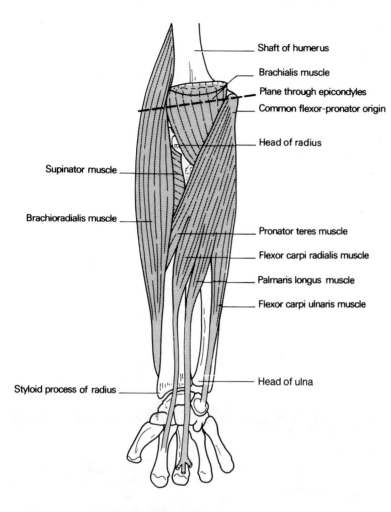

Figure 39-1.
Superficial muscles of the right forearm.

Shaft of humerus

Brachialis muscle

Plane through epicondyles

Common flexor-pronator origin

Head of radius

Supinator muscle

Brachioradialis muscle

Pronator teres muscle

Flexor carpi radialis muscle

Palmaris longus muscle

Flexor carpi ulnaris muscle

Head of ulna

Styloid process of radius

Nerve Supply. Radial nerve (C5, *C6,* and *C7*)

Action. Brachioradialis flexes the elbow; it is particularly effective if pronator teres has placed the radius in the midprone position.

Pronator Teres Muscle

This muscle forms the medial boundary of the cubital fossa (Fig. 39-1).

Proximal Attachments. Two heads: the common flexor origin on the medial epicondyle of the humerus and the coronoid process of the ulna.

Distal Attachment. The lateral border of the radius at its point of greatest curvature.

Nerve Supply. The median nerve (C6 and *C7*).

Action. Pronates the forearm and flexes the elbow.

CLINICAL NOTE

*The point of attachment of pronator teres is a common place for fractures (point of greatest curvature) of the radius. The upper fragment is wrapped in supinator muscle, which will **supinate** it while the lower frag-*

Figure 39-2.
Contents of the right cubital fossa.

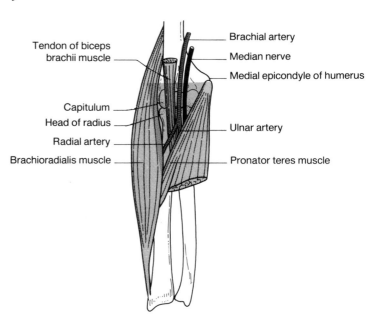

Tendon of biceps
brachii muscle

Capitulum
Head of radius
Radial artery
Brachioradialis muscle

Brachial artery
Median nerve
Medial epicondyle of humerus
Ulnar artery
Pronator teres muscle

*ment will be **pronated** by pronator teres. These axial rotations of the radial fragments are not always obvious on an x-ray film, and great care must be taken in reducing and immobilizing these fractures so that the pronated distal radius does not unite with the supinated proximal radius, that is, the lower forearm and hand must be placed in supination.*

Two Important Muscle Groups. Each humeral condyle gives attachment to an important group of muscles:

> ***Extensor-Supinator Group.*** These muscles of the forearm have a common attachment to the ***lateral epicondyle*** of the humerus and serve the extensor aspect of the forearm, wrist, and hand.
>
> ***Flexor-Pronator Group.*** These are muscles that have a common attachment to the ***medial epicondyle*** of the humerus. In general they pass to the anterior aspect of the forearm, wrist, and hand.

BONES OF THE FOREARM

The bones of the forearm are the ***radius*** and ***ulna*** (Fig. 39-4). The radius has the ability to rotate axially around the ulna. In full supination the radius lies parallel to the ulna, and in pronation the radius "crosses" the ulna. The radius carries the hand.

RADIUS

The upper end of the radius has already been described; the ***shaft*** is curved laterally and has a rounded ***lateral border***. The medial or ***interosseous border*** is sharp and gives attachment to the ***interosseous membrane*** (Fig. 39-4) through which the radius is connected to the ulna. The lower end of the radius has an ***articular facet*** for articulation with the small bones of the wrist. Laterally the lower end of the radius is prolonged into a point, the ***styloid process***. The posterior surface of the lower end bears the ***dorsal radial (Lister's) tubercle***. Medially the lower end of the radius shows the ***ulnar notch*** for articulation with the head of the ulna. (Note that for some arcane reason anatomists chose to call the lower end of the ulna the *head*.)

ULNA

The lower end of the ulna is slender and is expanded into the *head* of the ulna from which the *styloid* process projects. The head of the ulna does not participate in the wrist joint but is separated from the medial bones of the proximal row of the carpus by a triangular-shaped fibrocartilaginous *articular disc*. The apex of the articular disc attaches to the base of the ulnar styloid process, and its base attaches to the edge of the ulnar notch on the radius.

The *styloid processes* of the radius and ulna, the *head* of the ulna, and the *dorsal tubercle* of the radius are important palpable landmarks.

THE WRIST

There are eight bones of the wrist (*carpal bones*), divided into two rows (Fig. 39-4). The proximal row contains four bones, three of which may articulate with the radius and articular disc. They are, from lateral to me-

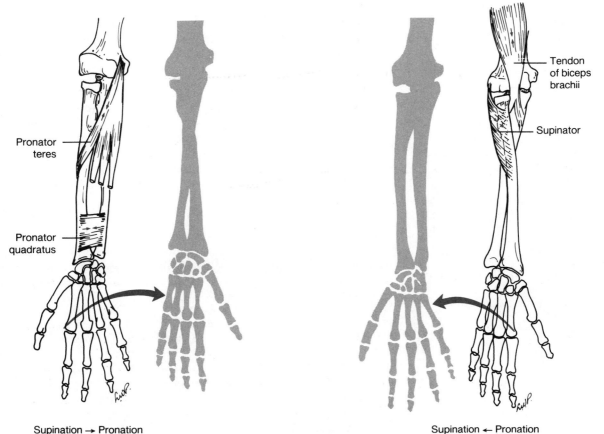

Pronator teres

Pronator quadratus

Tendon of biceps brachii

Supinator

Supination → Pronation

Supination ← Pronation

Figure 39-3.
Right forearm in pronation and supination.

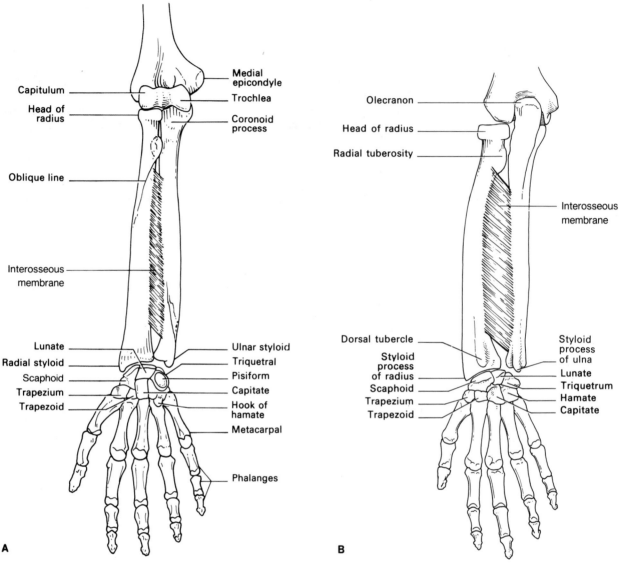

A

B

Figure 39-4.
Bones of the forearm and hand.
A. Anterior surface, right forearm.
B. Posterior surface, left forearm.

dial, the **scaphoid**, **lunate**, and **triquetral**. The **pisiform** is a small, easily palpable sesamoid bone that articulates with the anterior aspect of the triquetral and does not participate in the wrist joint.

The distal row comprises four bones, all of which articulate with the proximal row and with the metacarpals of the hand. These four bones, named from lateral to medial, are the **trapezium**, the **trapezoid**, the large **capitate**, and the **hamate**. The hamate bears a prominent anterior projection, the **hook**, which is palpable with some deep pressure in the hypothenar eminence of the hand distal to the pisiform.

CLINICAL NOTE

In about one third of individuals the blood supply to the scaphoid is from vessels that enter the bone at its distal portion, the proximal pole having no independent blood supply. In these people, if a fracture occurs through the waist of the scaphoid, the proximal pole of the scaphoid may undergo avascular necrosis unless careful and prolonged immobilization with close apposition of the fracture surfaces is achieved and maintained.

THE HAND

The skeleton of the hand consists of five **metacarpals**, which articulate with the distal row of bones of the carpus. The metacarpals articulate with the proximal **phalanges** of the digits. Each metacarpal and **phalanx** consists of a **base**, a **shaft**, and a **head**. Each finger has three phalanges; the thumb has only two.

INTEROSSEOUS MEMBRANE (Fig. 39-4)

The interosseous membrane of the forearm is a thin but strong membrane that attaches the **interosseous border** of the radius to the **interosseous border** of the ulna. The membrane separates the forearm into anterior (flexor) and posterior (extensor) compartments and also affords wide attachments for some muscles of the forearm.

CLINICAL NOTE

*The forearm, like the leg, is wrapped in a dense envelope of deep fascia and divided into anterior (flexor) and posterior (extensor) compartments by the forearm bones and the interosseous membrane. Trauma, such as fractures of the forearm bones with resulting bleeding into and edema of the tissues of the flexor compartment, can cause a **compartment compression syndrome.***

*The **symptoms** are extreme pain in the compartment and impaired sensation in the areas of distribution of the median and ulnar nerves in the hand and fingers. The **signs** are pallor and coldness of the fingers and severe pain on passive extension of the fingers which the patient will hold in the flexed position.*

*If untreated by surgical decompression, the result is ischemic necrosis of the muscles in the compartment. Later, when the necrosed muscles are replaced by scar tissue, the end result is a permanent flexion deformity of the wrist, metacarpophalangeal, and interphalangeal joints: **Volkman's ischemic contracture.** This condition is more common in children.*

FLEXOR RETINACULUM

The bones of the carpus form an arch with a forward concavity. The lateral pillar of this arch is formed by the scaphoid and trapezium, and the medial pillar is formed by the triquetral, pisiform, and hamate bones. The **flexor retinaculum,** which is a very dense thickening of the deep fascia of the region, extends from the pisiform and the hook of the hamate medially to the scaphoid and trapezium laterally. The attachment to the trapezium is split by the tendon of **flexor carpi radialis**. The function of the flexor retinaculum is to hold the long flexor tendons against the bones of the carpus so that they do not "bowstring" forward when the wrist is flexed (Fig. 39-5A).

Surface Marking. The distal skin **crease** on the anterior surface of the wrist marks the superior border of the flexor retinaculum.

Carpal Tunnel (Fig. 39-5). The tendons of muscles of the forearm that gain attachments to bones of the hand pass through a relatively rigid and

Figure 39-5.
Structures of the left wrist. **A.** Anterior
view illustrating levels of cross-sections
for B and C. **B.** Distal row of carpal
bones. **C.** Proximal row of carpal
bones.

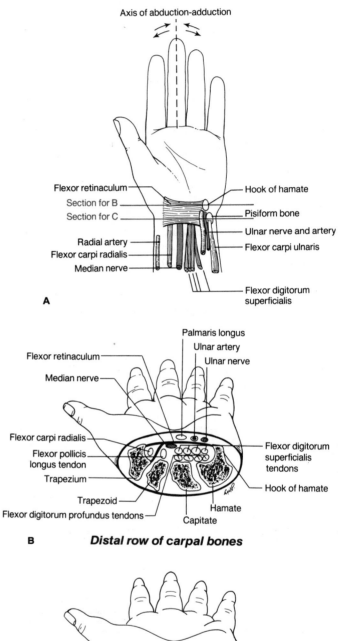

Axis of abduction-adduction

Flexor retinaculum
Section for B
Section for C

Radial artery
Flexor carpi radialis
Median nerve

Hook of hamate

Pisiform bone

Ulnar nerve and artery
Flexor carpi ulnaris

Flexor digitorum
superficialis

A

Palmaris longus
Ulnar artery
Ulnar nerve

Flexor retinaculum

Median nerve

Flexor carpi radialis

Flexor pollicis
longus tendon

Trapezium

Trapezoid

Flexor digitorum profundus tendons

Flexor digitorum
superficialis
tendons

Hook of hamate

Hamate

Capitate

B   *Distal row of carpal bones*

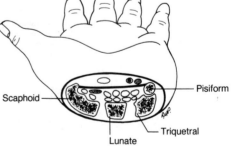

Scaphoid

Pisiform

Triquetral

Lunate

C   *Proximal row of carpal bones*

narrow space, the ***carpal tunnel***. The tendons' gliding action is facilitated by ***tendon sheaths*** that begin above the tunnel and extend into the palm of the hand. Competing for space in this narrow tunnel is the ***median nerve***.

CLINICAL NOTE: CARPAL TUNNEL SYNDROME

This extremely common syndrome results from increased pressure within the carpal tunnel. The causes of this increase in pressure may be disease of the tendon sheaths, wrist joint abnormalities, irregularities of the bones, fluid retention, or venous obstruction within the tunnel. The symptoms and signs may vary from pain and paresthesia (tingling) to anesthesia and motor paralysis in the territories supplied by the median nerve distal to the flexor retinaculum. The usually illustrated cross-sectional round shape of the median nerve in the tunnel is a myth. In the normal carpal tunnel the nerve looks more like a thick ribbon. When compressed in the tunnel the nerve becomes a flat band.

MUSCLES OF THE FRONT OF THE FOREARM

Brachioradialis and ***pronator teres*** have already been described. It is useful to remember that the ***lateral humeral epicondyle*** is the proximal point of attachment for the ***supinator-extensor*** group of muscles. These are primarily on the posterior aspect of the forearm and extend the fingers and wrist. The ***medial epicondyle*** provides proximal attachment for the ***pronator-flexor*** group of muscles. These are on the anterior aspect of the forearm, and, except for pronator teres, all have some action on the wrist joint; some also act on the digits. The muscles of the front of the forearm are divisible into several layers.

FIRST (SUPERFICIAL) LAYER

The superficial layer consists of four muscles that are either pronators of the forearm or flexors of the wrist and, to a lesser extent, the elbow (see Figs. 39-1 and 39-6). The position of the four muscles of the superficial layer can be matched by the four fingers of the hand if the thumb is placed behind and above the medial epicondyle and the fingers spread out on the

Figure 39-6.
A representation of the muscles of the superficial layer of the right forearm.

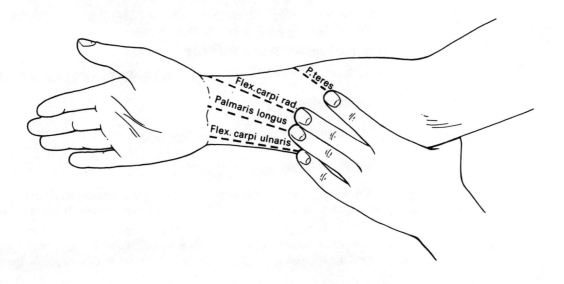

surface of the forearm (Fig. 39-6). The index finger then overlies pronator teres, the middle finger covers flexor carpi radialis, the ring finger covers palmaris longus, and the little finger overlaps the position of flexor carpi ulnaris.

Pronator Teres

This muscle was described on page 491.

Flexor Carpi Radialis

This is a long narrow muscle with a long narrow tendon.

Proximal Attachment. The common flexor origin (*i.e.*, the medial epicondyle).

Distal Attachments. The tendon passes in the groove of the trapezium to be attached to the bases of the second and third metacarpal bones.

Nerve Supply. The median nerve (C6 and *C7*).

Action. Flexion and abduction (radial deviation) of the wrist.

Palmaris Longus

This narrow "spare-part" muscle has an upper attachment to the common flexor origin, and its distal end attaches to the palmar aponeurosis, after passing superficial to and to some extent blending with the flexor retinaculum. The palmaris longus is often absent. It is supplied by the median nerve and is a weak flexor of the wrist. Its main claim to fame is its usefulness as the source of tendon tissue for tendon transplant operations.

Flexor Carpi Ulnaris

This muscle on the medial side of forearm is one of the one and one-half muscles of the region supplied by the ulnar nerve.

Proximal Attachment. The common flexor origin.

Distal Attachments. Its tendon attaches to the pisiform bone, and this attachment is transmitted via the pisohamate ligament to the hook of the hamate and the base of the fifth metacarpal.

Nerve Supply. The ulnar nerve (C7 and *C8*).

Action. Flexion and adduction (ulnar deviation) of the wrist.

SECOND LAYER

The second layer consists of a single muscle.

Flexor Digitorum Superficialis

The flexor digitorum superficialis has rather complicated upper attachments and attaches below into the four medial digits. It forms a layer in itself (Fig. 39-7).

Proximal Attachments. The ***humeroulnar head*** comes from the common flexor origin, from the capsule of the elbow joint, and from the ***coro-***

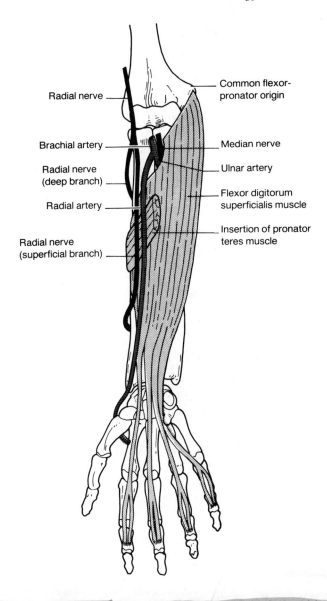

Radial nerve

Brachial artery

Radial nerve
(deep branch)

Radial artery

Radial nerve
(superficial branch)

Common flexor-
pronator origin

Median nerve

Ulnar artery

Flexor digitorum
superficialis muscle

Insertion of pronator
teres muscle

Figure 39-7.
Flexor digitorum superficialis muscle
and associated structures (second layer
of the right forearm).

noid process of the ulna. The **radial head** comes from the oblique line on the anterior surface of the radius, deep and medial to the attachment of the pronator teres. The **median nerve** and **ulnar artery** pass deep to the two heads to run down the forearm behind the flexor digitorum superficialis; in fact, the median nerve is usually adherent to the deep surface of the muscle.

Distal Attachments. Each of its four tendons attaches to the shaft of the middle phalanx of one of the medial four digits. (For more details on the digital attachments of flexor tendons, see Chapter 41, **The Hand**.)

Nerve Supply. The median nerve (C7, *C8*, and T1).

Action. The tendons flex all joints over which they travel, that is, flexion of the wrist, metacarpophalangeal, and proximal interphalangeal joints. With the wrist held in the **position of function** of slight extension at the

Figure 39-8.
Muscles of the third and fourth layers of
the flexor surface of the right forearm.

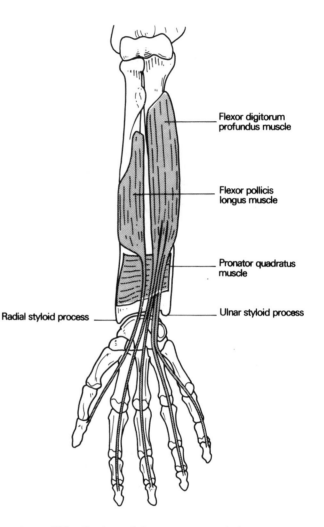

Flexor digitorum
profundus muscle

Flexor pollicis
longus muscle

Pronator quadratus
muscle

Ulnar styloid process

Radial styloid process

wrist, the action will be flexion of the metacarpophalangeal and proximal
interphalangeal joints.

THIRD LAYER

Flexor Digitorum Profundus

This is the only muscle that can flex the distal interphalangeal joints of the
fingers, and, because the muscle does not provide a separate muscle belly
for each finger, distal joint flexion may be impossible if the other distal
phalanges are forced into passive extension (Fig. 39-8). (Hold the joints
of the ring finger straight with your other hand and try flexing the distal
joint of the middle finger. You will probably fail, but there are individual
variations.)

Proximal Attachments. The upper three quarters of the anterior and
medial surfaces of the ulna and the medial half of the interosseous membrane.

Distal Attachments. Into the distal phalanges of the medial four digits
by passing through the split in the superficialis tendon. (See Chapter 41,
The Hand.)

Nerve Supply. The part of the muscle that sends tendons to the index and middle fingers is supplied by the ***anterior interosseous*** branch of the ***median nerve***; the part supplying the ring and little fingers is supplied by the ***ulnar nerve***. In ***both of these nerves***, the nerve fibers are derived from ***C8*** and T1.

Actions. Flexion of all the joints that it crosses, that is, wrist, metacarpophalangeal, and ***all*** the interphalangeal joints of the fingers. With the wrist in the ***position of function***, however, it flexes only the metacarpophalangeal and interphalangeal joints.

Flexor Pollicis Longus

The flexor pollicis longus acts to flex the joints of the thumb including the ***carpometacarpal joint*** (Fig. 39-8).

Proximal Attachments. To the anterior surface of radius, below the tuberosity and medial to the oblique line and from the lateral half of the anterior surface of the interosseous membrane.

Distal Attachment. To the distal phalanx of the thumb. The tendon is clothed in its own tendon sheath from superior to the flexor retinaculum to its attachment to the thumb.

Nerve Supply. The anterior interosseous branch of the median nerve (***C8***, T1).

Action. It flexes the interphalangeal joint of the thumb, the first metacarpophalangeal joint, the first carpometacarpal joint, and the wrist. ***Note*** that the first metacarpal bone is set at a right angle to the plane of the other four metacarpal bones so that flexion brings the thumb across the palm. The thumb carpometacarpal joint enjoys a great range of movement, which is what makes the human hand so versatile in the performance of skilled movements (see below).

FOURTH LAYER

The fourth layer of muscles in the front of the forearm consists of a single muscle, the pronator quadratus.

Pronator Quadratus

This is the only muscle in the forearm that confines its attachments to the radius and ulna (Fig. 39-8).

Medial Attachment. The lower quarter of the anterior surface of the ulna.

Lateral Attachment. The lower quarter of the anterior surface of the radius.

Nerve Supply. The anterior interosseous branch of the median nerve (***C8*** and T1).

Action. Pronation of the forearm.

It should be noted that all muscles in the anterior compartment of the forearm are supplied by either the ***median nerve*** or the ***ulnar nerve***. The median nerve supplies all but one and one-half muscles; the ***flexor carpi ulnaris*** and the medial half of the ***flexor digitorum profundus*** are supplied by the ulnar nerve.

All the digital flexors will also flex the wrist joint unless this joint is held in the position of function. If they are permitted to flex the wrist or if the wrist is forced into flexion, there will not be sufficient range of movement left in the flexors to provide a strong grip. This is why you can force someone to loosen his grip on an object by forcing his wrist into flexion.

NERVES

MEDIAN NERVE

The median nerve leaves the cubital fossa by passing between the humeral and the ulnar attachments of the pronator teres (see Fig. 39-2). Next it runs between the two heads of attachment of flexor digitorum superficialis (see Fig. 39-7), clinging to its deep surface, and becomes more superficial at the wrist, where it emerges from behind the lateral border of flexor digitorum superficialis to lie between the tendons of that muscle and the tendon of flexor carpi radialis. The nerve can be palpated at the wrist just medial to the last-named tendon. Next it enters the carpal tunnel deep to the flexor retinaculum (see Fig. 39-5).

In the forearm the median nerve gives off numerous muscular branches as well as the following two named branches.

1. *The anterior interosseous nerve* (Fig. 39-9) arises from the median nerve at the apex of the cubital fossa and, running deep to flexor digitorum superficialis, gains the interosseous membrane on which it descends between the flexor digitorum pollicis on its lateral side and flexor digitorum profundus on its medial side. In this course it supplies the flexor pollicis longus and the lateral half of flexor digitorum profundus. It passes deep to pronator quadratus, which it supplies, and terminates by supplying sensory fibers to the wrist joint.
2. *The palmar branch* leaves the median nerve a short distance proximal to the flexor retinaculum and, running superficial to the retinaculum, supplies the skin over the thenar eminence (the ball of the thumb) and the radial side of the palm.

ULNAR NERVE (Fig. 39-9)

The ulnar nerve reaches the forearm by passing *posterior* to the medial epicondyle, between it and the olecranon process. It enters the forearm by passing between the two heads of flexor carpi ulnaris and descends in the forearm on flexor digitorum profundus, with flexor carpi ulnaris overlapping it on the medial side. In the lower two thirds of the forearm the ulnar artery lies close to its lateral side. About 5 cm above the wrist the ulnar nerve gives off its *dorsal branch* and becomes superficial. It crosses the anterior aspect of the flexor retinaculum (see Fig. 39-5B) but is covered by some deep fascia in this region so that it has a superficial tunnel of its own (*Guyan's tunnel*). It continues into the palm, lateral to the pisiform bone and medial to the hook of the hamate, and divides into a superficial and a deep branch (see Fig. 39-9). Branches in the forearm include the following:

1. *Muscular branches* to the flexor carpi ulnaris and the medial half of flexor digitorum profundus.

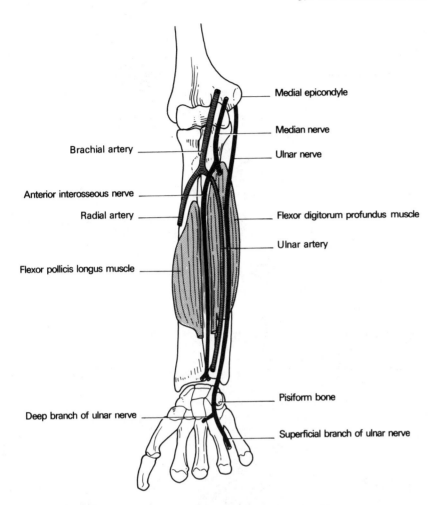

Figure 39-9.
Some vessels and nerves of the deeper regions of the right forearm (anterior view).

2. The ***palmar branch***, which leaves it near the middle of the forearm and, lying anterior to the ulnar artery, pierces the deep fascia above the wrist to supply the skin of the hypothenar eminence and the medial part of the palm of the hand (cutaneous).
3. The ***dorsal branch***, which passes deep to flexor carpi ulnaris tendon to supply the skin over the medial part of the posterior (dorsal) aspect of the hand.
4. ***Articular branches*** to the inferior radioulnar joint and the wrist joint.

RADIAL NERVE (Fig. 39-10)

The radial nerve enters the cubital fossa in a deep groove between brachialis and brachioradialis muscles and runs superficial to the supinator muscle in the floor of the cubital fossa. It divides into a ***deep branch***, which runs between the two layers of the supinator and emerges in the posterior compartment of the arm as the ***posterior interosseous nerve***, and a ***superficial branch***, which continues as a sensory nerve down the forearm overlapped by brachioradialis. Four centimeters above the wrist it passes deep to the tendon of brachioradialis and reaches the posterior aspect of

Figure 39-10.
The course of the radial nerve. The
forearm is pronated.

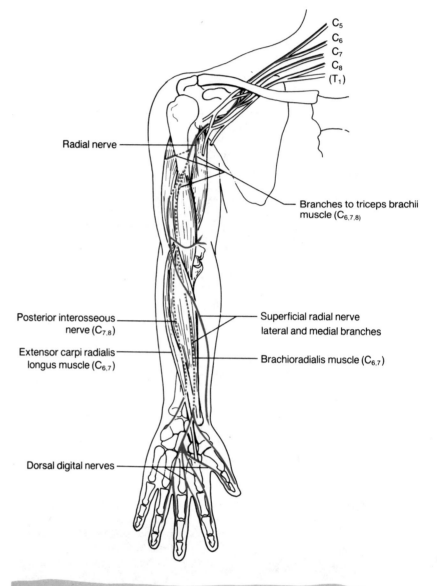

C₅
C₆
C₇
C₈
(T₁)

Radial nerve

Branches to triceps brachii
muscle ($C_{6,7,8}$)

Posterior interosseous
nerve ($C_{7,8}$)

Superficial radial nerve
lateral and medial branches

Extensor carpi radialis
longus muscle ($C_{6,7}$)

Brachioradialis muscle ($C_{6,7}$)

Dorsal digital nerves

the forearm to supply sensory fibers to the skin of the back of the hand and
fingers (for details, see p 522.). The superficial branch of the radial nerve
can be exposed from elbow to wrist without cutting any muscle or tendon,
and to the back of the hand by cutting only the brachioradialis tendon.

ARTERIES

BRACHIAL ARTERY (see Figs. 39-9 and 39-11)

The brachial artery terminates in the cubital fossa, at the level of the neck
of the radius, by dividing into the ***ulnar*** and ***radial*** arteries. Occasionally
it may bifurcate at a higher level, even in the axilla.

ULNAR ARTERY

The ulnar artery passes deep to the upper attachments of pronator teres and is accompanied by the median nerve as it passes deep to the two heads of flexor digitorum superficialis. Proximal to the middle of the forearm it crosses posterior to the median nerve to lie on the flexor digitorum profundus. At this level the ulnar nerve joins it to lie on its medial side. In the lower part of its course the ulnar artery is covered only by fascia and skin. It crosses in front of the flexor retinaculum, lateral to the pisiform and medial to the hook of the hamate (see Fig. 39-11).

CLINICAL NOTE

*The **ulnar pulse** can be taken at the wrist where the artery passes in front of the head of the ulna and lateral to the pisiform; it is generally not as prominent nor as easily palpable as the radial artery at the wrist (see Fig. 39-11).*

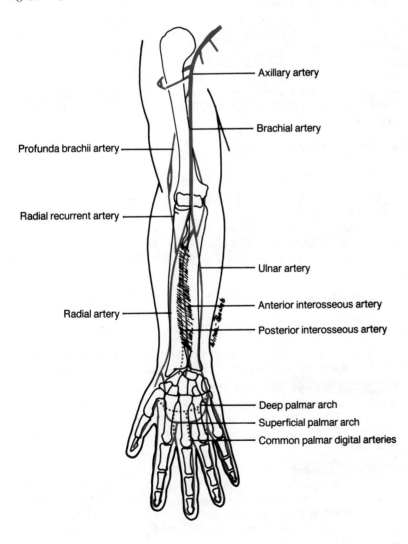

Figure 39-11.
Arterial supply of the upper limb.

Axillary artery

Brachial artery

Profunda brachii artery

Radial recurrent artery

Ulnar artery

Anterior interosseous artery

Posterior interosseous artery

Radial artery

Deep palmar arch

Superficial palmar arch

Common palmar digital arteries

Branches of the Ulnar Artery

There are several branches of the ulnar artery that merit mention (see Fig. 39-11).

1. Recurrent branches take part in the arterial anastomoses around the elbow.
2. The ***common interosseous artery*** arises in the first few centimeters of the ulnar artery and divides into ***anterior*** and ***posterior interosseous arteries***. They run on their respective surfaces of the interosseous membrane and terminate around the wrist.
3. Branches to the ***dorsal carpal network.***

RADIAL ARTERY

Everybody knows the radial artery: it is the pulse at the wrist. It begins at the neck of the radius as the other terminal branch of the brachial artery (see Figs. 39-9 and 39-11). It runs almost in a straight line toward the styloid process of the radius. In its upper course it is deep to brachioradialis, which can be pulled laterally to reveal the entire course of the artery in the forearm. In the middle third of the forearm the radial artery runs close and medial to the superficial branch of the radial nerve. At the lower part of the radius it lies against bone and is covered only by skin and fascia. In this superficial position, with the radius as a back-prop, it can be easily palpated as the ***radial pulse*** (see Fig. 41-19). Next, it passes deep to the tendons of abductor pollicis longus and extensor pollicis brevis to enter the posterior aspect of the wrist.

Branches of the Radial Artery

Several branches are given off in the forearm.

1. Branches that take part in the anastomosis around the elbow joint.
2. Muscular branches.
3. A branch to the ***dorsal carpal arch.***
4. The ***superficial palmar branch,*** which passes over or through the muscles of the thenar eminence to enter the palm.

STRUCTURES OF THE FRONT OF THE WRIST

SURFACE ANATOMY

The anterior aspect of the wrist contains several important structures that the student should be able to identify by observation or palpation.

Flexor Carpi Radialis Tendon. This tendon becomes visible in the lower third of the forearm if the wrist is flexed against resistance (see Fig. 39-5). If in doubt, place your fingers on this tendon and wiggle the fingers; the finger flexor tendons will move but flexor carpi radialis tendon will not.

Palmaris Longus Tendon. This tendon stands out if the wrist is flexed against resistance to 90 degrees and can be identified because it passes superficial to the flexor retinaculum (see Figs. 39-1 and 39-5B). Beware of anatomy examiners who do not have a palmaris longus tendon and invite you to demonstrate their tendon. Such lowlifes exist, but not in your school, of course!

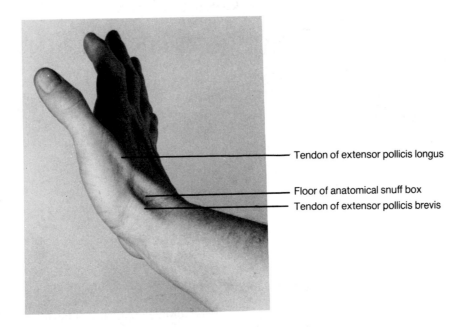

Tendon of extensor pollicis longus

Floor of anatomical snuff box
Tendon of extensor pollicis brevis

Flexor Digitorum Superficialis Tendons. The tendons immediately medial to the flexor carpi radialis tendon belong to flexor digitorum superficialis. They will move when the fingers are moved (see Fig. 39-5).

Flexor Carpi Ulnaris Tendon. When the wrist is flexed this tendon can be felt to attach to the pisiform bone (see Fig. 39-5).

Radial Artery. This artery and its pulse are palpable against the lower end of the radius lateral to flexor carpi radialis tendon (see Fig. 39-5A).

Ulnar Artery. The ulnar artery can be palpated against the head of the ulna just lateral to the tendon of flexor carpi ulnaris and the pisiform (see Fig. 39-5).

Crease of the Wrist. The distal flexion crease of the wrist corresponds to the position of the proximal border of the flexor retinaculum.

The Anatomical Snuffbox. The anatomical snuffbox is formed on the lateral surface of the wrist when the thumb is fully extended. It is bordered by three tendons: the lateral border is formed by the tendons of *abductor pollicis longus and extensor pollicis brevis*, whereas the medial border is the tendon of *extensor pollicis longus* (Fig. 39-12). The snuffbox is covered by skin and contains branches of the superficial radial nerve that can be rolled against the tendons, and the beginning of the *cephalic vein*. The floor is formed by the scaphoid and trapezium. The radial artery crosses the floor, and its pulsation can usually be felt in that location.

Figure 39-12.
Anatomical snuff box.

The Extensor-Supinator Surface of the Forearm

FASCIA

The **superficial fascia** of the forearm contains the **cephalic** and **basilic** veins, which begin at the extremities of the dorsal venous arch of the hand, as well as lymphatics and terminal branches of the radial and ulnar nerves.

The deep fascia envelops the forearm and gives attachment to the extensor muscles at and below the lateral humeral epicondyle. In the region of the wrist the deep fascia is thickened to form the **extensor retinaculum**, which keeps the tendons of the extensor muscles in place. Laterally the retinaculum is attached to the distal end of the radius, and medially it is attached to the styloid process of the ulna and to the triquetral and pisiform bones (Figs. 40-1 and 40-2).

MUSCLES

Brachioradialis and supinator were discussed with the cubital fossa.

SUPERFICIAL EXTENSOR MUSCLES (Fig. 40-1)

Extensor Carpi Radialis Longus

This is one of the two extensor muscles that also abduct (radially deviate) the wrist.

Proximal Attachments. The lateral supracondylar ridge of the humerus, the lateral epicondyle, and the associated intermuscular septum and deep fascia.

Distal Attachment. Its tendon runs through the anatomical snuffbox deep to the extensor retinaculum to attach to the posterior aspect of the base of the second metacarpal.

Nerve Supply. The radial nerve (C6 and C7).

Action. It extends and abducts the wrist.

Extensor Carpi Radialis Brevis

This is the other extensor and abductor of the wrist.

40

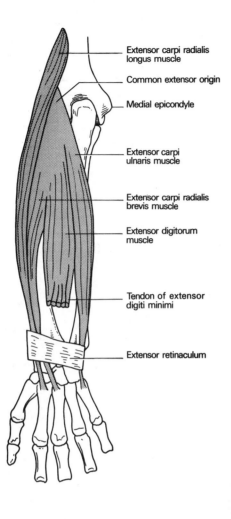

Extensor carpi radialis longus muscle

Common extensor origin

Medial epicondyle

Extensor carpi ulnaris muscle

Extensor carpi radialis brevis muscle

Extensor digitorum muscle

Tendon of extensor digiti minimi

Extensor retinaculum

Figure 40-1.
Superficial muscles of the extensor compartment of the left forearm.

Proximal Attachments. The common extensor tendon at the lateral epi-condyle, the lateral ligament of the elbow joint, and the deep fascia of the forearm.

Distal Attachment. Following a path that parallels that of the long radial extensor to the posterior aspect of the base of the third metacarpal.

Nerve Supply. The posterior interosseous nerve (*C7* and C8).

Action. Extension and abduction of the wrist.

Extensor Digitorum

This muscle is the principal extensor of the fingers.

Proximal Attachments. The common extensor origin, intermuscular septa, and deep fascia of the forearm.

Distal Attachments. By four tendons into the digital extensor expansions (see p. 514) of the four fingers. All the tendons pass in a common synovial sheath deep to the extensor retinaculum. On the back of the hand the tendons are interconnected by bands of connective tissue.

Nerve Supply. Posterior interosseous nerve (*C7* and C8).

Figure 40-2.
Superficial and deep (outcropping) muscles of the extensor compartment of the left forearm.

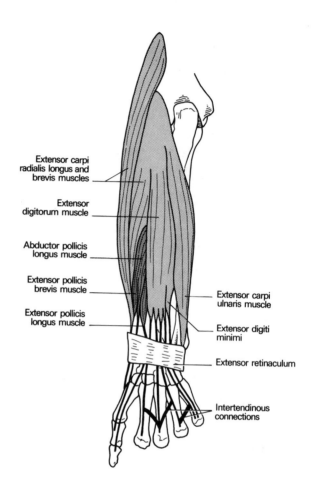

Extensor carpi radialis longus and brevis muscles

Extensor digitorum muscle

Abductor pollicis longus muscle

Extensor pollicis brevis muscle

Extensor pollicis longus muscle

Extensor carpi ulnaris muscle

Extensor digiti minimi

Extensor retinaculum

Intertendinous connections

Action. Extension of the radiocarpal (wrist) joint and of the metacarpophalangeal and interphalangeal joints of the fingers.

Extensor Digiti Minimi

This is an unimportant slip of muscle that gets its name simply because it is contained in its own synovial sheath.

Proximal Attachment. Common extensor origin.

Distal Attachment. Extensor expansion of the little finger.

Nerve Supply. Posterior interosseous nerve (**C7** and C8).

Action. Extends the metacarpophalangeal and interphalangeal joints of the little finger.

Extensor Carpi Ulnaris

This long, slender muscle acts as an extensor and adductor (ulnar deviator) of the wrist.

Proximal Attachments. The common extensor tendon, the posterior border of the ulna, and the antebrachial fascia.

Distal Attachment. After passing deep to the extensor retinaculum, to the posterior aspect of the base of the fifth metacarpal.

Nerve Supply. The posterior interosseous nerve (C7 and **C8**).

Action. Adducts and extends the wrist.

DEEP EXTENSOR (OUTCROPPING) MUSCLES (Figs. 40-2 and 40-3)

Four muscles on the extensor aspect of the forearm that have their upper attachments to the radius, ulna, and interosseous membrane lie deep to the above-mentioned muscles. Their tendons emerge from under cover of extensor digitorum at its lateral border to run superficial to extensor carpi radialis longus and brevis. Three of these muscles attach to the thumb and the fourth goes to the index finger.

Abductor Pollicis Longus

This muscle is important in the movements of the thumb.

Proximal Attachments. Radius, ulna, and interosseous membrane inferior to supinator.

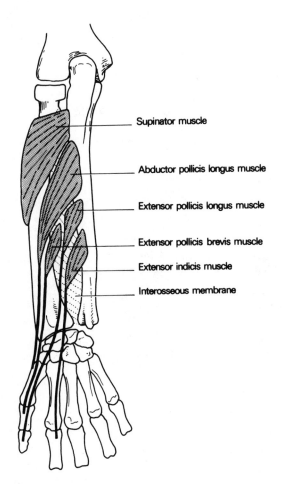

Supinator muscle

Abductor pollicis longus muscle

Extensor pollicis longus muscle

Extensor pollicis brevis muscle

Extensor indicis muscle

Interosseous membrane

Figure 40-3.
Deep (outcropping) and supinator muscles of the dorsum of the left forearm.

Distal Attachment. The tendon passes with that of extensor pollicis brevis deep to the extensor retinaculum in a common synovial sheath. It attaches to the base of the first metacarpal.

Nerve Supply. The posterior interosseous nerve (C7 and *C8*).

Action. Abduction and extension of the carpometacarpal joint of the thumb.

Extensor Pollicis Brevis

This relatively small muscle is important in movements of the thumb.

Proximal Attachments. The ulna and interosseous membrane inferior to abductor pollicis longus.

Distal Attachment. Its tendon runs with that of abductor pollicis longus deep to the extensor retinaculum and attaches to the base of the proximal phalanx of the thumb.

Nerve Supply. The posterior interosseous nerve (C7 and *C8*).

Action. Extension of the metacarpophalangeal and carpometacarpal joint of the thumb.

Extensor Pollicis Longus

This is the only muscle that can extend the interphalangeal joint of the thumb.

Proximal Attachments. The posterior surface of the ulna and interosseous membrane below abductor pollicis longus.

Distal Attachment. By a tendon that passes in its own tendon sheath deep to the extensor retinaculum, medial to the dorsal tubercle of the radius to the posterior surface of the base of the distal phalanx of the thumb.

Nerve Supply. The posterior interosseous nerve (C7 and *C8*).

Action. Extension of the carpometacarpal, metacarpophalangeal, and interphalangeal joints of the thumb.

Extensor Indices

This most medial of the deep extensor muscles is an extensor of the index finger.

Proximal Attachments. The ulna and interosseous membrane below the extensor pollicis longus.

Distal Attachment. By a tendon that passes deep to the extensor retinaculum to join the ulnar side of the tendon of extensor digitorum for the index finger.

Nerve Supply. Posterior interosseous nerve (C7 and *C8*).

Action. Extension of the index finger.

CLINICAL NOTE

Spontaneous rupture of the tendon of extensor pollicis longus at the point where it runs in a groove on the posterior aspect of the distal end of the

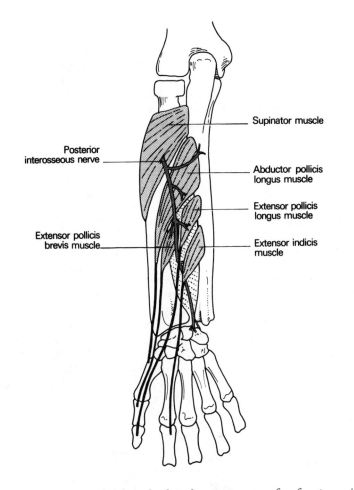

Figure 40-4.
Posterior interosseous nerve in the left
forearm.

Supinator muscle

Posterior
interosseous nerve

Abductor pollicis
longus muscle

Extensor pollicis
longus muscle

Extensor pollicis
brevis muscle

Extensor indicis
muscle

*radius, just medial to the dorsal tubercle, may occur after fractures in the
region or for no apparent reason. The cause is probably avascular necrosis; the tendon's upper blood supply comes from the anterior interosseous
artery, and its lower end is supplied by the radial artery. The two sources
of blood supply anastomose near the point where rupture usually occurs.*

NERVES

RADIAL NERVE

The radial nerve gives off its ***deep branch*** in the cubital fossa. This nerve
pierces the supinator muscle and becomes the ***posterior interosseous
nerve*** (Fig. 40-4) as it appears in the posterior region of the forearm. For
practical purposes it may be called the posterior interosseous nerve from
its beginning, thus reducing the confusion of adding "the deep branch of
the radial nerve" to our vocabulary. After its emergence from supinator
the nerve passes inferiorly between the extensor and (deep) outcropping
muscles. Next it goes deep to extensor pollicis longus to reach the interosseous membrane and ends by supplying the wrist joint. The posterior interosseous nerve supplies ***supinator, extensor digitorum, extensor digiti minimi, extensor carpi ulnaris***, and all the (deep) ***outcropping
muscles***.

ARTERIES

The posterior (dorsal) surface of the forearm is supplied by the posterior interosseous artery, which is a branch of the common interosseous from the ulnar artery, and by the dorsal carpal rete, which is formed by dorsal carpal branches from the *radial* and *ulnar* arteries (see Fig. 39-11). The dorsal carpal rete gives off three *dorsal metacarpal arteries* that anastomose with the palmar metacarpal arteries.

THE EXTENSOR EXPANSION

The *extensor expansion* (Fig. 40-5) is a special connective tissue arrangement by which tendons of the *long extensors* of the fingers and the *intrinsic muscles* of the hand insert on the phalanges. Basically, the extensor expansion consists of the expanded lower end of the tendons of *extensor digitorum*. The expansion is triangular in shape with the apex at the base of the distal phalanx and the base at the proximal end of the proximal phalanx. This triangular sheet is then wrapped around the digit so that the two proximal corners of the triangle come to lie on the sides of the proximal phalanx. Reference to Figure 40-5 will tell the story, where words fail.

The central fibers of the expansion insert into the base of the middle phalanx on its dorsal (posterior) surface. The lateral and medial fibers of the expansion pass distally to insert into the base of the distal phalanx.

At the proximal end of the expansion (the base of the triangle) there are attachments for the tendons of the appropriate *interosseous muscles*,

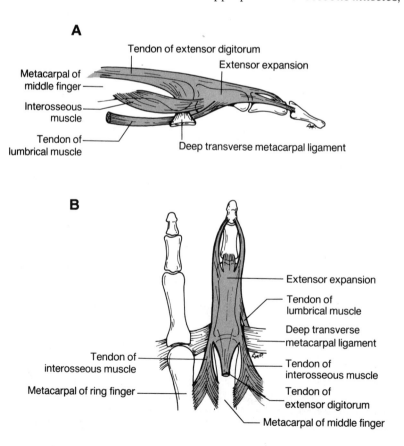

Figure 40-5.
Extensor expansion of the left middle finger. **A.** Lateral view. **B.** Dorsal view.

and on the radial side of the base of the expansion there is the insertion of the tendon of the **lumbrical muscle**.

The extensor expansion therefore gives insertion not only to the extensor digitorum and other appropriate extensors (*e.g.*, extensor indicis, extensor digiti minimi) but also to the interossei and lumbricals.

Note: The lumbricals pass **anterior** to the **deep transverse metacarpal ligament** and therefore are clearly anterior to the plane of the metacarpophalangeal joint (Fig. 40-5). Thus they can flex this joint. The interossei, passing posterior to the deep transverse metacarpal ligament, have less ability to flex the metacarpophalangeal joint. However, both the lumbricals and interossei are important in **extending** the **interphalangeal joints** when the metacarpophalangeal joint is held in flexion.

Extension of the interphalangeal joints with the metacarpophalangeal joints flexed is the movement that occurs when one makes an upstroke in writing. More details and an example of the delicate actions of the intrinsic muscles of the hand will be found in the following chapter on the hand.

The Hand

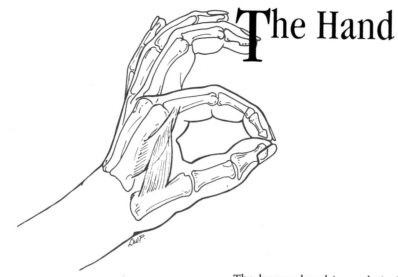

The human hand is a relatively undifferentiated replica of the distal appendage of a primitive pentadactyl limb. It is this essential lack of specialization or differentiation that makes the hand such a tremendous asset to man. The human hand differs from that of many primates by the shorter length of the thumb, the opposability of the thumb, and the fine discrimination of movements of the digits permitted by the intrinsic muscles.

The usefulness of the hand is enhanced by two other special features of the human body: the ability of the forearm to put the hand into pronation or supination, and the development of the trunk and lower limbs to enable man to function in the erect, bipedal position. Man's upright posture enables him to reach for the trees (for food); the pronated hand plucks the fruit, and the supinated hand puts the food into the mouth. How simple, and yet how complex. What would Adam and Eve have done without these movements?

The skin of the palmar surface of the hand and digits is raised in ridges (fingerprints and palm prints) and is kept moist by sweat glands to enhance the grip. In addition the hand's grasping facility is enhanced by the fusion of the palmar skin to the underlying palmar fascia, which prevents the skin from sliding.

The arrangement of the bones and joints of the hand is such that the hand and digits can accommodate themselves to objects of varying shapes. The greater the area of skin contact, the surer is the grip. Thus we can grasp the spherical surface of a tennis ball although at first glance our open hand appears rather flat. This accommodation is enabled by the opposability of the thumb and last finger and by some active and much passive rotation permitted at the metacarpophalangeal joints.

Consider the function of the hand in gripping a tool such as a hammer. The grip employed relies on the strength of the two most medial fingers, and the thumb and other two fingers help to guide and steady the hammer. As will be seen later the grip exerted by the two medial fingers and the guidance from the other fingers is largely a function of muscles innervated by the ulnar nerve. The recognition of objects and the proprioceptive information required for the fine guidance of a tool (*e.g.,* a pencil in drawing) are provided by the two lateral fingers and the thumb, which receive their sensory innervation by the median nerve. The median nerve has aptly been called "***the eyes of the hand.***"

41

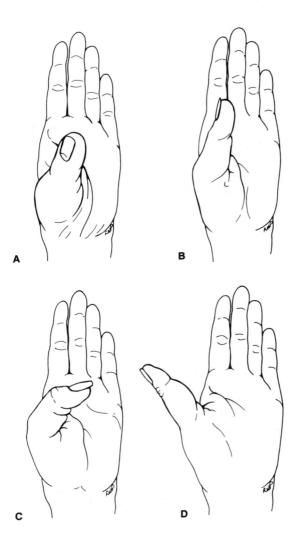

Figure 41-1.
Movements of the thumb.
A. Abduction. **B.** Adduction. **C.** Flexion.
D. Extension.

DIGITS

The five digits, named from the lateral side, are ***pollex, index, medius, annularis***, and ***minimus***. We shall call them thumb, index finger, middle finger, ring finger, and little finger (remember that in the anatomical position the hand is supinated, *i.e.*, it faces forward). Note again that the thumb and its metacarpal are rotated through 90 degrees in relation to the other digits. This enables the thumb to be opposed to all the other fingertips. Flexion-extension and abduction-adduction of the thumb thus occur at right angles to the similar movements of the fingers (Fig. 41-1).

PALMAR EMINENCES

There are two muscular thickenings in the palm. One is at the base of the thumb and is called the ***thenar eminence***; the other is at the base of the little finger and is called the ***hypothenar eminence*** (Fig. 41-2).

Figure 41-2.
Cutaneous innervation of the right hand.

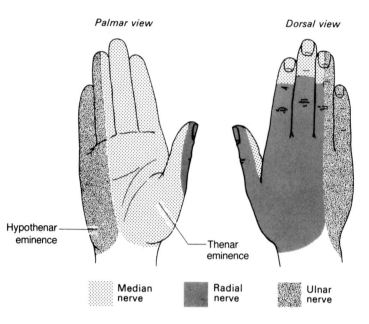

Palmar view

Dorsal view

Hypothenar eminence

Thenar eminence

Median nerve   Radial nerve   Ulnar nerve

JOINTS OF THE WRIST

The ***radiocarpal*** joint is usually referred to as the "wrist joint" (Fig. 41-3), but there are many other joints in the same area: ***midcarpal*** and ***intercarpal*** joints, ***carpometacarpal*** joints, and ***intermetacarpal*** joints. The number of bones and the way in which they articulate result in a relatively stable wrist but one that allows considerable movement.

RADIOCARPAL (WRIST) JOINT (Fig. 41-3)

Bony Surfaces. The lower end of the ***radius*** and the ***articular disc*** attaching it to the ulnar styloid articulate with the ***scaphoid***, ***lunate***, and ***triquetral*** bones (see Fig. 39-4). Together the three carpal bones form a

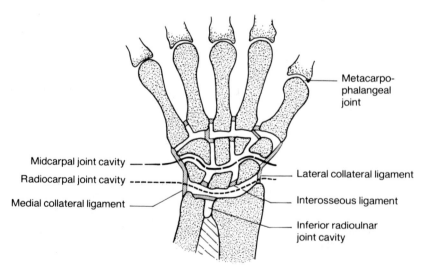

Metacarpophalangeal joint

Midcarpal joint cavity

Radiocarpal joint cavity

Medial collateral ligament

Lateral collateral ligament

Interosseous ligament

Inferior radioulnar joint cavity

Figure 41-3.
Coronal section of radiocarpal, midcarpal, carpometacarpal, and metacarpophalangeal joints of the right hand.

considerably larger articular surface than do the radius and the articular disc. This allows the movements of **adduction** (ulnar deviation) and **abduction** (radial deviation) and **flexion** and **extension** to occur at this joint. Note that the pisiform bone does not take part in this joint, except in the attachment of ligaments. The radial articular surface displays two concavities, and there is a similar depression on the articular disc. These concavities articulate with the scaphoid, lunate, and triquetral, respectively, when the wrist is in the neutral position.

Interosseous Ligaments. There are two interosseous ligaments: one joins scaphoid to lunate, the other joins lunate to triquetral. These two ligaments close off the joint cavity of the radiocarpal joint and form a continuous surface with the three carpal bones; thus they, as a unit, can articulate with the lower end of the radius and the articular disc.

Capsule. The capsule of the joint is unremarkable and relatively thin.

Synovial Membrane. The synovial membrane is confined to the radiocarpal joint and does not usually communicate with the inferior radioulnar or intercarpal joints.

Ligaments. The anterior and posterior radiocarpal ligaments are relatively weak. The **medial** (ulnar) **collateral ligament** runs from the styloid process of the ulna to the pisiform and the triquetral; it is relatively strong. The **lateral** (radial) **collateral ligament** runs from the radial styloid process to the scaphoid. It is a relatively strong ligament; the radial artery crosses superficial to it.

The articular surface of the radius is tilted anteriorly so that the joint in the position of function is closer to full extension than to full flexion. The movements possible at the radiocarpal joint are extension, flexion, abduction, and adduction. However, these are not **pure** movements; some of them also occur at the midcarpal joint (Fig. 41-4).

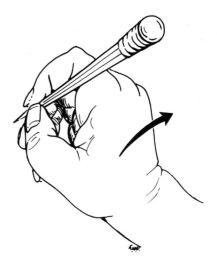

Figure 41-4.
"Position of function" for hand and fingers.

MIDCARPAL JOINT (see Fig. 41-3)

This is the joint between the carpal bones, primarily that between the proximal and distal rows of bones of the carpus, but transverse articulations between carpal bones are also included. They are plane joints. The proximal row consists of the scaphoid, lunate, and the triquetral. The distal row consists of the trapezium, trapezoid, capitate, and hamate. Interosseous ligaments join capitate to hamate and capitate to trapezoid. The posterior capsule is thicker than the anterior capsule; otherwise they are both unremarkable. The movements are extension, flexion, adduction, and abduction, but these movements cannot be entirely divorced from movements at the radiocarpal joint.

CARPOMETACARPAL JOINTS

The joints between the distal row of the carpus and the medial four metacarpals are rather featureless flat-shaped saddle (sellar) joints. They permit some gliding movements and some rotation which may be passive, as when the hand is "cupped" around a spherical object. The thumb carpometacarpal joint will be described below with the other joints and movements of the thumb.

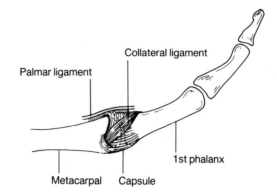

JOINTS OF THE HAND

Knuckles. These are the ***metacarpophalangeal*** joints (Fig. 41-5). Their unique feature is that the bone surfaces are oval (ellipsoid) in shape and therefore allow ***flexion*** and ***extension***, ***abduction*** and ***adduction***, and a small degree of rotation.

Flexion and extension of the metacarpophalangeal joints of the ***fingers*** occur in the ***sagittal plane***, flexion bringing the fingertips toward the palm. Abduction-adduction occurs in the ***coronal plane*** with the middle finger as the reference line. Abduction spreads the fingers away from that line, and adduction approximates them to that line. Note that the ***middle finger*** itself can deviate to the radial or ulnar side. Some ***opposition*** occurs at the fifth metacarpophalangeal joint.

Each joint has a capsule that is reinforced on the palmar surface by a dense plate of connective tissue, which assumes an almost cartilaginous appearance, the ***palmar ligament***. This fibrocartilaginous plate is loosely attached to the head of the metacarpal but densely attached to the base of the proximal phalanx; its palmar surface is grooved to accommodate the flexor tendons. The joints have two collateral ligaments which are tautest in the fully flexed position of the joint.

The medial four metacarpal heads are joined to each other by the ***deep transverse metacarpal ligaments***.

Interphalangeal Joints. These joints are small-scale versions of the metacarpophalangeal joints. They can flex and extend, and any rotation is passive accommodation to the shape of the object being held by the hand.

CLINICAL NOTE

*The fact that the collateral ligaments of the metacarpophalangeal and interphalangeal joints are tautest in full flexion is very important. An inflamed or traumatized joint will become surrounded by inflammatory exudate, which tends to mature into scar tissue. Scar tissue contracts over time. If a joint is to be immobilized it must be done in such a way that it will be able to resume its function. If a finger is splinted in extension the relaxed collateral ligaments will shorten, and subsequent flexion that requires the full length of the collateral ligaments will be impossible. **Never** splint a finger in extension.*

JOINTS OF THE THUMB

If we appear to place the thumb on a pedestal it is because of its unique structure and function. The base of the pedestal is the first carpometacarpal joint where the first metacarpal bone articulates with the trapezium. Remember that the first metacarpal is **rotated at 90 degrees** to the plane of the other metacarpals.

The carpometacarpal joint of the thumb is a saddle joint that functions like a universal joint, allowing flexion, extension, adduction, abduction, and rotation which, in this case, is **opposition**. This movement allows the pulp of the distal phalanx of the thumb to touch the tips of the other digits. The function of the thumb is so important and versatile that for insurance and pension purposes it accounts for 50% of the function of the hand.

The capsule of the thumb carpometacarpal joint is rather ordinary except for the fact that the palmar and dorsal fibers run obliquely to converge on the ulnar side of the base of the metacarpal. This provides an anchor around which the movement of opposition occurs.

The thumb has only two phalanges; in fact, the thumb metacarpal is morphologically a phalanx. This is suggested by the fact that its secondary center of ossification appears at its proximal end, as in the other phalanges.

An analysis of the movements of the hand will appear at the end of this chapter.

NERVES IN THE HAND (GENERAL PLAN) (Fig. 41-6)

The nerves in the hand are branches of the radial, ulnar, and median nerves. The following general plan needs to be understood.

Median Nerve. The median nerve gives off a **palmar cutaneous branch** above the flexor retinaculum; the branch runs superficial to the retinaculum to supply the skin of the lateral half of the palm. The median nerve also supplies the skin of the thumb, index finger, middle finger, and radial half of the ring finger on their palmar surfaces and sides and on the dorsum

Figure 41-6.
Nerves of the palmar surface of the right hand.

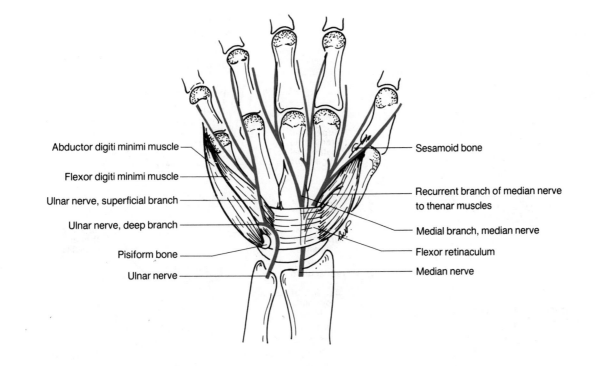

- Abductor digiti minimi muscle
- Flexor digiti minimi muscle
- Ulnar nerve, superficial branch
- Ulnar nerve, deep branch
- Pisiform bone
- Ulnar nerve

- Sesamoid bone
- Recurrent branch of median nerve to thenar muscles
- Medial branch, median nerve
- Flexor retinaculum
- Median nerve

of the terminal phalanx of the same digits. Its ***motor*** fibers supply the three ***thenar muscles*** and the lateral two ***lumbricals*** (see below).

Ulnar Nerve. This nerve gives off a ***palmar cutaneous branch*** that passes superficial to the flexor retinaculum and palmar aponeurosis to supply the skin of the medial half of the palm, and a ***dorsal cutaneous branch*** that supplies the medial half of the dorsum of the hand, the dorsum of the little finger, and the medial half of the dorsum of the ring finger. After passing superficial to the flexor retinaculum, the ulnar nerve divides into ***superficial*** and ***deep terminal*** branches. The superficial branch is mainly sensory, and the deep branch is mainly motor (see below).

Radial Nerve. The radial nerve has no motor function in the hand. It is sensory to the skin over the lateral two thirds of the dorsum of the hand and to the skin over the dorsum of the proximal and middle phalanges of the lateral three and one-half digits.

As might be expected these sensory (and to some extent motor) distributions are not engraved in stone, and variations and overlap are common.

THE PALMAR APONEUROSIS

The palmar aponeurosis (Fig. 41-7) is formed by the deep fascia of the hand. It is firmly attached to the skin of the palm and extends from the flexor retinaculum to split into four slips for the fingers. These digital slips blend with the ***fibrous flexor sheaths*** of the fingers. The central portion of the palmar aponeurosis is very thick and fades away over the thenar and hypothenar eminences. Two septa pass posteriorly to be attached to the first and fifth metacarpal bones, respectively. The medial septum lies lateral to the hypothenar muscles, and the lateral septum is medial to the

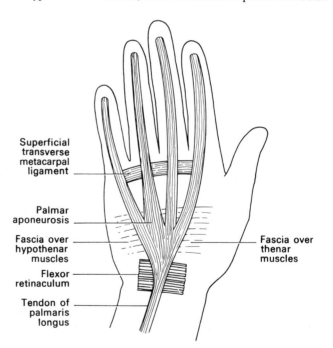

Superficial transverse metacarpal ligament

Palmar aponeurosis

Fascia over hypothenar muscles

Flexor retinaculum

Tendon of palmaris longus

Fascia over thenar muscles

Figure 41-7.
Fibrous structures of the palmar surface of the right hand (anterior view).

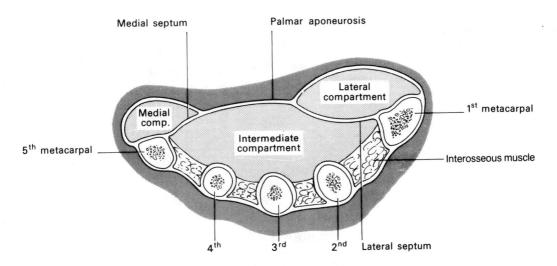

Medial septum — Palmar aponeurosis

Medial comp.

Lateral compartment

1st metacarpal

Intermediate compartment

5th metacarpal

Interosseous muscle

4th   3rd   2nd   Lateral septum

Figure 41-8.
Cross-section of the compartments of the palm. The metacarpal bones are joined by interosseous muscles.

thenar muscles. The palm is thus divided into three compartments (Fig. 41-8). The ***medial hypothenar compartment*** contains the hypothenar muscles, and the ***lateral thenar compartment*** contains the thenar muscles. The ***central compartment*** contains primarily the flexor tendons and their synovial sheath. The vessels and nerves that lie deep to the palmar aponeurosis reach their digits by passing between the digital slips of the palmar aponeurosis.

The middle compartment must not be thought of as a space in which the tendons and their sheaths rattle about. It is packed with loose connective tissue such that all spaces are filled and all movable structures are given the greatest freedom of motion. This loose connective tissue can become infected, for example, from a puncture wound, resulting in a middle palmar space abscess.

CLINICAL NOTE

*Infections in the palm of the hand generally remain confined to the compartments of the hand. Thus we can have thenar space infections, hypothenar space infections, or middle palmar space (sometimes called deep palmar space) infections. The thickness of the palmar aponeurosis prevents significant swelling in the palm of the hand when these spaces are full of pus; instead marked edema is evident in the loose subcutaneous tissue of the **dorsum** of the hand.*

TENDONS OF THE PALM

The freedom of movement of the long flexor tendons is assured by the ***synovial flexor sheaths*** (Fig. 41-9). As the deep and superficial flexor tendons pass deep to the flexor retinaculum they are surrounded by a synovial sheath that extends as far distally as the midregion of the palm. This sheath remains in continuity with the synovial covering of the flexor tendons to the little finger. The long flexor of the thumb has its own sheath that reaches from superior to the wrist to the insertion of the tendon. The tendons to the other three digits pick up another synovial sheath just proximal to the metacarpophalangeal joints and continue within that sheath to their distal attachments.

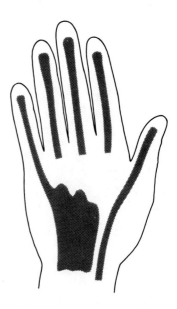

Figure 41-9.
Pattern of synovial sheaths of the right palm (anterior view).

Figure 41-10.
Schematic representation of the
osseofibrous tunnel of a typical digit
(cross-section).

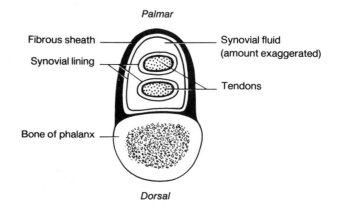

OSSEOFIBROUS TUNNEL

The palmar surfaces of the four fingers and the thumb are covered by a dense layer of connective tissue; together with the bones of the digits they form a tight, relatively unyielding tunnel in which there is sufficient space only to accommodate the flexor tendons, their synovial sheaths, and some slender blood vessels that nourish the tendons (Fig. 41-10). This tunnel arrangement is spoken of as the ***osseofibrous tunnel*** and is of great functional and clinical importance.

The tendons of ***flexor digitorum superficialis*** and ***profundus*** run on the palmar (ventral) aspects of the phalanges in these ***fibrous sheaths*** which are lined by the synovial sheaths mentioned above. Each tunnel consists of ***bone*** posteriorly and a ***fibrous arch*** on the sides and anteriorly. This fibrous tissue is thin in the regions of the joints so that the sheaths may bend when the fingers are bent.

Flexor Tendon Sheaths. The osseofibrous tunnel of each digit contains the two flexor tendons (only one for the thumb) wrapped in a synovial sheath that begins at the base of the proximal phalanx and is identifiable on the surface of the hand as extending from the distal transverse crease of the palm to the base of the distal phalanx.

Remember that the ***superficialis tendon*** splits into two slips that attach to each side of the middle phalanx and that the ***profundus tendon*** passes between these two slips to attach to the base of the distal phalanx (Fig. 41-11). As the flexor tendons approach their attachments they are connected to the dorsal part of their enclosing synovial sheaths by thread-

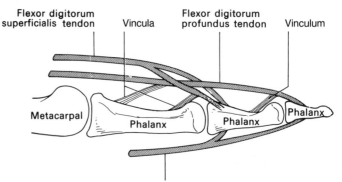

Figure 41-11.
Long tendons of a typical digit.

like bands of synovial tissue called ***vincula*** (singular, vinculum) brevia and longa. The names are unimportant, but they carry the ***blood supply*** to the tendons, and that ***is important***.

CLINICAL NOTE

Inflammation of the synovial sheaths results in a clinical condition known as **synovitis** *that produces severe pain. The patient does not tolerate passive extension of the affected digit, which is held in a position of partial flexion for comfort. If the little finger is affected, the infection can spread to the common synovial sheath in the palm and secondarily affect all four fingers. Similar remarks apply to the thumb, depending on whether its tendon sheath communicates with the common sheath.*

This condition requires emergency incision of the flexor sheath for drainage because the increased pressure, caused by the accumulation of inflammatory exudate within the flexor sheath, will soon obliterate the delicate blood vessels in the vincula and cause death of the tendons.

There is insufficient space within the osseofibrous tunnel to allow the repair of cut tendons within the tunnel. The resulting scar tissue would glue the tendon(s) to the walls of the tunnel, and a stiff finger would result. In these cases the tendons are **removed from the tunnel**, *and a suitable donor tendon (e.g., palmaris longus) is sutured to the proximal end of the cut tendon* **in the palm** *while the distal end is threaded through the tunnel and sutured to the base of the distal phalanx.* **Remember** *the territory from the distal palmar flexion crease to the base of the distal phalanx is* **no-man's-land**.

MUSCLES OF THE PALM (Fig. 41-12)

Thenar Muscles. As a group the thenar muscles act on the thumb or its metacarpal, or both. All three muscles are supplied by the recurrent branch

Figure 41-12.
Thenar and hypothernar muscles of the right palm.

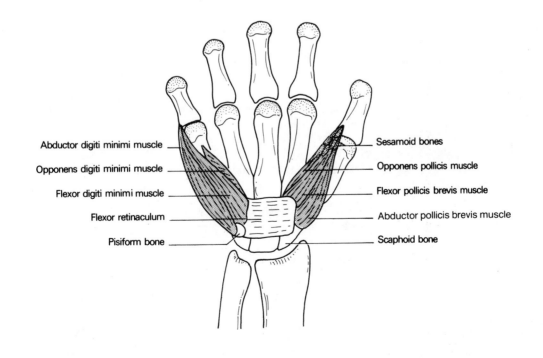

Abductor digiti minimi muscle	Sesamoid bones
Opponens digiti minimi muscle	Opponens pollicis muscle
Flexor digiti minimi muscle	Flexor pollicis brevis muscle
Flexor retinaculum	Abductor pollicis brevis muscle
Pisiform bone	Scaphoid bone

of the ***median nerve*** (see Fig. 41-6). No single action of these muscles can be considered in isolation; they act in unison.

CLINICAL NOTE

In median nerve paralysis the thenar muscles are inactive, and the usefulness of the thumb is largely lost because it can no longer be opposed.

Abductor Pollicis Brevis

This muscle lies on the radial side of the thenar eminence (Fig. 41-12).

Proximal Attachments. The flexor retinaculum, scaphoid, and trapezium.

Distal Attachment. Together with the tendon of flexor pollicis brevis to the radial side of the base of the proximal phalanx of the thumb.

Nerve Supply. The recurrent branch of median nerve (***C8***, T1).

Action. Abduction at the metacarpophalangeal and carpometacarpal joints together with some medial rotation, so that the thumb moves forward in a plane at right angles to the palm of the hand.

Flexor Pollicis Brevis

This muscle lies on the medial side of abductor pollicis brevis (Fig. 41-12).

Proximal Attachments. The flexor retinaculum and the trapezium.

Distal Attachment. With the abductor pollicis brevis to the radial side of the proximal phalanx of the thumb; the common tendon contains a ***sesamoid bone***.

Nerve Supply. The recurrent branch of the median nerve (***C8*** and T1).

Action. This muscle is an important flexor of the metacarpophalangeal joint of the thumb and assists in pinching the thumb against the tip of another finger (Fig. 41-12).

Opponens Pollicis

This muscle lies deep to abductor pollicis brevis and flexor pollicis brevis (Figs. 41-12 and 41-13).

Proximal Attachments. Flexor retinaculum and trapezium.

Distal Attachment. The lateral side of the shaft of the first metacarpal.

Nerve Supply. The recurrent branch of the median nerve (***C8*** and T1).

Action. Flexion and medial rotation, that is, opposition, of the first metacarpal.

Adductor Pollicis

This muscle lies deep to the flexor tendons in the palm of the hand (Fig. 41-13).

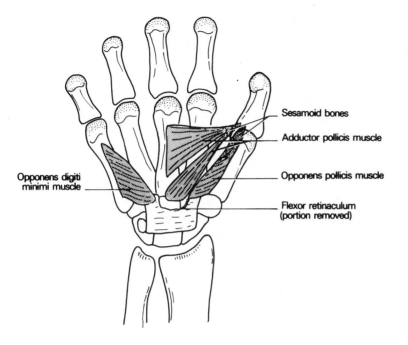

Figure 41-13.
Adductor pollicis and opponens
muscles of the right palm.

Sesamoid bones

Adductor pollicis muscle

Opponens pollicis muscle

Flexor retinaculum
(portion removed)

Opponens digiti
minimi muscle

Proximal Attachments. Two heads, a transverse head from the shaft of the third metacarpal and an oblique head from the capitate bone and adjacent ligaments.

Distal Attachment. By a tendon into the ulnar side of the base of the proximal phalanx of the thumb; the tendon contains a sesamoid bone.

Nerve Supply. The deep branch of the ***ulnar*** nerve (C8 and *T1*).

Action. Adduction of the thumb to bring it back to the palm.

Hypothenar Muscles

These are the muscles of the little finger. Their names are ***abductor digiti minimi***, ***flexor digiti minimi***, and ***opponens digiti minimi*** (Fig. 41-12). The proximal attachments are to the flexor retinaculum, the hamate bone, and adjacent structures, including the pisiform bone.

At their distal attachments the first two attach to the proximal phalanx of the finger, and the latter attaches to the shaft of the fifth metacarpal. Their actions are similar to those of the thenar muscles except that the fifth finger abducts in the plane of the hand. They are ***all*** supplied by the deep branch of the ***ulnar nerve*** (C8 and *T1*) (see Fig. 41-6).

Palmaris Brevis

This small cutaneous muscle wrinkles the skin over the ulnar border of the hand. It is supplied by the superficial branch of the ulnar nerve.

Lumbrical Muscles

Each lumbrical (lumbricus=earthworm) is attached to a tendon of flexor digitorum profundus, and some are attached to two adjacent tendons

Figure 41-14.
Lumbrical muscles of the right palm.

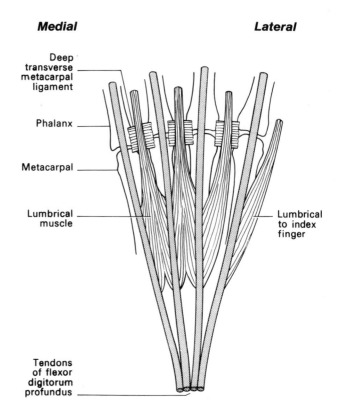

Medial   **Lateral**

Deep transverse metacarpal ligament

Phalanx

Metacarpal

Lumbrical muscle

Lumbrical to index finger

Tendons of flexor digitorum profundus

(Fig. 41-14). Their distal attachments are to the ***extensor expansions*** of the fingers.

Nerve Supply. The lateral two lumbricals are supplied by the ***median nerve*** and the medial two by the deep branch of the ***ulnar nerve*** (all from C8 and ***T1***).

Action. Flexion of the metacarpophalangeal joints and extension of the interphalangeal joints of the fingers. In other words, they place the fingers in the position that one assumes in holding a pen. Although these individual actions may be ascribed to the lumbricals, in general they are used for adjusting the motion imparted by the long tendons and placing the fingers exactly in the position required.

Also, if the flexor digitorum tendons have flexed the metacarpophalangeal joints, the lumbricals and interosseus muscles are the only muscles that can extend the interphalangeal joints.

Interosseus Muscles

There are three ***palmar interossei*** and four ***dorsal interossei***. Their attachments are shown in Figure 41-15. It should be noted that the ***d***orsal interossei ***ab***duct the fingers whereas the ***p***almar interossei ***ad***duct the fingers. More importantly, their distal attachments are similar to those of the lumbricals, except that their tendons pass posterior to the deep transverse carpal ligaments to attach to the extensor expansions of the fingers and into the bases of the proximal phalanges. Their actions are similar to those of the lumbricals.

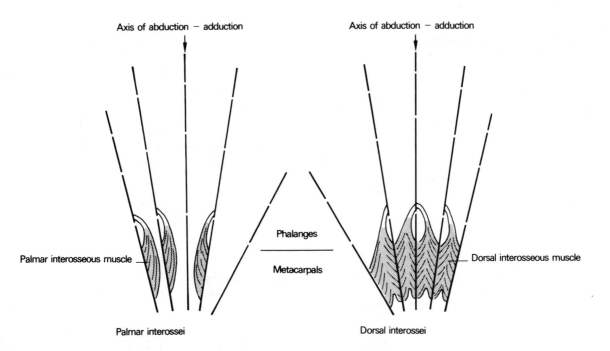

Axis of abduction − adduction

Axis of abduction − adduction

Phalanges

Metacarpals

Palmar interosseous muscle

Dorsal interosseous muscle

Palmar interossei

Dorsal interossei

Nerve Supply. All the interossei are supplied by the deep branch of the ulnar nerve (C8 and **T1**).

The lumbricals and interossei are referred to collectively as the **intrinsic muscles** of the hand. They **all** flex the metacarpophalangeal joints and extend the interphalangeal joints. If they are paralyzed (*e.g.*, in a **T1** lesion) the hand will adopt the opposite, **claw** position (main en griffe) (Fig. 41-16).

CLINICAL NOTE

If injury to the ulnar nerve is suspected, it may be tested by having the patient attempt to hold a piece of paper between the **extended** *and* **adducted** *fingers. Failure to hold the paper indicates ulnar nerve (or T1) damage.*

The first dorsal interosseous muscle is quite bulky and forms the "meat" of the dorsal aspect of the first interosseous space, which can be demonstrated by pinching the tip of the thumb against the tip of the index finger while trying to make the two digits describe the letter **O** *(Fig. 41-17).*

NERVES OF THE PALM

The distribution of the **radial nerve** to the dorsum of the hand and fingers has already been described.

ULNAR NERVE

The ulnar nerve (Fig. 41-18) enters the palm superficial to the flexor retinaculum, in its own little tunnel of deep fascia (Guyan's canal). It divides into **superficial** and **deep terminal branches**.

Figure 41-15.
Schematic representations of the interosseous muscles of the right hand.

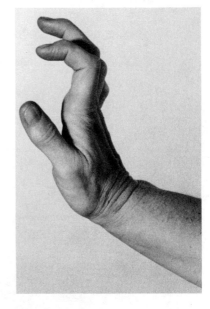

Figure 41-16.
Claw hand.

Figure 41-17.
Illustration of the function of the first
dorsal interosseous muscle.

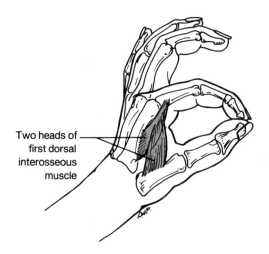

Two heads of
first dorsal
interosseous
muscle

Superficial Terminal Branch of Ulnar Nerve. This supplies palmaris
brevis and the skin over it. It then divides into two *palmar digital nerves*,
which supply all of the palmar surface of the little finger and the medial
half of the ring finger (see Fig. 41-6).

Deep Terminal Branch of Ulnar Nerve. This branch runs through the
hypothenar muscles, which it supplies, and then deep to the profundus
tendons crosses the palm to supply the *medial two lumbricals, the adductor pollicis and all the interossei* (Fig. 41-18).

MEDIAN NERVE

The median nerve (see Fig. 41-6) passes deep to the flexor retinaculum
between the tendons of the flexor digitorum superficialis and the tendon
of flexor carpi radialis. In the hand it is covered by skin, palmar aponeurosis, and the superficial palmar (arterial) arch.

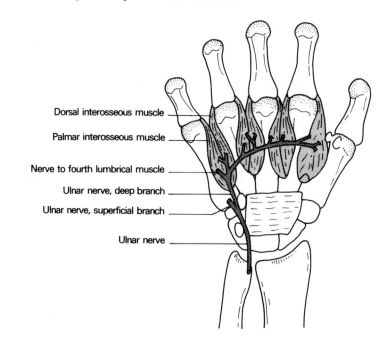

Dorsal interosseous muscle

Palmar interosseous muscle

Nerve to fourth lumbrical muscle

Ulnar nerve, deep branch

Ulnar nerve, superficial branch

Ulnar nerve

Figure 41-18.
Deep branch of the ulnar nerve of the
right hand. Anterior view.

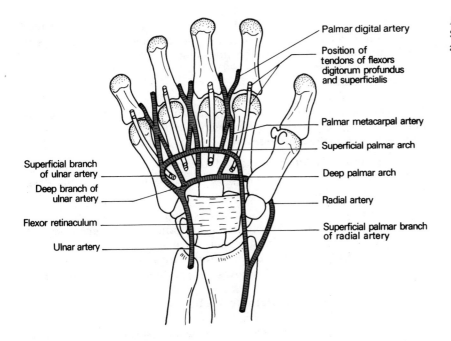

Palmar digital artery

Position of tendons of flexors digitorum profundus and superficialis

Palmar metacarpal artery

Superficial palmar arch

Deep palmar arch

Radial artery

Superficial palmar branch of radial artery

Superficial branch of ulnar artery

Deep branch of ulnar artery

Flexor retinaculum

Ulnar artery

Figure 41-19.
Superficial and deep palmar arterial arches of the right hand.

At the distal margin of the flexor retinaculum it gives off the important ***recurrent motor branch***, which supplies the muscles of the thenar eminence (abductor pollicis brevis, flexor pollicis brevis, and opponens pollicis) and then divides into medial and lateral ***palmar digital*** branches, which provide sensation to the entire palmar surfaces and sides of the thumb, index finger, middle finger, and lateral half of the ring finger and to the dorsum of the distal phalanges of these digits. It also supplies ***motor*** fibers to the lateral two lumbricals.

ARTERIES OF THE PALM

The palm is supplied by the radial and ulnar arteries (Fig. 41-19).

RADIAL ARTERY

The radial artery was last encountered at the wrist. Here it gives off a ***superficial palmar branch*** before winding around the lateral ligament of the wrist to enter the anatomical snuffbox.

Superficial Palmar Branch. This artery passes through the thenar muscle mass to cross the palm superficial to the flexor tendons. It joins the ***superficial palmar arch***, which is the continuation of the ulnar artery.

The radial artery continues around the wrist, deep to the tendons of abductor pollicis longus and extensor pollicis brevis. It then passes between the bases of the first two metacarpal bones, between the proximal heads of attachment of the first dorsal interosseous muscle, to enter the palm. It pierces the adductor pollicis muscle and joins the deep branch of the ulnar artery to form the ***deep palmar arch***, which is located deep to the long flexor tendons and lumbricals. The deep palmar arch gives off ***palmar metacarpal*** arteries, which anastomose with the ***palmar digital*** branches from the superficial palmar arch.

After piercing the first dorsal interosseous muscle the radial artery gives of the ***arteria princeps pollicis*** to supply branches to both sides of the thumb and also an artery to the index finger.

ULNAR ARTERY

As it passes lateral to the pisiform bone and anterior to the flexor retinaculum, the ulnar artery gives off a deep branch and then continues across the palm as the ***superficial palmar arch***.

Deep Branch. The deep branch of the ulnar artery passes through the hypothenar muscles to join the radial artery to complete the ***deep palmar arch***.

Superficial Branch. This branch passes laterally between the palmar aponeurosis and the long flexor tendons to form the ***superficial palmar arch***. It anastomoses with the superficial branch of the radial artery. The superficial arch gives off three ***palmar digital arteries***, which anastomose with the ***palmar metacarpal arteries*** and continue to supply the digits.

Surface Markings of the Palmar Arches

The deep arch runs along the bases of the metacarpals, which can be identified by palpation. The superficial arch lies about 1.25 cm distally.

Joints of the Upper Limb

In contrast to the joints of the lower limb, the upper limb joints can sacrifice some stability in favor of mobility. One might say that a major function of the non-weight-bearing upper limb is to provide a series of articulated levers for the manipulation of the hand into positions of optimum function.

STERNOCLAVICULAR JOINT

The sternoclavicular joint is the articulation between the manubrium sterni and the medial end of the clavicle. It is the only joint between the upper limb and the axial skeleton (Fig. 42-1).

The medial end of the clavicle is larger than the articular surface on the manubrium, and this end of the clavicle overlaps the joint to come into contact with the first costal cartilage. The medial end of the clavicle is covered by fibrocartilage, a reflection of the fact that the clavicle ossifies from membrane rather than from hyaline cartilage.

An *articular disc* divides the joint into two cavities. The disc is attached superiorly to the clavicle and inferiorly to the medial end of the first costal cartilage. This arrangement tends to prevent medial displacement of the clavicle on the manubrium. The force that results from leaning or falling on the outstretched hand tends to displace the clavicle medially, but this is resisted by the articular disc and by the *costoclavicular ligament* (Fig. 42-1).

Synovial Membrane. The synovial membrane lines the inside of the capsule and in the young child covers the articular disc. However, because of the wear produced by movement, it usually disappears from the articular disc as the individual matures.

Capsule. The fibrous capsule surrounds the joint. It is thickened anteriorly and posteriorly to form the *anterior sternoclavicular* and *posterior sternoclavicular ligaments*. An *interclavicular ligament* (Fig. 42-1) passes along the upper border of the manubrium to join the two clavicles together.

Extrinsic Ligament. The *costoclavicular ligament* joins the medial end of the clavicle to the first costal cartilage.

Movements. The joint acts as a pivot for the medial end of the clavicle. It allows the pectoral girdle to be moved in various directions around this fulcrum.

42

Figure 42-1.
Right sternoclavicular joint (coronal section).

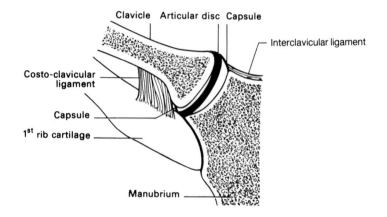

CLINICAL NOTE

The strength of the sternoclavicular joint is such that a fall on the point of the shoulder or onto the outstretched hand will fracture the clavicle (at the junction of its middle and lateral thirds) rather than dislocate the joint. Dislocations do occur but are usually the result of direct trauma applied to the anterior aspect of the medial end of the clavicle, such as hitting the chest wall against the steering column in a head-on automobile accident.

ACROMIOCLAVICULAR JOINT

The acromioclavicular joint is a plane synovial joint between the lateral end of the clavicle and the acromion process of the scapula (Fig. 42-2). Both joint surfaces are small, and the clavicle overlaps the acromion. The long axis of the joint is anteroposterior (its plane slopes inferiorly and medially). The joint contains an incomplete articular disc. The capsule is unremarkable except for some thickening on its superior aspect, which is called the acromioclavicular ligament.

Coracoclavicular Ligament. This **extrinsic** ligament is the strength of the joint. It is a very strong ligament that holds the clavicle to the coracoid

Figure 42-2.
Ligaments attaching to the acromion and coracoid processes of the right shoulder (anterior view).

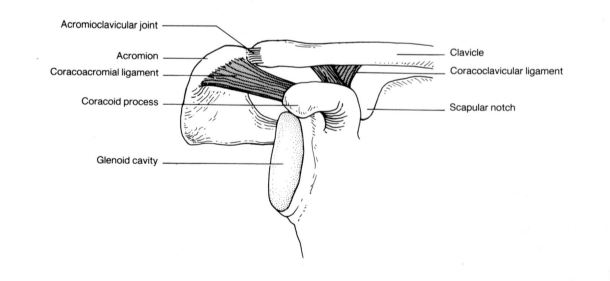

Table 42-1.
Movements of (Shoulder Girdle) Scapula

Movement	Muscles	Nerves	Spinal Cord Segments
Elevation	Trapezius (upper fibers)	Cranial n. XI	
	Levator scapulae	Nerve to levator scapulae and dorsal scapular	C3, C4, C5
	Serratus anterior (upper fibers)	Long thoracic	C5, C6, C7
Depression	Pectoralis major	Lateral and medial pectoral	C5, C6, C7, C8
	Latissimus dorsi	Thoracodorsal	C6, C7, C8
	Pectoralis minor	Medial pectoral	C5, C6, C7, C8
	Trapezius (lower fibers)	Cranial n. XI	
Superior rotation	Trapezius (upper and lower fibers)	Cranial n. XI	
	Serratus anterior	Long thoracic	C5, C6, C7
Inferior rotation	Levator scapulae	Nerve to levator scapulae and dorsal scapular	C3, C4, C5
	Rhomboids	Dorsal scapular	C4, C5
	Pectoralis major and minor	Lateral and medial pectoral	C5, C6, C7, C8
	Latissimus dorsi	Thoracodorsal	C6, C7, C8
Protraction	Serratus anterior	Long thoracic	C5, C6, C7
	Pectoralis minor	Medial pectoral	C5, C6, C7, C8
Retraction	Trapezius	Cranial n. XI	
	Rhomboids	Dorsal scapular	C4, C5
	Latissimus dorsi	Thoracodorsal	C6, C7, C8

process of the scapula (Fig. 42-2). A shoulder separation (acromioclavicular dislocation) cannot occur without rupture of the coracoclavicular ligament, and attempts to repair these separations are doomed to failure unless the integrity of the coracoclavicular bond is restored.

Movements of the Pectoral Girdle (Table 42-1). Movements of the pectoral girdle are most easily understood if they are considered in terms of movements of the scapula, with the clavicle acting as a strut that keeps the scapula propped away from the axial skeleton. These movements are elevation and depression, protraction (as in pushing the hand and arm forward), and retraction. Rotation of the scapula occurs when the arm is elevated above the shoulder. The pivot for these movements is the sternoclavicular joint, but some movement also occurs at the acromioclavicular articulation.

CLINICAL NOTE

Congenital absence of the clavicles is a rare birth defect and does not appear to hinder pectoral girdle movements and use of the upper limb in any significant way.

SHOULDER JOINT

The shoulder or **glenohumeral** joint (Fig. 42-3) is a very mobile ball and socket joint. However, mobility is achieved at the expense of stability, and dislocations of this joint are common. The capsule is relatively weak, the

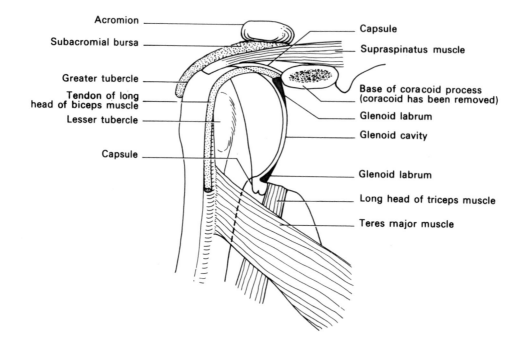

Figure 42-3.
Certain structures related to the right shoulder joint (anterior view with anterior portion of capsule removed).

normal relationship of the humerus to the glenoid being maintained largely by the ***rotator cuff muscles*** (see Fig. 37-9). Much of the apparent movement of the shoulder joint is actually produced by movements of the pectoral girdle.

Articular Surfaces. The articular surface of the glenoid cavity is much smaller than that of the head of the humerus. The joint cavity is deepened slightly by the fibrocartilaginous ***glenoid labrum***, which encircles the rim of the glenoid (Fig. 42-3). The glenoid cavity and the articular portion of the head of the humerus are both covered by articular (hyaline) cartilage.

Capsule. The capsule attaches to the scapula just proximal to the attachment of the glenoid labrum, and to the anatomical neck of the humerus (Fig. 42-4). The capsule is so lax that, in a shoulder with all muscles removed, if the humerus is abducted through perhaps 45 degrees, the head of the humerus may be pulled away from the glenoid cavity by as much as 2 cm. When the shoulder joint is adducted, as it is when the arm is at the side, the capsule is relatively tense superiorly, but inferiorly it is very lax, even redundant.

The capsule is reinforced by the blending with it of the fibers of humeral attachment of the ***rotator cuff*** muscles (see Fig. 37-9). The capsule is weakest inferiorly where it has no support from the rotator cuff muscles, and yet this is the very spot against which the head of the humerus is driven in forced abduction and lateral rotation of the shoulder that can produce dislocation.

Ligaments

Glenohumeral Ligaments. The glenohumeral ligaments are three thickenings in the capsule that are found anteriorly, running from the antero-superior margin of the glenoid to the lesser tubercle and anatomical neck of the humerus. They are best seen from within the joint cavity.

Coracohumeral Ligament (Fig. 42-4). This is a sturdy ligament passing from the base of the coracoid process of the scapula to the front of the greater tubercle. It is an intrinsic ligament and is tense in adduction of the joint.

Transverse Humeral Ligament (Fig. 42-4). This ligament spans the two tubercles and converts the intertubercular groove into a tunnel for the passage of the tendon of the long head of the biceps brachii muscle and its synovial tendon sheath.

Accessory Ligaments of the Scapula

Coracoacromial Ligament (see Fig. 42-2). This is a strong ligament that runs from the anterior edge of the acromion to the lateral border of the coracoid process. The coracoid process, the ligament, and the acromion process form the protective ***coracoacromial arch***, which tends to prevent upward displacement of the head of the humerus. The tendon of ***supraspinatus*** passes under this arch, being separated from it by the ***subacromial bursa*** (see Fig. 42-3).

Suprascapular Ligament. This small ligament bridges the scapular notch.

Glenoid Labrum. The ***glenoid labrum*** (see Fig. 42-3) is a fibrocartilaginous structure on the periphery of the glenoid cavity. It is triangular in cross-section with a free margin that is sharp and a base that attaches to the edge of the glenoid cavity.

Tendon. The tendon of the long head of ***biceps brachii*** passes through the joint. The tendon is covered by synovial membrane in its course through the joint from the supraglenoid tubercle to its exit deep to the transverse humeral ligament.

Synovial Membrane. The synovial membrane lines the inside of the capsule and any intra-articular parts of the humerus not covered by articular cartilage. It also covers the tendon of the long head of biceps brachii, where it lies in the intertubercular groove of the humerus. It may cover the whole, or just part, of the glenoid labrum.

Figure 42-4.
The right shoulder joint (anterior view).

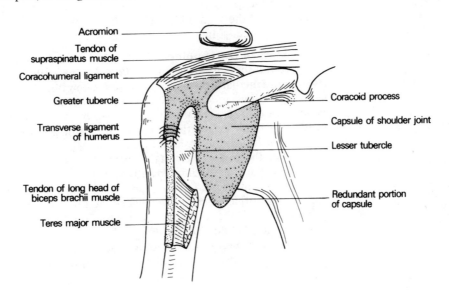

Acromion

Tendon of supraspinatus muscle

Coracohumeral ligament

Greater tubercle

Transverse ligament of humerus

Tendon of long head of biceps brachii muscle

Teres major muscle

Coracoid process

Capsule of shoulder joint

Lesser tubercle

Redundant portion of capsule

Table 42-2.
Shoulder Joint

Movement	Muscles	Nerves	Spinal Cord Segments
Flexion	Pectoralis major clavicular head	Lateral and medial pectoral	C5, C6
	Deltoid, anterior fibers	Axillary	C5, C6
	Coracobrachialis	Musculocutaneous	C6, C7
	Biceps brachii	Musculocutaneous	C5, C6
Extension	Latissimus dorsi	Thoracodorsal	C6, C7, C8
	Teres major	Lower subscapular	C5, C6
	Posterior fibers of deltoid	Axillary	C6
	Pectoralis major sternal head	Lateral and medial pectoral	C6, C7, C8
Abduction	Deltoid	Axillary	C5
	Supraspinatus	Suprascapular	C5
Adduction	Pectoralis major	Lateral and medial pectoral	C6, C7, C8
	Latissimus dorsi	Thoracodorsal	C6, C7, C8
Medial rotation	Pectoralis major	Lateral and medial pectoral	C6, C7, C8
	Deltoid, anterior fibers	Axillary	C6
	Latissimus dorsi	Thoracodorsal	C6, C7, C8
	Subscapularis	Upper and lower subscapular	C6, C7, C8
Lateral rotation	Deltoid, posterior fibers	Axillary	C5
	Teres minor	Axillary	C5
	Infraspinatus	Suprascapular	C5

Bursae. There are several bursae around the shoulder joint. The most important are the ***subacromial bursa*** (see Fig. 42-3), which was described with the ***supraspinatus muscle*** (see Chap. 37), and the bursa deep to ***subscapularis***, which protects its tendon where it runs over the anterior lip of the glenoid fossa. The latter bursa is usually connected to the glenohumeral joint cavity.

Movements of the Shoulder (Table 42-2). The shoulder permits movement around three axes. It may be ***flexed*** or ***extended***, ***abducted*** or ***adducted***, and medially or laterally ***rotated***. A combination of flexion, extension, abduction, and adduction describes a cone of movement called ***circumduction***. The relatively larger size of the articular surface of the head of the humerus and the laxity of the capsule give this joint the widest range of movement of any joint in the body.

When the arm is at rest in the anatomical position, the glenoid cavity faces just about equally laterally and anteriorly, and movements at the joint are described in relation to the plane of the scapula so that flexion brings the arm forward and across the front of the body, extension reverses that position, abduction brings the arm up laterally and forward, and adduction is the reverse of abduction.

The movements of the shoulder joint are enhanced by movements of the scapula, particularly rotation. For example, the average middle-aged adult can abduct the shoulder joint to about 100 degrees, but scapular rotation enables the hand to reach for the sky.

Principal Muscles Involved in Shoulder Movement and Main Spinal Cord Segments (in brackets) That Innervate These Muscles

Elevation of the Scapula. Trapezius, levator scapulae (Accessory (XI) [C3 and C4]).

Depression of the Scapula. Pectoralis minor, latissimus dorsi (C6, C7, and C8), and lower fibers of trapezius (accessory nerve XI).

Protraction of the Scapula. Pectoralis minor, serratus anterior (C6 and C7).

Retraction of the Scapula. Trapezius, rhomboids (Accessory (XI) [C3, C4, and C5]).

Rotation of the Scapula. So that glenoid cavity points upward (trapezius and lower fibers of serratus anterior [C3, C4, C5, and C6]) and so that glenoid points downward (gravity, pectoralis minor, levator scapulae, rhomboids [C3, C4, C5, and C6]).

Flexion of the Shoulder. Clavicular fibers of pectoralis major, anterior fibers of deltoid, coracobrachialis (C5 and C6).

Extension of the Shoulder. Sternal fibers of pectoralis major, posterior fibers of deltoid, latissimus dorsi (C7 and C8).

Abduction. Supraspinatus, deltoid (C5).

Adduction. Pectoralis major, latissimus dorsi (C6, C7, and C8).

Medial Rotation. Pectoralis major, anterior fibers of deltoid, latissimus dorsi, subscapularis (C6, C7, and C8).

Lateral Rotation. Posterior fibers of deltoid, teres minor, and infraspinatus (C5).

ELBOW AND PROXIMAL RADIOULNAR JOINT

Any description of the elbow joint must mention the proximal radioulnar joint because (1) the inferior two of the three bones making up the elbow joint (the radius and ulna) are in articulation with each other, (2) the joint cavity of the elbow is continuous with the joint cavity of the proximal radioulnar joint, and (3) the ligaments of the elbow are continuous with the ligaments of the proximal radioulnar joint (Fig. 42-5).

Bones. The trochlea and capitulum of the humerus articulate with the trochlear notch of the ulna and with the head of the radius, respectively. In addition, the head of the radius articulates with the radial notch of the ulna. The trochlea and trochlear notch are not completely congruent; the trochlea does not fit closely into the trochlear notch. This probably accounts for the fact that the head of the ulna (lower end) moves posteriorly in pronation and anteriorly in supination. The capitulum fits the radial head best when the elbow is semiflexed. The articular surfaces are covered by hyaline cartilage.

The periphery of the head of the radius articulates within the radial notch of the ulna to form the proximal radioulnar joint. The head of the radius pivots on the capitulum and moves against the radial notch.

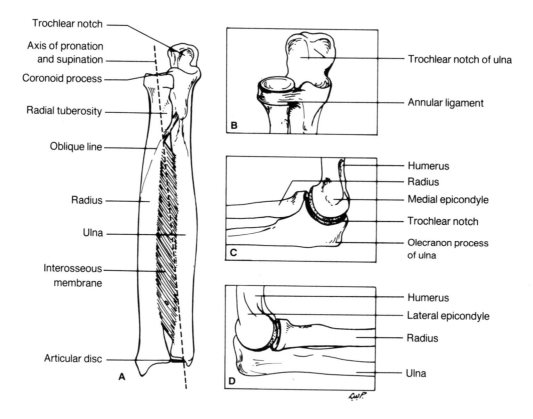

Figure 42-5.
Bones and ligaments of the right forearm. **A.** Anterior. **B.** Anterior. **C.** Medial. **D.** Lateral.

Capsule and Ligaments. Starting laterally the superior attachment of the **capsule** of the elbow joint to the humerus passes from the lateral epicondyle, superior to the radial and coronoid fossae to the medial epicondyle. Posteriorly it passes superior to the olecranon fossa to reach the lateral epicondyle. Inferiorly the capsule attaches to the annular ligament, the coronoid process, the medial surface of the ulna near the trochlear notch, and the olecranon.

The chief ligament of the superior radioulnar joint is the ***annular ligament***, which surrounds the head of the radius, that is, it forms a U (Fig. 42-5B). The ***quadrate ligament*** simply connects the two ulnar attachments of the annular ligament. The opening in the annular ligament is smaller than the head of the radius but fits the neck of the radius. The annular ligament is lined by synovial membrane.

CLINICAL NOTE

In the adult, the head of the radius usually cannot be dislocated without rupture of the annular ligament. However, in young children in whom the radial head is still small, an inferior dislocation can occur if the arm is pulled.

Lateral (Radial) Collateral Ligament (Fig. 42-6A). This triangular-shaped intrinsic ligament has its apex attached to the lateral epicondyle and its base attached to the annular ligament.

Medial (Ulnar) Collateral Ligament (Fig. 42-6B). The superior attachment of this ligament is to the medial epicondyle; inferiorly its anterior

Lateral view

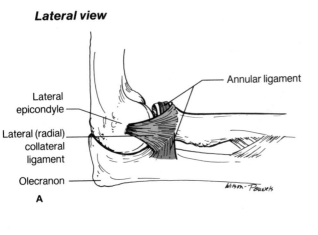

Annular ligament

Lateral epicondyle

Lateral (radial) collateral ligament

Olecranon

A

Figure 42-6.
Right elbow joint. **A.** Lateral (radial) collateral ligament. **B.** Medial (ulnar) collateral ligament.

Medial view

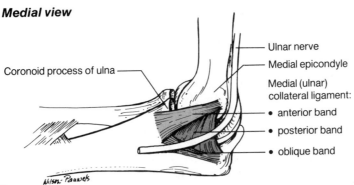

Coronoid process of ulna

Ulnar nerve

Medial epicondyle

Medial (ulnar) collateral ligament:

• anterior band

• posterior band

• oblique band

B

fibers attach to the coronoid process, and posteriorly its fibers pass to the olecranon. The two are linked by a weak oblique band. The ulnar nerve, as it passes behind the medial epicondyle, is in contact with the ulnar collateral ligament.

Movements of the Elbow and Radioulnar Joints (Tables 42-3 and 42-4). The elbow is only capable of *flexion* (biceps and brachialis [C5 and C6]) and *extension* (triceps [C7 and C8]). Extension is limited by the impingement of the olecranon process of the ulna on the olecranon fossa of the humerus.

Pronation. Pronation (pronators teres and quadratus [C7 and C8]) and *supination* (supinator and biceps brachii [C5 and C6]) occur at the radioulnar joints, but some ''rocking'' also occurs in the trochlear notch so that

Table 42-3.
Elbow Joint

Movement	Muscles	Nerves	Spinal Cord Segments
Flexion	Brachialis	Musculocutaneous	C5, C6
	Biceps brachii	Musculocutaneous	C5, C6
	Brachioradialis	Radial	C5, C6
Extension	Triceps brachii	Radial	C7, C8
	Anconeus	Radial	C7, C8

Table 42-4.
Radioulnar Joints

Movement	Muscles	Nerves	Spinal Cord Segments
Pronation	Pronator teres	Median	C7, C8
	Pronator quadratus	Anterior interosseous	C7, C8
Supination	Supinator	Posterior interosseous	C5, C6
	Biceps brachii	Musculocutaneous	C5, C6

the lower end of the ulna is pushed posteriorly in pronation and springs back anteriorly in supination (see Fig. 39-3).

Pronation and supination occur around an axis that passes through the head of the radius (and capitulum) and the base of the ulnar styloid process. The radius rotates around this axis.

Because the carpal bones articulate with the radius and the articular disc, the wrist and hand must move with the radius and not the ulna.

INFERIOR RADIOULNAR JOINT

The head of the ulna articulates with the ulnar notch of the radius. The articular surfaces are covered by hyaline (articular) cartilage and separated by a capillary layer of synovial fluid. The joint cavity is usually separated from the joint cavity of the radiocarpal joint, and the capsule of the joint is not remarkable. Joining the edges of the ulnar notch of the radius to the base of the ulnar styloid process is a fibrocartilaginous ***articular disc*** that provides the strongest attachment between the two bones at this joint (Fig. 42-7).

Interosseous Membrane. The interosseous membrane joining the interosseous borders of the radius and ulna provides a strong yet flexible attachment between the two bones as well as a surface for the attachment of muscles (Fig. 42-7).

Movements. ***The joints of the wrist and hand were described in the previous chapter*** (see Chap. 41). The movements are summarized here (Table 42-5).

Flexion of the Wrist. Flexor carpi ulnaris, flexor carpi radialis, and the long flexors of the digits (C6 and C7).

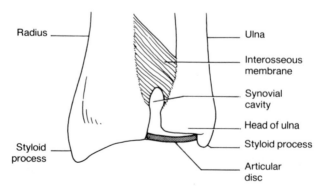

Figure 42-7.
Right inferior radioulnar joint (anterior view).

Table 42-5.
Wrist (Radiocarpal) Joint

Movement	Muscles	Nerves	Spinal Cord Segments
Flexion	Flexor carpi radialis	Median	C6, C7
	Flexor carpi ulnaris	Ulnar	C7, C8
	Palmaris longus	Median	C7, C8
	Long flexors of fingers and thumb	Median, anterior interosseous and ulnar	C7, C8, T1
Extension	Extensors carpi radialis longus and brevis	Radial	C6, C7, C8
	Extensor carpi ulnaris	Radial	C7, C8
	Extensor digitorum	Radial	C7, C8
Abduction	Extensors carpi radialis longus and brevis	Radial	C6, C7, C8
	Abductor pollicis longus	Posterior interosseous	C5, C6
	Flexor carpi radialis	Median	C6, C7
Adduction	Flexor carpi ulnaris	Ulnar	C7, C8
	Extensor carpi ulnaris	Radial	C7, C8

Extension of the Wrist. Extensor carpi radialis longus and brevis, extensor carpi ulnaris, and the extensors of the digits (C6 and C7).

Abduction of the Wrist. Flexor and extensor carpi radialis (C6 and C7) and abductor pollicis longus (C5 and C6).

Adduction of the Wrist. Flexor and extensor carpi ulnaris (C7 and C8).

Finger and Thumb Movements. The joints of the fingers and thumb were described in Chapter 41. Their movements are summarized here (Tables 42-6 and 42-7).

Finger Flexion and Extension (C7 and C8) (Table 42-7). Flexion of the metacarpophalangeal and interphalangeal joints is produced by flexors digitorum superficialis and profundus; flexion of the metacarpophalangeal joints with extension of the interphalangeal joints is a function of the lumbricals and interossei.

Table 42-6.
Movements of the Thumb (Including First Carpometacarpal Joint)

Movement	Muscles	Nerves	Spinal Cord Segments
Flexion	Flexor pollicis longus	Anterior interosseous	C8, T1
	Flexor pollicis brevis	Median	C8, T1
Extension	Extensors pollicis longus and brevis	Posterior interosseous	C5, C6
	Abductor pollicis longus	Posterior interosseous	C5, C6
Abduction	Abductor pollicis longus	Posterior interosseous	C5, C6
	Abductor pollicis brevis	Median	C8, T1
Adduction	Adductor pollicis	Ulnar	C8, T1
	1st dorsal interosseous	Ulnar	C8, T1
Opposition	Opponens pollicis	Median	C8, T1

Table 42-7.
Movements of the Fingers

Movement	Muscles	Nerves	Spinal Cord Segments
Flexion	Flexor digitorum superficialis	Median	C7, C8, T1
	Flexor digitorum profundus	Anterior interosseous and ulnar	C8, T1
Extension	Extensor digitorum longus	Radial	C7, C8
	Interossei and lumbricals	Median and ulnar	C8, T1
Abduction	Dorsal interossei	Ulnar	C8, T1
Adduction	Palmar interossei	Ulnar	C8, T1

Finger Abduction. The dorsal interossei.

Finger Adduction. The palmar interossei.

Flexion and Extension of the Thumb. Are functions of the long and short flexor and extensor muscles of the thumb.

Opposition of the Thumb. Produced by opponens pollicis.

Thumb Abduction. Abductors pollicis longus and brevis.

Thumb Adduction. Adductor pollicis and first dorsal interosseous.

Small Muscles of the Hand (T1). The small (intrinsic) muscles of the hand are the lumbricals and interossei, which flex the metacarpophalangeal joints and extend the interphalangeal joints (see Chap. 41).

Lymphatic System of the Upper Limb

The lymphatic system of the upper limb follows the general plan described in Chapter 2. The vessels are, in general, **superficial** and **deep**, and the **nodes** are primarily in the **axilla**.

LYMPH VESSELS

Superficial Vessels. Plexuses on the palmar surface of the fingers and hand pass to the dorsum of the hand; from there larger vessels pass superiorly. On the radial side of the hand and forearm the superficial vessels follow the line of the cephalic vein and terminate in the **axillary nodes**. The superficial vessels of the medial side of the hand pass to the ulnar side of the forearm and pick up superficial tributaries from the forearm and arm before they reach the **axillary** nodes. Some of the ulnar vessels pass first to the **cubital** nodes (Fig. 43-1).

Deep Vessels. These come from the periosteum, joints, muscles, and tendons and, running with the arteries, are joined by the superficial vessels above the elbow to pass to the **lateral group** of lymph nodes in the axilla.

LYMPH NODES

AXILLARY

The main group of lymph nodes of the upper limb is found in the axilla. The **axillary nodes** may be divided into five groups; the grouping is not necessarily distinct, and the location of individual nodes and groups of nodes varies (Fig. 43-2).

Apical Group. The apical lymph nodes form a small group around the axillary artery at the very apex of the axilla. This group receives afferents from all of the other axillary nodes. The efferents from the apical nodes join to form the **subclavian lymph trunk**, which joins the **jugular** and **bronchomediastinal trunk** to form the **right lymphatic duct** on the right side of the body. On the left side it joins the **thoracic duct**.

43

Figure 43-1.
Diagram of the superficial lymphatic
drainage of the upper limb.

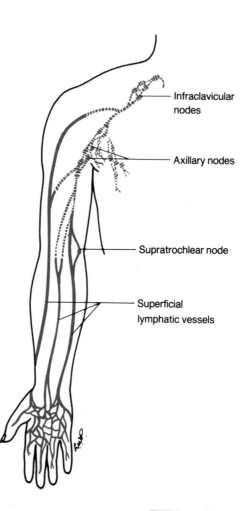

Infraclavicular
nodes

Axillary nodes

Supratrochlear node

Superficial
lymphatic vessels

Central Group. This group of nodes lies medial to the axillary artery below the pectoralis minor muscle. It receives afferents from the *lateral*, *pectoral*, and *subscapular* groups. Its efferents go to the *apical* group.

Lateral Group. These nodes are located lateral to the axillary artery near the lower border of teres major. They receive most of the lymph from the upper limb and drain into the *central* group.

Subscapular Group. This group is situated along the subscapular artery and receives afferents from the posterior surface of the chest wall and scapular region. Its efferents pass to the *central group*.

Pectoral Group. These nodes are found along the lateral thoracic artery at the lower border of pectoralis minor and are overlapped by the lower border of pectoralis major. They drain the anterior chest wall and the *breast*. Their efferents drain into the *central* and *apical* groups.

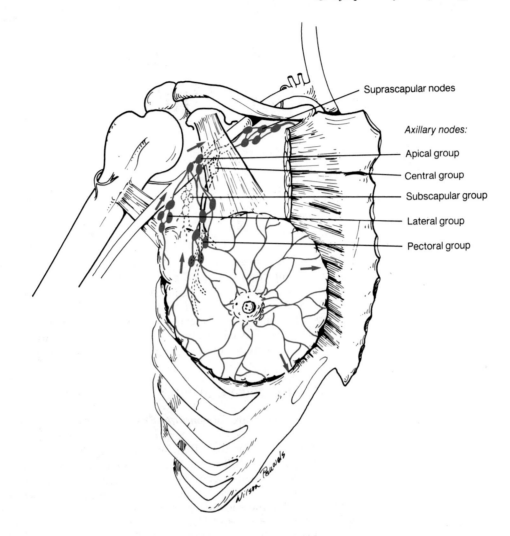

Suprascapular nodes

Axillary nodes:

Apical group

Central group

Subscapular group

Lateral group

Pectoral group

Figure 43-2.
Superficial and deep lymphatic drainage of the upper limb into axillary nodes.

CLINICAL NOTE

Because of the prevalence of malignancy of the breast, knowledge of the distribution of the axillary nodes is of great importance (see Chap. 2).

OTHER LYMPH NODES: UPPER EXTREMITY

A few other lymph nodes are found in different regions of the upper extremity. One group, the ***cubital***, is recognized. It is a small group of nodes located above the medial epicondyle; clinicians call this group the ***epitrochlear*** nodes.

A Review of the Nervous System of the Upper Limb

Injuries to the nerve supply of the upper limb by blunt or penetrating trauma are quite common, and the practitioner must be able to analyze the resulting deficits in order to diagnose the location of such lesions.

One cannot be too dogmatic in this analysis because variations in the makeup of the brachial plexus are common. The following represents an outline of the more usual innervation of the upper limb and of some of the more common nerve lesions.

AUTONOMIC NERVE SUPPLY

The upper limb requires a **sympathetic** nerve supply to regulate the caliber of its blood vessels, to control sweat gland secretion, and to provide a nerve supply to its arrector pili muscles (see Fig. 2-2). The source of the preganglionic fibers is the thoracic part of the spinal cord, and the postganglionic fibers arise in the second to fifth thoracic sympathetic ganglia. The postganglionic fibers pass to the brachial plexus by means of the first (sometimes also the second) thoracic spinal nerve and are then distributed through the plexus. Many postganglionic fibers travel with the branches of the plexus, but many others form a plexus around the axillary artery and are transmitted along its branches.

DERMATOMES

It may be remembered that a dermatome is an area of skin that receives its sensory nerve supply from a single specific segment of the spinal cord or area of the brain stem. Unfortunately there is considerable disagreement about the precise distribution of dermatomes throughout the body, and different areas will be mapped out by different investigators. The following is a rough guide to the dermatomes of the upper limb.

Remember that the brachial plexus contains the anterior primary rami of C5 to T1 with some variable contributions from C4 and T2, and that the thumb is on the cranial or preaxial side of the upper limb and the middle finger is the central digit of the limb, which drags C7 (the middle nerve) distally to supply the center of the back of the arm, forearm, hand, and middle finger. C5 supplies the lateral surface of the arm, C6 the lateral surface of the forearm, hand, and index finger, C8 the ring and little fingers, the medial side of the hand, and the forearm, and T1 the medial side of

44

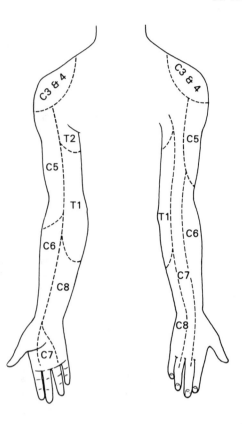

Figure 44-1.
Dermatomes (approximate) of anterior and posterior surfaces of upper limb.

the midforearm and arm. T2 (the intercostobrachial nerve) supplies the skin over the floor of the axilla and a small part of the medial aspect of the arm (Fig. 44-1).

MYOTOMES

A myotome is that portion of the muscle mass of the body that is supplied by the motor nerve from a single segment of the spinal cord or area of the brain stem. Again, different textbooks give different details, but the following may be taken as a guide to the myotomes and the joints they move.

> ***Shoulder***: Flexion, abduction, lateral rotation: C5. Extension,
> adduction and medial rotation: C6, C7, C8.
> ***Elbow***: Flexion: C5, C6. Extension: C7, C8.
> ***Forearm***: Pronation: C7, C8; Supination: C6
> ***Wrist***: Flexion-Extension: C6, C7.
> ***Fingers and Thumb***: Flexion-Extension: C7, C8.
> ***Hand***: Small muscles: C8, T1.

INJURIES TO THE NERVOUS SYSTEM

Injuries to nerves reveal themselves by ***paralysis*** or ***anesthesia***. The paralysis may be incomplete and difficult to detect, and anesthesia may be incomplete and show as ***altered*** sensation, for example, "pins and needles," which is called ***paresthesia***.

A nerve may be compressed so that its function is temporarily impaired, it may be crushed so that the axons are severed but the neurilemmal sheath remains intact, or it may be completely cut across.

Regeneration of axons from the proximal (cell) end will commence after a refractory period of 10 to 21 days, and the sprouting axons will try to grow down any neurilemmal sheath in the vicinity. If the nerve was injured by crushing with the neurilemmal sheaths intact, axons will find the correct passages to their destination. Axons grow at the rate of about 1 mm/day. If the nerve is severed, accurate surgical repair may result in varying degrees of restoration of function.

INJURIES TO THE BRACHIAL PLEXUS

Tearing or cutting injuries of the brachial plexus are almost impossible to repair because the complexity of the plexus makes accurate identification of nerve ends and reorientation of regenerating axons virtually impossible.

Certain limited paralyses are recognized and graced with eponyms of varying degrees of distinction and of uncertain historical and anecdotal accuracy.

Crutch Palsy. Prolonged use of a crutch that presses into the axilla has been known to injure the posterior cord or the radial nerve, resulting in varying degrees of extensor muscle paralysis such as wrist drop.

Saturday Night, Sleeper's, or Drunkard's Paralysis. Occurs when the sleeper's arm is draped over the back of a chair and the radial or median nerve is compressed.

Erb's Palsy. Erb's palsy is a birth injury that occurs most frequently when the baby's head is severely laterally flexed on the shoulder and pulled on during delivery. The upper trunk, involving C5 and C6, is damaged, and there is a loss of abduction, flexion, and lateral rotation at the shoulder and corresponding sensory loss over the lateral aspect of the limb.

Klumpke's Paralysis. Occurs when similar traction is exerted on the fetal shoulder during a breech delivery. In this case the lower part of the brachial plexus is pulled on and the lesion affects the lower trunk (C8, T1). There is impairment of wrist flexion and of the function of the intrinsic muscles of the hand.

PARALYSIS OF INDIVIDUAL NERVES

The functional loss that results from the cutting of an individual nerve will depend on the level of the lesion. Clearly structures supplied by the nerve and its branches above the lesion will escape unscathed, whereas structures supplied by the nerve distal to the lesion will be affected.

Median Nerve Injury

The median nerve contains fibers from all of the roots of the brachial plexus (Fig. 44-2).

Forearm Affected. That is, the lesion is at or above the elbow. There will be loss of pronation and appropriate sensory loss over the lateral part of the palm of the hand. In addition, distal dysfunction will occur as described for the appropriate levels below (*i.e.*, wrist and hand).

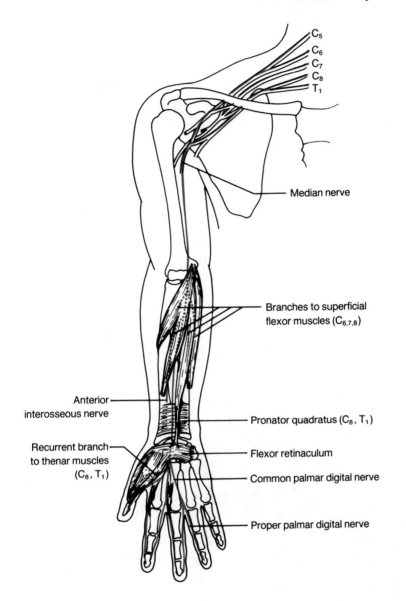

C₅
C₆
C₇
C₈
T₁

— Median nerve

Branches to superficial
flexor muscles (C₆,₇,₈)

Anterior
interosseous nerve

Pronator quadratus (C₈, T₁)

Recurrent branch
to thenar muscles
(C₈, T₁)

Flexor retinaculum

Common palmar digital nerve

Proper palmar digital nerve

Figure 44-2.
The course and branches of the median
nerve.

Wrist Affected. There will be severe weakening of wrist flexion, although some will still occur because of the action of flexor carpi ulnaris and the medial portion of flexor digitorum profundus. Adduction will be unimpaired and abduction will be weakened; there will be no loss of extension. The hand will be affected as described in the next paragraph.

Hand Affected. If the median nerve is severed at the wrist, a not uncommon site, the thenar muscles will be paralyzed with loss of opposition of the thumb and weakness at the first metacarpophalangeal joint (flexor pollicis brevis paralysis). The lateral three and one-half digits will be anesthetic on their palmar aspects and on the dorsum of their distal phalanges. The lateral two lumbricals will be affected, and there will thus be impairment of the fine movements of the first two fingers.

When paralysis has been present for some time, the affected muscles will atrophy. The thenar eminence will appear flat, and the thumb will be

Figure 44-3.
The course of the ulnar nerve.

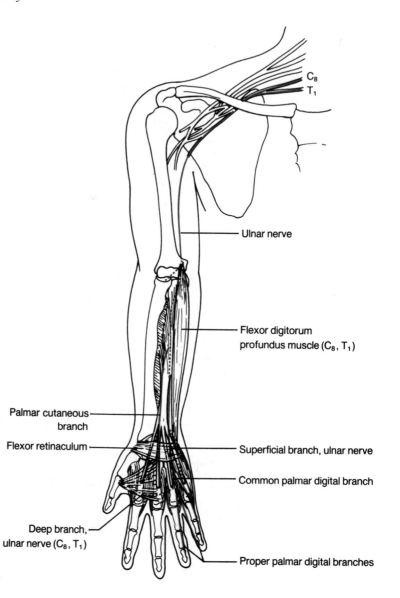

- C8
- T1
- Ulnar nerve
- Flexor digitorum profundus muscle (C8, T1)
- Palmar cutaneous branch
- Flexor retinaculum
- Superficial branch, ulnar nerve
- Common palmar digital branch
- Deep branch, ulnar nerve (C8, T1)
- Proper palmar digital branches

held close to the base of the lateral surface of the index finger. The two lateral fingers will show hyperextension at the metacarpophalangeal joint and some flexion at the interphalangeal joints.

Ulnar Nerve Injury

If the ulnar nerve (C8-T1) is cut in the arm the effect on the wrist is some impairment of flexion and impaired adduction (Fig. 44-3). In the medial two fingers there is poor flexion and grasp, and their terminal phalanges cannot be flexed. The ***interossei*** lose the power to abduct and adduct the fingers. The thumb cannot be adducted, and the medial two fingers will be held in hyperextension at the metacarpophalangeal joints and somewhat flexed at the proximal interphalangeal joints. The result is called a ***claw***

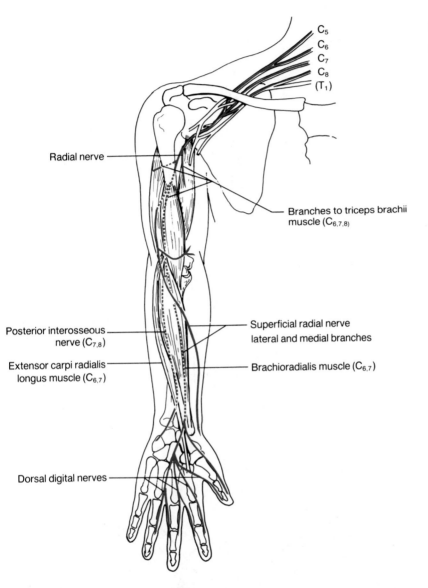

Figure 44-4.
The course of the radial nerve. The
forearm is pronated.

Radial nerve

Branches to triceps brachii
muscle ($C_{6,7,8}$)

Posterior interosseous
nerve ($C_{7,8}$)

Superficial radial nerve
lateral and medial branches

Extensor carpi radialis
longus muscle ($C_{6,7}$)

Brachioradialis muscle ($C_{6,7}$)

Dorsal digital nerves

hand. When muscle wasting occurs, there will be visible loss of substance
in the interosseous spaces between the metacarpal bones as viewed from
the dorsum of the hand.

   Sensation might be lost over the medial part of the hand and medial
one and one-half fingers.

Radial Nerve Injury

If the radial nerve ([C5] C6-C8 [T1]) is severed close to its origin, the *triceps muscle* will be paralyzed and the elbow cannot be extended against
gravity (Fig. 44-4). More frequently the radial nerve is injured in the *spiral
groove* of the humerus, in the region of the lateral humeral condyle, or
near the neck of the radius; if the injury occurs above the elbow, paralysis

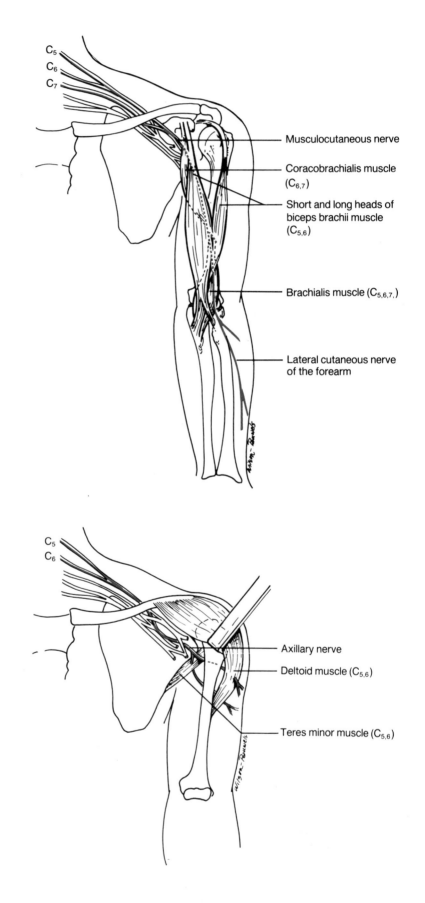

Figure 44-5.
The course of the musculocutaneous nerve.

C₅
C₆
C₇

Musculocutaneous nerve

Coracobrachialis muscle (C$_{6,7}$)

Short and long heads of biceps brachii muscle (C$_{5,6}$)

Brachialis muscle (C$_{5,6,7,}$)

Lateral cutaneous nerve of the forearm

C₅
C₆

Axillary nerve

Deltoid muscle (C$_{5,6}$)

Teres minor muscle (C$_{5,6}$)

Figure 44-6.
The course of the axillary nerve.

of the extensor muscles of the forearm will cause ***wrist drop*** and the thumb will be held in the adducted and flexed position. Some extension of the fingers will still be possible because of the actions of the interosseous and lumbrical muscles of the hand. Flexion of the fingers will be very ineffectual because the wrist extensors will not be able to stabilize the wrist in the position of function, and much of the flexor effort will be wasted flexing the wrist.

   Sensation seemingly should be lost over the posterior aspect of the forearm, hand, and proximal and middle phalanges of the lateral three and one-half digits, but the actual sensory loss is often surprisingly small and may be confined to an area involving the back of the thumb and adjacent dorsal aspect of the hand.

Musculocutaneous Nerve Injury

The musculocutaneous nerve (C5, C6, and C7) is not particularly prone to injury by indirect trauma, but, of course, a well-aimed stab or bullet will injure it (Fig. 44-5). If the injury occurs above its muscular branches there will be little flexion at the elbow joint and weak supination because biceps brachii and brachialis will be paralyzed. There may or may not be some anesthesia over the lateral aspect of the arm and forearm.

Axillary Nerve Injury

This nerve (C5 and C6) may be damaged in fractures through the surgical neck of the humerus or in dislocations of the shoulder joint (Fig. 44-6). Deltoid paralysis with loss of powerful abduction at the shoulder and some anesthesia of the skin covering the deltoid will result.

Glossary of Common Anatomical Terms

Abduction Movement away from the midline

Adduction Movement toward the midline

Anterior In front of

Aponeurosis Expanded tendon for attachment of muscle

Artery Blood vessel carrying blood from heart to tissues

Articulation A joint; connection between bones

Autonomic Nervous System Portion of nervous system to smooth muscle, cardiac muscle, glands, and viscera; consisting of sympathetic and parasympathetic components

Bone Rigid calcified tissue composing the skeleton

Capillary Microscopic-sized blood vessel connecting arterial to venous system

Cartilage Gristle; covers articular parts of bones; precursor of bone

Cell Structural and functional body unit of minute dimension

Central Nervous System Brain and spinal cord

Condyle Rounded articular ends of some bones

Contralateral Opposite side

Coronal A vertical plane at right angles to sagittal; drawn through the coronal suture of the skull

Crest A ridge or border; usually on a bone

Diaphysis Shaft of a long bone

Distal Farther away from origin or body's central axis

Dorsal Toward the back; also back of hand or superior aspect of foot

Eminence A low convexity; usually on a bone

Endocrine Gland A gland that secretes its hormone internally into the blood stream

Epicondyle Bony elevation or bump above a condyle

Epiphyseal Plate Cartilaginous plate between epiphysis and diaphysis of a bone within which bone growth occurs

Eversion Turning outward

Exocrine Gland A gland that secretes its hormones or enzymes by means of a duct system

Extension Straightening of a joint

Facet A small articular area on a bone

Fascia Dense fibrous envelope of muscles or other structures

Flexion Bending of a joint

Foramen Hole usually in a bone for passage of a blood vessel, nerve, or other structure

Fossa A shallow depression or pit; usually in bone

Ganglion Collection of nerve cells outside the central nervous system

Head Rounded end of a long bone

Inferior Below; away from the head end of the body

Insertion The relatively movable part of a muscular attachment

Inversion Turning inward

Ipsilateral Same side

Joint Articulation, junction between bones

Lateral Away from midline

Ligament Fibrous connection between bones, or fibrous bands or sheets holding muscles or tendons in place

Lymph Vessel Thin-walled tubes draining extracellular fluids to lymph nodes and ultimately into venous system

Medial Toward the midline

Median In the midline

Mesentery A double layer of peritoneum supporting an intra-abdominal or pelvic viscus

Motor Neuron Unit A nerve cell, its processes, and the muscle fibers supplied by it

Muscle Contractile tissue producing movement or change in shape of a structure

Nerve Specialized conducting fibers outside the central nervous system

Neuron A nerve cell and its processes

Nucleus An aggregation of nerve cell bodies within the central nervous system; the central functional and controlling part of any cell

Omentum A double layer of peritoneum connected to the stomach

Origin Relatively fixed part of a muscle attachment

Palmar Anterior, palm side of hand

Peripheral Nervous System Nerves and nerve cell bodies outside the central nervous system

Plantar Sole side of foot

Posterior Toward the back

Prone Body position when lying face down and palms down

Protruberance A visible or palpable swelling; usually on bone

Proximal Nearer to central axis of body

Ramus Plate-like piece of bone; branch of a nerve

Ramus Communicans A nerve branch connecting a sympathetic ganglion to a spinal nerve

Sagittal Vertical section or plane through sagittal suture of skull dividing the body into right and left halves

Shaft Body of a long bone

Sheath A protective covering, for example, nerve sheath or synovial sheath of a tendon

Superior Above, toward the head end of the body

Supine Lying on the back with palms facing upward

Suture Interlocking of bony edges, for example, skull

Symphysis A union of structures, for example, pubic symphysis

Tendon Fibrous tissue securing attachment of a muscle

Trochlea A pulley or spool-shaped articular surface

Tubercle A small bump, usually on a bone

Tuberosity A large bump, usually on a bone

Vein A blood vessel conducting blood to the heart

Volar Palmar side of the hand or finger

Bibliography and Suggested Reading

Some of the following books are referred to in the text. Others are added as useful general references although they have not been specifically mentioned.

Anderson JE: Grant's Atlas of Anatomy, 8th ed. Baltimore, Williams & Wilkins, 1983

Barr ML, Kiernan JA: The Human Nervous System, 5th ed. Philadelphia, JB Lippincott, 1988

Basmajian JV: Grant's Method of Anatomy, 10th ed. Baltimore, Williams & Wilkins, 1980

Clemente CD: Anatomy—A Regional Atlas of the Human Body, 3rd ed. Baltimore, Urban & Schwarzenberg, 1987

Cunningham's Textbook of Anatomy; see Romanes GJ

Gardner E, Gray DJ, O'Rahilly R: Basic Human Anatomy, 5th ed. Philadelphia, WB Saunders, 1986

Grant's Atlas of Anatomy; see Anderson JE

Gray's Anatomy; see Williams PL

Hollinshead WH: Anatomy for Surgeons, 3 vols., 3rd ed. Philadelphia, Harper & Row, 1982

Hollinshead WH, Rosse C: Textbook of Anatomy, 4th ed. Philadelphia, Harper & Row, 1985

Last JR: Anatomy—Regional and Applied, 7th ed. London, Churchill Livingstone, 1984

Lockhart RD: Living Anatomy, 5th ed. London, Faber and Faber, 1960

Michels NA: Blood Supply and Anatomy of the Upper Abdominal Organs. Philadelphia, JB Lippincott, 1955

Romanes GJ (ed): Cunningham's Textbook of Anatomy, 12th ed. London, Oxford University Press, 1981

Snell RS: Clinical Anatomy for Medical Students, 3rd ed. Boston, Little, Brown, 1986

Spence AP: Basic Human Anatomy, 2nd ed. Menlo Park, CA, Benjamin/Cummings, 1986

Williams PL, Warwick R (eds): Gray's Anatomy, 36th British ed. London, Churchill Livingstone, 1980

Wilson-Pauwels L, Akesson EJ, Stewart PA: Cranial Nerves. Burlington, ON, Canada, BC Decker, 1988

Woodburne RT: Essentials of Human Anatomy, 7th ed. London, Oxford University Press, 1983

Index

Anatomical parts are indexed under the general category to which they belong, for example, *Muscle(s)* and *Foramen*. The single exception is the bones, which are listed individually.

Numbers followed by an *f* indicate a figure; *t* following a page number indicates tabular material.